FINITE ELEMENT METHODS:
Parallel-Sparse Statics
and Eigen-Solutions

T0180569

FINITE ELEMENT METHODS:
Parallel-Sparse Statics and Eigen-Solutions

Duc Thai Nguyen

Old Dominion University
Norfolk, Virginia

 Springer

Prof. Duc Thai Nguyen
135 Kaufman
Old Dominion University
Department of Civil &
Environmental Engineering
Multidisc. Parallel-Vector Comp Ctr
Norfolk VA 23529

Finite Element Methods: Parallel-Sparse Statics and Eigen-Solutions

ISBN 978-1-4419-3985-2   e-ISBN 978-0-387-30851-7

Printed on acid-free paper.

Printed in the United States of America.

9 8 7 6 5 4 3 2 1

springer.com

To Dac K. Nguyen
Thinh T. Thai
Hang N. Nguyen
Eric N. D. Nguyen and Don N. Nguyen

Contents

Preface

Finite element methods (FEM) and associated computer software have been widely accepted as one of the most effective, general tools for solving large-scale, practical engineering and science applications. It is no wonder there is a vast number of excellent textbooks in FEM (not including hundreds of journal articles related to FEM) written in the past decades!

While existing FEM textbooks have thoroughly discussed different topics, such as linear/nonlinear, static/dynamic analysis, with varieties of 1-D/2-D/3-D finite element libraries, for thermal, electrical, contact, and electromagnetic applications, most (if not all) current FEM textbooks have mainly focused on the developments of "finite element libraries," how to incorporate boundary conditions, and some general discussions about the assembly process, solving systems of "banded" (or "skyline") linear equations. For implicit finite element codes, it is a well-known fact that efficient equation and eigen-solvers play critical roles in solving large-scale, practical engineering/science problems. Sparse matrix technologies have evolved and become mature enough that all popular, commercialized FEM codes have inserted sparse solvers into their software. Furthermore, modern computer hardware usually has multiple processors; clusters of inexpensive personal computers (under WINDOWS, or LINUX environments) are available for parallel computing purposes to dramatically reduce the computational time required for solving large-scale problems.

Most (if not all) existing FEM textbooks discuss the assembly process and the equation solver based on the "variable banded" (or "skyline") strategies. Furthermore, only limited numbers of FEM books have detailed discussions about Lanczos eigen-solvers or explanation about domain decomposition (DD) finite element formulation for parallel computing purposes.

This book has been written to address the concerns mentioned above and is intended to serve as a textbook for graduate engineering, computer science, and mathematics students. A number of state-of-the-art FORTRAN software, however, have been developed and explained with great detail. Special efforts have been made by the author to present the material in such a way to minimize the mathematical background requirements for typical graduate engineering students. Thus, compromises between rigorous mathematics and simplicities are sometimes necessary.

The materials from this book have evolved over the past several years through the author's research work and graduate courses (CEE715/815 = Finite Element I, CEE695 = Finite Element Parallel Computing, CEE711/811 = Finite Element II) at Old Dominion University (ODU). In Chapter 1, a brief review of basic finite element procedures for Linear/Statics/Dynamics analysis is given. One, two, and three-dimensional finite element types are discussed. The weak formulation is emphasized. Finite element general field equations are derived, isoparametric formulation is explained, and Gauss Quadrature formulas for efficient integration are discussed. In this chapter, only simple (non-efficient) finite element assembly procedures are explained. Chapter 2 illustrates some salient features offered by Message Passing Interface (MPI) FORTRAN environments. Unrolling techniques, efficient usage of computer cache memory, and some basic MPI/FORTRAN applications in matrix linear algebra operations are also discussed in this chapter. Different versions of direct, "SPARSE" equation solvers' strategies are thoroughly discussed in Chapter 3. The "truly sparse" finite element "assembly process" is explained in Chapter 4. Different versions of the Lanczos algorithms for the solution of generalized eigen-equations (in a sparse matrix environment) are derived in Chapter 5. Finally, the overall finite element domain decomposition computer implementation, which can exploit "direct" sparse matrix equation, eigen-solvers, sparse assembly, "iterative" solvers (for both "symmetrical" and "unsymmetrical" systems of linear equations), and parallel processing computation, are thoroughly explained and demonstrated in Chapter 6. Attempts have been made by the author to explain some difficult concepts/algorithms in simple language and through simple (hand-calculated) numerical examples. Many FORTRAN codes (in the forms of main program, and sub-routines) are given in Chapters 2 – 6. Several large-scale, practical engineering problems involved with several hundred thousand to over 1 million degree-of-freedoms (dof) have been used to demonstrate the efficiency of the algorithms discussed in this textbook.

This textbook should be useful for graduate students, practicing engineers, and researchers who wish to thoroughly understand the detailed step-by-step algorithms used during the finite element (truely sparse) assembly, the "direct" and "iterative" sparse equation and eigen-solvers, and incorporating the DD formulation for efficient parallel computation.

The book can be used in any of the following "stand-alone" courses:

(a) Chapter 1 can be expanded (with more numerical examples) and portions of Chapter 3 (only cover the sparse formats, and some "key components" of the sparse solver) can be used as a first (introductive type) course in finite element analysis at the senior undergraduate (or 1st year graduate) level.

(b) Chapters 1, 3, 4, and 5 can be used as a "stand-alone" graduate course such as "Special Topics in FEM: Sparse Linear Statics and Eigen-Solutions."

(c) Chapters 1, 2, 3, 4, and 6 can be used as a "stand-alone" graduate course, such as "Special Topics in FEM: Parallel Sparse Linear Statics Solutions."

(d) Chapters 2, 3, and 5, and portions of Chapter 6, can be used as a "stand-alone" graduate course such as "High Performance Parallel Matrix Computation."

The book also contains a limited number of exercises to further supplement and reinforce the presented concepts. The references for all chapters are listed at the end of the book.

The author would like to invite the readers to point out any errors they find. He also welcomes any suggestions or comments from readers.

Duc Thai Nguyen
Norfolk, Virginia

Acknowledgements

During the preparation of this book, I have received (directly and indirectly) help from many people. First, I would like to express my sincere gratitude to my colleagues at NASA Langley Research Center, Dr. Olaf O. Storaasli, Dr. Jaroslaw S. Sobieski, and Dr. Willie R. Watson, for their encouragement and support on the subject of this book during the past years.

The close collaborative work with Professor Gene Hou and Dr. J. Qin, in particular, has a direct impact on the writing of several sections in this textbook.

I am very grateful to Professors Pu Chen (China), S. D. Rajan (Arizona), B. D. Belegundu (Pennsylvania), J. S. Arora (Iowa), Dr. Brad Maker (California), Dr. Esmond Ng (California), Dr. Ed. D'Azevedo (Tennessee), and Mr. Maurice Sancer (California) for their enthusiasm and support of several topics discussed in this book.

My appreciation also goes to several of our current and former graduate students, such as Mrs. Shen Liu, Ms. N. Erbas, Mr. X. Guo, Dr. Yusong Hu, Mr. S. Tungkahotara, Mr. A.P. Honrao, and Dr. H. B. Runesha, who have worked with me for several years. Some of their research has been included in this book.

In addition, I would like to thank my colleagues at Old Dominion University (ODU) for their support, collaborative work, and friendship, among them, Professors Osman Akan, Chuh Mei, Hideaki Kaneko, Alex Pothen, Oktay Baysal, Bowen Loftin, Zia Razzaq, and Roland Mielke. The excellent computer facilities and consulting services provided by my ODU/OOCS colleagues (A. Tarafdar, Mike Sachon, Rusty Waterfield, Roland Harrison, and Tony D'Amato) over the past years are also deeply acknowledged.

The successful publication and smooth production of this book are due to Miriam I. Tejeda and Sue Smith (ODU office support staff members), and graduate students Mrs. Shen Liu, Ms. N. Erbas, Mr. S. Tungkahotara, and Mr. Emre Dilek. The timely support provided by Elaine Tham (editor), her colleagues, and staff members also gratefully acknowledged.

Special thanks go to the following publishers for allowing us to reproduce certain material for discussion in our textbook:

- Natalie David (N.David@elsevier.com) for reproducing some materials from "*Sparse Matrix Technology*", by Sergio Pissanetzky (pages 238 – 239, 263 – 264, 270, 282) for discussion in Chapters 3 and 4 of our textbook (see Tables 3.2, 3.3, 3.8, and 4.2).
- Michelle Johnson and Sabrina Paris (Sabrina.Paris@Pearsoned.com) for reproducing some materials from "*Finite Element Procedures*", by K.J. Bathe, 1st Edition, 1996 (pages 915 – 917, 924 – 927, 959) for discussion in Chapter 5 of our textbook (see examples on pages 293 – 297, 301 – 302; see sub-routine jacobiKJB and pages 344 – 348).

- Adam Hirschberg (AHirschberg@cambridge.org) for reproducing materials from *"Numerical Recipes"*, by W. H. Press, et.al. (pages 35 – 37, 366, 368, 374) for discussions in Chapter 5 of our textbook (see sub-routines LUBKSB, LUDCMP on pages 348 – 350; See tables 5.8, 5.9, 5.14).

Last but not least, I would like to thank my parents (Mr. Dac K. Nguyen and Mrs. Thinh T. Thai), my wife (Mrs. Hang N. Nguyen), and my sons (Eric N. D. Nguyen, and Don N. Nguyen) whose encouragement has been ever present.

Duc T. Nguyen
Norfolk, Virginia

Disclaimer of Warranty

1 A Review of Basic Finite Element Procedures

1.1 Introduction

Most (if not all) physical phenomena can be expressed in some form of partial differential equations (PDE), with appropriated boundary and/or initial conditions. Since exact, analytical solutions for complicated physical phenomena are not possible, approximated numerical procedures (such as Finite Element Procedures), have been commonly used. The focus of this chapter is to briefly review the basic steps involved during the finite element analysis [1.1–1.13]. This will facilitate several advanced numerical algorithms to be discussed in subsequent chapters.

1.2 Numerical Techniques for Solving Ordinary Differential Equations (ODE)

To simplify the discussion, rather than using a PDE example, a simple ODE (structural engineering) problem will be analyzed and solved in the subsequent sections.

Given the following ODE:

$$EI\frac{d^2y}{dx^2} = \frac{\omega x(L-x)}{2} \tag{1.1}$$

with the following boundary conditions:

$$y(@\ x = 0) = 0 = y(@\ x = L) \tag{1.2}$$

The above equations (1.1–1.2) represent a simply supported beam, subjected to a uniform load applied throughout the beam, as shown in Figure 1.1.

While several numerical methods are available (such as Variational, Galerkin, Colloquation, Minimize Residual methods), the Galerkin method is selected here due to its generality and simplicity. To facilitate the derivation and discussion of the Galerkin method, the ODE given in Eq.(1.1) can be re-casted in the following general form

$$Ly=f \tag{1.3}$$

L and f in Eq.(1.3) represent the "mathematical operator," and "forcing" function, respectively. Within the context of Eq.(1.1), the "mathematical operator" L, in this case, can be defined as:

$$L \equiv EI\frac{d^2}{dx^2}(\ ) \tag{1.4}$$

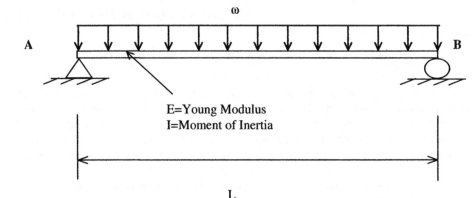

Figure 1.1 a Simply Supported Beam under Uniformly Applied Load

Hence

$$Ly = EI\frac{d^2(y)}{dx^2} \text{ and } f \equiv \frac{w\,x\,(L-x)}{2} \tag{1.5}$$

If we "pretend" the right-hand side "f" in Eq.(1.3) represents the "force," and the unknown function "y" represents the "beam deflection," then the virtual work can be computed as:

$$\delta w = \iiint_v (f)*\delta y\ dv \equiv \iiint_v (L\,y)*\delta y\ dv \tag{1.6}$$

In Eq.(1.6), δy represents the "virtual" displacement, which is consistent with (or satisfied by) the "geometric" boundary conditions (at the supports at joints A and B of Figure 1.1, for this example).

In the Galerkin method, the exact solution $y(x)$ will be replaced by the approximated solution $\tilde{y}(x)$, which can be given as follows:

$$\tilde{y}(x) = \sum_{i=1}^{N} a_i\phi_i(x) \tag{1.7}$$

where a_i are the unknown constant(s) and $\phi_i(x)$ are selected functions such that all "geometric boundary conditions" (such as given in Eq. [1.2]) are satisfied.

Substituting $\tilde{y}(x)$ from Eq.(1.7) into Eq.(1.6), one obtains:

$$\iiint_v (f)\delta y\ dv \neq \iiint_v (L\tilde{y})\,\delta y\ dv \tag{1.8}$$

However, we can adjust the values of a_i (for $\tilde{y}$), such that Eq.(1.8) will be satisfied, hence

$$\iiint_v (f)\delta y\, dv \approx \iiint_v (L\tilde{y})\,\delta y\, dv \qquad (1.9)$$

or

$$\iiint_v (L\tilde{y} - f)\delta y\, dv = 0 \qquad (1.10)$$

Based upon the requirements placed on the virtual displacement δy, one may take the following selection:

$$\delta y = \phi_i(x) \qquad (1.11)$$

Thus, Eq.(1.10) becomes the "Galerkin" equations, and is given as:

$$\iiint_v \underbrace{(L\tilde{y} - f)}_{\substack{\text{Residual} \\ \text{or error}}}\ \underbrace{\phi_i}_{\substack{\text{Weighting} \\ \text{Function}}}\ dv = 0 \qquad (1.12)$$

Thus, Eq.(1.12) states that the <u>summation</u> of the <u>weighting residual</u> is set to be zero. Substituting Eq.(1.7) into Eq.(1.12), one gets:

$$\iiint_v [L(\sum_{i=1}^{N} a_i\phi_i(x)) - f]\phi_i\, dv = 0 \qquad (1.13)$$

Eq.(1.13) will provide i=1,2,...,N equations, which can be solved simultaneously for obtaining a_i unknowns.

For the example provided in Figure 1.1, the "exact" solution y(x) can be easily obtained as:

$$y(x) = \left(\frac{1}{EI}\right) * \left(\frac{2\omega L x^3 - \omega x^4 - \omega L^3 x}{24}\right) \qquad (1.14)$$

assuming a two-term approximated solution for $\tilde{y}(x)$ is sought, then (from Eq. [1.7]):

$$\tilde{y} = (A_1 + A_2 x) * \phi_1(x) \qquad (1.15)$$

or

$$\tilde{y} = A_1\phi_1(x) + A_2 x\phi_1(x) \qquad (1.16)$$

or

$$\tilde{y} = A_1\phi_1(x) + A_2\phi_2(x) \qquad (1.17)$$

where

$$\phi_2(x) = x * \phi_1(x) \qquad (1.18)$$

Based on the given geometric boundary conditions, given by Eq.(1.2), the function $\phi_1(x)$ can be chosen as:

$$\phi_1(x) = \sin\left(\frac{\pi x}{L}\right) \tag{1.19}$$

Substituting Eqs.(1.18 – 1.19) into Eq.(1.17), one has:

$$\tilde{y} = A_1 * \sin\left(\frac{\pi x}{L}\right) + A_2 * x \sin\left(\frac{\pi x}{L}\right) \tag{1.20}$$

Substituting the given differential equation (1.1) into the approximated Galerkin Eq.(1.12), one obtains:

$$\int_{x=0}^{L} [EI\tilde{y}'' - \frac{\omega x(L-x)}{2}] * \phi_i(x)dx = 0 \tag{1.21}$$

For i = 1, one obtains from Eq.(1.21):

$$\int_{x=0}^{L} [EI\tilde{y}'' - \frac{\omega x(L-x)}{2}] * [\phi_1(x) = \sin(\frac{\pi x}{L})]dx = 0 \tag{1.22}$$

For i = 2 one obtains from Eq.(1.21):

$$\int_{x=0}^{L} [EI\tilde{y}'' - \frac{\omega x(L-x)}{2}] * [\phi_2(x) = x\sin(\frac{\pi x}{L})]dx = 0 \tag{1.23}$$

Substituting Eq.(1.20) into Eqs.(1.22 – 1.23), and performing the integrations using MATLAB, one obtains the following two equations:

$$[2EIA_1 * \pi^5 + EIA_2 * \pi^5 * L + 8\omega L^4] = 0 \tag{1.24}$$

$$[3EIA_1 * \pi^5 + 3EIA_2 * \pi^3 * L + 2EIA_2 * \pi^5 * L + 12\omega L^4] = 0 \tag{1.25}$$

or, in the matrix notations:

$$(EI) * \begin{bmatrix} 2\pi^5 & \pi^5 L \\ 3\pi^5 & 3\pi^3 L + 2\pi^5 L \end{bmatrix} * \begin{Bmatrix} A_1 \\ A_2 \end{Bmatrix} = \begin{Bmatrix} -8\omega L^4 \\ -12\omega L^4 \end{Bmatrix} \tag{1.26}$$

Using MATLAB, the solution of Eq.(1.26) can be given as:

$$A_1 = \frac{-4\omega L^4}{\pi^5 EI} = \frac{(-4)*(5)\omega L^4}{(5)\pi^5 EI} = \frac{-5\omega L^4}{(\frac{5}{4})\pi^5 EI} = \frac{-5\omega L^4}{382.523 EI} \tag{1.27}$$

$$A_2 = 0 \tag{1.28}$$

Thus, the approximated solution $\tilde{y}(x)$, from Eq.(1.20), becomes:

$$\tilde{y}(x) = \left(\frac{-5\omega L^4}{382.523EI}\right) * \sin(\frac{\pi x}{L}) \tag{1.29}$$

At $x = \dfrac{L}{2}$, the exact solution is given by Eq.(1.14):

$$y = \frac{-5\omega L^4}{384EI} \qquad (1.30)$$

At $x = \dfrac{L}{2}$, the approximated solution is given by Eq.(1.29):

$$\tilde{y} = \frac{-5\omega L^4}{382.523EI} \qquad (1.31)$$

Remarks

(a) The selected function $\phi_1(x)$ can also be easily selected as a polynomial, which also satisfies the geometrical boundary conditions (see Eq.[1.2])

$$\phi_1(x) = (x - 0)(x - L) \qquad (1.32)$$

and, therefore we have:

$$\tilde{y}(x) = (A_1 + A_2 x) * \phi_1(x) \qquad (1.33)$$

or

$$\tilde{y}(x) = A_1 \phi_1(x) + A_2 \phi_2(x) \qquad (1.34)$$

where

$$\phi_2(x) = x\phi_1(x) \qquad (1.35)$$

(b) If the function $\phi_1(x)$ has to satisfy the following "hypothetical" boundary conditions,

$$y(@\, x = 0) = 0 \qquad (1.36)$$
$$y'(@\, x = 0) = 0 \qquad (1.37)$$
$$y(@\, x = L) = 0 \qquad (1.38)$$
$$y''(@\, x = 0) = 0 \qquad (1.39)$$

then a possible candidate for $\phi_1(x)$ can be chosen as:

$$\phi_1(x) = (x - 0)^2 (x - L)^3 \qquad (1.40)$$

The first and second terms of Eq.(1.40) are raised to the power 2 and 3 in order to satisfy the "slope" and "curvature" boundary conditions, as indicated in Eq.(1.37), and Eq.(1.39), respectively.

(c) In practical, real-life problems, the "exact" analytical solution (such as the one shown in Eq.[1.14]) is generally unknown. Thus, one way to evaluate the quality of the approximated solution (or check if the approximated solution is already converged or not) is to keep increasing the number of unknown constant terms used in Eq.(1.33), such as:

$$\tilde{y}(x) = (A_1 + A_2 x + A_3 x^2 + A_4 x^3 + \dots) * \phi_1(x) \qquad (1.41)$$

or

$$\tilde{y}(x) = A_1\phi_1(x) + A_2\phi_2(x) + A_3\phi_3(x) + A_4\phi_4(x) + ... \qquad (1.42)$$

where

$$\phi_2(x) = x * \phi_1(x)$$
$$\phi_3(x) = x * \phi_2(x) \qquad (1.43)$$
$$\phi_4(x) = x * \phi_3(x) \qquad (1.44)$$

Convergence is achieved when the current (more unknown terms) approximated solution does not change much, as compared to the previous (less unknown terms) approximated solution (evaluated at some known, discretized locations).

1.3 Identifying the "Geometric" versus "Natural" Boundary Conditions

For the "Structural Engineering" example depicted in Figure 1.1, it is rather easy to recognize that "geometric" boundary conditions are usually related to the "deflection and/or slope" (the vertical deflections at the supports A and B are zero), and the "natural" boundary conditions are usually related to the "shear force and/or bending moment" (the moments at the simply supports A and B are zero).

For "non-structural" problems, however, one needs a more general approach to distinguish between the "geometric" versus "natural" boundary conditions. The abilities to identify the "geometrical" boundary conditions are crucially important since the selected function $\phi_1(x)$ has to satisfy all these geometrical boundary conditions. For this purpose, let's consider the following "beam" equation:
$$EIy'''' = \omega(x) \qquad (1.45)$$

Our objective here is to identify the possible "geometric" boundary conditions from the above 4$\underline{th}$ order, ordinary differential equations (ODE). Since the highest order of derivatives involved in the ODE (1.45) is four, one sets:
$$2n = 4 \qquad (1.46)$$
hence
$$n - 1 = 1 \qquad (1.47)$$

Based on Eq.(1.47), one may conclude that the function itself (= y), and all of its derivatives up to the order "n−1" (= 1, in this case, such as y') are the "geometrical" boundary conditions. The "natural" boundary conditions (if any) will involve derivatives of the order $n(= 2), n + 1(= 3),...$ and higher (such as $y'', y''',...$).

1.4 The Weak Formulations

Let's consider the following differential equation
$$-\frac{d}{dx}\left[a(x)\frac{dy}{dx}\right] = b(x) \quad \text{for } 0 \le x \le L \qquad (1.48)$$

subjected to the following boundary conditions

$$y(@ \ x = 0) = y_0 \tag{1.49}$$

$$\left(a \frac{dy}{dx} \right)_{@ \ x=L} = Q_0 \tag{1.50}$$

In Eqs.(1.48 – 1.50), $a(x)$, $b(x)$ are known functions, y_0 and Q_0 are known values, and L is the length of the 1-dimensional domain.

The approximated N-term solution can be given as:

$$y(x) \approx \tilde{y}(x) = \sum_{i=1}^{N} A_i \phi_i(x) + \phi_0(x) \tag{1.51}$$

Based on the discussion in Section 1.3, one obtains from Eq.(1.48):

$$2n = 2 \ (= \text{the highest order of derivative}) \tag{1.52}$$

Hence

$$n - 1 = 0 \tag{1.53}$$

Thus, Eq.(1.49) and Eq.(1.50) represent the "geometrical" and "natural" boundary conditions, respectively.

The <u>non-homogeneous</u> "geometrical" (or "essential") boundary conditions can be satisfied by the function $\phi_0(x)$ such as $\phi_0(@ \ x_0) = y_0$. The functions $\phi_i(x)$ are required to satisfy the <u>homogeneous</u> form of the same boundary condition $\phi_i(@ \ x_0) = 0$.

If all specified "geometrical" boundary conditions are homogeneous (for example, $y_0=0$), then $\phi_0(x)$ is taken to be zero and $\phi_i(x)$ must still satisfy the same boundary conditions (for example, $\phi_i(@ \ x_0) = 0$).

The "weak formulation," if it exists, basically involves the following three steps:

Step 1

Applying the Galerkin Eq.(1.12) into the given differential equation (1.48), one obtains:

$$\int_0^L [\ \underbrace{-\frac{d}{dx}\left\{ a(x)\frac{dy}{dx} \right\} - b(x)}_{dv} \ ] * \underbrace{W_i(x)}_{u} dx = 0 \tag{1.54}$$

Step 2

While it is quite possible that the N selected weighting function W_i (shown in Eq.1.54) will help us solve for N unknown constants A_1, A_2,..., A_N, it requires that the derivatives up to the highest order specified by the original differential equation are available. However, if the differentiation is distributed between the approximated solution $\tilde{y}(x)$ and the weighting function(s) $W_i(x)$, then the resulting integral form (see Eq.1.12) will require weaker continuity condition on $\phi_i(x)$, hence the weighted integral statement is called the "weak form." Two major advantages are associated with the weak formulation:

(a) It requires weaker (or less) continuity of the dependent variable, and it often results in a "symmetrical" set of algebraic equations.
(b) The natural boundary conditions are included in the weak form, and therefore the approximated solution $\tilde{y}(x)$ is required to satisfy only the "geometrical" (or "essential") boundary conditions of the problem.

In this second step, Eq.(1.54) will be integrated by parts to become:

$$-\left[W_i(x)*a(x)\frac{dy}{dx}\right]_0^L + \int_0^L [a(x)\frac{dy}{dx}*\frac{dW_i(x)}{dx} - b(x)*W_i(x)]\,dx = 0 \quad (1.55)$$

The coefficient of the weighting function $W_i(x)$ in the boundary term can be recognized as:

$$a(x)\frac{dy}{dx}*n_x \equiv Q \equiv \text{"secondary" variable (= heat, for example)} \quad (1.56)$$

In Eq.(1.56), n_x is defined as the cosine of the angle between the x-axis and the outward normal to the boundary for a 1-D problem, $n_x \equiv \cos(0°) = 1$ at the right end $(x = L)$ and $n_x \equiv \cos(180°) = -1$ at the left end $(x = 0)$ of the beam, as indicated in Figure 1.2.

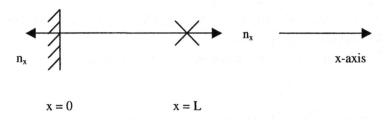

Figure 1.2 Normal (= n_x) to the Boundary of the Beam

Using the new notations introduced in Eq.(1.56), Eq.(1.55) can be re-written as:

$$0 = -[W_i(x)*a(x)\frac{dy}{dx}]^{@\,x=L} + [W_i(x)*a(x)\frac{dy}{dx}]_{@\,x=0}$$
$$+ \int_0^L [a(x)\frac{dy}{dx}\frac{dW_i(x)}{dx} - b(x)W_i(x)]dx$$

$$0 = -[W_i(x)*a(x)\frac{dy}{dx}*n_x]^{@\ x=L} - [W_i(x)*a(x)\frac{dy}{dx}*n_x]_{@\ x=0}$$
$$+ \int_0^L [a(x)\frac{dy}{dx}\frac{dW_i(x)}{dx} - b(x)W_i(x)]dx$$

or

$$0 = \int_0^L [a(x)\frac{dy}{dx}\frac{dW_i(x)}{dx} - b(x)W_i(x)]dx$$
$$-[W_i(x)*Q]^{@\ x=L} + [W_i(x)*Q]_{@\ x=0}$$

(1.57)

It should be noted that in the form of Eq.(1.54), the "primary" variable y(x) is required to be <u>twice</u> differentiable but only <u>once</u> in Eq.(1.57).

<u>Step 3</u>

In this last step of the weak formulation, the actual boundary conditions of the problem are imposed. Since the weighting function $W_i(x)$ is required to satisfy the homogeneous form of the specified geometrical (or essential) boundary conditions, hence

$$W_i(@\ x=0) = 0 ; \text{ because } y(@x=0)=y_0$$

(1.58)

Eq.(1.57) will reduce to:

$$0 = \int_0^L [a(x)*\frac{dw_i}{dx} - b(x)*W_i(x)]\, dx - W_i(@\ x=L)*Q(@\ x=L)$$

or, using the notation introduced in Eq.(1.50), one has:

$$0 = \int_0^L [a(x)*\frac{dw_i}{dx} - b(x)*W_i(x)]\, dx - W_i(@\ L)*Q_0$$

(1.59)

Eq.(1.59) is the weak form equivalent to the differential equation (1.48) and the natural boundary condition equation (1.50).

1.5 Flowcharts for Statics Finite Element Analysis

The finite element Galerkin method is quite similar to the Galerkin procedures described in the previous sections. The key difference between the above two approaches is that the former will require the domain of interests to be divided (or discretized) into a finite number of sub-domains (or finite elements), and the selected "shape" functions $\phi_i(x)$ are selected to satisfy all essential boundary conditions associated with a particular finite element only.

The unknown primary function (say, deflection function) $f(x_i)$ at any location within a finite element can be computed in terms of the element nodal "displacement" vector $\{r'\}$ as:

$$\{f(x_i)\}_{3x1} = [N(x_i)]_{3xn} * \{r'\}_{nx1} \tag{1.60}$$

In Eq.(1.60), n represents the number of "displacements" (or degree-of-freedom≡dof) per finite element, and $N(x_i)$ represents the known, selected shape function(s). For the most general 3-D problems, the shape functions $[N(x_i)]$ are functions of x_1, x_2, and x_3 (or x, y, and z). The left-hand side of Eq.(1.60) represents the three components of the displacement along the axis x_1, x_2, and x_3. The strain-displacement relationships can be obtained by taking the partial derivatives of displacement function Eq.(1.60), with respect to the independent variables x_i, as follows:

$$\{\varepsilon\} = \frac{\partial}{\partial x_i}[N(x_i)] * \{r'\} \tag{1.61}$$

or

$$\{\varepsilon\} = [B(x_i)] * \{r'\} \tag{1.62}$$

where

$$[B(x_i)] \equiv \frac{\partial}{\partial x_i}[N(x_i)] \tag{1.63}$$

The internal virtual work can be equated with the external virtual work, therefore

$$\int_V \delta \varepsilon^T * \sigma \, dV = \{\varepsilon r'\}^T * \{p'\} \tag{1.64}$$

In Eq.(1.64), the superscript "T" represents the transpose of the virtual strain $\delta\varepsilon$ and virtual nodal displacement $\delta r'$, whereas σ and p' represent the stress and nodal loads, respectively.

The stress-strain relationship can be expressed as:

$$\{\sigma\} = [D]\{\varepsilon\} \tag{1.65}$$

where in the above equation, [D] represents the material matrix. For a 1-D structural problem, Eq.(1.65) reduces to the following simple Hooke's law scalar equation

$$\sigma_{xx} = E \, \varepsilon_{xx} \tag{1.66}$$

where E represents the material Young Modulus.

From Eq.(1.62), one obtains:

$$\{\varepsilon\}^T = \{r'\}^T [B]^T \tag{1.67}$$

Hence

$$\{\delta\varepsilon\}^T = \{\delta r'\}^T [B]^T \tag{1.68}$$

substituting Eqs.(1.65, 1.62, 1.68) into the virtual work equation (1.64), one obtains:

$$\int_V \{\delta r'\}^T [B]^T * [D]\{\varepsilon = B\,r'\}\,dV - \{\delta r'\}^T \{p'\} = 0 \tag{1.69}$$

or

$$\{\delta r'\}^T * \left[\int_V [B]^T [D][B]dV * \{r'\} - \{p'\} \right] = 0 \tag{1.70}$$

The virtual displacements $\{\delta r\}$ can be arbitrarily selected as long as they are consistent with (or satisfied by) the geometrical boundary constraints, hence in general $\{\delta r'\} \neq 0$. Therefore, from Eq.(1.70), one requires:

$$[k'] * \{r'\} - \{p'\} = \{0\} \tag{1.71}$$

where the "element" local stiffness matrix [k'] in Eq.(1.71) is defined as:

$$[k'] \equiv \int_V [B]^T [D][B]dV \tag{1.72}$$

Since the element local coordinate axis in general will NOT coincide with the system global coordinate axis, the following transformation need to be done (see Figure 1.3):

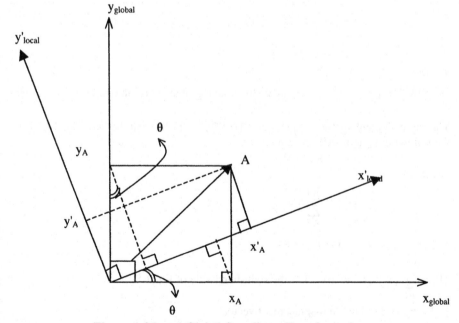

Figure 1.3 Local-Global Coordinate Transformation

$$\{r'\} = \begin{Bmatrix} x'_A \\ y'_A \end{Bmatrix} = \begin{Bmatrix} x_A \cos(\theta) + y_A \sin(\theta) \\ y_A \cos(\theta) - x_A \sin(\theta) \end{Bmatrix} = \begin{bmatrix} \cos(\theta) & \sin(\theta) \\ -\sin(\theta) & \cos(\theta) \end{bmatrix} \begin{Bmatrix} x_A \\ y_A \end{Bmatrix} \qquad (1.73)$$

or

$$\{r'\} = [\lambda] \begin{Bmatrix} x_A \\ y_A \end{Bmatrix} \equiv [\lambda]\{r\} \qquad (1.74)$$

where

$$[\lambda] \equiv \begin{bmatrix} \cos(\theta) & \sin(\theta) \\ -\sin(\theta) & \cos(\theta) \end{bmatrix}; \text{ for 2-D problems} \qquad (1.75)$$

Similarly, one has:

$$\{p'\} = [\lambda]\{p\} \qquad (1.76)$$

Substituting Eqs.(1.74, 176) into Eq.(1.71), one has:

$$[k']*[\lambda]\{r\} = [\lambda]\{p\} \qquad (1.77)$$

Pre-multiplying both sides of Eq.(1.77) by $[\lambda]^T$, one gets:

$$[\lambda]^T[k'][\lambda]*\{r\} = [\lambda]^T[\lambda]\{p\} \qquad (1.78)$$

Since $[\lambda]^T[\lambda] = [I]$, thus $[\lambda]^T = [\lambda]^{-1}$, hence Eq.(1.78) becomes:

$$[\lambda]^T[k'][\lambda]*\{r\} = \{p\} \qquad (1.79)$$

or

$$[k]*\{r\} = \{p\} \qquad (1.80)$$

where

$$[k] = [\lambda]^T[k'][\lambda] \equiv [k^{(e)}] = \text{element stiffness in global coordinate reference} \qquad (1.81)$$

The finite <u>element</u> stiffness equations, given by Eq.(1.80), can be assembled to form the following <u>system</u> stiffness equations:

$$[K]*\{R\} = \{P\} \qquad (1.82)$$

where

$$[K] = \sum_{e=1}^{NEL} [k^{(e)}] \qquad (1.83)$$

and $[k^{(e)}]$ is defined in Eq.(1.81)

$$\{P\} = \sum_{e=1}^{NEL} \{p^{(e)}\}; \text{ and } \{p^{(e)}\} \text{ is the right-hand-side of Eq.(1.80)}$$

$\{R\}$ = system unknown displacement vector

For structural engineering applications, the system matrix [K] is usually a sparse, symmetrical, and positive definite matrix (after imposing appropriated boundary conditions). At this stage, however [K] is still a singular matrix. After imposing the appropriated boundary conditions, [K] will be non-singular, and the unknown vector {R} in Eq.(1.82) can be solved.

1.6 Flowcharts for Dynamics Finite Element Analysis

The kinetic energy (= 1/2*mass*velocity^2) can be computed as:

$$\text{K.E.} = \frac{1}{2} \int_V \{\dot{f}\}^T * \{\dot{f}\} * \rho dV \tag{1.84}$$

In Eq.(1.84) "ρdV," and $\{\dot{f}\} \equiv \dfrac{\partial f}{\partial t}$ represents the "mass" and "velocity," respectively.

Substituting Eq.(1.60) into Eq.(1.84), one obtains:

$$\text{K.E.} = \frac{1}{2} \{\dot{r}'\}^T \int_V [N(x_i)]^T [N(x_i)] \rho dV * \{\dot{r}'\} \tag{1.85}$$

The kinetic energy can also be expressed as:

$$\text{K.E.} = \frac{1}{2} \{\dot{r}'\}^T [m'] \{\dot{r}'\} \tag{1.86}$$

Comparing Eq.(1.85) with Eq.(1.86), the <u>local element mass matrix</u> [m'] can be identified as:

$$[m'] \equiv \int_V [N(x_i)]^T [N(x_i)] \rho dV \equiv [m'^{(e)}] \tag{1.87}$$

The <u>global</u> element mass matrix [m] can be obtained in a similar fashion as indicated in Eq.(1.81), hence

$$[m] = [\lambda]^T [m'^{(e)}][\lambda] \equiv [m^{(e)}] \tag{1.88}$$

The system mass matrix can be assembled as:

$$[M] = \sum_{e=1}^{NEL} [m^{(e)}] \tag{1.89}$$

where $[m^{(e)}]$ has already been defined in Eq.(1.88). The system dynamical equilibrium equations, therefore, can be given as:

$$[M]\{\ddot{R}\} + [K]\{R\} = \{P(t)\} \tag{1.90}$$

If a "proportional" damping matrix [C] is used, then

$$[C] = \alpha_1[K] + \alpha_2[M] \tag{1.91}$$

where α_1 and α_2 are constant coefficients, and Eq.(1.90) can be generalized to:

$$[M]\{\ddot{R}\} + [C]\{\dot{R}\} + [K]\{R\} = \{P(t)\} \tag{1.92}$$

with the following initial conditions:

$$@\, t = 0, \{R\} = \{R^{(0)}\} \text{ and } \{\dot{R}\} = \{\dot{R}^{(0)}\} \tag{1.93}$$

For undamped, free vibration, Eq.(1.92) will be simplified to:

$$[M]\{\ddot{R}\} + [K]\{R\} = \{0\} \tag{1.94}$$

Let $\{R\} = \{y\}\sin(\omega t)$ (1.95)

where $\omega \equiv$ natural frequency of the structure. Then

$$\{\dot{R}\} = \{y\}\omega\cos(\omega t) \tag{1.96}$$

$$\{\ddot{R}\} = -\{y\}\omega^2\sin(\omega t) = -\omega^2\{R\} \tag{1.97}$$

Substituting Eqs.(1.95, 1.97) into Eq.(1.94), one gets:

$$-[M]\{y\}\omega^2 + [K]\{y\} = \{0\} \tag{1.98}$$

or

$$[K]\{y\} = \omega^2[M]\{y\} = \lambda[M]\{y\} \tag{1.99}$$

Thus, Eq.(1.99) can be recognized as the <u>Generalized Eigen-Value Problem</u>. The eigen-values of Eq.(1.99) can be solved from:

$$[K - \lambda M]\{y\} = \{0\} \tag{1.100}$$

In Eq.(1.100), λ and $\{y\}$ represent the eigen-values and the corresponding eigen-vectors, respectively.

1.7 Uncoupling the Dynamical Equilibrium Equations

To facilitate the discussion, the following system stiffness and mass matrices are given (assuming proper boundary conditions are already included):

$$[K] = \begin{bmatrix} 2 & -1 \\ -1 & 1 \end{bmatrix} \tag{1.101}$$

$$[M] = \begin{bmatrix} 2 & 0 \\ 0 & 1 \end{bmatrix} \tag{1.102}$$

In order to have a "non-trivial" solution for Eq.(1.100), one requires that

$$Det|K - \lambda M| = Det\begin{vmatrix} 2 - 2\lambda & -1 \\ -1 & 1 - \lambda \end{vmatrix} = 0 \tag{1.103}$$

The solution for Eq.(1.103) can be given as:

$$\lambda = \frac{2 \mp \sqrt{2}}{2}$$

Hence

$$\lambda_1 = \frac{2 - \sqrt{2}}{2} \tag{1.104}$$

and

$$\lambda_2 = \frac{2 + \sqrt{2}}{2} \tag{1.105}$$

Substituting the 1st eigen-value ($= \lambda_1$) into Eq.(1.100), one obtains:

$$\left(\begin{bmatrix} 2 & -1 \\ -1 & 1 \end{bmatrix} - \left(\frac{2 - \sqrt{2}}{2} \right) * \begin{bmatrix} 2 & 0 \\ 0 & 1 \end{bmatrix} \right) * \begin{Bmatrix} y_1^{(1)} \\ y_2^{(1)} \end{Bmatrix} = \begin{Bmatrix} 0 \\ 0 \end{Bmatrix} \tag{1.106}$$

In Eq.(1.106), the superscripts of y represent the 1st eigen-vector (associated with the 1st eigen-value λ_1) while the subscripts of y represent the components of the 1st eigen-vector. There are two unknowns in Eq.(1.106), namely $y_1^{(1)}$ and $y_2^{(1)}$ however, there is only one linearly independent equation in (1.106). Hence, if we let

$$y_1^{(1)} = 1 \tag{1.107}$$

then, one can solve for

$$y_2^{(1)} = \sqrt{2} \tag{1.108}$$

The 1st eigen-vector $\phi^{(1)}$ (associated with the 1st eigen-value λ_1), therefore, is given as:

$$\phi^{(1)} = \begin{Bmatrix} 1 \\ \sqrt{2} \end{Bmatrix} \tag{1.109}$$

Similar procedures can be used to obtain the 2nd eigen-vector $\phi^{(2)}$ (associated with the 2nd eigen-value λ_2), and can be obtained as:

$$\phi^{(2)} = \begin{Bmatrix} 1 \\ -\sqrt{2} \end{Bmatrix} \tag{1.110}$$

The eigen-vector $\phi^{(1)}$, defined in Eq.(1.109), can be normalized with respect to the mass matrix [M], according to the following two steps:

Step 1 Calculate $c_1 = \phi^{(1)^T} [M] \phi^{(1)} = 4$ \hfill (1.111)

Step 2 Compute the 1st normalized eigen-vector $\phi_N^{(1)}$ as:

$$\phi_N^{(1)} = \frac{\phi^{(1)}}{\sqrt{c_1}} = \begin{Bmatrix} 1/2 \\ \sqrt{2}/2 \end{Bmatrix} \tag{1.112}$$

Similarly, one obtains:

$$\phi_N^{(2)} = \begin{Bmatrix} 1/2 \\ -\sqrt{2}/2 \end{Bmatrix} \tag{1.113}$$

Thus, the normalized eigen-matrix $[\Phi]$ can be assembled as:

$$[\Phi] = \left[\phi_N^{(1)} \phi_N^{(2)}\right] = \begin{bmatrix} \dfrac{1}{2} & \dfrac{1}{2} \\ \dfrac{\sqrt{2}}{2} & -\dfrac{\sqrt{2}}{2} \end{bmatrix} \tag{1.114}$$

Now, let's define a new variable vector $\{\Lambda\}$ as:

$$\{R\} = [\Phi] * \{\Lambda\} \tag{1.115}$$

Hence, Eq.(1.92) becomes (damping term is neglected):

$$[M] * [\Phi] * \{\ddot{\Lambda}\} + [K] * [\Phi] * \{\Lambda\} = \{P(t)\} \tag{1.116}$$

Pre-multiplying both sides of Eq.(1.116) by $[\Phi]^T$, one obtains:

$$[\Phi]^T [M][\Phi]\{\ddot{\Lambda}\} + [\Phi]^T [K][\Phi]\{\Lambda\} = [\Phi]^T \{P(t)\} \tag{1.117}$$

The above equation can be represented as:

$$[M^*]\{\ddot{\Lambda}\} + [K^*]\{\Lambda\} = \{P^*(t)\} \tag{1.118}$$

where

$$[M^*] \equiv [\Phi]^T [M][\Phi] = [I] = \text{Identity Matrix} \tag{1.119}$$

$$[K^*] \equiv [\Phi]^T [K][\Phi] = \begin{bmatrix} \lambda_1 & 0 \\ 0 & \lambda_2 \end{bmatrix} = \text{Diagonal (E\_value) Matrix} \tag{1.120}$$

$$\{P^*(t)\} \equiv [\Phi]^T \{P(t)\} \tag{1.121}$$

Since both $[M^*]$ and $[K^*]$ are diagonal matrices, the dynamical equilibrium equations, Eq.(1.118), are uncoupled! The initial conditions associated with the new variable $\{\Lambda\}, \{\ddot{\Lambda}\}$ can be computed as follows:

Taking the first derivative of Eq.(1.115) with respect to time, one gets:

$$\{\dot{R}\} = [\Phi]\{\dot{\Lambda}\} \tag{1.122}$$

Pre-multiplying Eq.(1.122) with $[\Phi]^T [M]$, one has:

$$[\Phi]^T [M]\{\dot{R}\} = \{\dot{\Lambda}\} \tag{1.123}$$

Similarly, one obtains:

$$[\Phi]^T [M]\{R\} = \{\Lambda\} \tag{1.124}$$

Using the initial conditions shown in Eq.(1.93), one has:

$$@ \, t = 0, \text{ then } \{\dot{\Lambda}\} = [\Phi]^T [M]\{\dot{R}^{(0)}\} \equiv \{\dot{\Lambda}^{(0)}\} \tag{1.125}$$

and

$$\{\Lambda\} = [\Phi]^T [M] \{R^{(0)}\} \equiv \{\Lambda^{(0)}\} \tag{1.126}$$

Detailed descriptions of efficient "sparse eigen-solution" algorithms will be discussed in Chapter 5.

1.8 One-Dimensional Rod Finite Element Procedures

The axially loaded rod member is shown in Figure 1.4:

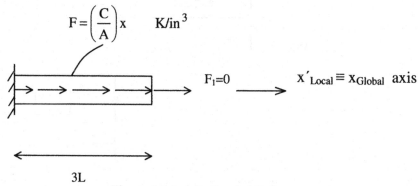

Figure 1.4 Axially Loaded Rod

The governing differential equation for the rod shown in Figure 1.4 can be given as:

$$-EA\frac{\partial^2 u}{\partial x^2} = q(x) \tag{1.127}$$

where E, A, q(x), and u represent Material Young modulus, cross-sectional area of the rod, axially distributed load (where $q(x) = cx^{K/in}$) and the rod's axial displacement, respectively. The geometric (or essential) boundary condition from Figure 1.4 can be given as:

$$u(@ \ x = 0) = u_0 = 0 \tag{1.128}$$

and the natural boundary condition can be expressed as:

$$EA\frac{\partial u}{\partial x}\bigg|_{@ \ x=3L} = F_1 \ (\equiv \text{Axial Force}), \tag{1.129}$$

assuming the rod (shown in Figure 1.4) is divided into three rod finite elements as indicated in Figure 1.5.

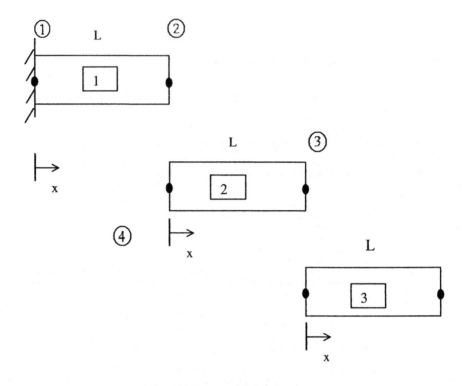

Figure 1.5 an Axially Loaded Rod with three Finite Elements

1.8.1 One-Dimensional Rod Element Stiffness Matrix

Since each finite element axial rod is connected by two nodes, and each node has only one axial unknown displacement (or 1 degree-of-freedom ≡ 1 dof), therefore, the axial displacement within a rod element #1 $u(x)$ can be expressed as:

$$u(x) = a_1 + a_2 x = [1, x] * \begin{Bmatrix} a_1 \\ a_2 \end{Bmatrix} \qquad (1.130)$$

where the unknown constants a_1 and a_2 can be solved by invoking the two geometric boundary conditions:

$$\text{At } x = 0; u = u_1 = a_1 + a_2(x = 0) \qquad (1.131)$$
$$\text{At } x = L; u = u_2 = a_1 + a_2(x = L) \qquad (1.132)$$

To assure that the approximated solution will converge to the actual ("exact") solution as the number of finite elements is increased, the following requirements need to be satisfied by the approximated solution $u(x)$, shown in Eq.(1.130):
1. It should be a complete polynominal (including all lower order terms up to the highest order used).
2. It should be continuous and differentiable (as required by the weak form).
3. It should be an interpolant of the primary variables at the nodes of the finite element.

Eqs.(1.131 – 1.132) can be expressed in matrix notations as:

$$\begin{Bmatrix} u_1 \\ u_2 \end{Bmatrix} = \begin{bmatrix} 1 & 0 \\ 1 & L \end{bmatrix} * \begin{Bmatrix} a_1 \\ a_2 \end{Bmatrix} \tag{1.133}$$

Eq.(1.133) can also be expressed in a more compact notations as:

$$\{u\} = [A]\{a\} \tag{1.134}$$

where

$$\{u\} \equiv \begin{Bmatrix} u_1 \\ u_2 \end{Bmatrix} \tag{1.135}$$

$$[A] \equiv \begin{bmatrix} 1 & 0 \\ 1 & L \end{bmatrix} \tag{1.136}$$

$$\{a\} \equiv \begin{Bmatrix} a_1 \\ a_2 \end{Bmatrix} \tag{1.137}$$

From Eq.(1.134), one obtains:

$$\{a\} = [A]^{-1} * \{u\} \tag{1.138}$$

where

$$[A]^{-1} = \left(\frac{1}{L}\right) \begin{bmatrix} L & 0 \\ -1 & 1 \end{bmatrix} \tag{1.139}$$

Substituting Eq.(1.138) into Eq.(1.130), one gets:

$$u(x) = [1, x] * [A]^{-1} * \{u\} \tag{1.140}$$

or

$$u(x) = [1, x] * \begin{bmatrix} 1 & 0 \\ \dfrac{-1}{L} & \dfrac{1}{L} \end{bmatrix} * \begin{Bmatrix} u_1 \\ u_2 \end{Bmatrix} \tag{1.141}$$

Eq.(1.141) has exactly the same form as described earlier by Eq.(1.160), where

$$\{f(x_i)\} \equiv u(x) \tag{1.142}$$

$$[N(x)] \equiv [1, x] * \begin{bmatrix} 1 & 0 \\ \dfrac{-1}{L} & \dfrac{1}{L} \end{bmatrix} = \{1 - \frac{x}{L}, \frac{x}{L}\} \tag{1.143}$$

$$\{r'\} \equiv \begin{Bmatrix} u_1 \\ u_2 \end{Bmatrix} \tag{1.144}$$

The shape (or interpolation) functions $[N(x)] = [\phi_1^e(x) = 1 - \dfrac{x}{L} ; \phi_2^e(x) = \dfrac{x}{L}]$,

shown in Eq.(1.143), have the following properties:

$$\phi_i^e(@ \, x_j) = \delta_{ij} = \begin{cases} 0, & \text{if } i \neq j \\ 1, & \text{if } i = j \end{cases} \tag{1.143}$$

$$\sum_{i=1}^{n} \phi_i(x) = 1, \; therefore \;\; \sum_{i=1}^{n} \frac{d\phi_i(x)}{dx} = 0 \tag{1.144}$$

$$where \; n \equiv the \; number \; of \; dof \; per \; element$$

From Eq.(1.63), the strain displacement matrix [B(x)] can be found as:

$$[B(x)] = \frac{\partial}{\partial x}[N(x)] = \left\{ \frac{-1}{L}, \frac{1}{L} \right\} \tag{1.145}$$

The element stiffness matrix [k '] in the local coordinate reference can be found from Eq.(1.72) as:

$$[k'] = \int_0^L [B]^T (D)[B]dx \tag{1.146}$$

$$[k'] = \int_0^L \left\{ \begin{matrix} -\dfrac{1}{L} \\ \dfrac{1}{L} \end{matrix} \right\} (D) \left[-\dfrac{1}{L}, \dfrac{1}{L} \right] dx \tag{1.147}$$

$$[k'] = \frac{(D)}{L} \begin{bmatrix} 1 & -1 \\ -1 & 1 \end{bmatrix} = \frac{(AE)}{L} \begin{bmatrix} 1 & -1 \\ -1 & 1 \end{bmatrix} \tag{1.148}$$

The above element stiffness matrix [k'] and material matrix [D] can also be obtained from the energy approach, as described in the following steps:

Step 1

Strain energy density U_0 can be computed as:

$$U_0 = \frac{1}{2} \varepsilon_x^T E \varepsilon_x \tag{1.149}$$

Strain energy U can be computed as:

$$U = \iiint U_0 dV = \int_0^L U_0 (Adx) \tag{1.150}$$

$$U = \int_0^L \frac{1}{2} \varepsilon_x^T E \varepsilon_x A dx \tag{1.151}$$

Step 2

Substituting Eqs.(1.62, 1.65) into Eq.(1.151), one gets:

$$U = \int_0^L \frac{1}{2} \{r'\}^T [B]^T AE[B]\{r'\} dx \tag{1.152}$$

or

$$U = \frac{1}{2}\{r'\}^T * \int_0^L [B]^T (AE)[B]dx * \{r'\} \tag{1.153}$$

or

$$U = \frac{1}{2}\{r'\}^T *[k']*\{r'\} \tag{1.154}$$

From Eq.(1.153), the element stiffness matrix [k'] can be easily identified as it was given earlier in Eq.(1.146), and the material matrix can be recognized as:

$$[D]_{1x1} \equiv AE \tag{1.155}$$

1.8.2 Distributed Loads and Equivalent Joint Loads

The given traction force per unit volume (see Figure 1.4)

$$F = \frac{cx}{A} \quad K/in^3 \tag{1.156}$$

can be converted into the distributed load as:

$$q = F*A = cx \quad K/in \tag{1.157}$$

The work done by the external loads acting on finite element rod #1 can be computed as:

$$W = \iiint_V F * u * dV = \int_{x=0}^L FuAdx \tag{1.158}$$

or

$$W = \int_0^L (cx)*([N]\{r'\})dx \tag{1.159}$$

or

$$W = \int_0^L (cx)*\{r'\}^T [N]^T dx = \{r'\}^T \int_0^L q*[N]^T dx \tag{1.160}$$

or

$$W = \{r'\}^T * \int_0^L (cx) \begin{bmatrix} 1-\dfrac{x}{L} \\ \dfrac{x}{L} \end{bmatrix} dx \tag{1.161}$$

$$\text{Let } \{F_{equiv}\} \equiv \int_0^L (cx) \begin{bmatrix} 1-\dfrac{x}{L} \\ \dfrac{x}{L} \end{bmatrix} dx \tag{1.162}$$

Hence

$$\{F^1_{equiv}\} = \frac{cL^2}{6}*\begin{Bmatrix} 1 \\ 2 \end{Bmatrix} \tag{1.163}$$

Thus, Eq.(1.161) becomes:

$$W = \{r'\}^T * \{F_{equiv}\} \tag{1.164}$$

In Eq.(1.164), since $\{r'\}^T$ represents the nodal displacement, hence $\{F_{equiv}\}$ represents the "equivalent nodal loads" vector. In other words, the distributed axial load $(f = cx)$ can be converted into an equivalent nodal loads vector by Eq.(1.162), or it can be expressed in a more general case as:

$$\{F_{equiv}\} \equiv \int_0^L (q)*[N(x)]^T dx \tag{1.165}$$

Similarly, for finite element rod #2 and #3, one obtains (see Figure 1.4).

$$\{F^2_{equiv}\} \equiv \int_0^L c(L+x)[N]^T dx \tag{1.166}$$

$$\{F^2_{equiv}\} = \left(\frac{cL^2}{6}\right)\begin{Bmatrix} 4 \\ 5 \end{Bmatrix} \tag{1.167}$$

and

$$\{F^3_{equiv}\} \equiv \int_0^L c(2L+x)[N]^T dx = \left(\frac{cL^2}{6}\right)\begin{Bmatrix} 7 \\ 8 \end{Bmatrix} \tag{1.168}$$

1.8.3 Finite Element Assembly Procedures

The element stiffness matrix [k'] in the "local" coordinate system, shown in Eq.(1.148), is the same as the element stiffness matrix [k] (see Eq.[1.81]) in the "global" coordinate. This observation is true, due to the fact that the local axis x'_{Local} of the rod element coincides with the global x_{Global} axis (see Figure 1.4). Thus, in this case the transformation matrix $[\lambda]$ (see Eq.[1.75]) becomes an identity matrix, and therefore from Eq.(1.81), one gets:

$$[k] = [k']$$

The system stiffness matrix can be assembled (or added) from element stiffness matrices, as indicated in Eq.(1.83)

$$[K]_{4x4} = \sum_{e=1}^{NEL=3} [k^{(e)}] \tag{1.169}$$

or, using Eq.(1.148), one obtains:

$$[K]_{4x4} = \left(\frac{AE}{L}\right) * \left(\begin{bmatrix} 1 & -1 & 0 & 0 \\ -1 & 1 & 0 & 0 \\ 0 & 0 & 0 & 0 \\ 0 & 0 & 0 & 0 \end{bmatrix} + \begin{bmatrix} 0 & 0 & 0 & 0 \\ 0 & 1 & -1 & 0 \\ 0 & -1 & 1 & 0 \\ 0 & 0 & 0 & 0 \end{bmatrix} + \begin{bmatrix} 0 & 0 & 0 & 0 \\ 0 & 0 & 0 & 0 \\ 0 & 0 & 1 & -1 \\ 0 & 0 & -1 & 1 \end{bmatrix} \right)$$

or

$$[K]_{4x4} = \left(\frac{AE}{L}\right) * \begin{bmatrix} 1 & -1 & 0 & 0 \\ -1 & 2 & -1 & 0 \\ 0 & -1 & 2 & -1 \\ 0 & 0 & -1 & 1 \end{bmatrix} \tag{1.170}$$

Similarly, the system nodal load vector can be assembled from its elements' contributions:

$$\{P\}_{4x1} = \sum_{e=1}^{NEL=3} \{p^{(e)}\} \equiv \sum_{e=1}^{NEL=3} \{F_{Equiv.}^{(e)}\} \tag{1.171}$$

Utilizing Eqs.(1.163, 1.167, 1.168), one obtains:

$$\{P\}_{4x1} = \left(\frac{cL^2}{6}\right) * \left(\begin{Bmatrix} 1 \\ 2 \\ 0 \\ 0 \end{Bmatrix} + \begin{Bmatrix} 0 \\ 4 \\ 5 \\ 0 \end{Bmatrix} + \begin{Bmatrix} 0 \\ 0 \\ 7 \\ 8 \end{Bmatrix} \right) \tag{1.172}$$

or

$$\{P\}_{4x1} = \left(\frac{cL^2}{6}\right) * \begin{Bmatrix} 1 \\ 6 \\ 12 \\ 8 \end{Bmatrix} \tag{1.173}$$

Thus, the system matrix equilibrium equations is given as:

$$\left(\frac{AE}{L}\right) \begin{bmatrix} 1 & -1 & 0 & 0 \\ -1 & 2 & -1 & 0 \\ 0 & -1 & 2 & -1 \\ 0 & 0 & -1 & 1 \end{bmatrix} * \begin{Bmatrix} u_1 \\ u_2 \\ u_3 \\ u_4 \end{Bmatrix} = \left(\frac{cL^2}{6}\right) \begin{Bmatrix} 1 \\ 6 \\ 12 \\ 8 \end{Bmatrix} \tag{1.174}$$

or

$$[K]_{4x4} * \{D\}_{4x1} = \{P\}_{4x1} \tag{1.175}$$

Detailed descriptions of efficient "sparse assembly" algorithms will be discussed in Chapter 4.

1.8.4 Imposing the Boundary Conditions

The system stiffness matrix equations, given by Eq.(1.174), are singular, due to the fact that the boundary conditions (such as $u_1 = 0$) have not yet been incorporated. Physically, it means that the rod (shown in Figure 1.4) will be "unstable" without proper boundary condition(s) imposed (such as shown in Eq.[1.128]).

To make the discussion more general, let's assume that the boundary condition is prescribed at node 1 as:

$$u_1 = \alpha_1 \text{ (where } \alpha_1 = \text{known value)} \qquad (1.176)$$

Let $F_{unknown1}$ be defined as the unknown axial "reaction" force at the supported node 1, then Eq.(1.174) after imposing the boundary condition(s) can be symbolically expressed as:

$$\begin{bmatrix} K_{11} & K_{12} & K_{13} & K_{14} \\ K_{21} & K_{22} & K_{23} & K_{24} \\ K_{31} & K_{32} & K_{33} & K_{34} \\ K_{41} & K_{42} & K_{43} & K_{44} \end{bmatrix} * \begin{Bmatrix} u_1 = \alpha_1 \\ u_2 \\ u_3 \\ u_4 \end{Bmatrix} = \begin{Bmatrix} F_{unknown1} \\ P_2 \\ P_3 \\ P_4 \end{Bmatrix} \qquad (1.177)$$

Eq.(1.177) is <u>equivalent</u> to the following matrix equation

$$\begin{bmatrix} 1 & 0 & 0 & 0 \\ 0 & K_{22} & K_{23} & K_{24} \\ 0 & K_{32} & K_{33} & K_{34} \\ 0 & K_{42} & K_{43} & K_{44} \end{bmatrix} * \begin{Bmatrix} u_1 \\ u_2 \\ u_3 \\ u_4 \end{Bmatrix} = \begin{Bmatrix} \alpha_1 \\ P_2 - K_{21}*\alpha_1 \\ P_3 - K_{31}*\alpha_1 \\ P_4 - K_{41}*\alpha_1 \end{Bmatrix} \qquad (1.178)$$

or

$$[K_{bc}]*\{D\} = \{P_{bc}\} \qquad (1.179)$$

Eq.(1.178) is more preferable as compared to Eq.(1177), due to the following reasons:
  (i) The modified system stiffness matrix $[K_{bc}]$ is non-singular
  (ii) The modified right-hand-side vector $\{P_{bc}\}$ is completely known

Thus, the unknown displacement vector $\{D\} = \begin{Bmatrix} u_1 \\ u_2 \\ u_3 \\ u_4 \end{Bmatrix}$ can be solved by the existing efficient algorithms and software, which are available in different computer platforms. Detailed descriptions of "sparse equation solution" algorithms [1.9] will be discussed in Chapter 3.

Having obtained all the unknown global displacement vector $\{D\} = \begin{Bmatrix} u_1 \\ u_2 \\ u_3 \\ u_4 \end{Bmatrix}$ by solving Eq.(1.178), the unknown axial reaction force $F_{unknown1}$ can be easily solved from Eq.(1.177).

Furthermore, element nodal load (or stress) vector $\{p\}$ (in the <u>global</u> coordinate axis) can be easily re-covered by Eq.(1.80) since the element nodal displacement vector $\{r\}$ is simply a sub-set of the (known) system nodal displacement vector $\{D\}$ (shown in Eq. [1.179]).

1.8.5 Alternative Derivations of System of Equations from Finite Element Equations

From a given ODE, shown in Eq.(1.127), it can be applied for a typical e^{th} element (shown in Figure 1.5) as follows:

Step 1 Setting the integral of weighting residual to zero

$$0 = \int_{x_A}^{x_B} w \left[R = -EA \frac{\partial^2 u}{\partial x^2} - q(x) \right] dx \tag{1.179a}$$

Step 2 Integrating by parts once

$$0 = \left[w * (-EA \frac{\partial u}{\partial x}) \right]_{x_A}^{x_B} - \int_{x_A}^{x_B} (-EA \frac{\partial u}{\partial x}) * \frac{\partial w}{\partial x} dx$$

$$- \int_{x_A}^{x_B} w * q(x) dx \tag{1.179b}$$

Let $Q \equiv +EA \dfrac{\partial u}{\partial x} n_x$ (for the boundary terms) $\tag{1.179c}$

where n_x has already been used/defined in Eq.(1.56)

Hence Eq.(1.179b) becomes:

$$0 = [-w(@ x_B) * Q(@ x_B)] - [w(@ x_A) * Q(@ x_A)]$$

$$+ \int_{x_A}^{x_B} (EA \frac{\partial u}{\partial x}) * \frac{\partial w}{\partial x} dx - \int_{x_A}^{x_B} w * q(x) dx \tag{1.179d}$$

Step 3 Imposing "actual" boundary conditions

$$0 = \int_{x_A}^{x_B} (EA \frac{\partial u}{\partial x}) * \frac{\partial w}{\partial x} dx - w(x_A) * Q_A - w(x_B) * Q_B$$

$$- \int_{x_A}^{x_B} w * q(x) dx \tag{1.179e}$$

Step 4 Finite Element Equations

Let $w \equiv \phi_i^e(x)$ $\tag{1.179f}$

Let $u \equiv \displaystyle\sum_{j=1}^{n} u_j^e \, \phi_j^e(x)$ $\tag{1.179g}$

For the case $n = 2$, $\phi_i^e(x)$ and $\phi_j^e(x)$ have already been identified as the shape functions $[N(x)]$ shown in Eq.(1.143), where $i, j = 1,2$.

Substituting Eqs.(1.179f, 1.179g) into Eq.(1.179e), one obtains:

$$0 = \int_{x_A}^{x_B} EA \frac{\partial}{\partial x}(\sum_{j=1}^{n} u_j^e \, \phi_j^e) * \frac{\partial \phi_i^e}{\partial x} \, dx - \sum_{j=1}^{n} \phi_i^e(@ \, x_j) * Q_j$$
$$- \int_{x_A}^{x_B} \phi_i^e * q(x) \, dx \qquad (1.179h)$$

It should be noted that the last two (boundary) terms in Eq.(1.179e) have been replaced by the last (summation) term in Eq.(1.179h), so that an axially loaded finite element that has more than (or equal to) two nodes can be handled by Eq.(1.179h).

Using the properties of the finite element shape (or interpolation) functions (see Eqs.[1.143, 1.144]), one obtains:

$$\sum_{j=1}^{n} \phi_i^e(@ \, x_j) * Q_j = \sum_{j=1}^{n} \delta_{ij} * Q_j \equiv Q_i^e \qquad (1.179i)$$

Utilizing Eq.(1.179i), Eq.(1.179h) can be expressed as:

$$0 = \int_{x_A}^{x_B} EA \frac{\partial}{\partial x}(\sum_{j=1}^{n} u_j^e \, \phi_j^e) * \frac{\partial \phi_i^e}{\partial x} \, dx - Q_i^e$$
$$- \int_{x_A}^{x_B} \phi_i^e * q(x) \, dx \qquad (1.179j)$$

or

$$0 = \left(\sum_{j=1}^{n} [k_{ij}^e] * u_j^e \right) - f_i^e - Q_i^e \qquad (1.179k)$$

where $i = 1, 2, ..., n$ (= number of dof per element)

and $[k_{ij}^e] \equiv \int_{x_A}^{x_B} (EA \frac{\partial \phi_i^e}{\partial x} * \frac{\partial \phi_j^e}{\partial x}) \, dx \equiv$ element "stiffness" matrix $\qquad (1.179l)$

$$\{f_i^e\} \equiv \int_{x_A}^{x_B} q(x) * \phi_i^e \, dx \equiv \text{"Equivalent" nodal load vector} \qquad (1.179m)$$

$\{Q_i^e\}$ = "Internal" nodal load vector   (1.179n)

Equation (1.179k) can be further re-arranged as:

$$[k^e]*\{u^e\} = \{f^e\} + \{Q^e\}$$ (1.179o)

Remarks

(1) Referring to Figure 1.5, a typical e^{th} finite element will have:
- 2 equations (see Eq.[1.179k], assuming n = 2)
- 4 unknowns (say, for the 2nd finite element), which

can be identified as $u_2^2, u_3^2, Q_2^2, and \ Q_3^2$

(2) Thus, for the "entire" domain (which contains all 3 finite elements, as shown in Figure 1.5), one has:

- $2\dfrac{equations}{element}*3\,elements = 6\,equations$

- 12 unknowns, which can be identified as:

$u_1^1, u_2^1, Q_1^1 \ and \ Q_2^1, \cdots, u_3^3, u_4^3, Q_3^3, Q_4^3$

(3) The additional (6, in this example) equations can be obtained from:
- System, geometrical boundary condition(s), such as:

$u_1 = 0$

- Displacement compatibility requirements at "common" nodes (between "adjacent" finite elements), such as:

$u_2^1 = u_2^2$

$u_3^2 = u_3^3$

- Applied "nodal" loads, such as:

$Q_2^1 + Q_2^2 = 0$

$Q_3^2 + Q_3^3 = 0$

$Q_4^3 \quad = F_1$ (see Figures 1.4 and 1.5)

1.9 Truss Finite Element Equations

The formulation derived in Section 1.8 for the axially loaded rod element can be trivially expanded to analyze 2-D truss structures such as the one shown in Figure

1.6:

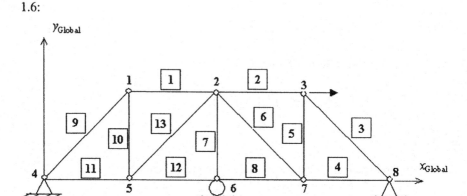

Figure 1.6 a Typical 2-D Truss Structure

A typical truss member, such as member #$\boxed{13}$ of Figure 1.6, is shown separately with its own local axis (x' and y') in Figure 1.7:

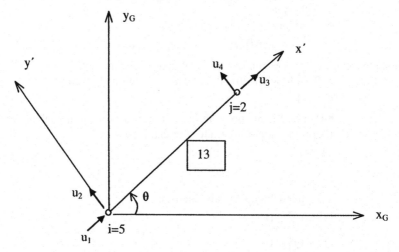

Figure 1.7 a Typical 2-D Truss Member with Its Nodal Displacements

Since a truss member can only carry axial force and have only "axial" displacements, such as u_1 and u_3 in Figure 1.7, the "fictitious" local displacements u_2 and u_4 are indicated in the same figure. Thus, the local element stiffness matrix [k'] of the rod (see Eq.[1.148]) can be slightly modified to become the 4x4 local element stiffness matrix of a 2-D e^{th} truss element.

$$[k^{1(e)}] = \left(\frac{AE}{L}\right) \begin{bmatrix} 1 & 0 & -1 & 0 \\ 0 & 0 & 0 & 0 \\ -1 & 0 & 1 & 0 \\ 0 & 0 & 0 & 0 \end{bmatrix} \qquad (1.180)$$

The 4x4 global element stiffness matrix of a 2-D truss element can be computed from Eq.(1.81) as:

$$[k^{(e)}]_{4x4} = \begin{bmatrix} [\lambda] & [0] \\ [0] & [\lambda] \end{bmatrix}^T \begin{bmatrix} 1 & 0 & -1 & 0 \\ 0 & 0 & 0 & 0 \\ -1 & 0 & 1 & 0 \\ 0 & 0 & 0 & 0 \end{bmatrix} \begin{bmatrix} [\lambda] & [0] \\ [0] & [\lambda] \end{bmatrix} \left(\frac{AE}{L} \right) \qquad (1.181)$$

The 2x2 submatrix $[\lambda]$ in the above equation has already been defined in Eq.(1.75)

1.10 Beam (or Frame) Finite Element Equations

A typical beam element, its local/global coordinate axis, and its local displacements (or degree -of- freedom, dof) are shown in Figure 1.8. Axial dof is usually neglected and there are only 4 dof associated with a 2-D beam (or frame) element:

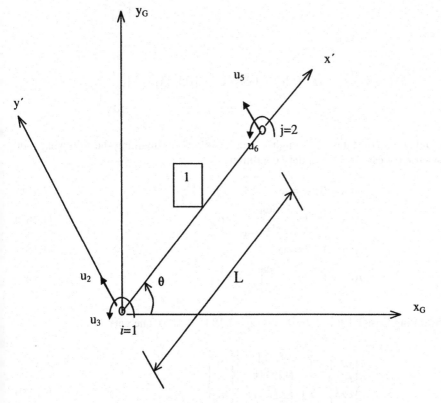

Figure 1.8 a Typical 2-D Beam Element without Axial dof

In the Euler-Bernoulli beam theory, it is assumed that a plane cross-section remains perpendicular to the beam axis even after deformation. The transverse deflection of the beam $\omega(x)$ is governed by the following 4th order differential equation

$$\frac{d^2}{dx^2}\left(D\frac{d^2\omega}{dx^2}\right) = f(x), \quad for\ 0 \le x \le L \tag{1.182}$$

where f(x) is the transverse distributed loads acting on a beam and the material matrix $[D]_{1x1}$ is defined as:

$$[D]_{1x1} = EI \tag{1.183}$$

The transverse displacement field $\omega(x)$ within a 4 dof 2-D beam element can be given as:

$$\omega(x) = a_1 + a_2 x + a_3 x^2 + a_4 x^3 = [1, x, x^2, x^3]\begin{Bmatrix} a_1 \\ a_2 \\ a_3 \\ a_4 \end{Bmatrix} \tag{1.184}$$

Hence

$$\omega'(x) = \frac{d\omega}{dx} = a_2 + 2a_3 x + 3a_4 x^2 = [0,1,2x,3x^2]\begin{Bmatrix} a_1 \\ a_2 \\ a_3 \\ a_4 \end{Bmatrix} \tag{1.185}$$

The unknown constants a_1 through a_4 can be solved by invoking the following four geometric (or essential) boundary conditions:

$$At\ x = 0,\ \ \omega = \omega_1 \tag{1.186}$$

$$At\ x = 0,\ \ \omega' = \frac{d\omega}{dx}\bigg|_{x=0} = \theta_2 \tag{1.187}$$

$$At\ x = L,\ \ \omega = \omega_3 \tag{1.188}$$

$$At\ x = 0,\ \ \omega' = \frac{d\omega}{dx}\bigg|_{x=L} = \theta_4 \tag{1.189}$$

Substituting Eqs.(1.186 – 1.189) into Eqs.(1.184 – 1.185), one gets:

$$\begin{Bmatrix} \omega 1 \\ \theta 2 \\ \omega 3 \\ \theta 4 \end{Bmatrix} = \begin{bmatrix} 1,0,0,1 \\ 0,1,0,0 \\ 1, L, L^2, L^3 \\ 0,1,2L,3L^2 \end{bmatrix}\begin{Bmatrix} a_1 \\ a_2 \\ a_3 \\ a_4 \end{Bmatrix} \tag{1.190}$$

The above matrix equation can also be expressed in a compacted notation as:

$$\{u\}_{4x1} = [A]_{4x4}\{a\}_{4x1} \tag{1.191}$$

Hence

$$\{a\} = [A]^{-1}\{u\} \tag{1.192}$$

Substituting Eq.(1.192) into Eq.(1.184), one obtains:

$$\omega(x) = [1, x, x^2, x^3]_{1x4}[A]_{4x4}^{-1}\{u\}_{4x1} \tag{1.193}$$

The product of the first two terms of the right-hand side of Eq.(1.193) can be recognized as the beam element shape functions N(x), hence

$$\omega(x) = [N(x)]_{1x4} * \{u\}_{4x1} \tag{1.194}$$

where the shape functions [N(x)] are given as:

$$[N(x)]_{1x4} = [1, x, x^2, x^3]\begin{bmatrix} 1 & 0 & 0 & 0 \\ 0 & 1 & 0 & 0 \\ 1 & L & L^2 & L^3 \\ 0 & 1 & 2L & 3L^2 \end{bmatrix}^{-1} \tag{1.195}$$

$$[N(x)] = \left\{ \frac{L^3 - 3Lx^2 + 2x^3}{L^3}, \frac{x(L^2 - 2Lx + x^2)}{L^2}, \frac{-x^2(-3L + 2x)}{L^3}, \frac{x^2(-L + x)}{L^2} \right\} \tag{1.196}$$

The distributed load $f(x)=f_0$ K/in acting on the beam member can be converted into the equivalent joint loads by the following formula:

$$\{F_{equiv}\} = \int_0^L (f_0) * [N(x)]^T \, dx \tag{1.197}$$

or

$$\{F_{equiv}\} = \left\{ \begin{array}{c} \dfrac{f_0 * L}{2} \\ \dfrac{f_0 * L^2}{12} \\ \dfrac{f_0 * L}{2} \\ -\dfrac{f_0 * L^2}{12} \end{array} \right\} \tag{1.198}$$

1.11 Tetrahedral Finite Element Shape Functions

The governing 3-D Poisson equation can be given as:

$$-\frac{\partial}{\partial x}\left(c_1 \frac{\partial \omega}{\partial x}\right) - \frac{\partial}{\partial y}\left(c_2 \frac{\partial \omega}{\partial y}\right) - \frac{\partial}{\partial z}\left(c_3 \frac{\partial \omega}{\partial z}\right) = f \text{ in } \Omega \tag{1.199}$$

with the following geometric boundary condition(s):

$$\omega = \omega_0 \text{ on } \Gamma_1$$

and the natural boundary condition(s):

$$c_1 \frac{\partial \omega}{\partial x} n_x + c_2 \frac{\partial \omega}{\partial y} n_y + c_3 \frac{\partial \omega}{\partial z} n_z = q_0 \text{ on } \Gamma_2 \qquad (1.200)$$

where $c_i = c_i(x,y,z)$ and $f = f(x,y,z)$ are given functions on the boundaries Γ_1 and Γ_2, respectively.

The weak formulation can be derived by the familiar three-step procedures:

Step 1
Setting the weighted residual of the given differential equation to be zero, thus

$$0 = \int_{\Omega^e} W \left[-\frac{\partial}{\partial x} \left(c_1 \frac{\partial \omega}{\partial x} \right) - \frac{\partial}{\partial y} \left(c_2 \frac{\partial \omega}{\partial y} \right) - \frac{\partial}{\partial z} \left(c_3 \frac{\partial \omega}{\partial z} \right) - f \right] d\Omega \qquad (1.201)$$

where $d\Omega \equiv dxdydz$ (1.202)

Step 2
Eq.(1.201) can be integrated by parts once, to give:

$$0 = \int_{\Gamma^e} W \left[-c_1 \frac{\partial \omega}{\partial x} n_x - c_2 \frac{\partial \omega}{\partial y} n_y - c_3 \frac{\partial \omega}{\partial z} n_z \right]$$

$$- \int_{\Omega^e} \left[-c_1 \frac{\partial \omega}{\partial x} \frac{\partial W}{\partial x} - c_2 \frac{\partial \omega}{\partial y} \frac{\partial W}{\partial y} - c_3 \frac{\partial \omega}{\partial z} \frac{\partial W}{\partial z} + Wf \right] d\Omega \qquad (1.203)$$

Step 3
Let

$$q_n \equiv c_1 \frac{\partial \omega}{\partial x} n_x + c_2 \frac{\partial \omega}{\partial y} n_y + c_3 \frac{\partial \omega}{\partial z} n_z \qquad (1.204)$$

Then, Eq.(1.203) can be re-written as:

$$0 = \int_{\Omega^e} \left[c_1 \frac{\partial \omega}{\partial x} \frac{\partial W}{\partial x} + c_2 \frac{\partial \omega}{\partial y} \frac{\partial W}{\partial y} + c_3 \frac{\partial \omega}{\partial z} \frac{\partial W}{\partial z} - Wf \right] d\Omega - \oint_{\Gamma^e} W q_n \, d\Gamma \qquad (1.205)$$

The primary dependent function ω can be assumed as:

$$\omega = \sum_{j=1}^{n} \omega_j N_j^e(x, y, z) \equiv [N(x, y, z)]_{1xn} * \begin{Bmatrix} \omega_1 \\ \omega_2 \\ \vdots \\ \omega_n \end{Bmatrix}_{nx1} \qquad (1.206)$$

In Eq.(1.206), n, ω_j, and N_j represent the number of dof per element, element nodal displacements, and element shape functions, respectively.

For a 4-node tetrahedral element (see Figure 1.9) n = 4, the assumed field can be given as:

$$\omega(x, y, z) = a_1 + (a_2x + a_3y + a_4z) \tag{1.207}$$

or

$$\omega(x, y, z) = [1, x, y, z] * \begin{Bmatrix} a_1 \\ a_2 \\ a_3 \\ a_4 \end{Bmatrix} \tag{1.208}$$

For an 8-node brick element (see Figure 1.9), n = 8, the assumed field can be given as:

$$\omega(x,y,z) = a_1 + (a_2x + a_3y + a_4z) + (a_5xy + a_6yz + a_7zx) + (a_8xyz) \tag{1.209}$$

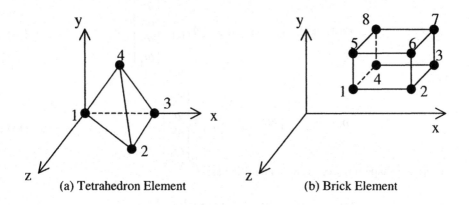

(a) Tetrahedron Element   (b) Brick Element

Figure 1.9 Three-Dimensional Solid Elements

The shape functions for the 4-node tetrahedral element can be obtained by the same familiar procedures. The geometric boundary conditions associated with an eth element are given as:

$$\left.\begin{array}{l} \text{At node 1: } x = x_1;\ y = y_1;\ z = z_1, \text{ then } \omega = \omega_1 \\ \vdots \\ \text{At node 4: } x = x_4;\ y = y_4;\ z = z_4, \text{ then } \omega = \omega_4 \end{array}\right\} \tag{1.210}$$

Substituting Eq.(1.210) into Eq.(1.208), one obtains:

$$\begin{Bmatrix} \omega_1 \\ \omega_2 \\ \omega_3 \\ \omega_4 \end{Bmatrix} = \begin{bmatrix} 1 & x_1 & y_1 & z_1 \\ 1 & x_2 & y_2 & z_2 \\ 1 & x_3 & y_3 & z_3 \\ 1 & x_4 & y_4 & z_4 \end{bmatrix} \begin{Bmatrix} a_1 \\ a_2 \\ a_3 \\ a_4 \end{Bmatrix} \tag{1.211}$$

In a more compacted notation, Eq.(1.211) can be expressed as:

$$\vec{\omega} = [A]_{4x4} \{a\}_{4x1} \tag{1.212}$$

From Eq.(1.212), one gets:

$$\{a\} = [A]^{-1} \vec{\omega} \tag{1.213}$$

Substituting Eq.(1.213) into Eq.(1.208), one obtains:

$$\omega(x, y, z) = [1, x, y, z]_{1x4} * [A]_{4x4}^{-1} * \begin{Bmatrix} \omega_1 \\ \omega_2 \\ \omega_3 \\ \omega_4 \end{Bmatrix} \tag{1.214}$$

or

$$\omega(x, y, z) = [N(x, y, z)]_{1x4} * \begin{Bmatrix} \omega_1 \\ \omega_2 \\ \omega_3 \\ \omega_4 \end{Bmatrix} \tag{1.215}$$

where the shape functions can be identified as:

$$[N(x, y, z)]_{1x4} \equiv [1, x, y, z] * [A]^{-1} \tag{1.216}$$

Let $W = N_i(x, y, z)$, for $i = 1, 2, 3, 4$ (tetrahedral) $\tag{1.217}$

and substituting Eq.(1.215) into Eq.(1.205), one obtains the following (finite) element equations:

$$0 = \int_{\Omega^e} \left[c_1 \frac{\partial(N_j \omega_j)}{\partial x} \frac{\partial N_i}{\partial x} + c_2 \frac{\partial(N_j \omega_j)}{\partial y} \frac{\partial N_i}{\partial y} + c_3 \frac{\partial(N_j \omega_j)}{\partial z} \frac{\partial N_i}{\partial z} \right] d\Omega$$
$$- \int_{\Omega^e} N_i f d\Omega - \oint_{\Gamma^e} N_i q_n d\Gamma \tag{1.218}$$

or

$$\int_{\Omega^e} \left[c_1 \frac{\partial N_j}{\partial x} \frac{\partial N_i}{\partial x} + c_2 \frac{\partial N_j}{\partial y} \frac{\partial N_i}{\partial y} + c_3 \frac{\partial N_j}{\partial z} \frac{\partial N_i}{\partial z} \right] d\Omega * \{\omega_j\}$$
$$= \int_{\Omega^e} N_i f d\Omega + \oint_{\Gamma^e} N_i q_n d\Gamma \tag{1.219}$$

or

$$[k_{ij}^{(e)}]_{4 \times 4} * \{\omega_j^{(e)}\}_{4 \times 1} = \{F_i^{(e)}\} \tag{1.220}$$

where

$$[k_{ij}^{(e)}] \equiv \int_{\Omega^e} \left[c_1 \frac{\partial N_j}{\partial x} \frac{\partial N_i}{\partial x} + c_2 \frac{\partial N_j}{\partial y} \frac{\partial N_i}{\partial y} + c_3 \frac{\partial N_j}{\partial z} \frac{\partial N_i}{\partial z} \right] d\Omega$$

$$\{F_i^{(e)}\}_{4 \times 1} = \int_{\Omega^e} N_i f d\Omega + \oint_{\Gamma^e} N_i q_n d\Gamma \tag{1.221}$$

The first term on the right side of Eq.(1.221) represents the equivalent joint loads due to the distributed "body" force f, while the second term represents the equivalent joint loads due to the distributed "boundary" force q_n.

1.12 Finite Element Weak Formulations for General 2-D Field Equations

The two-dimensional time-dependent field equation can be assumed in the following form:

$$c_1 \frac{\partial^2 u}{\partial x^2} + c_2 \frac{\partial^2 u}{\partial y^2} + c_3 \frac{\partial^2 u}{\partial x \partial y} + c_4 \frac{\partial u}{\partial x} + c_5 \frac{\partial u}{\partial y} + c_6 u^2 + c_7 u + c_8$$
$$+ c_{11} ctg(u) \frac{\partial u}{\partial x} + c_{12} ctg(x) \frac{\partial u}{\partial x} = c_9 \frac{\partial^2 u}{\partial t^2} + c_{10} \frac{\partial u}{\partial t} \tag{1.222}$$

where $c_i, i = 1 - 12$ are constants; $u = u(x, y, t)$

It should be noted that the terms associated with constants c_{11} and c_{12} are included for handling other special applications [1.14].

The weighted residual equation can be established by the familiar procedure

$$\iint_{\Omega^e} w(c_1 \frac{\partial^2 u}{\partial x^2} + c_2 \frac{\partial^2 u}{\partial y^2} + c_3 \frac{\partial^2 u}{\partial x \partial y} + c_4 \frac{\partial u}{\partial x} + c_5 \frac{\partial u}{\partial y} + c_6 u^2 + c_7 u + c_8$$
$$- c_9 \frac{\partial^2 u}{\partial t^2} - c_{10} \frac{\partial u}{\partial t} + c_{11} ctg(u) \frac{\partial u}{\partial x} + c_{12} ctg(x) \frac{\partial u}{\partial x}) dx dy = 0 \tag{1.223}$$

where $w \equiv$ Weighting functions.

The following relationships can be established through integration by parts:

$$c_1 \iint_{\Omega^e} w \frac{\partial^2 u}{\partial x^2} dxdy$$

$$= c_1 \iint_{\Omega^e} w \frac{\partial}{\partial x}(\frac{\partial u}{\partial x}) dxdy$$

$$= c_1 \iint_{\Omega^e} [\frac{\partial}{\partial x}\left(w \frac{\partial u}{\partial x}\right) - \frac{\partial w}{\partial x}\frac{\partial u}{\partial x}] dxdy \qquad (1.224)$$

$$= c_1 \oint_{\Gamma^e} w \frac{\partial u}{\partial x} n_x ds - c_1 \iint_{\Omega^e} \frac{\partial w}{\partial x}\frac{\partial u}{\partial x} dxdy ;$$

$$c_2 \iint_{\Omega^e} w \frac{\partial^2 u}{\partial y^2} dxdy$$

$$= c_2 \iint_{\Omega^e} w \frac{\partial}{\partial y}(\frac{\partial u}{\partial y}) dxdy$$

$$= c_2 \iint_{\Omega^e} [\frac{\partial}{\partial y}\left(w \frac{\partial u}{\partial y}\right) - \frac{\partial w}{\partial y}\frac{\partial u}{\partial y}] dxdy \qquad (1.225)$$

$$= c_2 \oint_{\Gamma^e} w \frac{\partial u}{\partial y} n_y ds - c_2 \iint_{\Omega^e} \frac{\partial w}{\partial y}\frac{\partial u}{\partial y} dxdy ;$$

$$c_3 \iint_{\Omega^e} w \frac{\partial^2 u}{\partial x \partial y} dxdy$$

$$= \frac{c_3}{2} \iint_{\Omega^e} w \frac{\partial}{\partial x}(\frac{\partial u}{\partial y}) dxdy + \frac{c_3}{2} \iint_{\Omega^e} w \frac{\partial}{\partial y}(\frac{\partial u}{\partial x}) dxdy$$

$$= \frac{c_3}{2} \iint_{\Omega^e} [\frac{\partial}{\partial x}\left(w \frac{\partial u}{\partial y}\right) - \frac{\partial w}{\partial x}\frac{\partial u}{\partial y}] dxdy + \frac{c_3}{2} \iint_{\Omega^e} [\frac{\partial}{\partial y}\left(w \frac{\partial u}{\partial x}\right) - \frac{\partial w}{\partial y}\frac{\partial u}{\partial x}] dxdy$$

$$= \frac{c_3}{2} \oint_{\Gamma^e} w \frac{\partial u}{\partial y} n_x ds + \frac{c_3}{2} \oint_{\Gamma^e} w \frac{\partial u}{\partial x} n_y ds - \frac{c_3}{2} \iint_{\Omega^e} \frac{\partial w}{\partial x}\frac{\partial u}{\partial y} dxdy - \frac{c_3}{2} \iint_{\Omega^e} \frac{\partial w}{\partial y}\frac{\partial u}{\partial x} dxdy;$$

$$(1.226)$$

Substituting Eqs.(1.224 – 1.226) into Eq.(1.223), one gets:

$$\iint_{\Omega^e}(-c_1\frac{\partial w}{\partial x}\frac{\partial u}{\partial x}-c_2\frac{\partial w}{\partial y}\frac{\partial u}{\partial y}-\frac{c_3}{2}\frac{\partial w}{\partial x}\frac{\partial u}{\partial y}-\frac{c_3}{2}\frac{\partial w}{\partial y}\frac{\partial u}{\partial x}-c_4 w\frac{\partial u}{\partial x}-c_5 w\frac{\partial u}{\partial y}$$

$$+c_6 wu^2+c_7 wu+c_8 w-c_9 w\frac{\partial^2 u}{\partial t^2}-c_{10} w\frac{\partial u}{\partial t}$$

$$+c_{11} wctg(u)\frac{\partial u}{\partial x}+c_{12} wctg(x)\frac{\partial u}{\partial x})dxdy$$

$$+\oint_{\Gamma^e} w[n_x\left(c_1\frac{\partial u}{\partial x}+\frac{c_3}{2}\frac{\partial u}{\partial y}\right)+n_y\left(c_2\frac{\partial u}{\partial y}+\frac{c_3}{2}\frac{\partial u}{\partial x}\right)]ds=0$$

$$(1.227)$$

Let

$$n_x\left(c_1\frac{\partial u}{\partial x}+\frac{c_3}{2}\frac{\partial u}{\partial y}\right)+n_y\left(c_2\frac{\partial u}{\partial y}+\frac{c_3}{2}\frac{\partial u}{\partial x}\right)\equiv q_n \qquad (1.228)$$

Then Eq.(1.227) becomes:

$$\iint_{\Omega^e}(-c_1\frac{\partial w}{\partial x}\frac{\partial u}{\partial x}-c_2\frac{\partial w}{\partial y}\frac{\partial u}{\partial y}-\frac{c_3}{2}\frac{\partial w}{\partial x}\frac{\partial u}{\partial y}-\frac{c_3}{2}\frac{\partial w}{\partial y}\frac{\partial u}{\partial x}-c_4 w\frac{\partial u}{\partial x}$$

$$-c_5 w\frac{\partial u}{\partial y}+c_6 wu^2+c_7 wu+c_8 w-c_9 w\frac{\partial^2 u}{\partial t^2}-c_{10} w\frac{\partial u}{\partial t}$$

$$+c_{11} wctg(u)\frac{\partial u}{\partial x}+c_{12} wctg(x)\frac{\partial u}{\partial x})dxdy+\oint_{\Gamma^e} wq_n ds=0$$

$$(1.229)$$

The dependent variable field u(x,y,t) is assumed to be in the following form:

$$u(x,y,t)\approx\sum_{j=1}^{n}u_j^e(t)\psi_j^e(x,y), \text{ where } n\equiv \text{ the dof per element} \qquad (1.230)$$

Let the weighting function $w=\psi_i^e(x,y)$, (see Eq.[1.223]), then Eq.(1.229) becomes:

$$\iint_{\Omega^e} \left(\begin{array}{l} -c_1 \dfrac{\partial \psi_i}{\partial x}(\sum_{j=1}^{n} u_j \dfrac{\partial \psi_j}{\partial x}) - c_2 \dfrac{\partial \psi_i}{\partial y}(\sum_{j=1}^{n} u_j \dfrac{\partial \psi_j}{\partial y}) \\[2mm] -\dfrac{c_3}{2}\dfrac{\partial \psi_i}{\partial x}(\sum_{j=1}^{n} u_j \dfrac{\partial \psi_j}{\partial y}) - \dfrac{c_3}{2}\dfrac{\partial \psi_i}{\partial y}(\sum_{j=1}^{n} u_j \dfrac{\partial \psi_j}{\partial x}) \\[2mm] -c_4 \psi_i(\sum_{j=1}^{n} u_j \dfrac{\partial \psi_j}{\partial x}) - c_5 \psi_i(\sum_{j=1}^{n} u_j \dfrac{\partial \psi_j}{\partial y}) + c_6 \psi_i(\sum_{j=1}^{n} u_j \psi_j)u \\[2mm] +c_7 \psi_i(\sum_{j=1}^{n} u_j \psi_j) + c_8 \psi_i + c_{11}\psi_i ctg(u)(\sum_{j=1}^{n} u_j \dfrac{\partial \psi_j}{\partial x}) \\[2mm] +c_{12}\psi_i ctg(x)(\sum_{j=1}^{n} u_j \dfrac{\partial \psi_j}{\partial x}) - c_9 \psi_i(\sum_{j=1}^{n} \dfrac{d^2 u}{dt^2}\psi_j) \\[2mm] -c_{10}\psi_i(\sum_{j=1}^{n} \dfrac{du}{dt}\psi_j) \end{array} \right) dxdy$$

$$+ \iint_{\Gamma^e} w q_n ds = 0$$

$$(1.231)$$

Eq.(1.231) can also be expressed as:

$$\sum_{j=1}^{n} \iint_{\Omega^e} \left[\begin{array}{l} (-c_1 \dfrac{\partial \psi_i}{\partial x}\dfrac{\partial \psi_j}{\partial x} - c_2 \dfrac{\partial \psi_i}{\partial y}\dfrac{\partial \psi_j}{\partial y} - \dfrac{c_3}{2}\dfrac{\partial \psi_i}{\partial x}\dfrac{\partial \psi_j}{\partial y} \\[2mm] -\dfrac{c_3}{2}\dfrac{\partial \psi_i}{\partial y}\dfrac{\partial \psi_j}{\partial x} - c_4 \psi_i \dfrac{\partial \psi_j}{\partial x} - c_5 \psi_i \dfrac{\partial \psi_j}{\partial y} + c_6 u \psi_i \psi_j \\[2mm] +c_7 \psi_i \psi_j + c_{11}\psi_i ctgu \dfrac{\partial \psi_j}{\partial x} + c_{12}\psi_i ctgx \dfrac{\partial \psi_j}{\partial x})u_j \\[2mm] -c_9 \psi_i \psi_j \dfrac{d^2 u_j}{dt^2} - c_{10}\psi_i \psi_j \dfrac{du_j}{dt}) \end{array} \right] dxdy$$

$$+ \iint_{\Omega^e} c_8 \psi_i dxdy + \oint_{\Gamma^e} w q_n ds = 0$$

$$(1.232)$$

In matrix form, Eq.(1.232) becomes:

$$[K^e]\{u^e\} + [C^e]\{\dot{u}^e\} + [M^e]\{\ddot{u}^e\} = \{f^e\} + \{Q^e\} \qquad (1.233)$$

where

$$K_{ij}^e = \iint\limits_{\Omega^e} [(-c_1 \frac{\partial \psi_i}{\partial x}\frac{\partial \psi_j}{\partial x} - c_2 \frac{\partial \psi_i}{\partial y}\frac{\partial \psi_j}{\partial y} - \frac{c_3}{2}\frac{\partial \psi_i}{\partial x}\frac{\partial \psi_j}{\partial y} - \frac{c_3}{2}\frac{\partial \psi_i}{\partial y}\frac{\partial \psi_j}{\partial x} - c_4 \psi_i \frac{\partial \psi_j}{\partial x}$$

$$-c_5 \psi_i \frac{\partial \psi_j}{\partial x} + c_6 u \psi_i \psi_j + c_7 \psi_i \psi_j + c_{11}ctg(u)\psi_i \frac{\partial \psi_j}{\partial x} + c_{12}ctg(x)\psi_i \frac{\partial \psi_j}{\partial x}]dxdy$$

$$(1.234)$$

$$C_{ij}^e = -\iint\limits_{\Omega^e} c_{10}\psi_i\psi_j dxdy \qquad (1.235)$$

$$M_{ij}^e = -\iint\limits_{\Omega^e} c_9\psi_i\psi_j dxdy \qquad (1.236)$$

$$f_i^e = -\iint\limits_{\Omega^e} c_8\psi_i dxdy \qquad (1.237)$$

$$Q_i^e = -\oint\limits_{\Gamma^e} \psi_i q_n ds \qquad (1.238)$$

Note that matrix K is <u>nonlinear</u> when $c_6 \neq 0$ or $c_{11} \neq 0$.

For a 3-node triangular element (see Figure 1.10),

Let $\alpha_i = x_j y_k - x_k y_j$, $\beta_i = y_j - y_k$, $\gamma_i = -(x_j - x_k)$ $\qquad (1.239)$

where $i \neq j \neq k$; and i, j, and k permute in a natural order. Then the triangular area can be computed as:

$$A = (\alpha_1 + \alpha_2 + \alpha_3) \qquad (1.240$$

$$N_i \equiv \psi_i^e = \frac{1}{A}(\alpha_i^e + \beta_i^e x + \gamma_i^e y) \quad (i = 1,2,3) \qquad (1.241)$$

In local coordinates, suppose that the triangular element has its dimensions, (base and height) a and b, respectively (see Figure 1.10)

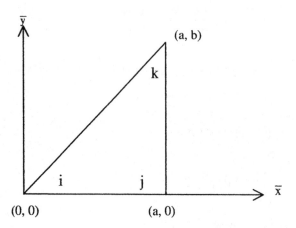

Figure 1.10 a 2-D Triangular Element with Its Local Coordinates

Then Eq.(1.241) becomes:

$$\psi_1^e(\overline{x},\overline{y}) = 1 - \frac{\overline{x}}{a}, \quad \psi_2^e(\overline{x},\overline{y}) = \frac{\overline{x}}{a} - \frac{\overline{y}}{b}, \quad \psi_3^e(\overline{x},\overline{y}) = \frac{\overline{y}}{b} \qquad (1.242)$$

For the 4-node rectangular element (see Figure 1.11), the corresponding shape functions in the local coordinates can be given as:

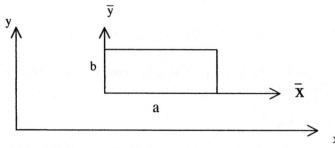

Figure 1.11 a 2-D Rectangular Element with Its Local Coordinates

$$\psi_1^e = (1-\frac{\overline{x}}{a})(1-\frac{\overline{y}}{b}), \quad \psi_2^e = \frac{\overline{x}}{a}(1-\frac{\overline{y}}{b}), \quad \psi_3^e = \frac{\overline{x}}{a}\frac{\overline{y}}{b}, \quad \psi_4^e = (1-\frac{\overline{x}}{a})\frac{\overline{y}}{b} \qquad (1.243)$$

To facilitate the computation of Eq.(1.234), K_{ij}^e can be expressed as the sum of several basic matrices $[S^{\alpha\beta}]$ $(\alpha,\beta = 0,1,2)$.

Thus

$$\left[K^e\right] = -c_1\left[S^{11}\right] - c_2\left[S^{22}\right] - \frac{c_3}{2}\left[S^{12}\right] - \frac{c_3}{2}\left[S^{12}\right]^T - c_4\left[S^{01}\right]$$
$$- c_5\left[S^{02}\right] + c_6 u\left[S^{00}\right] + c_7\left[S^{00}\right] + c_{11}ctg(u)\left[S^{01}\right] + c_{12}ctg(x)\left[S^{01}\right] \tag{1.244}$$

$$[C^e] = -c_{10}[S^{00}] \tag{1.245}$$

$$[M^e] = -c_9[S^{00}] \tag{1.246}$$

where $[\ ]^T$ denotes the transpose of the enclosed matrix or vector, and

$$S_{ij}^{\alpha\beta} = \int\limits_{\Omega^e} \psi_{i,\alpha}\psi_{j,\beta}dxdy \tag{1.247}$$

with

$$\psi_{i,\alpha} = \frac{\partial\psi_i}{\partial x_\alpha}, x_1 = x, x_2 = y; \psi_{i,0} = \psi_i, \psi_{j,0} = \psi_j \tag{1.248}$$

Using the Linear Triangular Element and let

$$I_{mn} = \int\limits_{\Delta} x^m y^n dxdy \tag{1.249}$$

Then

$$I_{00} = A, I_{10} = A\hat{x}, \hat{x} = \frac{1}{3}\sum_{i=1}^{3} x_i, \ I_{01} = A\hat{y}, \hat{y} = \frac{1}{3}\sum_{i=1}^{3} y_i \tag{1.250}$$

$$I_{11} = \frac{A}{12}\left(\sum_{i=1}^{3} x_i y_i + 9\hat{x}\hat{y}\right), I_{20} = \frac{A}{12}\left(\sum_{i=1}^{3} x_i^2 + 9\hat{x}^2\right), I_{02} = \frac{A}{12}\left(\sum_{i=1}^{3} y_i^2 + 9\hat{y}^2\right) \tag{1.251}$$

From Eq.(1.247), one obtains:

$$S_{ij}^{11} = \frac{1}{4A}\beta_i\beta_j, \ S_{ij}^{12} = \frac{1}{4A}\beta_i\gamma_j, \ S_{ij}^{22} = \frac{1}{4A}\gamma_i\gamma_j \tag{1.252}$$

$$S_{ij}^{01} = \frac{1}{4A}(\alpha_i\beta_j + \beta_i\beta_j\hat{x} + \gamma_i\beta_j\hat{y}), S_{ij}^{02} = \frac{1}{4A}(\alpha_i\gamma_j + \beta_i\gamma_j\hat{x} + \gamma_i\gamma_j\hat{y}) \tag{1.253}$$

$$S_{ij}^{00} = \frac{1}{4A}[\alpha_i\alpha_j + (\alpha_i\beta_j + \alpha_j\beta_i)\hat{x} + (\alpha_i\gamma_j + \alpha_j\gamma_i)\hat{y}] + \frac{1}{4A^2}[I_{20}\beta_i\beta_j$$
$$+ I_{11}(\gamma_i\beta_j + \gamma_j\beta_i) + I_{02}\gamma_i\gamma_j] \tag{1.254}$$

Using the special cases (triangular element local coordinates; see Fig. 1.10), one has:

$$\left[S^{11}\right] = \frac{b}{2a}\begin{bmatrix} 1 & -1 & 0 \\ -1 & 1 & 0 \\ 0 & 0 & 0 \end{bmatrix}, \left[S^{22}\right] = \frac{a}{2b}\begin{bmatrix} 0 & 0 & 0 \\ 0 & 1 & -1 \\ 0 & -1 & 1 \end{bmatrix}, \left[S^{12}\right] = \frac{1}{2}\begin{bmatrix} 0 & 1 & -1 \\ 0 & -1 & 1 \\ 0 & 0 & 0 \end{bmatrix} \tag{1.255}$$

$$\left[S^{01}\right] = \frac{b}{6}\begin{bmatrix} -1 & 1 & 0 \\ -1 & 1 & 0 \\ -1 & 1 & 0 \end{bmatrix}, \left[S^{02}\right] = \frac{a}{6}\begin{bmatrix} 0 & -1 & 1 \\ 0 & -1 & 1 \\ 0 & -1 & 1 \end{bmatrix}, \left[S^{00}\right] = \frac{ab}{24}\begin{bmatrix} 2 & 1 & 1 \\ 1 & 2 & 1 \\ 1 & 1 & 2 \end{bmatrix} \tag{1.256}$$

$$f_i^e = -\iint_{\Omega^e} c_8 \psi_i dxdy = -c_8 \frac{ab}{6}\{1 \quad 1 \quad 1\}^T \tag{1.257}$$

Similarly, using rectangular element (local coordinates), one has:

$$\left[S^{11}\right] = \frac{b}{6a}\begin{bmatrix} 2 & -2 & -1 & 1 \\ -2 & 2 & 1 & -1 \\ -1 & 1 & 2 & -2 \\ 1 & -1 & -2 & 2 \end{bmatrix}, \left[S^{22}\right] = \frac{a}{6b}\begin{bmatrix} 2 & 1 & -1 & -2 \\ 1 & 2 & -2 & -1 \\ -1 & -2 & 2 & 1 \\ -2 & -1 & 1 & 2 \end{bmatrix} \tag{1.258}$$

$$\left[S^{12}\right] = \frac{1}{4}\begin{bmatrix} 1 & 1 & -1 & -1 \\ -1 & -1 & 1 & 1 \\ -1 & -1 & 1 & 1 \\ 1 & 1 & -1 & -1 \end{bmatrix}, \left[S^{01}\right] = \frac{b}{12}\begin{bmatrix} -2 & 2 & 1 & -1 \\ -2 & 2 & 1 & -1 \\ -1 & 1 & 2 & -2 \\ -1 & 1 & 2 & -2 \end{bmatrix} \tag{1.259}$$

$$\left[S^{02}\right] = \frac{a}{12}\begin{bmatrix} -2 & -1 & 1 & 2 \\ -1 & -2 & 2 & 1 \\ -1 & -2 & 2 & 1 \\ -2 & -1 & 1 & 2 \end{bmatrix}, \left[S^{00}\right] = \frac{ab}{36}\begin{bmatrix} 4 & 2 & 1 & 2 \\ 2 & 4 & 2 & 1 \\ 1 & 2 & 4 & 2 \\ 2 & 1 & 2 & 4 \end{bmatrix} \tag{1.260}$$

$$f_i^e = -\iint_{\Omega^e} c_8 \psi_i dxdy = -c_8 \frac{ab}{4}\{1 \quad 1 \quad 1 \quad 1\}^T \tag{1.261}$$

The following cases are considered for the purposes of comparing the above derived formulas with [1.1]:

Case 1 Let's define

$$c_1 = -a_{11}, c_2 = -a_{22}, \frac{c_3}{2} = -a_{12}, \frac{c_3}{2} = -a_{21}, c_7 = a_{00},$$
$$c_8 = -f, c_4 = c_5 = c_6 = c_9 = c_{10} = c_{11} = c_{12} = 0 \tag{1.262}$$

Then, Eq.(1.222) will be simplified to J. N. Reddy's Eq.(8.1) [1.1]. Substituting Eq.(1.262) into Eqs.(1.234, 1.237), one obtains:

$$K_{ij}^e = \iint_{\Omega^e} [(a_{11} \frac{\partial \psi_i}{\partial x} \frac{\partial \psi_j}{\partial x} + a_{22} \frac{\partial \psi_i}{\partial y} \frac{\partial \psi_j}{\partial y} + a_{12} \frac{\partial \psi_i}{\partial x} \frac{\partial \psi_j}{\partial y} + a_{21} \frac{\partial \psi_i}{\partial y} \frac{\partial \psi_j}{\partial x} + a_{00} \psi_i \psi_j] dxdy$$

$$= \iint_{\Omega^e} [\frac{\partial \psi_j}{\partial x} (a_{11} \frac{\partial \psi_i}{\partial x} + a_{12} \frac{\partial \psi_i}{\partial y}) + \frac{\partial \psi_i}{\partial y} (a_{22} \frac{\partial \psi_j}{\partial y} + a_{21} \frac{\partial \psi_j}{\partial x}) + a_{00} \psi_i \psi_j] dxdy$$

$$(1.263)$$

$$f_i^e = -\iint_{\Omega^e} c_8 \psi_i dxdy = \iint_{\Omega^e} f \psi_i dxdy \qquad (1.264)$$

Again, Eqs.(1.263 – 1.264) are the same as J. N. Reddy's Eq.(8.14b) [1.1].

Case 2 Let's define

$$c_1 = -a_{11}, c_2 = -a_{22}, c_3 = c_4 = c_5 = c_6 = c_{11} = c_{12} = 0,$$
$$c_7 = a_0, c_8 = -f, c_9 = 0, c_{10} = -c \qquad (1.265)$$

Then, Eq.(1.222) will be simplified to J. N. Reddy's Eq.(8.154) [1.1]. Substituting Eq.(1.265) into Eqs.(1.234, 1.235, 1.237), one obtains:

$$K_{ij}^e = \iint_{\Omega^e} [(a_{11} \frac{\partial \psi_i}{\partial x} \frac{\partial \psi_j}{\partial x} + a_{22} \frac{\partial \psi_i}{\partial y} \frac{\partial \psi_j}{\partial y} + a_0 \psi_i \psi_j] dxdy \qquad (1.266)$$

$$C_{ij}^e = -\iint_{\Omega^e} c_{10} \psi_i \psi_j dxdy = \iint_{\Omega^e} c \psi_i \psi_j dxdy \qquad (1.267)$$

$$f_i^e = -\iint_{\Omega^e} c_8 \psi_i dxdy = \iint_{\Omega^e} f \psi_i dxdy \qquad (1.268)$$

Eqs.(1.266 – 1.268) are the same as J. N. Reddy's Eq.(8.159c)[1.1].

Case 3 Let's define

$$c_1 = -a_{11}, c_2 = -a_{22}, c_3 = c_4 = c_5 = c_6 = 0,$$
$$c_7 = a_0, c_8 = -f, c_9 = -c, c_{10} = c_{11} = c_{12} = 0 \qquad (1.269)$$

Then, Eq.(1.222) will be simplified to J. N. Reddy's Eq.(8.167a)[1.1]. Substituting Eq.(1.268) into Eqs.(1.234, 1.236, 1.237), one obtains:

$$K_{ij}^e = \iint_{\Omega^e} [(a_{11} \frac{\partial \psi_i}{\partial x} \frac{\partial \psi_j}{\partial x} + a_{22} \frac{\partial \psi_i}{\partial y} \frac{\partial \psi_j}{\partial y} + a_0 \psi_i \psi_j] dxdy \qquad (1.270)$$

$$M_{ij}^e = - \iint_{\Omega^e} c_9 \psi_i \psi_j dxdy = \iint_{\Omega^e} c \psi_i \psi_j dxdy \qquad (1.271)$$

$$f_i^e = - \iint_{\Omega^e} c_8 \psi_i dxdy = \iint_{\Omega^e} f \psi_i dxdy \qquad (1.272)$$

Eqs.(1.270 – 1.272) are identical to J. N. Reddy's Eq.(8.159c)[1.1].

1.13 The Isoparametric Formulation

Isoparametric formulation is very useful in modeling arbitrary shapes and curve boundaries. The natural coordinate system (ξ, η, ϑ) is used commonly, which can be related to the global (x, y, z) system through a transformation (Jacobian) matrix [J]. Furthermore, for non-rectangular elements (see Figure 1.12), numerical integration rather than analytical integration is normally done.

Displacements or coordinates can be interpolated from their nodal values, hence

$$[\xi \ \eta \ \vartheta]^T = [\ N\ ][d] \qquad (1.273)$$
$$[x \ y \ z]^T = [\ \overline{N}\ ][c] \qquad (1.274)$$

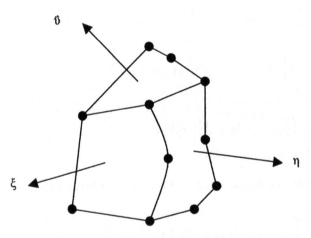

Figure 1.12 Quadratic Solid Element with Some Linear Edges

In Eqs.(1.273 and 1.274), if the shape functions $[N]$ and $[\overline{N}]$ are identical, then the element is an isoparametric element.

CASE 1 Isoparametric Bar Element

Consider a straight, 3-node element with its natural coordinates, ξ, as shown in Figure 1.13:

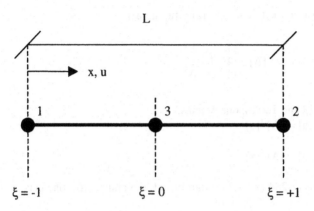

Figure 1.13 a Three-node Bar Element with Its Natural Coordinates ξ.

Utilizing Eqs.(1.273 and 1.274), one has:

$$u = a_1 + a_2\xi + a_3\xi^2 = [1 \ \ \xi \ \ \xi^2][a_1 \ \ a_2 \ \ a_3]^T \qquad (1.275)$$

$$x = a_4 + a_5\xi + a_6\xi^2 = [1 \ \ \xi \ \ \xi^2][a_4 \ \ a_5 \ \ a_6]^T \qquad (1.276)$$

where the a_i are generalized coordinates.
Applying the boundary conditions (see Figure 1.13) into Eqs.1.275 and 1.276 one has:

$$
\begin{aligned}
u_1 &= a_1 + a_2(-1) + a_3(-1)^2 \\
u_2 &= a_1 + a_2(1) \ \ + a_3(1)^2 \\
u_3 &= a_1 + a_2(0) \ \ + a_3(0)^2
\end{aligned}
\qquad (1.277)
$$

and

$$
\begin{aligned}
x_1 &= a_1 + a_2(-1) + a_3(-1)^2 \\
x_2 &= a_1 + a_2(1) \ \ + a_3(1)^2 \\
x_3 &= a_1 + a_2(0) \ \ + a_3(0)^2
\end{aligned}
\qquad (1.278)
$$

Eqs.(1.277 and 1.278) can be expressed in matrix notations

$$
[u_1 \ \ u_2 \ \ u_3]^T = \begin{bmatrix} 1 & -1 & 1 \\ 1 & 1 & 1 \\ 1 & 0 & 0 \end{bmatrix} \times [a_1 \ \ a_2 \ \ a_3]^T \qquad (1.279)
$$

or

$$[u] = [A][a] \qquad (1.279a)$$

and

$$[x] = [A][\bar{a}] \tag{1.280}$$

where

$$[x] = [x_1 \ x_2 \ x_3]^T \quad \text{and} \quad [\bar{a}] = [a_4 \ a_5 \ a_6]^T \tag{1.281}$$

$$[u] = [u_1 \ u_2 \ u_3]^T \quad \text{and} \quad [a] = [a_1 \ a_2 \ a_3]^T \tag{1.282}$$

and

$$[A] = \begin{bmatrix} 1 & -1 & 1 \\ 1 & 1 & 1 \\ 1 & 0 & 0 \end{bmatrix} \tag{1.283}$$

From Eqs.(1.279 and 1.280), one obtains:

$$[a] = [A]^{-1}[u] \tag{1.284}$$

$$[\bar{a}] = [A]^{-1}[x] \tag{1.285}$$

Substituting Eqs.(1.284 and 1.285) into Eqs.(1.275 and 1.276), one gets:

$$u = [1 \ \xi \ \xi^2][A]^{-1}[u] \tag{1.286}$$
$$x = [1 \ \xi \ \xi^2][A]^{-1}[x] \tag{1.287}$$

or

$$u = [N]_{1x3}[u]_{3x1} \tag{1.288}$$

$$x = [N]_{1x3}[x]_{3x1} \tag{1.289}$$

where the "shape function" [N] can be defined as:

$$[N] \equiv N_i = [\tfrac{1}{2}(-\xi + \xi^2) \quad \tfrac{1}{2}(\xi + \xi^2) \quad (1 - \xi^2)] \tag{1.290}$$

Strain-displacement relationship for axial members can be given as:

$$E_{xx} = du/dx = d\ ([N][u])\ /dx = (d[N(\xi)]/dx)[u_1 \ u_2 \ u_3]^T$$
$$= [B][u] \tag{1.291}$$

where

$$d/dx = (d/d\xi)(d\xi/dx) \quad \text{and} \quad d[N(\xi)]/dx \equiv [B] \tag{1.292}$$

From Eq.(1.356), one obtains:

$$dx/d\xi \equiv J = (d/d\xi)[N][x_1 \ x_2 \ x_3]^T = (d[N]/d\xi)[x_1 \ x_2 \ x_3]^T \tag{1.293}$$

$$dx/d\xi \equiv J = [\tfrac{1}{2}(-1 + 2\xi) \quad \tfrac{1}{2}(1 + 2\xi) \quad (-2\xi)][x_1 \ x_2 \ x_3]^T \tag{1.294}$$

where J is called the "Jacobian."
Since $dx/d\xi \equiv J$, hence

$$d\xi/dx = 1/J \tag{1.295}$$

Element stiffness matrix is given by the following familiar expression:

$$[k] = \int_v [B]^T[D][B] \ dv = \int_{x=0}^{L} [B]^T[D \equiv E][B]A \ dx \tag{1.296}$$

$$[k] = \int_{\xi = -1}^{1} [B]^T (AE)[B] J \, d\xi \qquad (1.297)$$

where

$$[B] \equiv (d[N(\xi)]/dx) = (d[N(\xi)]/d\xi)(d\xi/dx = 1/J) \qquad (1.298)$$

CASE 2 Isoparametric Linear Plane Element

Consider a straight-line quadrilateral element as shown in Figure 1.14.

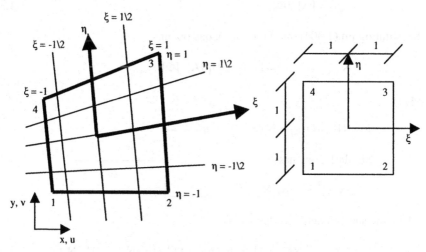

Figure 1.14 Four-Node Plane Isoparametric Element in the x - y and $\xi - \eta$
Space, Respectively.

Case 2.1

Assuming there is only one dof per node (scalar field problem), one has:

$$\Phi = a_1 + a_2\xi + a_3\eta + a_4\xi\eta \qquad (1.299)$$

or

$$\Phi = [1 \; \xi \; \eta \; \xi\eta][a_1 \; a_2 \; a_3 \; a_4]^T \qquad (1.300)$$

Applying the four boundary conditions:

At node 1: $(\xi, \eta) = (-1, -1)$; $\Phi = \Phi_1$
At node 2: $(\xi, \eta) = (1, -1)$; $\Phi = \Phi_2$ (1.301)
At node 3: $(\xi, \eta) = (1, 1)$; $\Phi = \Phi_3$
At node 4: $(\xi, \eta) = (-1, 1)$; $\Phi = \Phi_4$

Then, Eq.(1.300) becomes:

$$[\Phi_1 \quad \Phi_2 \quad \Phi_3 \quad \Phi_4]^T = \begin{bmatrix} 1 & -1 & -1 & 1 \\ 1 & 1 & -1 & -1 \\ 1 & 1 & 1 & 1 \\ 1 & -1 & 1 & -1 \end{bmatrix} \times [a_1 \quad a_2 \quad a_3 \quad a_4]^T \qquad (1.302)$$

or

$$[\Phi] = [A][a] \qquad (1.303)$$

Hence

$$[a] = [A]^{-1}[\Phi] \qquad (1.304)$$

Substituting Eq.(1.304) into Eq.(1.300), one obtains:

$$\Phi = [1 \quad \xi \quad \eta \quad \xi\eta][A]^{-1}[\Phi] \qquad (1.305)$$

or

$$\Phi = [N]_{1 \times 4}[\Phi]_{4 \times 1} = N_i \Phi_i$$

Similarly, one has: $\qquad\qquad\qquad\qquad\qquad$ (1.306)

$$[x \quad y]^T = [N_i x_i \quad N_i y_i]^T$$

where the shape function is defined as:

$$[N]_{1 \times 4} = [1 \quad \xi \quad \eta \quad \xi\eta]_{1 \times 4}[A]^{-1}_{4 \times 4} = [N_1 \quad N_2 \quad N_3 \quad N_4]_{1 \times 4} \qquad (1.307)$$

and

$$\begin{aligned} N_1 &= \tfrac{1}{4}(1 - \xi)(1 - \eta) \\ N_2 &= \tfrac{1}{4}(1 + \xi)(1 - \eta) \\ N_3 &= \tfrac{1}{4}(1 + \xi)(1 + \eta) \\ N_4 &= \tfrac{1}{4}(1 - \xi)(1 + \eta) \end{aligned} \qquad (1.308)$$

Element stiffness matrix [k] is, again, given by Eq.(1.296):

$$[k] = \int_x \int_y [B]^T[D][B] \, t \, dx \, dy \qquad (1.309)$$

where

$$[B] = [(\partial N_i / \partial x \quad \partial N_i / \partial y)]^T = \begin{bmatrix} N_{1,x} & N_{2,x} & N_{3,x} & N_{4,x} \\ N_{1,y} & N_{2,y} & N_{3,y} & N_{4,y} \end{bmatrix} \qquad (1.310)$$

[D], t = material properties and thickness of an element, respectively. (1.311)

In the natural coordinate $\xi - \eta$ system, Eq.(1.310) becomes:

$$[k] = \int_{\xi=-1}^{1} \int_{\eta=-1}^{1} [B]^T [D][B] \, t \, J \, d\xi \, d\eta \qquad (1.312)$$

From the chain rule derivative, one has:

$$\begin{bmatrix} \partial\Phi/\partial x \\ \partial\Phi/\partial y \end{bmatrix} = \begin{bmatrix} (\partial\Phi/\partial\xi)(\partial\xi/\partial x) + (\partial\Phi/\partial\eta)(\partial\eta/\partial x) \\ (\partial\Phi/\partial\xi)(\partial\xi/\partial y) + (\partial\Phi/\partial\eta)(\partial\eta/\partial y) \end{bmatrix} = \begin{bmatrix} \xi_{,x} & \eta_{,x} \\ \xi_{,y} & \eta_{,y} \end{bmatrix} \times \begin{bmatrix} \Phi_{,\xi} & \Phi_{,\eta} \end{bmatrix}^T \qquad (1.313)$$

or

$$[\Phi_{,x} \ \Phi_{,y}]^T = [\Gamma]_{2x2}[\Phi_{,\xi} \ \Phi_{,\eta}]^T = \begin{bmatrix} \Gamma_{11} & \Gamma_{12} \\ \Gamma_{21} & \Gamma_{22} \end{bmatrix} \times [\Phi_{,\xi} \ \Phi_{,\eta}]^T \qquad (1.314)$$

where

$$[\Gamma] \equiv \begin{bmatrix} \xi_{,x} & \eta_{,x} \\ \xi_{,y} & \eta_{,y} \end{bmatrix} \qquad (1.315)$$

Since the partial derivatives of ξ and η with respect to x and y are <u>not</u> directly available, Eq.(1.314) can be re-written as:

$$\partial\Phi/\partial\xi = (\partial\Phi/\partial x)(\partial x/\partial\xi) + (\partial\Phi/\partial y)(\partial y/\partial\xi) \qquad (1.316)$$

$$\partial\Phi/\partial\eta = (\partial\Phi/\partial x)(\partial x/\partial\eta) + (\partial\Phi/\partial y)(\partial y/\partial\eta) \qquad (1.317)$$

or

$$[\Phi_{,\xi} \ \Phi_{,\eta}]^T = \begin{bmatrix} x_{,\xi} & y_{,\xi} \\ x_{,\eta} & y_{,\eta} \end{bmatrix} \times [\Phi_{,x} \ \Phi_{,y}]^T \qquad (1.318)$$

Thus the "Jacobian" matrix [J] can be identified as:

$$[J] = \begin{bmatrix} J_{11} & J_{12} \\ J_{21} & J_{22} \end{bmatrix} = \begin{bmatrix} x_{,\zeta} & y_{,\zeta} \\ x_{,\eta} & y_{,\eta} \end{bmatrix} \qquad (1.319)$$

Since $x = N_i(\xi,\eta)x_i$, for $i = 1,2,3,4$ (see Eq.[1.306]), hence

$$[J] = \begin{bmatrix} N_{i,\xi}x_i & N_{i,\xi}y_i \\ N_{i,\eta}x_i & N_{i,\eta}y_i \end{bmatrix} = [N_{i,\xi}] * \begin{bmatrix} x_1 & y_1 \\ x_2 & y_2 \\ x_3 & y_3 \\ x_4 & y_4 \end{bmatrix}_{4x2} \qquad (1.320)$$

$[N_{i,\eta}]_{2x4.}$

Comparing Eq.(1.314 and 1.318), one recognizes that:

$$[\Gamma] = [J]^{-1} \text{ or } [J] = [\Gamma]^{-1} \qquad (1.321)$$

Using Eq.(1.308), one obtains:

$$N_{i,\xi} = \tfrac{1}{4}[-(1-\eta) \quad (1-\eta) \quad (1+\eta) \quad -(1+\eta)] \qquad (1.322)$$

$$N_{i,\eta} = \tfrac{1}{4}[-(1-\xi) \quad -(1-\xi) \quad (1+\xi) \quad (1+\xi)] \qquad (1.323)$$

The "Jacobian" J (see Eq. 1.312) is the determinant of the "Jacobian" matrix [J], hence

$$J = J_{11}J_{22} - J_{21}J_{12} \qquad (1.324)$$

Case 2.2

If there are two dof per node (plane stress element), then

$$u = \Sigma N_i(\xi,\eta)u_i \quad \text{and} \quad v = \Sigma N_i(\xi,\eta)v_i \qquad (1.325)$$

and the strain-displacement relationship is given as:

$$[\varepsilon] \equiv \begin{bmatrix} \varepsilon_{xx} \\ \varepsilon_{yy} \\ \gamma_{xy} = 2\varepsilon_{xy} \end{bmatrix} = \begin{bmatrix} \partial u/\partial x \\ \partial v/\partial y \\ (\partial u/\partial y)+(\partial v/\partial x) \end{bmatrix} = \begin{bmatrix} 1 & 0 & 0 & 0 \\ 0 & 0 & 0 & 1 \\ 0 & 1 & 1 & 0 \end{bmatrix} \times \begin{bmatrix} u_{,x} \\ u_{,y} \\ v_{,x} \\ v_{,y} \end{bmatrix}$$

Utilizing Eq.(1.314), one gets:

$$\begin{bmatrix} u_{,x} \\ u_{,y} \\ v_{,x} \\ v_{,y} \end{bmatrix} = \begin{bmatrix} \Gamma_{11} & \Gamma_{12} & 0 & 0 \\ \Gamma_{21} & \Gamma_{22} & 0 & 0 \\ 0 & 0 & F_{11} & \Gamma_{12} \\ 0 & 0 & \Gamma_{21} & \Gamma_{22} \end{bmatrix} \times \begin{bmatrix} u_{,\xi} \\ u_{,\eta} \\ v_{,\xi} \\ v_{,\eta} \end{bmatrix} \qquad (1.326)$$

Utilizing Eq.(1.325), one obtains:

$$\begin{bmatrix} u_{,\xi} \\ u_{,\eta} \\ v_{,\xi} \\ v_{,\eta} \end{bmatrix} = \begin{bmatrix} N_{1,\xi} & 0 & N_{2,\xi} & 0 & N_{3,\xi} & 0 & N_{4,\xi} & 0 \\ N_{1,\eta} & 0 & N_{2,\eta} & 0 & N_{3,\eta} & 0 & N_{4,\eta} & 0 \\ 0 & N_{1,\xi} & 0 & N_{2,\xi} & 0 & N_{3,\xi} & 0 & N_{4,\xi} \\ 0 & N_{1,\eta} & 0 & N_{2,\eta} & 0 & N_{3,\eta} & 0 & N_{4,\eta} \end{bmatrix} \times \begin{bmatrix} u_1 \\ v_1 \\ u_2 \\ v_2 \\ u_3 \\ v_3 \\ u_4 \\ v_4 \end{bmatrix} \qquad (1.327)$$

Substituting Eqs.(1.326 and 1.327) into Eq.(1.325), one obtains:

$$[\varepsilon] = \begin{bmatrix} \varepsilon_{xx} \\ \varepsilon_{yy} \\ \gamma_{xy} \end{bmatrix} = \begin{bmatrix} 1 & 0 & 0 & 0 \\ 0 & 0 & 0 & 1 \\ 0 & 1 & 1 & 0 \end{bmatrix} \times \begin{bmatrix} \Gamma_{11} & \Gamma_{12} & 0 & 0 \\ \Gamma_{21} & \Gamma_{22} & 0 & 0 \\ 0 & 0 & \Gamma_{11} & \Gamma_{12} \\ 0 & 0 & \Gamma_{21} & \Gamma_{22} \end{bmatrix}$$

$$\times \begin{bmatrix} N_{1,\xi} & 0 & N_{2,\xi} & 0 & N_{3,\xi} & 0 & N_{4,\xi} & 0 \\ N_{1,\eta} & 0 & N_{2,\eta} & 0 & N_{3,\eta} & 0 & N_{4,\eta} & 0 \\ 0 & N_{1,\xi} & 0 & N_{2,\xi} & 0 & N_{3,\xi} & 0 & N_{4,\xi} \\ 0 & N_{1,\eta} & 0 & N_{2,\eta} & 0 & N_{3,\eta} & 0 & N_{4,\eta} \end{bmatrix} \times [d] \qquad (1.328)$$

or

$$[\varepsilon]_{3x1} = [B]_{3x8}[d]_{8x1} \tag{1.329}$$

Finally, the element stiffness matrix $[k]_{8x8}$ can be given as:

$$[k]_{8x8} = \int\int [B]^T_{8x3}[D]_{3x3}[B]_{3x8} \ t \ dx \ dy = \int_{\xi=-1}^{1}\int_{\eta=-1}^{1}[B]^T[D][B] \ t \ J \ d\xi \ d\eta \tag{1.330}$$

The "equivalent" joint load can be given as:

$$[f_{equiv}]_{8x1} = \int_{-1}^{1}\int_{-1}^{1}[N]^T_{8x2}[F]_{2x1} \ t \ J \ d\xi \ d\eta \tag{1.331}$$

where [F] = distributed loads acting on the member.

1.14 Gauss Quadrature

Using the isoparametric formulation, the element stiffness matrices (shown in Eqs.[1.312 and 1.330]) need to be numerically integrated since a close-formed, analytical solution is, in general, not possible.

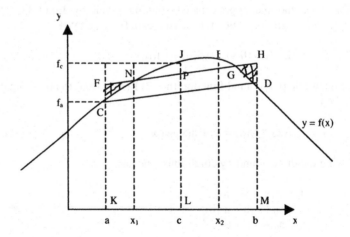

Figure 1.15 Integration of y = f(x) Between x = a and x = b

The approximate numerical integration for the closed-formed solution of

$$I = \int_{a}^{b} f(x) \ dx \tag{1.332}$$

can be given as:

$$I_n \approx \int_a^b f_n(x) \, dx \qquad (1.333)$$

where

$$f_n(x) = a_0 + a_1x + a_2x^2 + \ldots + a_nx^n \qquad (1.334)$$

The order (= n) of the polynomial function will determine, in general, the quality of the numerical approximation. For n = 1, one has:

$$f_1(x) = a_0 + a_1x^1 \qquad (1.335)$$

Applying the two boundary conditions, one obtains:

$$f_a = a_0 + a_1(a)$$
$$f_b = a_0 + a_1(b) \qquad (1.336)$$

Solving Eq.(1.336) simultaneously for a_0 and a_1 then substituting into Eq.(1.335), one gets:

$$f_1(x) = f_a (x-b)/(a-b) + f_b (x-a)/(b-a) \qquad (1.337)$$

It can be seen from Figure 1.15 that Eq.(1.337) is a straight line (or linear) function of line CD, and the approximated solution given by Eq.(1.333) represents the "Trapezoid Integration Rule" for the (trapezoid) area CDMK.

$$I \approx I_1 = (b-a) * 0.5(f(a) + f(b)) \qquad (1.338)$$

If the quadratic polynomial function (n = 2) $f_2(x)$ is used to approximate the function f(x), then

$$f(x) \approx f_2(x) = a_0 + a_1x + a_2x^2 \qquad (1.339)$$

Applying the three boundary conditions, one has:

$$\left. \begin{array}{l} f_a = a_0 + a_1a + a_2(a)^2 \\ f_b = a_0 + a_1b + a_2(b)^2 \\ f_c = a_0 + a_1c + a_2(c)^2 \end{array} \right\} \longrightarrow \qquad (1.340)$$

The unknown constants a_0, a_1, and a_2 can be solved from Eq.(1.340) and then are substituted into Eq.(1.339) to obtain:

$$f(x) \approx f_2(x) = f_a ((x-b)/(x-c))/((a-b)/(a-c)) + f_b ((x-a)/(x-c))/((b-a)/(b-c))$$
$$+ f_c ((x-a)/(x-b))/((c-a)/(c-b)) \qquad (1.341)$$

Thus, Eq.(1.333) gives:

$$I \approx I_2 = \int_a^b f_2(x)\, dx = (b-a)(f_a + 4f_c + f_b)/6 \tag{1.342}$$

Equation (1.342) represents the Simpson's Integration Rule.

The Trapezoid Integration Rule, given by Eq.(1.338), is based on the area under the straight line CD which passes through "two end-points" associated with the integration limit. The error produced by the Trapezoid rule is represented by the area CNJEGD. However, if one finds the appropriated "two interior points" N and G on the curve $y = f(x)$, shown in Figure 1.15, then the area under the "extended" straight line NG between the two end-points ($x = a$ and $x = b$) will produce much less error as compared to the "exact" integration.

The main idea is to identify the appropriated locations and the weighting factors associated with the "two interior points" N and G, as indicated in Figure 1.15. For this purpose, the Trapezoid Rule (see Eq.[1.339]) will be re-derived by using the following method of "undetermined coefficients." Let the exact integration $I = \int_a^b f(x)\, dx$ be expressed as:

$$I = c_1 f(a) + c_2 f(b) = \int_{-0.5(b-a)}^{0.5(b-a)} f(x)\, dx \tag{1.343}$$

Since the Trapezoid Rule will give an exact solution when the function being integrated is a constant, or a straight line (see Figure 1.16a and Figure 1.16b), Eq.(1.343) can be expressed as:

$$c_1 f(a) + c_2 f(b) = \int_{-0.5(b-a)}^{0.5(b-a)} 1\, dx \tag{1.344}$$

$$c_1 f(a) + c_2 f(b) = \int_{-0.5(b-a)}^{0.5(b-a)} x\, dx \tag{1.345}$$

Using Figure 1.16a, Eq.(1.344) becomes:

$$c_1(1) + c_2(1) = b - a \tag{1.346}$$

Using Figure 1.16b, Eq.(1.345) becomes:

$$c_1 f(a) + c_2(-f(a)) = 0 \tag{1.347}$$

or

$$c_1 = c_2 \tag{1.348}$$

Solving Eqs.(1.346 and 1.348) simultaneously, one obtains:

$$c_1 = c_2 = 0.5(b-a) \tag{1.349}$$

Therefore, Eq.(1.343) becomes:

$$I = (c_1 = 0.5(b-a)) * f(a) + (c_2 = 0.5(b-a)) * f(b) \tag{1.350}$$

As expected, Eq.(1.350) is identical to the Trapezoid Eq.(1.338). The above method of "undetermined coefficients" can be extended for derivation of the "Two-Point Gauss Quadrature" formula as follows:

Let the exact integration be approximated by:

$$I = \int_a^b f(x)\,dx \approx c_1 f(x_1) + c_2 f(x_2) \tag{1.351}$$

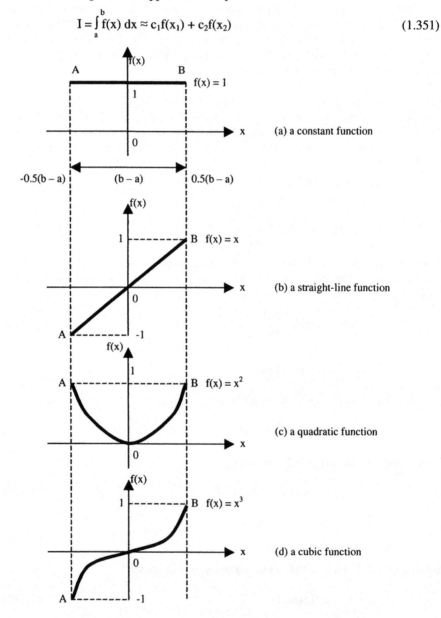

Figure 1.16 Integration of Some Basic Polynomial Functions.

The key difference between Eq.(1.351) and Eq.(1.343) is that the former has four unknowns (c_1, c_2, x_1, and x_2) while the latter (method of undetermined coefficients) has only two unknowns (c_1 and c_2). The four unknowns can be solved from the following four equations, which utilize Figure 1.16 and Eq.(1.351):

$$\int_{-1}^{1} 1 \, dx = c_1(f(x_1) = 1) + c_2(f(x_2) = 1) \tag{1.352}$$

$$\int_{-1}^{1} x \, dx = c_1(f(x_1) = x_1) + c_2(f(x_2) = x_2) \tag{1.353}$$

$$\int_{-1}^{1} x^2 \, dx = c_1(f(x_1) = x_1^2) + c_2(f(x_2) = x_2^2) \tag{1.354}$$

$$\int_{-1}^{1} x^3 \, dx = c_1(f(x_1) = x_1^3) + c_2(f(x_2) = x_2^3) \tag{1.355}$$

In Eqs.(1.352 – 1.355), it has been assumed that the integration limits are $\{-1 \quad +1\}$, instead of $\{a \quad b\}$, or $\{-0.5(b-a), 0.5(b-a)\}$.

Solving Eqs. (1.352 – 1.355) simultaneously, one obtains:
$$c_1 = c_2 = 1 \tag{1.356}$$
$$x_1 = -1/\sqrt{3} = -0.577350269\ldots \tag{1.357}$$
$$x_2 = +1/\sqrt{3} \tag{1.358}$$

Thus, the "Two-Point Gauss Quadrature" Eq.(1.351) becomes:

$$I = \int_{-1}^{1} f(x) \, dx \approx (c_1 = 1) * f(x_1 = -1/\sqrt{3}) + (c_2 = 1) * f(x_2 = +1/\sqrt{3}) \tag{1.359}$$

The graphical interpretation of Eq.(1.359) can be expressed as the area under the "extended straight line NG" (line FNGH in Figure 1.15), where the x-coordinates of points N and G are given by Eqs.(1.357) and (1.358), respectively. The area KFHMK, given by the "Two-Point Gauss Quadrature" formula (Eq.[1.359]) is much closer to the "exact" area as compared to the approximated area KCDMK, given by the Trapezoid Eq.(1.338).

For the "Three-Point Gauss Quadrature" formula, one obtains:

$$I = \int_{-1}^{1} f(x) \, dx \approx c_1 f(x_1) + c_2 f(x_2) + c_3 f(x_3) \tag{1.360}$$

The six unknowns (c_1, c_2, c_3, x_1, x_2, x_3) can be similarly obtained from the following six equations:

$$\int_{-1}^{1} 1 \, dx = c_1*1 + c_2*1 + c_3*1 \tag{1.361}$$

$$\int_{-1}^{1} x \, dx = c_1 x_1 + c_2 x_2 + c_3 x_3 \tag{1.362}$$

-1

$$\int_{-1}^{1} x^2 \, dx = c_1 x_1^2 + c_2 x_2^2 + c_3 x_3^2 \qquad (1.363)$$

$$\int_{-1}^{1} x^3 \, dx = c_1 x_1^3 + c_2 x_2^3 + c_3 x_3^3 \qquad (1.364)$$

$$\int_{-1}^{1} x^4 \, dx = c_1 x_1^4 + c_2 x_2^4 + c_3 x_3^4 \qquad (1.365)$$

$$\int_{-1}^{1} x^5 \, dx = c_1 x_1^5 + c_2 x_2^5 + c_3 x_3^5 \qquad (1.366)$$

The results for up to "Six-Point Gauss Quadrature" formulas are summarized in Table 1.1.

Table 1.1 Gauss Quadrature Formulas

Points	Weighting Factors	Gauss Quadrature Locations
2	$c_1 = 1.000000000$	$x_1 = -0.577350269$
	$c_2 = 1.000000000$	$x_2 = 0.577350269$
3	$c_1 = 0.555555556$	$x_1 = -0.774596669$
	$c_2 = 0.888888889$	$x_2 = 0.0$
	$c_3 = 0.555555556$	$x_3 = 0.774596669$
4	$c_1 = 0.347854845$	$x_1 = -0.861136312$
	$c_2 = 0.652145155$	$x_2 = -0.339981044$
	$c_3 = 0.652145155$	$x_3 = 0.339981044$
	$c_4 = 0.347854845$	$x_4 = 0.861136312$
5	$c_1 = 0.236926885$	$x_1 = -0.906179846$
	$c_2 = 0.478628670$	$x_2 = -0.538469310$
	$c_3 = 0.568888889$	$x_3 = 0.0$
	$c_4 = 0.478628670$	$x_4 = 0.538469310$
	$c_5 = 0.236926885$	$x_5 = 0.906179846$
6	$c_1 = 0.171324492$	$x_1 = -0.932469514$
	$c_2 = 0.360761573$	$x_2 = -0.661209386$
	$c_3 = 0.467913935$	$x_3 = -0.238619186$
	$c_4 = 0.467913935$	$x_4 = 0.238619186$
	$c_5 = 0.360761573$	$x_5 = 0.661209386$
	$c_6 = 0.171324492$	$x_6 = 0.932469514$

Remarks about Integration Limits

The evaluation of the exact integration:

$$I = \int_{z=a}^{b} \bar{f}(z)\, dz \qquad (1.367)$$

can be transformed into the following form:

$$I = \int_{x=-1}^{+1} f(x)\, dx \qquad (1.368)$$

through the following simple transformation of variables:

$$z = a_0 + a_1 x \qquad (1.369)$$

Corresponding to the limit values $z = (a, b)$, one has $x = (-1, 1)$, hence Eq.(1.369) can be written as:

$$a = a_0 + a_1(-1) \qquad (1.370)$$
$$b = a_0 + a_1(1) \qquad (1.371)$$

The two unknowns a_0 and a_1 in Eq.(1.369) can be obtained by solving Eqs.(1.370) and (1.371) simultaneously:

$$a_0 = 0.5(b + a) \qquad (1.372)$$
$$a_1 = 0.5(b - a) \qquad (1.373)$$

Thus, Eq.(1.369) becomes:

$$z = 0.5(b + a) + 0.5(b - a)x \equiv g(x) \qquad (1.374)$$

Hence
$$dz/dx = 0.5(b - a)$$
or
$$dz = 0.5(b - a)\, dx \qquad (1.375)$$

Substituting Eqs.(1.374 − 1.375) into Eq.(1.367), one gets:

$$I = \int_{z=a}^{b} \bar{f}(z)\, dz = \int_{x=-1}^{+1} \underbrace{f(g(x)) * 0.5(b - a)}_{f(x)}\, dx \qquad (1.376)$$

or

$$I = \int_{x=-1}^{+1} f(x)\, dx \qquad (1.377)$$

Example

$$\text{Evaluate } I = \int_{z=0}^{0.8} \overline{f(z)} \, dz \qquad (1.378)$$

$$\text{where } \overline{f(z)} \equiv 0.2 + 20z^2 - 90z^4 \qquad (1.379)$$

Applying Eq.(1.376) with $a = 0.0$ and $b = 0.8$, Eq.(1.378) becomes:

$$I = \int_{x=-1}^{1} [0.2 + 20\{0.5(0.8 + 0.0) + 0.5(0.8 - 0.0)x\}^2 - 90(0.4 + 0.4x)^4] \, 0.4 \, dx \qquad (1.380)$$

or

$$I = \int_{x=-1}^{1} f(x) \, dx \qquad (1.381)$$

where

$$f(x) \equiv [0.2 + 20\{0.5(0.8 + 0.0) + 0.5(0.8 - 0.0)x\}^2 - 90(0.4 + 0.4x)^4] \, 0.4 \qquad (1.382)$$

Gauss Quadrature formulas, shown in Table 1.1, can be used to efficiently evaluate the integral Eq.(1.381).

Gauss Quadratures for 2D and 3D Problems

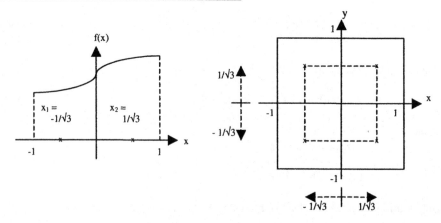

Figure 1.17 2-Point Gauss Quadrature Locations for 1D Problem   **Figure 1.18** 2x2 Point Gauss Quadrature Locations for 2D Problem

For a 1D problem, shown in Figure 1.17, Eq.(1.360) can be generalized to:

$$\overset{N = \# \text{ of Gauss points in the x-direction}}{I_1 = \sum_{i=1}^{N} c_i \, f(x_i)} \qquad (1.383)$$

2D problems, shown in Figure 1.18, and 3D problems, Eq.(1.383), can be further generalized to:

$$I_2 = \sum_{j=1}^{Ny} \sum_{i=1}^{Nx} c_{ij} f(x_i, y_j), \text{ where one usually selects } N_x = N_y = N \quad (1.384)$$

$$I_3 = \sum_{k=1}^{Nz} \sum_{j=1}^{Ny} \sum_{i=1}^{Nx} c_{ijk} f(x_i, y_j, z_k), \text{ where one usually selects}$$
$$N_x = N_y = N_z = N \quad (1.385)$$

1.15 Summary

The basic finite element procedures have been reviewed in this chapter. Rayleigh Ritz, and Galerkin methods are emphasized for finite element solutions of the Partial Differential Equations (PDE). One-, two-, and three-dimensional engineering problems can be treated, isoparametric formulation has also been explained. Unified step-by-step procedures for deriving the "element interpolation (or shape) function" have been discussed. Both the finite element assembly process (and how to handle the geometric and natural boundary conditions) and efficient numerical integrations have also been explained.

1.16 Exercises

1.1 Obtain the 1-term approximated displacement equation, $\tilde{y}(x)$, for the simply supported beam, governed by the following ordinary differential equation (ODE):

$$EI \frac{d^2 y}{dx^2} = \frac{M_o x}{L}$$

where E, I, M_o, L represents the Young modulus, moment of inertia, applied moment at $x = L$, and length of the beam, respectively.
with the boundary conditions:
$$y(@ x = 0) = 0$$
$$y(@ x = L) = 0$$
Using the 1-term trial solutions discussed in section 1.2,
$$\tilde{y}(x) = A_1(x - 0)(x - L)$$
Also, compare the approximated displacement at the center of the beam, $\tilde{y}(@ x = \dfrac{L}{x})$, with the theoretical, exact value $y_{ex}(@ x = \dfrac{L}{x})$.

1.2 Re-do problem 1.1, by using a different (1-term) trial solution:

$$\tilde{y}(x) = A_1 \sin(\frac{\pi x}{L})$$

1.3 Re-do problem 1.1, by using a 2-term approximated displacement equation.

1.4 Re-do problem 1.2, by using a 2-term approximated displacement equation.

1.5 Consider the following ODE:

$$x^2 \frac{d^2y}{dx^2} + 2x \frac{dy}{dx} = 6x$$

with the given boundary conditions:
$$y(@ \, x = 1) = 0$$
$$y(@ \, x = 2) = 0$$

(a) Which of the above boundary conditions is "geometrical," or "natural" boundary conditions.

(b) Using the "weak formulation," discussed in section 1.4, find the 1-term approximated solution, $\tilde{y}(x)$.

(c) The exact solution, $y_{ex}(x)$, for the above ODE is given as:
$$y_{ex} = \frac{6}{x} + 3x - 9$$

Compare your approximated solution $\tilde{y}(@ \, x = 1.5)$ with the exact solution $y_{ex}(@ \, x = 1.5)$.

1.6 Find the 3-term trial solution $\tilde{y}(x)$ that will satisfy all the following boundary conditions:
$$at \; x = 2, y = 0$$
$$at \; x = 4, y = 0, y' = 0$$
$$at \; x = 3, y = 0, y'' = 0$$

1.7 Consider the following ODE:
$$x^3 \frac{d^4y}{dx^4} + 6x^2 \frac{d^3y}{dx^3} + 6x \frac{d^2y}{dx^2} - 10x = 0$$

with the following boundary conditions:
$$y(@ \, x = 1) = 0$$
$$y'(@ \, x = 1) = 0$$
$$y(@ \, x = 3) = 0$$
$$y'(@ \, x = 3) = 0$$

Using the "weak formulation," find the 2-term approximated solution for the above ODE.

1.8 A clamped (or fixed) rectangular plate having dimensions 2a×2b (on the x-y plane) is subjected to a uniform load distribution q_0 (acting along the z-direction). The plate equilibrium equation can be described by:
$$\nabla^4 w = \frac{q_0}{D}$$

where $w(x, y)$, and $D\left(=\dfrac{E h^3}{12(1-\upsilon^2)}\right)$ represents the lateral deflection (in the z-direction) and bending rigidity of the plate, respectively.

The following boundary conditions are given:

$$w = 0 = \frac{\partial w}{\partial x} \quad \text{at } x = \pm a$$

$$w = 0 = \frac{\partial w}{\partial y} \quad \text{at } y = \pm b$$

Using the 1-term Galerkin trial solution, $\tilde{w}(x, y) = c_1\left(x^2 - a^2\right)^2 \left(y^2 - b^2\right)^2$, find the approximated solution $\tilde{w}(x, y)$.

1.9 Re-do problem 1.8 using the 4-term Galerkin trial solution:

$$\tilde{w}(x, y) = \left(c_1 + c_2 x + c_3 y + c_4 xy\right)\left(x^2 - a^2\right)^2 \left(y^2 - b^2\right)^2$$

1.10 Using the finite element procedures,

    (a) find all nodal displacements
    (b) find all element (axial) forces
    (c) find all support reactions

for the following 2-D Truss structure:

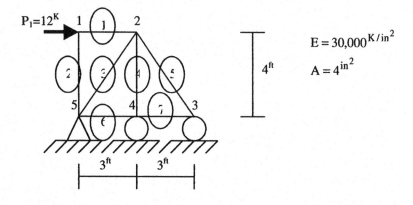

1.11 Derive the element interpolation (or shape) function for a 4-node "axially loaded" element as shown in the following figure:

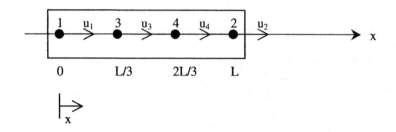

1.12 Evaluate the following integral:

$$I_1 = \int_1^4 (x^2 - 2x + \frac{1}{x} - 2e^{-3x})\, dx$$

(a) by using a 2 point Gauss Quadrature formula
(b) by using a 3 point Gauss Quadrature formula

1.13 Re-do problem 1.12 for the following integral:

$$I_2 = \int_{x=1}^4 \int_{y=1}^4 (x^2 y - 2xy^2 + 6xy - 8)\, dy\, dx$$

2 Simple MPI / FORTRAN Applications

2.1 Introduction

The numerical algorithms discussed throughout this textbook can be implemented by using Message Passing Interface (MPI) either in FORTRAN or C++ language. MPI/FORTRAN is essentially the same as the "regular" FORTRAN 77 and/or FORTRAN 90, with some additional, "special FORTRAN" statements for parallel computation, sending (or broadcasting), receiving and merging messages among different processors. Simple and most basic needs for these "special" FORTRAN statements can be explained and illustrated through two simple examples, which will be described in subsequent sections.

2.2 Computing Value of π by Integration

The numerical value of π (≈ 3.1416) can be obtained by integrating the following function:

$$f(x) = \frac{4}{1+x^2}, \quad \text{for } x = [0, 1] \tag{2.1}$$

Since

$$\int \frac{1}{a+bx^2} dx = \frac{1}{\sqrt{ab}} \tan^{-1}\left(\frac{x\sqrt{ab}}{a}\right)$$

hence

$$\int_0^1 \frac{4}{1+x^2} dx = 4\left[\tan^{-1}(x)\right]_0^1 = \pi$$

Integrating the function given in Eq.(2.1) can be approximated by computing the area under the wave $f(x)$, as illustrated in Figure 2.1.

In Figure 2.1, assuming the domain of interests, $x = 0 \rightarrow 1$, is divided into $n = 8$ segments. For example, segment 1 will correspond to $x = 0$ and $2/16$, segment 4 corresponds to $x = 6/16$ and $8/16$, etc. Thus, the integral size $h = (1-0)/8 = 1/8$. It is further assumed that the number of processors (= numprocs) to be used for parallel computation is 4 (= processors P_0, P_1, P_2, and P_3). The values of "myid" in Table 2.1 (see loop 20) are given by 0, 1, 2, and 3, respectively.

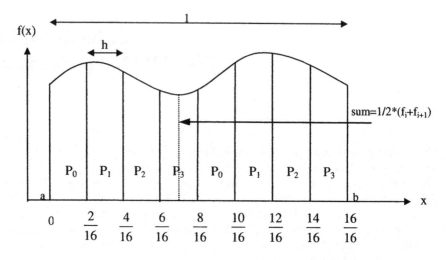

Figure 2.1 Computing π by Integrating $f(x) = \dfrac{4}{1+x^2}$

According to Figure 2.1, a typical processor (such as P_3) will be responsible to compute the areas of two segments. Thus, the total areas computed by processor P_3 are:

$$(\text{Area})_{P_3} = \frac{1}{2}\left(f\big|_{x=\frac{6}{16}} + f\big|_{x=\frac{8}{16}}\right)h + \frac{1}{2}\left(f\big|_{x=\frac{14}{16}} + f\big|_{x=\frac{16}{16}}\right)h \tag{2.2}$$

$$= (\text{Area of segment 4}) + (\text{Area of segment of 8}) \tag{2.3}$$

or

$$(\text{Area})_{P_3} = (h * \text{Average height of segment 4}) + (h * \text{Average height of segment 8}) \tag{2.4}$$

$$= h * (\text{Average height of segment 4} + \text{Average height of segment 8}) \tag{2.5}$$

$$= h * \left\{f\left(@\ x = \tfrac{7}{16}\right) + f\left(@\ x = \tfrac{15}{16}\right)\right\} \tag{2.6}$$

$$(\text{Area})_{P_3} = h * \{\text{sum}\} = \text{mypi} \tag{2.7}$$

The values of {sum} and mypi are computed inside and outside loop 20, respectively (see Table 2.1).

Since there are many comments in Table 2.1, only a few remarks are given in the following paragraphs to further clarify the MPI/FORTRAN code:

Remark 1. In a parallel computer environment, and assuming four processors (= P_0, P_1, P_2, and P_3) are used, then the same code (such as Table 2.1) will be executed by each and every processor.

Remark 2. In the beginning of the main parallel MPI/FORTRAN code, "special" MPI/FORTRAN statements, such as include "mpif.h", call MPI_INIT, ..., and call MPI_COMM_SIZE (or near the end of the main program, such as call MPI_FINALIZE), are always required.

Remark 3. The variable "myid" given in Table 2.1 can have integer values as 0, 1, 2, ... (# processors −1) = 3, since # processors = numprocs = 4 in this example.

Remark 4. The "If-Endif" block statements, shown near the bottom of Table 2.1, are only for printing purposes, and therefore, these block statements can be executed by any one of the given processors (such as by processor P_0 or myid = 0), or by processor P_3 (or myid = 3). To be on the safe side, however, myid = 0 is used, so that the parallel MPI code will not give an error message when it is executed by a single processor!

Remark 5. The parameters given inside the call MPI_BCAST statement have the following meanings:

1st parameter = n = The "information" that needs to broadcast to all processors. This "information" could be a scalar, a vector, or a matrix, and the number could be INTEGER, or REAL, etc.

In this example, n represents an INTEGER scalar.

2nd parameter = 1 (in this example), which represents the "size" of the 1st parameter.

3rd parameter = MPI_INTEGER, since in this example the 1st parameter is an integer variable.

4th parameter = 0, which indicates that the 1st parameter will be broadcasting to all other processors.

Remark 6. Depending on the processor number (= myid), each processor will be assigned to compute a different segment of rectangular areas (see do-loop 20, in Table 2.1).

Remark 7. After executing line "mypi=h*sum," each and every processor has already computed its own (or local) net areas (= area of 2 rectangular segments, as shown in Figure 2.1), and is stored under variable "mypi." Therefore, the final value for π can be computed by adding the results from each processor. This task is done by "call MPI_REDUCE." The important (and less obvious) parameters involved in " call MPI_REDUCE," shown in Table 2.1 are explained in the following paragraphs:

1st parameter = mypi = net (rectangular) areas, computed by <u>each</u> processor.

2nd parameter = pi = total areas (= approximate value for π), computed by adding (or summing) each processor's area.

3rd parameter = 1 (= size of the 1st parameter)

4th parameter = MPI_DOUBLE_PRECISION = double precision is used for summation calculation.

5th parameter = MPI_SUM = performing the "summing" operations from each individual processor's results (for some application, "merging" operations may be required).

6th parameter = 0

Remark 8. Before ending an MPI code, a statement, such as call MPI_FINALIZE (see Table 2.1), needs to be included.

Table 2.1 Compute π by Parallel Integration [2.1]

```
c   pi.f - compute pi by integrating f(x) = 4/(1 + x**2)
c
c   Each node:
c    1) receives the number of rectangles used in the approximation.
c    2) calculates the areas of it's rectangles.
c    3) Synchronizes for a global summation.
c   Node 0 prints the result.
c
c   Variables:
c
c   pi   the calculated result
c   n    number of points of integration.
c   x        midpoint of each rectangle's interval
c   f        function to integrate
c   sum,pi    area of rectangles
c   tmp       temporary scratch space for global summation
c   i        do loop index
c++++++++++++++++++++++++++++++++++++++++++++++++++++++++++++++++++++++++++
      program main
      include 'mpif.h'
      double precision PI25DT
      parameter    (PI25DT = 3.141592653589793238462643d0)
```

```
      double precision  mypi, pi, h, sum, x, f, a,time1,time2
      integer n, myid, numprocs, i, rc
c                        function to integrate
      f(a) = 4.d0 / (1.d0 + a*a)
      call MPI_INIT( ierr )
      call MPI_COMM_RANK( MPI_COMM_WORLD, myid, ierr )
      call MPI_COMM_SIZE( MPI_COMM_WORLD, numprocs, ierr )
      print *, 'Process ', myid, ' of ', numprocs, ' is alive'
c----------------------------------------------------------------------
      time1=MPI_Wtime()
c----------------------------------------------------------------------
      sizetype  = 1
      sumtype   = 2
 10   if ( myid .eq. 0 ) then
c        write(6,98)
c 98     format('Enter the number of intervals: (0 quits)')
c        read(5,99) n
c 99     format(i10)
         read(101,*) n
c        n=10000000
      endif
      call MPI_BCAST(n,1,MPI_INTEGER,0,MPI_COMM_WORLD,ierr)
c     check for quit signal
      if ( n .le. 0 ) goto 30
c     calculate the interval size
      h = 1.0d0/n
      sum  = 0.0d0
      do 20 i = myid+1, n, numprocs
        x = h * (dble(i) - 0.5d0)
        sum = sum + f(x)
 20   continue
      mypi = h * sum
c     collect all the partial sums
      call MPI_REDUCE(mypi,pi,1,MPI_DOUBLE_PRECISION,MPI_SUM,0,
     $    MPI_COMM_WORLD,ierr)
c     node 0 prints the answer.
      if (myid .eq. 0) then
         write(6, 97) pi, abs(pi - PI25DT)
 97      format(' pi is approximately: ', F18.16,
     +        ' Error is: ', F18.16)
      endif
c     goto 10
 30   call MPI_FINALIZE(rc)
c----------------------------------------------------------------------
      time2=MPI_Wtime()
      write(6,*) 'MPI elapse time= ',time2-time1
c----------------------------------------------------------------------
      stop
      end
```

2.3 Matrix-Matrix Multiplication

The objective of this section is to again illustrate the usage of few basic parallel MPI/FORTRAN statements for simple applications such as multiplying 2 matrices. To further simplify the discussion, it is assumed that 4 processors (numtasks = number of processors = 4, see line #021 in Table 2.2) are used for parallel multiplication of 2 square matrices A (1000, 1000) and B (1000, 1000). The main steps involved in this example are summarized as follows:

Step 1: Input (or generate) matrices [A] and [B] by a master program (= processor 0). This task is done by lines # 048 – 058 in Table 2.2.

Step 2: The master processor will distribute certain columns of matrix [B] to its workers (= numworkers = numtasks – 1 = 4 - 1 = 3 workers, see line #022 in Table 2.2). Let NCB = Number of Columns of [B] = 1,000. The average number of columns to be received by each "worker" processor (from a "master" processor) is:

Avecol = NCB/numworkers = 1,000/3 = 333 (see line #060).

The "extra" column(s) to be taken care of by certain "worker" processor(s) is calculated as:

Extra = mod (NCB, numworkers) = 1 (see line #061).

In this example, we have a "master" processor (taskid = 0, see line #020) and 3 "worker" processors (taskid = 1, 2, or 3). Each "worker" processor has to carry a minimum of 333 columns of [B]. However, those worker processor number(s) that are less than or equal to "extra" (= 1), such as worker processor 1, will carry:

Cols = avecol + 1 = 334 columns (see line #066).

The parameter "offset" is introduced and initialized to 1 (see line #062). The value of "offset" is used to identify the "starting column number" in the matrix [B] that will be sent (by a MASTER processor) to each "worker" processor. Its value is updated as shown at line #079. For this particular example, the values of "offset" are 1, 335, and 668 for "worker processor" number 1, 2, and 3 (or processor numbers P_1, P_2, and P_3), respectively.

In other words, worker processor P_1 will receive and store 334 columns of [B], starting from column 1 (or offset = 1), whereas worker processors P_2 (and P_3) will receive and store 333 columns of [B], starting from column 335 (and 668), respectively.

Inside loop 50 (see lines #064 – 080), the "master" processor (P_0) will send the following pieces of information to the appropriate "worker processors":

(a) The value of "offset"   (see line #071)
(a) The number of columns of [B]   (see line #073)
(b) The entire matrix [A]   (see line #075)
(c) The specific columns of [B]   (see line #077)

The parameters involved in MPI_ SEND are explained as follows:

1st parameter = name of the variable (could be a scalar, a vector, a matrix, etc.)
        that needs to be sent

2nd parameter = the size (or dimension) of the 1st parameter (expressed in
        words, or numbers)

3rd parameter = the type of variable (integer, real, double precision, etc.)

4th parameter = dest = destination, which processor will receive the message (or
        information)

5th parameter = mtype = 1 (= FORM_MASTER, also see lines #013, 063)

6th parameter = as is, unimportant

7th parameter = as is, unimportant

Having sent all necessary data to "worker processors," the "master processor" will wait to receive the results from its workers (see lines # 081 – 091), and print the results (see lines # 092 – 105).

Step 3: Tasks to be done by each "worker" processor.

(a) Receive data from the "master" processor (see lines #106 – 118).
(b) Compute the matrix-matrix multiplications for [entire A]*[portion of columns B] ≡ [portion of columns C], see lines #119 – 125.
(c) Send the results (= [portion of columns C]) to the "master" processor (see lines #126 – 136).

It should be noted here that the parameters involved in MPI_RCV are quite similar (nearly identical to the ones used in MPI_SEND).

Table 2.2 Parallel Computation of Matrix Times Matrix [2.1]

program mm	! 001
c--	! 002
c.This file is stored under ~/cee/MPI-oduSUNS/matrix_times_matrix_mpi.f	! 003
c---	! 004
include 'mpif.h'	! 005
c NRA : number of rows in matrix A	! 006
c NCA : number of columns in matrix A	! 007
c NCB : number of columns in matrix B	! 008
parameter (NRA = 1000)	! 009

```
      parameter (NCA = 1000)                                          ! 010
      parameter (NCB = 1000)                                          ! 011
      parameter (MASTER = 0)                                          ! 012
      parameter (FROM_MASTER = 1)                                     ! 013
      parameter (FROM_WORKER = 2)                                     ! 014
      integer        numtasks,taskid,numworkers,source,dest,mtype,    ! 015
     &         cols,avecol,extra, offset,i,j,k,ierr                   ! 016
      integer status(MPI_STATUS_SIZE)                                 ! 017
      real*8a(NRA,NCA), b(NCA,NCB), c(NRA,NCB)                        ! 018
      call MPI_INIT( ierr )                                           ! 019
      call MPI_COMM_RANK( MPI_COMM_WORLD, taskid, ierr )              ! 020
      call MPI_COMM_SIZE( MPI_COMM_WORLD, numtasks, ierr )            ! 021
      numworkers = numtasks-1                                         ! 022
      print *, 'task ID= ',taskid                                     ! 023
c+++++++++++++++++++++++++++++++++++++++++++++++++                    ! 024
c     write(6,*) 'task id = ',taskid                                  ! 025
c     call system ('hostname')                                        ! 026
c     ! to find out WHICH computers run this job                      ! 027
c     ! this command will SLOW DOWN (and make UNBALANCED             ! 028
c     ! workloads amongst processors)                                 ! 029
c+++++++++++++++++++++++++++++++++++++++++++++++++                    ! 030
      if(numworkers.eq.0) then                                        ! 031
      time00 = MPI_WTIME()                                            ! 032
C     Do matrix multiply                                              ! 033
      do 11 k=1, NCB                                                  ! 034
       do 11 i=1, NRA                                                 ! 035
        c(i,k) = 0.0                                                  ! 036
        do 11 j=1, NCA                                                ! 037
         c(i,k) = c(i,k) + a(i,j) * b(j,k)                            ! 038
 11    continue                                                       ! 039
      time01 = MPI_WTIME()                                            ! 040
      write(*,*) ' C(1,1)   : ', c(1,1)                               ! 041
      write(*,*) ' C(nra,ncb): ',C(nra,ncb)                           ! 042
      write(*,*)                                                      ! 043
      write(*,*) ' Time  me=0: ', time01 - time00                     ! 044
      go to 99                                                        ! 045
      endif                                                           ! 046
C ************************** master task **************************    ! 047
      if (taskid .eq. MASTER) then                                    ! 048
      time00 = MPI_WTIME()                                            ! 049
C     Initialize A and B                                              ! 050
      do 30 i=1, NRA                                                  ! 051
       do 30 j=1, NCA                                                 ! 052
        a(i,j) = (i-1)+(j-1)                                          ! 053
 30    continue                                                       ! 054
      do 40 i=1, NCA                                                  ! 055
       do 40 j=1, NCB                                                 ! 056
        b(i,j) = (i-1)*(j-1)                                          ! 057
 40     continue                                                      ! 058
C     Send matrix data to the worker tasks                            ! 059
      avecol = NCB/numworkers                                         ! 060
      extra = mod(NCB, numworkers)                                    ! 061
      offset = 1                                                      ! 062
```

```
      mtype = FROM_MASTER                                              ! 063
      do 50 dest=1, numworkers                                         ! 064
        if (dest .le. extra) then                                      ! 065
         cols = avecol + 1                                             ! 066
        else                                                           ! 067
         cols = avecol                                                 ! 068
        endif                                                          ! 069
        write(*,*)' sending',cols,' cols to task',dest                 ! 070
        call MPI_SEND( offset, 1, MPI_INTEGER, dest, mtype,            ! 071
     &           MPI_COMM_WORLD, ierr )                                ! 072
        call MPI_SEND( cols, 1, MPI_INTEGER, dest, mtype,              ! 073
     &           MPI_COMM_WORLD, ierr )                                ! 074
        call MPI_SEND( a, NRA*NCA, MPI_DOUBLE_PRECISION, dest, mtype,  ! 075
     &           MPI_COMM_WORLD, ierr )                                ! 076
        call MPI_SEND( b(1,offset), cols*NCA, MPI_DOUBLE_PRECISION,    ! 077
     &           dest, mtype, MPI_COMM_WORLD, ierr )                   ! 078
        offset = offset + cols                                         ! 079
50    continue                                                         ! 080
C     Receive results from worker tasks                                ! 081
      mtype = FROM_WORKER                                              ! 082
      do 60 i=1, numworkers                                            ! 083
        source = i                                                     ! 084
        call MPI_RECV( offset, 1, MPI_INTEGER, source,                 ! 085
     &           mtype, MPI_COMM_WORLD, status, ierr )                 ! 086
        call MPI_RECV( cols, 1, MPI_INTEGER, source,                   ! 087
     &           mtype, MPI_COMM_WORLD, status, ierr )                 ! 088
        call MPI_RECV( c(1,offset), cols*NRA, MPI_DOUBLE_PRECISION,    ! 089
     &           source, mtype, MPI_COMM_WORLD, status, ierr )         ! 090
60    continue                                                         ! 091
      time01  = MPI_WTIME()                                            ! 092
C     Print results                                                    ! 093
c     do 90 i=1, NRA                                                   ! 094
c       do 80 j = 1, NCB                                               ! 095
c        write(*,70)c(i,j)                                             ! 096
c 70      format(2x,f8.2,$)                                            ! 097
c 80     continue                                                      ! 098
c       print *, ' '                                                   ! 099
c 90   continue                                                        ! 100
      write(*,*) ' C(1,1)   : ', c(1,1)                                ! 101
      write(*,*) ' C(nra,ncb): ',C(nra,ncb)                            ! 102
      write(*,*)                                                       ! 103
      write(*,*) ' Time  me=0: ', time01 - time00                     ! 104
      endif                                                            ! 105
C ********************** worker task **************************        ! 106
      if (taskid.gt.MASTER) then                                       ! 107
      time11  = MPI_WTIME()                                            ! 108
C     Receive matrix data from master task                            ! 109
      mtype = FROM_MASTER                                              ! 110
      call MPI_RECV( offset, 1, MPI_INTEGER, MASTER,                   ! 111
     &           mtype, MPI_COMM_WORLD, status, ierr )                 ! 112
      call MPI_RECV( cols, 1, MPI_INTEGER, MASTER,                     ! 113
     &           mtype, MPI_COMM_WORLD, status, ierr )                 ! 114
      call MPI_RECV( a, NRA*NCA, MPI_DOUBLE_PRECISION, MASTER,         ! 115
```

```
 &              mtype, MPI_COMM_WORLD, status, ierr )            ! 116
      call MPI_RECV( b, cols*NCA, MPI_DOUBLE_PRECISION, MASTER,   ! 117
 &              mtype, MPI_COMM_WORLD, status, ierr )            ! 118
C   Do matrix multiply                                          ! 119
      do 100 k=1, cols                                          ! 120
       do 100 i=1, NRA                                          ! 121
        c(i,k) = 0.0                                            ! 122
        do 100 j=1, NCA                                         ! 123
         c(i,k) = c(i,k) + a(i,j) * b(j,k)                      ! 124
 100   continue                                                 ! 125
C   Send results back to master task                            ! 126
      mtype = FROM_WORKER                                       ! 127
      call MPI_SEND( offset, 1, MPI_INTEGER, MASTER, mtype,     ! 128
 &            MPI_COMM_WORLD, ierr )                            ! 129
      call MPI_SEND( cols, 1, MPI_INTEGER, MASTER, mtype,       ! 130
 &            MPI_COMM_WORLD, ierr )                            ! 131
      call MPI_SEND( c, cols*NRA, MPI_DOUBLE_PRECISION, MASTER, ! 132
 &            mtype, MPI_COMM_WORLD, ierr )                     ! 133
      time22 = MPI_WTIME()                                      ! 134
      write(*,*) ' Time  me=',taskid,': ', time22 - time11      ! 135
      endif                                                     ! 136
 99   call MPI_FINALIZE(ierr)                                   ! 137
      end                                                       ! 138
```

2.4 MPI Parallel I/O

Some basics in Input/Output (I/O) operations under MPI environments are explained and demonstrated in Table 2.3

Table 2.3 I/O Operations under MPI Environments

```
      program mpiio
      include 'mpif.h'
c      implicit real*8(a-h,o-z)    ! will get error if use this "implicit" stmt
c*********************************************************************
c   Purposes:  Using MPI parallel i/o for writing & reading by different
c              processors, on "different segments" of the "same" file
c   Person(s): Todd and Duc T. Nguyen
c   Latest Date:  April 10, 2003
c   Stored at:  cd ~/cee/mpi_sparse_fem_dd/mpi_io.f
c   Compile ??:  Just type
c              tmf90 -fast -xarch=v9 mpi_io.f -lmpi
c   Execute ??:  Just type (assuming 2 processors are used)
c              bsub -q hpc-mpi-short -I -n 2 a.out
c   Output:    Stored at files 301+(me= processors 0, 1, etc...)
c*********************************************************************
      real*8 a, b
      real*8 t1,t2,t3
      parameter (nns = 10000000)
      integer op,nns,nsz,myfh
      dimension a(nns),b(nns)
      integer (kind=mpi_offset_kind) idisp
```

```
      call MPI_INIT(ierr)
      call MPI_COMM_RANK(MPI_COMM_WORLD, me, ierr)
      call MPI_COMM_SIZE(MPI_COMM_WORLD, np, ierr)
      ns = nns/np
      ip = 101
      op = 301+me
c     kind = MPI_OFFSET_KIND
c++++++++++++++++++++++++++++++++++++++++++++
c Initial testing arrays
c++++++++++++++++++++++++++++++++++++++++++++
      do i = 1,ns
       a(i) = i+ns*me
       b(i) = 0
      enddo
c++++++++++++++++++++++++++++++++++++++++++++
c Open the file
c++++++++++++++++++++++++++++++++++++++++++++
      call MPI_FILE_OPEN(MPI_COMM_WORLD, 'test',
     & MPI_MODE_rdwr+mpi_mode_create,MPI_INFO_NULL,myfh,ierr)
c++++++++++++++++++++++++++++++++++++++++++++
c Set view point for each process
c++++++++++++++++++++++++++++++++++++++++++++
      idisp = ns*8*me
      write(op,*) '----------------------'
      write(op,*) 'me,kind',me,kind
      write(op,*) 'idisp',idisp
      call MPI_FILE_SET_VIEW(myfh,idisp,MPI_double_precision,
     & MPI_double_precision,'native',MPI_INFO_NULL,ierr )
c     call mpi_file_seek(myfh,idisp,mpi_seek_set,ierr)
      nsz = ns
      write(op,*) 'nsz',nsz
c++++++++++++++++++++++++++++++++++++++++++++
c Write array to the disk
c++++++++++++++++++++++++++++++++++++++++++++
c     write(op,*) 'a',(a(i),i=1,nsz)
      t1 = MPI_WTIME()
      call MPI_FILE_WRITE(myfh,a,nsz,MPI_double_precision,
     & mpi_status_ignore,ierr)
      t2 = MPI_WTIME()
      write(op,*) 'Time to write(MPI) to the disk',t2-t1
c++++++++++++++++++++++++++++++++++++++++++++
c Read file
c++++++++++++++++++++++++++++++++++++++++++++
      idisp = ns*8*(me+1)
      if (me .eq. np-1) then
       idisp = 0
      endif
      call MPI_FILE_SET_VIEW(myfh,idisp,MPI_double_precision,
     & MPI_double_precision,'native',MPI_INFO_NULL,ierr )
      call MPI_FILE_READ(myfh,b,nsz,MPI_double_precision,
     & mpi_status_ignore,ierr)
      call MPI_FILE_CLOSE(myfh,ierr)
      t3 = MPI_WTIME()
```

```
      write(op,*) 'Time to read(MPI) from the disk',t3-t2
c     write(op,*) 'b',(b(i),i=1,nsz)
c     call MPI_FILE_CLOSE(myfh,ierr)
999   call MPI_FINALIZE(ierr)
      stop
      end
```

Preliminary timing for READ/WRITE by a different number of processors is reported in Table 2.4.

Table 2.4 Preliminary Performance of MPI Parallel I/O

(Using HELIOS, but compiled and executed from CANCUN, SUN_10,000 computer)

NP	1	2	4	Comments
Write	83.04 sec	46.34 sec (me = 0)	29.75 sec	$\left\{\begin{array}{l}\text{using MPI wall clock}\\ \text{time subroutine}\end{array}\right.$
			30.02 sec	
		46.68 sec (me = 1)	37.55 sec	{ write is slower than read
			58.38 sec	
Read	15.39 sec	18.34 sec (me = 0)	11.77 sec	$\left\{\begin{array}{l}\text{Parallel \underline{write} seems to offer}\\ \text{some speed}\end{array}\right.$
			28.35 sec	
		19.68 sec (me = 1)	22.36 sec	$\left\{\begin{array}{l}\text{Parallel \underline{read} seems to be}\\ \text{worse}\end{array}\right.$
			8.79 sec	
	Each processor write 10 million double procession words, and read 10 Million words	Each processor write and read 5 million double precision words (in parallel)	Correspond to 4 processors $me = \left\{\begin{array}{l}0\\1\\2\\3\end{array}\right.$ each processor write and read $2.5*10^6$ words	

2.5 Unrolling Techniques [1.9]

To enhance the performance of the codes, especially on a single vector processor, unrolling techniques are often used to reduce the number of "load and store" movements (to and from the CPU), and, therefore, code performance can be improved (up to a factor of 2, on certain computer platforms).

To illustrate the main ideas, consider the simple task of finding the product of a given matrix [A] and a vector {x}, where

$$[A] = \begin{bmatrix} 1 & 2 & 3 & 4 \\ 5 & 6 & 7 & 8 \\ 9 & 10 & 11 & 12 \\ 13 & 14 & 15 & 16 \end{bmatrix} \text{ and } x = \begin{Bmatrix} 2 \\ 0 \\ 1 \\ 1 \end{Bmatrix} \tag{2.8}$$

Algorithm 1: Dot product operations

In this algorithm, each row of a matrix [A] operates on a vector {x} to obtain the

"final" solution. The final answer for $b_1 = \begin{Bmatrix} 1 \\ 2 \\ 3 \\ 4 \end{Bmatrix} \cdot \begin{Bmatrix} 2 \\ 0 \\ 1 \\ 1 \end{Bmatrix} = 9$, which involves the dot

product operations of two vectors. The FORTRAN code for this algorithm is given in Table 2.5.

Table 2.5 Matrix Times Vector (Dot Product Operations)

```
       Do 1 I = 1, N (say = 4)
       Do 1 J = 1, N
1         b(I) = b(I) + A(I, J) * x(J)
```

Algorithm 2: Dot product operations with unrolling techniques

We can group a few (say NUNROL = 2) rows of [A] and operate on a vector {x}. Thus, algorithm 1 can be modified and is given in Table 2.6.

Table 2.6 Matrix Times Vector (Dot Product Operations, with Unrolling Level 2)

```
       Do 1 I = 1, N, NUNROL (= 2)
       Do 1 J =1, N
          b(I) = b(I) + A(I, J) * x(J)
1         b(I+1) = b(I+1) + A(I+1, J) * x(J)
```

Algorithm 3: Saxpy operations

In this algorithm, each column of matrix [A] operates on an appropriate component of vector $\{x\}$ to get a "partial, or incomplete" solution vector $\{b\}$. Thus, the first "partial" solution for $\{b\}$ is given as:

$$\{b\}_{incomplete} = \begin{Bmatrix} 1 \\ 5 \\ 9 \\ 13 \end{Bmatrix} * 2 = \begin{Bmatrix} 2 \\ 10 \\ 18 \\ 26 \end{Bmatrix} \tag{2.9}$$

and eventually, the "final, complete" solution for $\{b\}$ is given as:

$$\{b\}_{final} = \begin{Bmatrix} 2 \\ 10 \\ 18 \\ 26 \end{Bmatrix} + \begin{Bmatrix} 2 \\ 6 \\ 10 \\ 14 \end{Bmatrix} * (0) + \begin{Bmatrix} 3 \\ 7 \\ 11 \\ 15 \end{Bmatrix} * (1) + \begin{Bmatrix} 4 \\ 8 \\ 12 \\ 16 \end{Bmatrix} * (1) = \begin{Bmatrix} 9 \\ 25 \\ 41 \\ 61 \end{Bmatrix} \tag{2.10}$$

The FORTRAN code for this algorithm is given in Table 2.7.

Table 2.7 Matrix Times Vector (Saxpy Operations)

```
       Do 1 I = 1, N
       Do 1 J = 1, N
         b(J)  =  b(J)  +  A(J, I)  *  x(I)
1        Continue
```

The operation inside loop 1 of Table 2.7 involves a summation of a scalar $\underline{a}$ (= x (I), in this case, since x(I) is independent from the innermost loop index J) times a vector $\underline{x}$ (= A(J, I), in this case, since the Ith column of [A] can be considered as a vector) plus a previous vector $\underline{y}$ (= b(J), in this case), hence the name SAXPY!

Algorithm 4: Saxpy operations with unrolling techniques

We can group a few (say NUNROL = 2) columns of matrix [A] and operate on appropriate components of vector $\{x\}$. Thus, algorithm 3 can be modified as shown in Table 2.8.

Table 2.8 Matrix Times Vector (Saxpy Operation with Unrolling Level 2)

```
       Do 1 I = 1, N, NUNROL (= 2)
       Do 1 J = 1, N
         b(J)  =  b(J)  +  A(J, I)  *  x(I)  +  A(J, I+1)  *  x(I+1)
1        Continue
```

Remarks: Unrolling dot product operations are different from unrolling saxpy operations in several aspects:

(a) The former (unrolling dot product) gives a "final, or complete" solution, whereas the latter (unrolling saxpy) gives a "partial, or incomplete" solution.

(b) In FORTRAN coding, the former requires "several" FORTRAN statements inside the innermost loop (see Table 2.6), whereas the latter requires "one long" FORTRAN statement (see Table 2.8).

(c) The former involves the operation of the dot product of two given vectors, whereas the latter involves the operation of a CONSTANT times a VECTOR, plus another vector.

(d) On certain computer platforms (such as Cray 2, Cray YMP and Cray – C90, etc.), saxpy operations are faster than dot product operations. Thus, algorithms and/or solution strategies can be tailored to specific computer platforms to improve the numerical performance.

(e) In Tables 2.6 and 2.8, if N is "not" a multiple of "NUNROL", then the algorithms have to be modified slightly in order to take care of a few remaining rows (or columns) of matrix [A], as can be seen in Exercises 2.4 and 2.6.

2.6 Parallel Dense Equation Solvers

In this section, it will be demonstrated that simple, yet highly efficient parallel strategies can be developed for solving a system of dense, symmetrical, positive definite equations. These strategies are based upon an efficient matrix times matrix subroutine, which also utilizes the "unrolling" techniques discussed in the previous sections.

2.6.1 Basic Symmetrical Equation Solver

Systems of linear, symmetrical equations can be represented as:

$$A \cdot x = b \tag{2.11}$$

One way to solve Eq.(2.11) is to first decompose the coefficient matrix A into the product of two triangular matrices

$$A = U^T U \tag{2.12}$$

Where U is an upper-triangular matrix which can be obtained by

when $i \neq j$, then
$$u_{ij} = \frac{a_{ij} - \sum_{k=1}^{i-1} u_{ki} u_{kj}}{u_{ii}} \tag{2.13}$$

when $i = j$, then
$$u_{ii} = \sqrt{a_{ii} - \sum_{k=1}^{i-1} u_{ki}^2} \tag{2.14}$$

Then the unknown vector x can be solved through the forward/backward elimination, such as:

$$U^T y = b \qquad (2.15)$$

for y, with

$$y_j = \frac{b_j - \sum\limits_{i=1}^{j-1} u_{ij} y_i}{u_{jj}} \qquad (2.16)$$

and to solve

$$Ux = y \qquad (2.17)$$

for x, with

$$x_j = \frac{y_j - \sum\limits_{i=j+1}^{n} u_{ji} x_i}{u_{jj}} \qquad (2.18)$$

The efficiency of an equation solver on massively parallel computers with distributed memory is dependent on both its vector and/or cache performance and its communication performance. In this study, we have decided to adopt a skyline column storage scheme to exploit dot product operations.

2.6.2 Parallel Data Storage Scheme

Assuming 3 processors are used to solve a system of 15 matrix equations, the matrix will be divided into several parts with 2 columns per blocks (ncb = 2). Therefore, from Figure 2.2, processor 1 will handle columns 1, 2, 7, 8, 13, and 14. Also, processor 2 will handle columns 3, 4, 9, 10, and 15, and processor 3 will handle columns 5, 6, 11, and 12. The columns that belong to each processor will be stored in a one-dimensional array in a column-by-column fashion. For example, the data in row 4 and column 7 will be stored by processor 1 at the 7th location of one-dimensional array A of processor 1. Likewise, each processor will store only portions of the whole matrix [A]. The advantage of this storage scheme is that the algorithm can solve a much bigger problem.

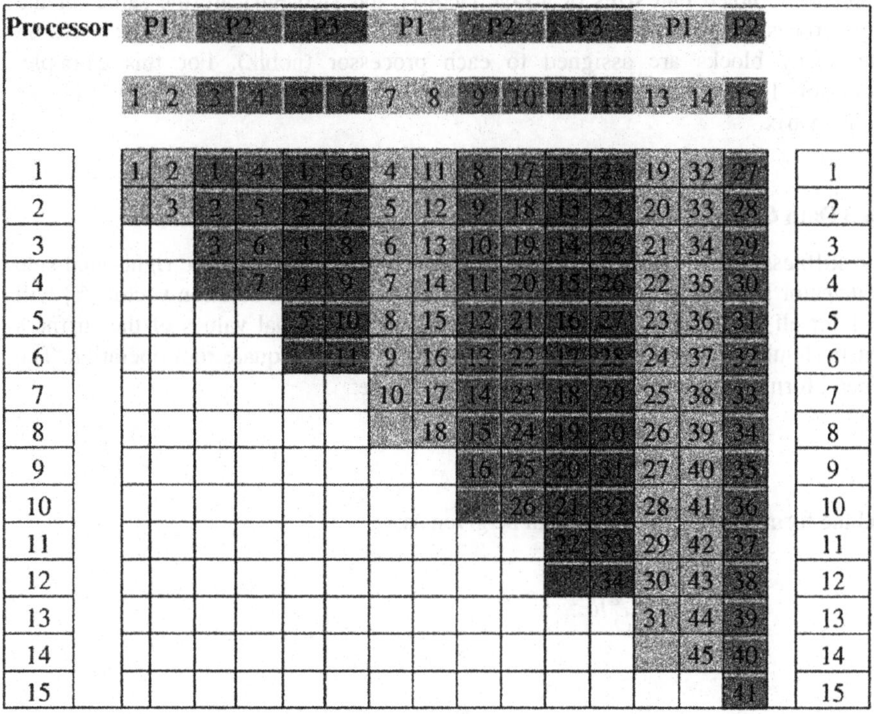

Processor	P1		P2		P3		P1		P2		P3		P1		P2	
	1	2	3	4	5	6	7	8	9	10	11	12	13	14	15	
1	1	2	1	4	1	6	4	11	8	17	12	21	19	32	27	1
2		3	2	5	2	7	5	12	9	18	13	24	20	33	28	2
3			3	6	3	8	6	13	10	19	14	25	21	34	29	3
4				7	4	9	7	14	11	20	15	26	22	35	30	4
5					5	10	8	15	12	21	16	27	23	36	31	5
6						11	9	16	13	22	17	28	24	37	32	6
7							10	17	14	23	18	29	25	38	33	7
8								18	15	24	19	30	26	39	34	8
9									16	25	20	31	27	40	35	9
10										26	21	32	28	41	36	10
11											22	33	29	42	37	11
12												34	30	43	38	12
13													31	44	39	13
14														45	40	14
15															41	15

Figure 2.2 Block Columns Storage Scheme

Two more small arrays are also needed to store the starting columns (icolst) and ending columns (icolend) of each block.

Therefore, from this example:

$$icolst_1 := \begin{pmatrix} 1 \\ 7 \\ 13 \end{pmatrix} \qquad icolend_1 := \begin{pmatrix} 2 \\ 8 \\ 14 \end{pmatrix}$$

$$icolst_2 := \begin{pmatrix} 3 \\ 9 \\ 15 \end{pmatrix} \qquad icolend_2 := \begin{pmatrix} 4 \\ 10 \\ 15 \end{pmatrix}$$

$$icolst_3 := \begin{pmatrix} 5 \\ 11 \end{pmatrix} \qquad icolend_3 := \begin{pmatrix} 6 \\ 12 \end{pmatrix}$$

It should be noted that sizes of arrays {icolst} and {icolend} are the same for the same processor, but may be different from other processors. In fact, it depends on how many blocks are assigned to each processor (noblk). For this example, processors 1 and 2 each stores 3 blocks of the matrix, and processor 3 stores 2 blocks of the matrix.

2.6.3 Data Generating Subroutine

The stiffness matrix, stored in a one-dimensional array {A} and the right-hand-side load vector {b} will be automatically generated such that the solution vector {x} will be 1 for all values of {x}. Also, from Eq.2.14, the diagonal values of the stiffness matrix should be large to avoid the negative values in the square root operation. The general formulas to generate the stiffness matrix are:

$$a_{i,i} = 50000 \cdot (i+i)$$

$$a_{i,j} = i+j$$

and the formula for RHS vector can be given as:

$$b_i = \sum_{j=1}^{n} a_{i,j} \, .$$

2.6.4 Parallel Choleski Factorization [1.9]

Assuming the first four rows of the matrix A (see Figure 2.2) have already been updated by multiple processors, and row 5 is currently being updated, according to Figure 2.2, terms such as $u_{5,5}... u_{5,6}$ and $u_{5,11}... u_{5,12}$ are processed by processor 3. Similarly, terms such as $u_{5,7}... u_{5,8}$ and $u_{5,13}... u_{5,14}$ are handled by processor 1 while terms such as $u_{5,9}, u_{5,10}$ and $u_{5,15}$ are executed by processor 2.

As soon as processor 3 completely updated column 5 (or more precisely, updated the diagonal term $u_{5,5}$ since the terms $u_{1,5}$ $u_{2,5}$ ... $u_{4,5}$ have already been factorized earlier), it will send the entire column 5 (including its diagonal term) to all other processors. Then processor 3 will continue to update its other terms of row 5. At the same time, as soon as processors 1 and 2 receive column 5 (from processor 3), these processors will immediately update their own terms of row 5.

To enhance the computational speed, by using the optimum available cache, the "scalar" product operations (such as $u_{ki} * u_{kj}$) involved in Eq.(2.13) can be replaced by "sub-matrix" product operations. Thus, Eq.(2.13) can be re-written as:

$$[u_{ij}] = [u_{ii}]^{-1} \cdot \left([a_{ij}] - \sum_{k=1}^{i-1} [u_{ki}]^T [u_{kj}] \right) \qquad (2.19)$$

Similar "sub-matrix" expressions can be used for Eqs.(2.14, 2.16, 2.18).

2.6.5 A Blocked And Cache Based Optimized Matrix-Matrix Multiplication

Let's consider matrix $C(m,n)$ to be the product of two dense matrices A and B of dimension (m,l) and (l,n), respectively.

$$C = A \cdot B \qquad\qquad (2.20)$$

A basic matrix-matrix multiplication algorithm consists of triple-nested do loops as follows:

```
Do i=1,m
  Do j=1,n
    Do k=1,l
       c(i,j)=c(i,j)+a(i,k)*b(k,j)
    ENDDO
  ENDDO
ENDDO
```

Further optimization can be applied on this basic algorithm to improve the performance of the multiplication of two dense matrices. The following optimization techniques are used for the matrix-matrix multiplication sub-routine.

- Re-ordering of loop indexes
- Blocking and strip mining
- Loop unrolling on the considered sub-matrices
- Stride minimization
- Use of temporary array and leftover computations

2.6.5.1 Loop Indexes And Temporary Array Usage

Fortran uses column-major order for array allocation. In order to get a "stride" of one on most of the matrices involved in the triple do loops of the matrix-matrix multiplication, one needs to interchange the indices as follows:

```
Do j=1,n
  Do i=1,m
    Do k=1,l
       c(i,j)=c(i,j)+a(i,k)*b(k,j)
    ENDDO
  ENDDO
ENDDO
```

Note that in the above equation, the order of the index loop has been re-arranged to get a stride of 1 on matrix C and B. However, the stride on matrix A is m. In order to have a stride of 1 during the computation on matrix A, a temporary array is used to load a portion of matrix A before computation.

2.6.5.2 Blocking and Strip Mining

Blocking is a technique to reduce cache misses in nested array processing by calculating in blocks or strips small enough to fit in the cache. The general idea is that an array element brought in should be processed as fully as possible before it is flushed out.

To use the blocking technique in this basic matrix-matrix multiplication, the algorithm can be re-written by adding three extra outer loops (JJ, II, and KK) with an increment equal to the size of the blocks.

```
Do JJ=1,n,jb
  Do II=1,m,ib
    Do KK=1,l,kb
      Do J=JJ,min(n,JJ+jb-1)
        Do I=II,min(m,II+ib-1)
          Do K=KK,min(l,KK+kb-1)
            c(i,j)=c(i,j)+a(i,k)*b(k,j)
          ENDDO
        ENDDO
      ENDDO
    ENDDO
  ENDDO
ENDDO
```

In the above code-segment, ib, jb, and kb are the block sizes for i, j, and k, respectively, and they are estimated function of the available cache size for different computer platforms.

2.6.5.3 Unrolling of Loops

Unrolling of loops is considered at various stages. To illustrate the idea, let's consider the inner do loop (see the index k) where the actual multiplication is performed. Here a 4 by 4 sub-matrix is considered at the time, and the inner do loop of Eqs.(2.13 – 2.14) can be unrolled in the following fashion:

```
DO j=1, number of J blocks
  DO i=1, number of I blocks
    DO k=1, number of K blocks
      C{i,j}   = C{i,j}   + T{1,1} * B{k,j}
      C{i+1,j} = C{i+1,j} + T{1,2} * B{k,j}
      C{i,j+1} = C{i,j+1} + T{1,1} * B{k,j+1}
```

```
                C{i+1,j+1} = C{i+1,j+1} + T{1,2} * B{k,j+1}
                C{i,j+2} = C{i,j+2} + T{1,1} * B{k,j+2}
                C{i+1,j+2} = C{i+1,j+2} + T{1,2} * B{k,j+2},
                C{i,j+3} = C{i,j+3} + T{1,1} * B{k,j+3}
                C{i+1,j+3} = C{i+1,j+3} + T{1,2} * B{k,j+3}
                C{i+2,j} = C{i+2,j} + T{1,1} * B{k,j}
                C{i+3,j} = C{i+3,j} + T{1,2} * B{k,j}
                C{i+2,j+1} = C{i+2,j+1} + T{1,1} * B{k,j+1}
                C{i+3,j+1} = C{i+3,j+1} + T{1,2} * B{k,j+1}
                C{i+2,j+2} = C{i+2,j+2} + T{1,1} * B{k,j+2}
                C{i+3,j+2} = C{i+3,j+2} + T{1,2} * B{k,j+2}
                C{i+2,j+3} = C{i+2,j+3} + T{1,1} * B{k,j+3}
                C{i+3,j+3} = C{i+3,j+3} + T{1,2} * B{k,j+3}
        ENDDO
      ENDDO
    ENDDO
```

In the above code-segment, T is the temporary 2-D array, which contains portions of the matrix [A].

2.6.6 Parallel "Block" Factorization

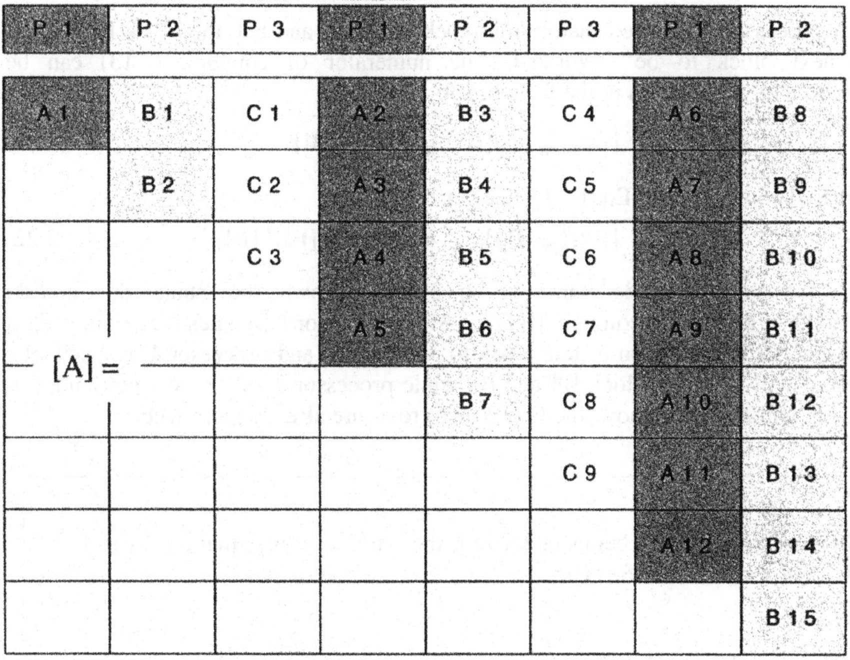

Figure 2.3 Parallel Block Factorization

The procedure starts at factorizing the first block of processor 1 by using basic sequential Choleski factorization. Then, processor 1 will send its first block (A1) to processor 2 and 3. After that, processor 1 will factorize A2 and A6 blocks. At the same time, processor 2 is factorizing B1, B3, and B8 blocks, and processor 3 is factorizing C1 and C4 blocks. Up to now, the first ncb rows are already factorized.

Note:

Suppose the block A2 is being factorized, the formula that can be used for factorization is

$$[A2]_{new} = (\,[A1]^{T}\,)^{-1}\,[A2]_{old}$$

Pre-multiply both side with $[A1]^{T}$, we get

$$[A1]^{T}\,[A2]_{new} = [A2]_{old}$$

$[A2]_{new}$, therefore, can be solved by performing the forward operations for ncb times.

After processor 2 finished factorizing blocks B1, B3, and B8, the block B2 will be the next block to be factorized. The numerator of equation (2.13) can be "symbolically" rewritten in the form of sub-matrix as:

$$[B2]_{new} = \sqrt{[B2]_{old} - [B1]^{T}[B1]} \tag{2.21}$$

"Squared" both sides of Eq.(6.21), one gets:

$$[B2]_{new}^{T}[B2]_{new} = [B2]_{old} - [B1]^{T}[B1] \tag{2.22}$$

Since the right-hand-side matrix of Eq.(2.22) is known, the "standard" Choleski method can be used to compute $[B2]_{new}$. After processor 2 finishes factorizing B2, it will send the information of this block to processor 3 and processor 1, respectively. Then, it will factorize block B4 and B9 while processor 3 and 1 are factorizing C2, C5, A3, and A7. Up to now, the first 2 (ncb) rows are already factorized.

Note:

Suppose block C2 is being factorized, the numerator of equation 3 can be re-written in sub-matrix form as:

$$[C2]_{new} = [C2]_{old} - [B1]^{T}\,[C1]$$

Then, the denominator term will be processed like the step that has been used to factorize block A2.

Now, processor 3 will be the master processor (i.e., the processor that will factorize the diagonal block), and the step mentioned above will be repeated for subsequent steps.

2.6.7 "Block" Forward Elimination Subroutine

From Section 2.6.1, the system of Eq. (2.11) can be rewritten as:

$$[U]^T \, \vec{y} = \vec{b}$$

or

In this forward substitution phase, three different types of processors are identified:

1. The master processor is the processor that calculates the "final" solution $\{x_i\}$ for this phase (i.e., processor 1 at starting point).
2. The next master processor is the processor that will calculate the final $\{x_i\}$ in the next loop of operation (i.e., processor 2 at starting point).
3. The workers are the processors that will calculate the updated value of $\{x_i\}$ in the next loop of operation (i.e., processor 3 at starting point).

The procedure from the given example starts at the forward substitution for the final solution $\{y_i\}$ by doing forward substitution of block A1 by processor 1, the master processor. Then, processor 1 will send the solution of $\{y_1\}$ to processor 2 (the next master processor) and 3 (the worker). After that, processor 1 will "partially" update $\{y4\}$ and $\{y7\}$ by doing forward of substitution blocks A2 and A6, respectively. For the next master processor, processor 2, it will update and calculate the final $\{y_2\}$ by doing forward substitution of blocks B1 and B2 respectively. After that, processor 2 will send the solution of $\{y_2\}$ to processor 3 (the next master processor) and 1 (the worker). At the same time, the worker, processor 3 will do forward substitution of block C1 to update $\{y_3\}$ and C4 to update $\{y_6\}$. The next step will be that processor 3 is the master processor, and the procedure will be repeated until all intermediate unknowns are solved.

2.6.8 "Block" Backward Elimination Subroutine

In this backward substitution phase, the unknown vector $\{x\}$, shown in Eq.(2.11), will be solved.

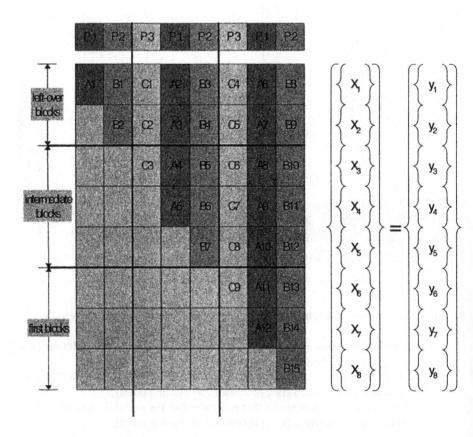

Figure 2.4 Backward Elimination Phase

From Figure 2.4, processor 2 will be the master of the starting process. The main ideas of the algorithm are to divide the upper-factorized matrix into a sub-group of block matrices, which have, normally ncb x np rows. Also, there are 3 types of sub-group blocks, which are the first blocks, the intermediate blocks, and the last blocks or leftover block. Then each processor will do backward substitution block by block from bottom to top and, if possible, from right to left. After doing backward substitution, each processor will send the solution (partially updated or final) to its left processor (i.e. processor 2 will send the solution to processor 1, processor 1 will send the solution to processor 3, and processor 3 will send the solution to processor 2). From the given example, processor 2 will do backward substitution of block B15 to find the final solution of X_8. Then, it will send X_8 to the adjacent (left) processor, processor 1, and start to do backward substitution of block B14 to find the partially updated solution of X_7. Then, processor 2 will send X_7 to processor 1. Now, processor 1 will do backward substitution of block A12 to find the final solution of X_7, while processor 2 is doing backward substitution of block B13 to find the partial updated solution of X_6. To clearly understand the steps involved in this phase, see Table 2.9.

Table 2.9 Processor Tasks in Backward Substitution Subroutine

Processor 1	Processor 2 (Master)	Processor 3	Remark
Idle (Waiting until B14 is finished)	B15	Idle (Waiting until A11 is finished)	Final solution $\{X_8\}$ is found
Idle	B14	Idle	$\{X_7\}_{partial}$
A12	B13	Idle	$\{X_7\}_{final},$ $\{X_6\}_{partial}$
A11	B12	Idle	$\{X_6\}_{partial},$ $\{X_5\}_{partial}$
A10	B11	C9	$\{X_6\}_{final},$ $\{X_5\}_{partial},$ $\{X_4\}_{partial}$
A9	B10	C8	$\{X_5\}_{partial},$ $\{X_4\}_{partial},$ $\{X_3\}_{partial}$
A8	B7	C7	$\{X_5\}_{final},$

			$\{X_4\}_{\text{partial}},$ $\{X_3\}_{\text{partial}}$
Idle (Waiting until B6 is finished)	B6	C6	$\{X_4\}_{\text{partial}},$ $\{X_3\}_{\text{partial}}$
A5	B5	Idle (Waiting until A4 is finished)	$\{X_4\}_{\text{final}},$ $\{X_3\}_{\text{partial}}$
A4	B9	Idle	$\{X_3\}_{\text{partial}},$ $\{X_2\}_{\text{partial}}$
A7	B8	C3	$\{X_3\}_{\text{final}},$ $\{X_2\}_{\text{partial}},$ $\{X_1\}_{\text{partial}}$
A6	Idle (Waiting until C5 is finished)	C5	$\{X_2\}_{\text{partial}},$ $\{X_1\}_{\text{partial}}$
Idle (Waiting until B4 is finished)	B4	C4	$\{X_2\}_{\text{partial}},$ $\{X_1\}_{\text{partial}}$
A3	B3	Idle (Waiting until A3 is finished)	$\{X_2\}_{\text{partial}},$ $\{X_1\}_{\text{partial}}$
A2	Idle (Waiting until C2 is finished)	C2	$\{X_2\}_{\text{partial}},$ $\{X_1\}_{\text{partial}}$
Idle (Waiting until B1 is finished)	B2	C1	$\{X_2\}_{\text{final}},$ $\{X_1\}_{\text{partial}}$
Idle	B1	Done	$\{X_1\}_{\text{partial}}$
A1	Done	Done	$\{X_1\}_{\text{final}}$

2.6.9 "Block" Error Checking Subroutine

After the solution of the system of equations has been obtained, the next step that should be considered is error checking. The purpose of this phase is to evaluate how good the obtained solution is. There are four components that need to be considered:

1. X_{max} is the X_i that has the maximum absolute value (i.e., the maximum displacement, in structural engineering application).
2. Absolute summation of X_i

3. Absolute error norm of $[A] \cdot \vec{x} - \vec{b}$
4. Relative error norm is the ratio of absolute error norm and the norm of RHS vector $\vec{b}$.

P 1	P 2	P 3	P 1	P 2	P 3	P 1	P 2			
A1	B1	C1	A2	B3	C4	A6	B8	x_1		b_1
	B2	C2	A3	B4	C5	A7	B9	x_2		b_2
		C3	A4	B5	C6	A8	B10	x_3		b_3
			A5	B6	C7	A9	B11	x_4	$=$	b_4
				B7	C8	A10	B12	x_5		b_5
					C9	A11	B13	x_6		b_6
						A12	B14	x_7		b_7
							B15	x_8		b_8

Figure 2.5 Error Norms Computation (Upper Triangular Portion)

The first two components can be found without any communication between each processor because the solution {x} is stored in every processor. However, the next two components, absolute error norm and relative error norm, require communication between each processor because the stiffness matrix of the problem was divided in several parts and stored in each processor. Therefore, the parallel algorithm for error norm checking will be used to calculate these error norms. The key concepts of the algorithm are that every processor will calculate its own $[A] \cdot \vec{x} - \vec{b}$, and the result will be sent to the master processor, processor 1.

The procedure starts at each processor, which will calculate $[A]\{X\}_I$, which corresponds to $\{b_i\}$ for the lower part of the stiffness matrix. From Figure 2.6, processor 1 will partially calculate $[A]\{X\}_1$ by multiplying the lower part of $[A1]$ with $\{X_1\}$.

Similarly, Processor 2 will partially calculate $[A]\{X\}_2$ by multiplying the lower part of $[B1]$ and $[B2]$ with $\{X_1\}$ and $\{X_2\}$, respectively. The step will be repeated until every processor finishes calculating the lower part of the stiffness matrix. In this step, there is no communication between processors.

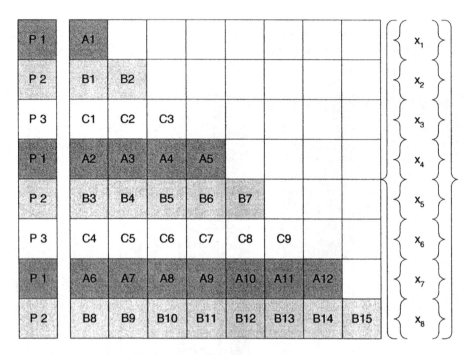

Figure 2.6 Error Norms Computation (Lower Triangular Portion)

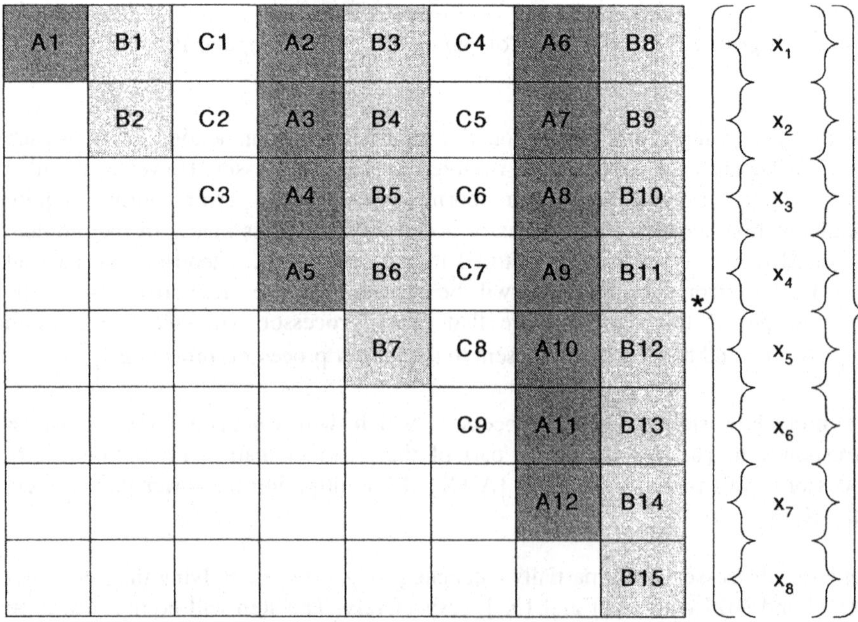

Figure 2.7 Error Norms Computation (Upper Triangular Portion)

Once the calculating for the lower part of the stiffness matrix is done, the next step will be the calculation of the diagonal and the upper part of the stiffness matrix. This procedure starts with the same concept as the calculation of the lower part. After the diagonal and the upper part is done, every processor will have its own $[A]\{X\}_i$.

Then, the worker processors will send $[A]\{X\}_i$ to the master processor. Therefore, processor 1 will have the complete values of $[A]\{X\}$. Next, processor 1 will find the difference between $[A]\{X\}$ and $\{b\}$, and the norm of this difference will be the absolute error norm. The absolute error norm will then be divided by the norm of the RHS vector to give the relative error norm.

2.6.10 Numerical Evaluation

Several full-size (completely dense) symmetrical matrix equations are solved, and the results (on different computer platforms) are compared and tabulated in Tables 2.10 – 2.12. All reported times are expressed in seconds. Performance of the "regular" versus "cache-based" matrix-matrix multiplication sub-routines is presented in Table 2.13. Results for a parallel SGI dense solver (on NASA LaRC SGI/Origin 2000, Amber Computer, 40 GB RAM), as compared to our developed software, are given in Tables 2.14 – 2.15.

It is interesting to note the "major difference" between our developed parallel dense solver (on the Sun 10k computer) as compared to the parallel SGI dense solver (on the NASA SGI/ORIGIN-2000 computer).

In particular, let's look at Tables 2.14 and 2.15, paying attention to the "time ratio" of "Forward-Backward / Factorization."

For 5,000 dense equations, using 8 (SGI Amber) processors, and with the near optimal block size ncb = 64, our developed parallel solver has the "time ratio" = (Forward & Backward / Factorization) = (3.5 / 19.1) = 18.32% as indicated in Table 2.15.

However, for the same problem size (= 5,000 equations), using 8 (NASA SGI/ORIGIN-2000, Amber) processors, the SGI Parallel Dense Solver library sub-routine has the "time ratio" = (Forward & Backward / Factorization) = (4.8 / 15.8) = 30.38%, as indicated in Table 2.14.

Thus, our developed software is much more efficient than the well-respected SGI software, especially for the Forward & Backward solution phases.

Finally, we show the results from porting the software to run on commodity computer systems. The software system using Windows 2000 workstation OS was developed using Compaq FORTRAN compiler (Version 6.1) and uses MPI-

Softtech's MPI library (Version 1.6.3). The same numerical experiments used with the two earlier hardware platforms yielded ncb = 32 and 64 showing somewhat similar performances. However, ncb = 64 was used as the block size. Moreover, the dense equations used in evaluating the performance came from actual finite element models. Table 2.16 shows the results for 4,888 equations while Table 2.17 is for 10,368 equations.

The performance of the commodity cluster is better than both the workstation platforms. When the matrices fit into memory (Table 2.16), the speedup is sub-linear. The speedup, however, is super-linear with larger problems and can be misleading.

This is because the matrices do not fit into memory and a larger number of page faults occur with the single processor version. Nevertheless, the overall throughput and performance is impressive.

Table 2.10 Two Thousand (2,000) Dense Equations (Sun 10k)

n = 2000	Factorization time				Forward Substitution				Backward Substitution			
	ncb=16	32	64	128	ncb=16	32	64	128	ncb=16	32	64	128
1 processor	16.243	9.896	8.552	8.337	0.176	0.173	0.173	0.175	0.110	0.097	0.098	0.097
2	7.637	5.170	4.873	5.108	0.087	0.040	0.066	0.261	0.236	0.083	0.048	0.039
4	3.736	3.028	3.010	3.548	0.028	0.034	0.044	0.194	0.109	0.046	0.028	0.026
8	2.634	2.076	2.470	3.186	0.025	0.029	0.047	0.210	0.057	0.027	0.020	0.021
16	1.766	1.569	1.723	✕	0.034	0.049	0.095	✕	0.038	0.020	0.016	✕

Table 2.11 Five Thousand (5,000) Dense Equations (Sun 10k)

n = 5000	Factorization time				Forward Substitution				Backward Substitution			
	ncb=16	32	64	128	ncb=16	32	64	128	ncb=16	32	64	128
1 processor	277.5	182.2	143.3	130.4	1.108	1.110	1.105	1.153	0.775	0.784	0.784	0.776
2	140.0	93.8	76.8	72.2	0.582	0.554	0.545	0.620	1.553	0.826	0.623	0.582
4	73.4	51.3	42.4	44.3	0.300	0.287	0.293	0.338	0.781	0.424	0.319	0.336
8	40.8	27.6	25.5	29.4	0.170	0.162	0.208	0.435	0.445	0.209	0.164	0.183
16	42.4	23.0	25.08	22.0	2.685	0.760	0.260	0.284	0.737	0.114	0.084	0.125

Table 2.12 Ten Thousand (10,000) Dense Equations (Sun 10k)

n = 10000	Factorization time				Forward Substitution				Backward Substitution			
	ncb=16	32	64	128	ncb=16	32	64	128	ncb=16	32	64	128
1 processor	2284.2	1548.6	1203.9	1059.2	4.565	4.881	4.541	4.548	3.317	3.560	3.323	3.302
2	1178.8	774.3	596.8	548.4	2.450	2.360	2.240	2.244	6.410	3.578	2.590	2.385
4	595.2	397.4	316.7	306.2	1.244	1.187	1.133	1.229	3.310	1.764	1.354	1.264
8	307.1	211.3	176.0	180.6	0.640	0.600	0.643	0.839	1.759	0.920	0.700	0.687
16	161.3	114.0	105.6	120.2	0.395	0.401	0.610	1.060	0.989	0.490	0.386	0.416

Table 2.13 Matrix-Matrix Multiplication Subroutines

	LIONS (Helios)			LIONS(Cancun)	NASA SGI/Origin-2000
	Regular version (seconds)	Cache version (seconds)	Cache Version (seconds)	Cache Version (seconds)	Cache Version (seconds)
1000x1000	177.405[*]	6.391[*]	6.420[†]	10.9[†]	9.5[†]
2000x2000	1588.03[*]	53.13[*]	54.02[†]	N/A	N/A

[*] using MPI_WTIME( )

[†] using etime( )

Table 2.14 Five Thousand (5,000) Dense Equations Timings Using SGI Parallel (SGI Amber) (seconds)

# of Processes	Factorization Time	Forward& Backward Substitution	Total Time	Speedup
1	87.4	4.2	91.6	1
2	48.8	4.6	53.4	1.7
4	23.8	4.4	28.2	3.2
8	15.8	4.8	20.6	4.4

Table 2.15 Five Thousand (5,000) Dense Equations Timings (SGI Amber) (seconds)

# of Processes	Factorization Time	Forward& Backward Substitution	Total Time	Speedup
1	103.1	3.4	106.5	1
2	55.9	2.8	58.7	1.8
4	31.8	1.7	33.5	3.2
8	19.1	3.5	22.6	4.7
10	16.2	4.4	20.6	5.2
16	12.9	5.4	18.3	5.8

Table 2.16 Five Thousand (4,888) Dense Equations Timings (Intel FEM) (seconds)

# of Processes	Factorization Time	Forward& Backward Substitution	Total Time	Speedup
1	61.1	0.2	61.3	1
2	33.7	0.7	34.4	1.8
3	24.2	0.5	24.7	2.5
4	19.3	0.5	19.8	3.1
5	16.6	0.4	17.0	3.6
6	14.6	0.3	14.9	4.1

Table 2.17 Ten Thousand (10,368) Dense Equations Timings (Intel FEM) (seconds)

# of Processes	Factorization Time	Forward& Backward Substitution	Total Time	Speedup
1	1258	1.1	1259.1	1
2	305	3.3	308.3	4.1
3	210	2.2	212.2	5.9
4	166	1.6	167.6	7.5
5	137	1.5	138.5	9.1
6	120	1.6	121.6	10.4
7	106	1.4	107.4	11.7

2.6.11 Conclusions

Detailed numerical algorithms and implementation for solving a large-scale system of dense equations have been described. Numerical results obtained from the developed MPI dense solver, and the SGI parallel dense solvers on different parallel computer platforms have been compared and documented.

The following conclusions, therefore, can be made:

1. The developed parallel dense solver is simple, efficient, flexible, and portable.
2. Using the developed parallel dense solver, a large-scale dense matrix can be stored across different processors. Hence, each processor only has to store "small portions" of the large dense matrix.
3. From Tables 2.10 – 2.12, it seems that when the problem size is getting bigger, the optimized number of columns per block that fits in a system cache will be approximately 64 columns per block.
4. Optimum performance of the developed parallel dense solver can be achieved by fine-tuning the block size on different computer platforms.

2.7 Developing/Debugging Parallel MPI Application Code on Your Own Laptop

A Brief Description of MPI/Pro

MPI/Pro is a commercial MPI middleware product. MPI/Pro optimizes the time for parallel processing applications.

MPI/Pro supports the full Interoperable Message Passing Interface (IMPI). IMPI allows the user to create heterogeneous clusters, which gives the user added flexibility while creating the cluster.

Verari Systems Software offers MPI/Pro on a wide variety of operating systems and interconnects, including Windows, Linux, and Mac OS X, as well as Gigabit Ethernet, Myrinet, and InfiniBand.

Web site

http://www.mpi-softtech.com/products/

Contact Information

Telephone:
        Voice: (205) 397-3141
        Toll Free: (866) 851-5244
        Fax: (205) 397-3142

Sales support

sales@mpi-softtech.com

Technical support

mpipro-sup@mpi-softtech.com

Cost

- $100 per processor (8 processor minimum), one time payment
- Support and Maintenance are required for the first year at a rate of 20% per annum of the purchase price. Support ONLY includes: NT4SP5 and higher, 2000 Pro, XP Pro, and Windows 2003 Server.

Steps to Run Mpi/Pro on Windows OS

MPI/Pro requires Visual Studio 98 to run parallel FORTRAN code on Windows OS. Visual Studio 98 supports both FORTRAN and C programming languages. MPI is usually written in C programming language; therefore Visual Studio is sufficient for the user's needs.

A. Open Fortran Compiler Visual Studio

B. Open Project

Open a new "project" as a *Win 32 Console Application*. Give a name to your project, e.g., *project1*. This will create a folder named "project1."

C. Create a Fortran File

Create a new Fortran file under the project folder you have just created, e.g., *project1.for*, and when you create a file, click on the option *"Add to Project."*

Then, type your program in this file. If you have other files or sub-routines to link to this file

- Create a file or files under the folder *"project1,"*.
- Then, from the pulldown menu, go to Project/ Add to Project and select File.
- Select the files to be linked one by one and add them to your project.

D. Parallelizing Your Own Program With Mpi/Pro

There are a total of four call statements to type in a program running sequentially. But, first of all, one has to type the following line at the very beginning of the code:

include 'mpif.h'

Then, right after defining the data types and character strings, the following three call statements need to be typed:

> *call MPI_INIT (ierr)*
> *call MPI_COMM_RANK (MPI_COMM_WORLD, me,ierr)*
> *call MPI_COMM_SIZE (MPI_COMM_WORLD, np,ierr)*

where

me   = the ID of each processor (*e.g., me = 0, 1, 2, 3, 4, 5 if six processors are used*)
np   = total number of processors
ierr   = error message

In order to finalize the procedure of parallel programming, another call statement has to be placed at the end of the program:

> *call MPI_FINALIZE (ierr)*

E. A Simple Example of Parallel MPI/FORTRAN Code

The entire list of the "trivial" MPI code is shown below:

```
implicit real*8(a-h,o-z)
include 'mpif.h'
real*8 a(16)

call MPI_INIT (ierr)
call MPI_COMM_RANK(MPI_COMM_WORLD, me, ierr)
call MPI_COMM_SIZE(MPI_COMM_WORLD, np, ierr)

if (me .eq. 0) then
open(5,file='rectdata.txt')
read(5,*) b,h
write(6,*) me,' b,h',b,h
endif

do i = 1,np
a(i) = (me+i)*2
enddo

write(6,*) 'me=',me,' a(-)',(a(i),i=1,np)
```

```
call sub1(me)
call sub2(me)

call MPI_FINALIZE(ierr)

stop
end
```

As you can see, there are two subroutine calls in the trivial MPI code. These sub-routines are added to the project as described in Section C. The sub-routines created for this example are listed below:

```
subroutine sub1(me)
write(6,*) me,' is in sub1 subroutine'
return
end

subroutine sub2(me)
write(6,*) me,' is in sub2 subroutine'
return
end
```

F. Before Compilation

A user makes some modifications in the FORTRAN compiler before he/she compiles his/her code. These modifications include defining the paths for the directory in which the file *'mpif.h'* is stored for the library files required by MPI/Pro.

1. From the pulldown menu, go to Tools then click on Options.

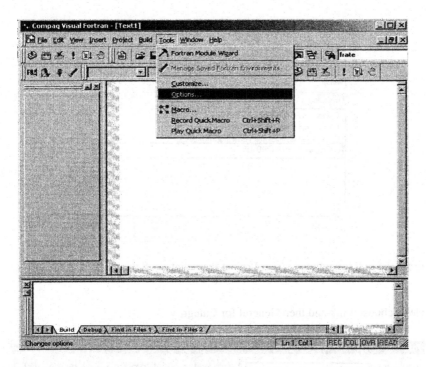

It will open up the Options window; click on Directories. Under the Show directories for sub-menu, the path for MPI/Pro include file is defined.

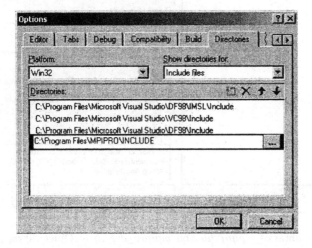

The same procedure is repeated for MPI/Pro library files:

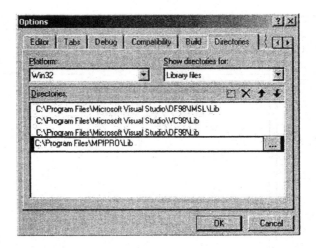

2. From the pulldown menu, go to Project then click on Settings.

Here, choose Link and then General for Category.

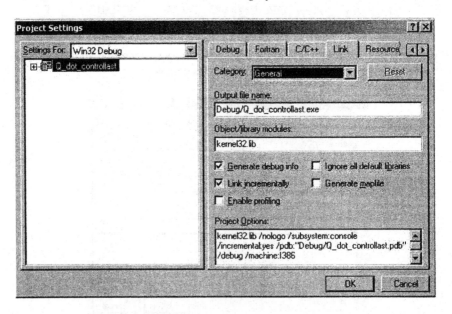

Add MPI/Pro libraries *mpipro.lib* and *mpipro_dvf.lib* to Object/libraries modules. Then, switch the Category to Input.

Type *libc.lib* in the Ignore libraries section as shown below:

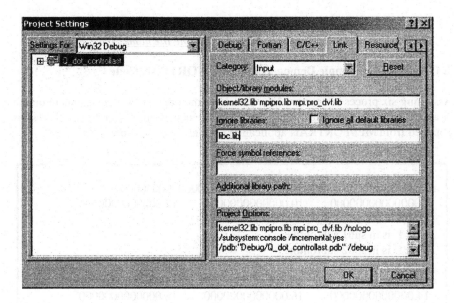

G. Compilation

As the next step, the user is supposed to compile his/her program and create the executable file. To do this:

- From the pulldown menu, go to Build, then select Compile to compile the MPI code. When you do this, a folder named "Debug" will be created.
- Go to Build again, and select Build to create the executable of the MPI code. The executable file will be located under the folder "Debug."

H. Specifying the Computer Name(s)

In order to specify the machine name, create a file named *"Machines"* under the same directory where the executable file exists. Type in the computer name as your processor for parallel processing.

I. Running Parallel

1. Open up the window for a command prompt. To do this, go to Windows Start/Run and type "cmd" or "cmd.exe." This will run the command prompt.

2. In command prompt, go to the folder "Debug" where you have created the executable file.

3. Then type:

mpirun –np <u>number of processor</u> <u>executable file name</u>

J. Outputs of the Simple Demonstrated MPI/FORTRAN Code

Assuming six processors are used to "simulate" the parallel MPI environment on the "Laptop" personal computer, using the MPI code discussed in Section E. The results obtained from MPI/FORTRAN application code are shown below:

```
me=       1  a(-)  4.00000000000000      6.00000000000000
    8.00000000000000      10.0000000000000      12.0000000000000
    14.0000000000000
        1  is in sub1 subroutine
        1  is in sub2 subroutine

me=       4  a(-)  10.0000000000000      12.0000000000000
    14.0000000000000      16.0000000000000      18.0000000000000
    20.0000000000000
        4  is in sub1 subroutine
        4    is in sub2 subroutine

me=       2  a(-)  6.00000000000000      8.00000000000000
    10.0000000000000      12.0000000000000      14.0000000000000
    16.0000000000000
        2  is in sub1 subroutine
        2  is in sub2 subroutine

me=       3  a(-)  8.00000000000000      10.0000000000000
    12.0000000000000      14.0000000000000      16.0000000000000
    18.0000000000000
        3  is in sub1 subroutine
        3  is in sub2 subroutine

me=       5  a(-)  12.0000000000000      14.0000000000000
    16.0000000000000      18.0000000000000      20.0000000000000
    22.0000000000000
        5  is in sub1 subroutine
        5  is in sub2 subroutine
        0  b,h  6.00000000000000      8.00000000000000

me=       0  a(-)  2.00000000000000      4.00000000000000
    6.00000000000000      8.00000000000000      10.0000000000000
    12.0000000000000
        0  is in sub1 subroutine
        0  is in sub2 subroutine
```

2.8 Summary

In this chapter, a brief summary of how to write simple parallel MPI/FORTRAN codes has been presented and explained. Through two simple examples (given in Sections 2.2 – 2.3), it is rather obvious that writing parallel MPI/FORTRAN codes are essentially the same as in the sequential mode of FORTRAN 77, or FORTRAN 90, with some specially added parallel MPI/FORTRAN statements. These special MPI/FORTRAN statements are needed for the purposes of parallel computation and processors' communication (such as sending/receiving messages, summing /merging processor's results, etc.). Powerful unrolling techniques associated with "dot-product" and "saxpy" operations have also been explained. These unrolling strategies should be incorporated into any application codes to substantially improve the computational speed.

2.9 Exercises

2.1 Write a <u>sequential</u> FORTRAN (or C++) code to find the product $\{b\} = [A] * \{x\}$, using <u>dot-product</u> operations, where

$$[A]_{n \times n} = A_{i,j} = i + j$$

$$\{x\}_{n \times 1} = x_j = j$$

$$n = 5000$$

2.2 Re-do problem 2.1 using <u>saxpy</u> operations.

2.3 Re-do problem 2.1 using <u>unrolling level 10</u>.

2.4 Re-do problem 2.1 using <u>unrolling level 7</u>.

2.5 Re-do problem 2.2 using <u>unrolling level 10</u>.

2.6 Re-do problem 2.1 using <u>unrolling level 7</u>.

2.7 Assuming 4 processors are available (NP = 4), write a <u>parallel MPI/FORTRAN</u> (or C++) program to re-do problem 2.1.

2.8 Given a real array A(-) that contains "N" random values 0.00 and 1.00 (say, N = 10^6 numbers), and assuming there are "NP" processors available (say, NP = 4), write a MPI_parallel FORTRAN (or C^{++}) that will print the above "N" numbers in "increasing" order.

3 Direct Sparse Equation Solvers

3.1 Introduction

In the finite element procedures, the most time-consuming portions of the computer code usually occur when solving a system of discretized, linear simultaneous equations. While truly sparse equation solvers have already penetrated through powerful (and popular) finite element commercial software (such as MSC-NASTRAN, ABAQUS, etc.), to the best of the author's knowledge, most (if not all) past and currently available finite element books still discuss and explain either skyline and/or variable bandwidth (profiled) equation solution strategies. It should be emphasized here that while sparse equation solvers have been well documented in the mathematical and computer science communities, very few have been thoroughly explained and reported by the engineering communities [1.9].

The focus of this chapter, therefore, is to discuss important computational issues for solving large-scale engineering system of sparse, linear equations, which can be represented in the matrix notation as:

$$[A]\vec{x} = \vec{b} \tag{3.1}$$

3.2 Sparse Storage Schemes

Considering a $NR \times NC$ (= 5×5) unsymmetrical sparse matrix as shown in Eq.(3.1):

$$[A] = \quad \begin{array}{c|c|c|c|c|c|}
 & 1 & 2 & 3 & 4 & 5 \\
\hline
1 & d_1 & & a_1 & a_2 & \\
\hline
2 & & d_2 & & a_3 & \\
\hline
3 & b_1 & & d_3 & a_4 & \\
\hline
4 & b_2 & b_3 & b_4 & d_4 & \\
\hline
5 & & & & & d_5 \\
\hline
\end{array} \tag{3.2}$$

Where NR and NC represent Number of Rows, and Number of Columns of matrix [A], respectively.

Sparse matrix Eq.(3.2) can be represented in the compact row storage as:

$$
IA \begin{pmatrix} 1 \\ 2 \\ 3 \\ 4 \\ 5 \\ NR+1=6 \end{pmatrix} = \begin{Bmatrix} 1 \\ 4 \\ 6 \\ 9 \\ 13 \\ 14 \end{Bmatrix} = \begin{array}{c} \text{starting locations of the } 1^{st} \text{ non} - \text{zero} \\ \text{term for each row} \end{array}
\tag{3.3}
$$

$$
JA \begin{pmatrix} 1 \\ 2 \\ 3 \\ 4 \\ 5 \\ 6 \\ 7 \\ 8 \\ 9 \\ 10 \\ 11 \\ 12 \\ 13 = NCOEF1 \end{pmatrix} = \begin{Bmatrix} 1 \\ 3 \\ 4 \\ 2 \\ 4 \\ 1 \\ 3 \\ 4 \\ 1 \\ 2 \\ 3 \\ 4 \\ 5 \end{Bmatrix} = \begin{array}{c} \text{column numbers associated with} \\ \text{non} - \text{zero terms of each row} \end{array}
\tag{3.4}
$$

$$
AN (1, 2, 3, 4 \ldots NCOEF1)^T = \{ d_1, a_1, a_2, d_2, \ldots, d_5 \}^T
\tag{3.5}
$$

In Eq.(3.4), NCOEF1 represents the total non-zero terms (before factorization, see section 3.3) and can be calculated as:

$$
NCOEF1 = IA (NR+1) - 1 = 14 - 1 = 13
\tag{3.6}
$$

If matrix [A] is <u>symmetrical</u> (such as $a_1 = b_1, \ldots, a_4 = b_4$), then it can be represented in the compact row storage formats as:

$$IA \begin{pmatrix} 1 \\ 2 \\ 3 \\ 4 \\ 5 \\ NR+1=6 \end{pmatrix} = \begin{Bmatrix} 1 \\ 3 \\ 4 \\ 5 \\ 5 \\ 5 \end{Bmatrix} = \begin{matrix} \text{starting location of the 1}^{st}\text{ nonzero,} \\ \text{off} - \text{diagonal term for each row} \end{matrix} \qquad (3.7)$$

$$JA \begin{pmatrix} 1 \\ 2 \\ 3 \\ NCOEF1=4 \end{pmatrix} = \begin{Bmatrix} 3 \\ 4 \\ 4 \\ 4 \end{Bmatrix} = \begin{matrix} \text{column numbers associated} \\ \text{with non} - \text{zero, off} - \text{diagonal} \\ \text{terms for each row} \end{matrix} \qquad (3.8)$$

$$AD \begin{pmatrix} 1 \\ 2 \\ 3 \\ 4 \\ NR=5 \end{pmatrix} = \begin{Bmatrix} d_1 \\ d_2 \\ d_3 \\ d_4 \\ d_5 \end{Bmatrix} = \begin{matrix} \text{values of diagonal terms} \\ \text{of the matrix [A]} \end{matrix} \qquad (3.9)$$

$$AN \begin{pmatrix} 1 \\ 2 \\ 3 \\ NCOEF1=4 \end{pmatrix} = \begin{Bmatrix} a_1 \\ a_2 \\ a_3 \\ a_4 \end{Bmatrix} = \begin{matrix} \text{values of non} - \text{zero, off} - \text{diagonal} \\ \text{terms of the matrix [A]} \end{matrix} \qquad (3.10)$$

The unsymmetrical sparse matrix [A], shown in Eqs.(3.2 – 3.5), can also be stored in the compact column storage formats [3.1]:

$$IA_{SGI} \begin{pmatrix} 1 \\ 2 \\ 3 \\ 4 \\ 5 \\ NR+1=6 \end{pmatrix} = \begin{Bmatrix} 1 \\ 4 \\ 6 \\ 9 \\ 13 \\ 14 \end{Bmatrix} = \begin{array}{l} \text{starting locations of the 1}^{st} \\ \text{non}-\text{zero term for} \\ \text{each column} \end{array} \qquad (3.11)$$

$$JA_{SGI} \begin{pmatrix} 1 \\ 2 \\ 3 \\ 4 \\ 5 \\ 6 \\ 7 \\ 8 \\ 9 \\ 10 \\ 11 \\ 12 \\ 13 = NCOEF1 \end{pmatrix} = \begin{Bmatrix} 1 \\ 3 \\ 4 \\ 2 \\ 4 \\ 1 \\ 3 \\ 4 \\ 1 \\ 2 \\ 3 \\ 4 \\ 5 \end{Bmatrix} = \text{Row numbers} \qquad (3.12)$$

$$AN_{SGI} (1, 2, 3, 4, \ldots, NCOEF1)^{T} = \{d_1, b_1, b_2, d_2, \ldots, d_5\}^{T} \qquad (3.13)$$

Remarks

(1) In this particular example, matrix [A] (shown in Eq.3.2) is unsymmetrical in numerical "values," however, the non-zero terms are still symmetrical in "locations." Thus, Eqs.(3.11–3.12) are identical to Eqs.(3.3–3.4). This observation will <u>not</u> be true for the most general case where <u>both</u> "values and locations" of the matrix [A] are unsymmetrical!

(2) The unsymmetrical sparse solver MA28 [3.2] will store the matrix [A], shown in Eq.(3.2), according to the following fashion:

$$IA_{row, MA28} \begin{pmatrix} 1 \\ 2 \\ 3 \\ 4 \\ 5 \\ 6 \\ 7 \\ 8 \\ 9 \\ 10 \\ 11 \\ 12 \\ 13 = NCOEF1 \end{pmatrix} = \begin{Bmatrix} 1 \\ 1 \\ 1 \\ 2 \\ 2 \\ 3 \\ 3 \\ 3 \\ 4 \\ 4 \\ 4 \\ 4 \\ 5 \end{Bmatrix} = \text{row numbers} \tag{3.14}$$

$$JA_{col, MA28} \begin{pmatrix} 1 \\ 2 \\ 3 \\ 4 \\ 5 \\ 6 \\ 7 \\ 8 \\ 9 \\ 10 \\ 11 \\ 12 \\ 13 = NCOEF1 \end{pmatrix} = \begin{Bmatrix} 1 \\ 3 \\ 4 \\ 2 \\ 4 \\ 1 \\ 3 \\ 4 \\ 1 \\ 2 \\ 3 \\ 4 \\ 5 \end{Bmatrix} = \text{column numbers} \tag{3.15}$$

$$AN_{MA28}(1,2,\ldots,NCOEF1=13)^T = \{ \boxed{d_1}, a_1, a_2, \boxed{d_2}, a_3, \ldots, \boxed{d_5} \}^T \tag{3.16}$$

3.3 Three Basic Steps and Re-ordering Algorithms

A system of simultaneous linear equations $[A]\vec{x} = \vec{b}$ can be solved in three basic steps, regardless of which direct sparse algorithm is implemented. The Choleski method is preferred as compared to the LDL^T method when matrix $[A]$ is symmetrical and positive definite. If matrix $[A]$, however, is non-positive definite (and symmetrical), then the LDL^T algorithm can be utilized. If matrix $[A]$ is unsymmetrical, then the LDU algorithm can be used.

(a) Choleski Algorithm

Step 1: Factorization phase

In this step, matrix $[A]$ is factorized (or decomposed) as:

$$[A] = [U]^T [U] \tag{3.17}$$

where

[U] is an upper-triangular matrix.

Step 2: Forward solution phase

Substituting Eq.(3.17) into Eq.(3.1), one obtains:

$$[U]^T [U]\vec{x} = \vec{b} \tag{3.18}$$

Renaming the product of $[U]\vec{x}$ as:

$$[U]\vec{x} = \vec{y} \tag{3.19}$$

Hence, Eq.(3.18) becomes:

$$[U]^T \vec{y} = \vec{b} \tag{3.20}$$

Equation (3.20) can be easily solved for $\vec{y} = \{y_1, y_2, ..., y_n\}^T$ according to the orders y_1, y_2,... then i_n (hence the name "forward" solution phase), where n = the size of the coefficient matrix $[A]$ (also = NR).

Step 3: Backward solution phase

The original unknown vector $\vec{X} = \{x_1, x_2, \ldots x_n\}^T$ can be solved from Eq.(3.19) according to the orders x_n, $x_{n-1}, \ldots$ then x_1 (hence the name "backward" solution phase).

Explicit formulas for the upper-triangular matrix [U] and vectors $\{y\}$ and $\{x\}$ can be derived by the following simple example.

$$[A] = \begin{bmatrix} a_{11} & a_{12} & a_{13} \\ a_{21} & a_{22} & a_{23} \\ a_{31} & a_{32} & a_{33} \end{bmatrix} = \begin{bmatrix} u_{11} & 0 & 0 \\ u_{12} & u_{22} & 0 \\ u_{13} & u_{23} & u_{33} \end{bmatrix} \begin{bmatrix} u_{11} & u_{12} & u_{13} \\ 0 & u_{22} & u_{23} \\ 0 & 0 & u_{33} \end{bmatrix} \quad (3.21)$$

Multiplying the right-hand side of Eq.(3.21), and equating with the upper-triangular portion of the left-hand side of Eq.(3.21), one obtains:

$$u_{11} = \sqrt{a_{11}} \quad (3.22)$$

$$u_{12} = \frac{a_{12}}{u_{11}} \quad (3.23)$$

$$u_{13} = \frac{a_{13}}{u_{11}} \quad (3.24)$$

$$u_{22} = \sqrt{a_{22} - u_{12}^2} \quad (3.25)$$

$$u_{23} = \frac{a_{23} - u_{12}u_{13}}{u_{22}} \quad (3.26)$$

$$u_{33} = \sqrt{a_{33} - u_{13}^2 - u_{23}^2} \quad (3.27)$$

The requirement of "positive definite" matrix [A] assures that the numerical values under the square root, shown in Eqs.(3.22, 3.25 and 3.27), will be "positive."

For the general $n \times n$ matrix [A], the factorized matrix $[U] = u_{ij}$ can be given as:

$$u_{i,i} = \sqrt{a_{i,i} - \sum_{k=1}^{i-1} u_{k,j}^2} \qquad (3.28)$$

$$u_{i,j} = \frac{a_{i,j} - \sum_{k=1}^{i-1} u_{k,i} u_{k,j}}{u_{i,i}}, \text{ where } j>i \qquad (3.29)$$

(b) LDL^T Algorithm

In this case, the symmetrical matrix [A] can be factorized as:

$$[A] = \begin{bmatrix} a_{11} & a_{12} & a_{13} \\ a_{21} & a_{22} & a_{23} \\ a_{31} & a_{32} & a_{33} \end{bmatrix} = \begin{bmatrix} 1 & 0 & 0 \\ l_{21} & 1 & 0 \\ l_{31} & l_{32} & 1 \end{bmatrix} \begin{bmatrix} d_1 & 0 & 0 \\ 0 & d_2 & 0 \\ 0 & 0 & d_3 \end{bmatrix} \begin{bmatrix} 1 & l_{21} & l_{31} \\ 0 & 1 & l_{32} \\ 0 & 0 & 1 \end{bmatrix} \qquad (3.30)$$

By equating the upper-triangular portions of the left-hand side, and the right-hand side of Eq. (3.30), one can obtain the explicit formulas for the lower-triangular matrix [L], and the diagonal matrix [D], respectively. The "forward" solution phase can be done as follows:

- solving $\vec{Z} = \begin{Bmatrix} Z_1 \\ Z_2 \\ \vdots \\ Z_n \end{Bmatrix}$ from $[L]\vec{Z} = \vec{b}$ $\qquad (3.31)$

- solving $\vec{\omega} = \begin{Bmatrix} \omega_1 \\ \omega_2 \\ \vdots \\ \omega_n \end{Bmatrix}$ from $[D]\vec{\omega} = \vec{Z}$ $\qquad (3.32)$

Finally, the "backward" solution phase can be done by solving for $\vec{x}$ from:

$$[L]^T \vec{x} = \vec{\omega} \qquad (3.33)$$

Assuming matrix [A] is completely full, then the factorized matrices $[L]^T$ or $[U]$, and [D] can be obtained as shown in Table 3.1.

Table 3.1 FORTRAN Code for LDL Factorization of a Full Matrix

```
        implicit real*8(a-h,o-z)                              ! 001
        dimension u(3,3)                                      ! 002
        n=6                                                   ! 004
        do 5 i=1,n                                            ! 005
        read(5,*) (u(i,j),j=i,n)                              ! 006
        write(6,*) 'upper [A] = ',(u(i,j),j=i,n)              ! 007
5       continue                                              ! 008
        write(6,*) '+++++++++++++++++'                        ! 009
c,,,,,,Factorizing row 1 will be done at line # 022           ! 010
c......Factorizing subsequent rows                            ! 011
        do 1 i=2,n                                            ! 012
c......Current factorized i-th row requires previous rows     ! 013
        do 2 k=1,i-1                                          ! 014
c......Compute multiplier(s), where [U] = [L] transpose       ! 015
        xmult=u(k,i)/u(k,k)                                   ! 016
        do 3 j=i,n                                            ! 017
c......Partially (incomplete) factorize i-th row              ! 018
c......Thus, using SAXPY operations                           ! 019
            u(i,j)=u(i,j)-xmult*u(k,j)                        ! 020
3           continue                                          ! 021
        u(k,i)=xmult                                          ! 022
2       continue                                              ! 023
1       continue                                              ! 024
c
        do 6 i=1,n                                            ! 025
        write(6,*) 'upper [U] = ',(u(i,j),j=i,n)              ! 026
6       continue                                              ! 027
        stop                                                  ! 028
        end                                                   ! 029
```

For a simple 3×3 example shown in Eq.(3.30), the LDL^T algorithm presented in Table 3.1 can also be explained as:

    (a) Row #2$_{new}$ = Row #2$_{original}$- (xmult = $u_{1,2}/u_{1,1}$)*Row #1$_{new}$

    (b) Row #3$_{new}$ = Row #3$_{original}$- (xmult1 = $u_{1,3}/u_{1,1}$)*Row #1$_{new}$

                   -(xmult2 = $u_{2,3}/u_{2,2}$)*Row #2$_{new}$

    etc.

A detailed explanation of Table 3.1 is given in the following paragraphs:

Line 004: Input the size of the coefficient matrix [A], say n = 6.

Lines 005 – 007: Input the numerical values of the upper-triangular portions of the symmetrical matrix [A], according to the row-by-row fashion. Since the original matrix [A] will be overwritten by its factorized matrix [U] for saving computer memories, matrix [U] is also used for this input phase.

$$[U] = \begin{bmatrix} 112.0 & 7.0 & 0.0 & 0.0 & 0.0 & 2.0 \\ \times & 110.0 & 5.0 & 4.0 & 3.0 & 0.0 \\ O & \times & 88.0 & 0.0 & 0.0 & 1.0 \\ O & \times & O & 66.0 & 0.0 & 0.0 \\ O & \times & O & O & 44.0 & 0.0 \\ \times & O & \times & O & O & 11.0 \end{bmatrix} \qquad (3.34)$$

Line 012: The current i^{th} row of [U] is factorized.

Line 014: All previously factorized k^{th} rows will be used to "partially" factorize the current I^{th} row.

Line 020: The ith row is "partially" factorized by the previous k^{th} row. This operation involves a constant (= -xmult) times a vector (= column j, corresponding to row "k" of [U]), and then adds the result to another vector (= column j, corresponding to row "i" of [U]).

Lines 25 – 27: Output the upper-triangular portions of the factorized matrix [U], according to the row-by-row fashion.

$$[U]_{factorized} = \begin{bmatrix} 112.0000 & 0.0625 & 0.0000 & 0.0000 & 0.0000 & 0.0179 \\ & 109.5625 & 0.0456 & 0.0365 & 0.0274 & -0.0011 \\ & & 87.7718 & -0.0021 & -0.0016 & 0.0115 \\ & & & 65.8536 & -0.0017 & 0.0001 \\ & & & & 43.9175 & 0.0001 \\ & & & & & 10.9526 \end{bmatrix} (3.35)$$

Remarks

There are only 6 non-zero (ncoef1 = 6), off-diagonal terms of the "original" upper-triangular matrix [U] as can be seen from Eq.(3.34). However, there are 12 non-zero (ncoef2 = 12), off-diagonal terms of the "factorized" upper-triangular matrix [U] as shown in Eq(3.35). These extra non-zero terms are referred to as "fill-in" terms (= ncoef2 - ncoef1 = 12 - 6 = 6).

(c) LDU Algorithm

If the matrix [A] is <u>unsymmetrical</u>, then it can be factorized as:

$$[A] = \begin{bmatrix} a_{11} & a_{12} & a_{13} \\ a_{21} & a_{22} & a_{23} \\ a_{31} & a_{32} & a_{33} \end{bmatrix} = \begin{bmatrix} 1 & 0 & 0 \\ l_{21} & 1 & 0 \\ l_{31} & l_{32} & 1 \end{bmatrix} \begin{bmatrix} d_1 & 0 & 0 \\ 0 & d_2 & 0 \\ 0 & 0 & d_3 \end{bmatrix} \begin{bmatrix} 1 & u_{12} & u_{13} \\ 0 & 1 & u_{23} \\ 0 & 0 & 1 \end{bmatrix} \quad (3.36)$$

Equating both sides of Eq.(3.36), one will obtain nine equations for nine unknowns (three unknowns for [L], three unknowns for [D], and three unknowns for [U], respectively). Equation (3.36) can be generalized to any matrix size n×n.

It should be noted here that if matrix [A] is symmetrical, then $[U] = [L]^T$, and therefore, Eq.(3.36) is identical to Eq.(3.30) of the LDL^T algorithm.

(d) Re-ordering Algorithm

The extra non-zero "fill-in" terms occur during the factorization phase for solving Eq.(3.1) need to be minimized, so that both computer memory requirement and computational time can be reduced. While Reversed CutHill-Mckee and Gippspool Stockmeiger algorithms can be used to minimize the bandwidth, or skyline of the coefficient matrix [A]; these algorithms will not be optimized for "truly sparse" solution strategies discussed in this chapter. For truly sparse algorithms, re-ordering algorithms such as Nested Disection (ND), Multiple Minimum Degree (MMD), or METiS [3.3 – 3.4] are often used since these algorithms will provide the output "mapping" integer array {IPERM}, which can be utilized to minimize the number of fill-in terms.

Most (if not all) sparse re-ordering algorithms will require two adjacency integer arrays ({ixadj} and {iadj}) in order to produce the output integer array {IPERM}. These two adjacency arrays can be obtained as soon as the "finite element connectivity" information is known.

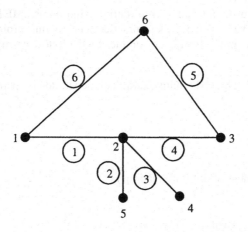

Figure 3.1 Simple Finite Element Model

As an example, the finite element connectivity information is shown in Figure 3.1. This finite element model has 6 nodes, and 6 one-dimensional (line) elements. Assuming each node has one unknown degree-of-freedom (dof), then the coefficient (stiffness) matrix will have the non-zero patterns as shown in Eq.(3.34). The two-adjacency arrays associated with Figure 3.1 are given as:

$$\{IADJ\}^T = \{\boxed{2,6}, \boxed{1,3,4,5}, \boxed{2,6}, \boxed{2}, \boxed{2}, \boxed{1,3}\}^T \tag{3.37}$$

Basically, Eq.(3.37) represents the node(s) that "connect" node 1,2,..., 6, respectively.

$$\{IXADJ\}^T = \{1,3,7,9,10,11,13\}^T \tag{3.38}$$

Equation (3.38) gives the "starting location" corresponding to each node (or dof) of the finite element model (shown in Figure 3.1).

Remarks

(1) Equations (3.34, 3.37 – 3.38) play "nearly identical" roles as the unsymmetrical sparse storage scheme explained in Eqs.(3.2 – 3.4). The only (minor) difference is diagonal terms "included" in Eq.(3.3), whereas they are "excluded" in Eq.(3.38). Furthermore, even though the original (stiffness) matrix shown in Eq.(3.34) is symmetrical, both its lower and upper-triangular portions are included in Eq.(3.37).

(2) Among different sparse re-ordering algorithms, METiS [3.3] seems to perform very well for most practical engineering problems. The user can easily integrate METiS sub-routines by the following sub-routine calls:

```
call METIS_NodeND (ndof, ixadj, iadj, ncont, itemp, iperm, invp)

where :

ndof= number of dof (or equations)
ncont= 1
itemp(i)= 0 ; (working array, with i=0,1,2,...,7)
ixadj(ndof+1)= see Eq. (3.38) for explanation ;
```

> iadj(k)=see Eq.(3.37) for explanation ; k= ixadj(ndof+1)-1
>
> iperm(ndof)= see Eq. (3.39) for explanation ;
>
> invp(old dof numbers)= {new dof numbers}

(3) For the original coefficient (stiffness) matrix data shown in Eq.(3.34), upon exiting from a sparse re-ordering algorithm (such as MMD, ND, or METiS, etc.), an output mapping integer array {IPERM} is produced, such as:

$$\text{IPERM} \begin{pmatrix} 1 \\ 2 \\ 3 \\ 4 \\ 5 \\ 6 \end{pmatrix} = \begin{Bmatrix} 6 \\ 5 \\ 4 \\ 3 \\ 2 \\ 1 \end{Bmatrix} \qquad (3.39)$$

or IPERM(new dof numbers) = {old dof numbers}   (3.40)

Using the information provided by Eq.(3.39), the original stiffness matrix given by Eq.(3.34) can be re-arranged to:

$$[U] = \begin{bmatrix} 11.0 & 0.0 & 0.0 & 1.0 & 0.0 & 2.0 \\ & 44.0 & 0.0 & 0.0 & 3.0 & 0.0 \\ & & 66.0 & 0.0 & 4.0 & 0.0 \\ & & & 88.0 & 5.0 & 0.0 \\ & \text{symmetrical} & & & 110.0 & 7.0 \\ & & & & & 112.0 \end{bmatrix} \qquad (3.41)$$

The diagonal term u(2,2) in the "old" dof numbering system (= 110.0) has been moved to the "new" location u(5,5) as shown in Eq.(3.41).

Similarly, the off-diagonal term u(1,2) in the "old" dof numbering system (= 7.0, see Eq.3.34) has been moved to the "new" location u(6,5), or u(5,6) due to a symmetrical property as shown in Eq.(3.41).

Now, the factorized matrix [U], produced by Table 3.1, is given as:

$$[U]_{\text{factorized}} = \begin{bmatrix} 11.0000 & 0.0000 & 0.0000 & 0.0909 & 0.0000 & 0.1818 \\ & 44.0000 & 0.0000 & 0.0000 & 0.0682 & 0.0000 \\ & & 66.0000 & 0.0000 & 0.0606 & 0.0000 \\ & & & 87.9091 & 0.0569 & -0.0021 \\ & & & & 109.2686 & 0.0642 \\ & & & & & 111.1862 \end{bmatrix} \quad (3.42)$$

From Eq.(3.42), one can clearly see that there is only one extra, non-zero fill-in term at location u(4,6). This is much better than six non-zero fill-in terms shown in Eq.(3.35)!

Thus, instead of solving Eq.(3.1) directly, one should first go through the re-ordering process (to minimize the total number of fill-in terms), which will solve the following system for $\vec{x}'$:

$$[A']\vec{x}' = \vec{b}' \qquad (3.43)$$

In Eq.(3.43), $[A']$ can be obtained simply by moving row(s) and column(s) of $[A]$ to different locations (by using the mapping array {IPERM} shown in Eq.3.39). Vectors $\vec{b}'$ and $\vec{x}'$ simply represent the rows interchanged from their corresponding vector $\vec{b}$ and $\vec{x}$ (see Eq.(3.1), respectively.

3.4 Symbolic Factorization with Re-ordering Column Numbers

(a) <u>Sparse Symbolic Factorization</u> [1.9]

The purpose of symbolic factorization is to find the locations of all non-zero (including "fill-in" terms), off-diagonal terms of the factorized matrix [U] (which has NOT been done yet!). Thus, one of the major goals in this phase is to predict the required computer memory for subsequent numerical factorization. The outputs from this symbolic factorization phase will be stored in the following two integer arrays (assuming the stiffness matrix data shown in Eq.3.41 is used).

$$\text{JSTARTROW} \begin{pmatrix} 1 \\ 2 \\ 3 \\ 4 \\ 5 \\ 6 \\ 7 = N+1 \end{pmatrix} = \begin{Bmatrix} 1 \\ 3 \\ 4 \\ 5 \\ 7 \\ 8 \\ 8 \end{Bmatrix} \qquad (3.44)$$

$$
\text{JCOLNUM} = \begin{pmatrix} 1 \\ 2 \\ 3 \\ 4 \\ 5 \\ 6 \\ 7 = \text{NCOEF2} \end{pmatrix} = \begin{Bmatrix} 4 \\ 6 \\ 5 \\ 5 \\ 5 \\ 6 \\ 6 \end{Bmatrix} \tag{3.45}
$$

The following "new" definitions are used in Eqs.(3.44 – 3.45):

NCOEF2 $\equiv$ The number of non-zero, off-diagonal terms of the factorized matrix [U]

JSTARTROW(i) $\equiv$ Starting location of the first non-zero, off-diagonal term for the ith row of the factorized matrix [U]. The dimension for this integer array is $N + 1$.

JCOLNUM(j) $\equiv$ Column numbers associated with each non-zero, off-diagonal term of [U] (in a row-by-row fashion). The dimension for this integer array is NCOEF2. Due to "fill-in" effects, NCOEF2>>NCOEF1.

As a rule of thumb for most engineering problems, the ratio of NCOEF2/NCOEF1 will be likely in the range between 7 and 20 (even after using re-ordering algorithms such as MMD, ND, or METiS).

The key steps involved during the symbolic phase will be described in the following paragraphs:

Step 1: Consider each ith row (of the original stiffness matrix [K]).

Step 2: Record the locations (such as column numbers) of the original non-zero, off-diagonal terms.

Step 3: Record the locations of the "fill-in" terms due to the contribution of some (not all) appropriated, previous rows j (where $1 \le j \le i-1$). Also, consider if current i^{th} row will have any immediate contribution to a "future" row.

Step 4: Return to Step 1 for the next row.

A simple, but highly inefficient way to accomplish Step 3 (of the symbolic phase) would be to identify the non-zero terms associated with the ith column. For example, there will be no "fill-in" terms on row 3 (using the data shown in Eq.3.41) due to "no contribution" of the previous rows 1 and 2. This fact can be easily realized by observing that the associated 3rd column of [K] (shown in Eq.3.41) has no non-zero terms.

On the other hand, if one considers row 4 in the symbolic phase, then the associated 4th column will have one non-zero term (on row 1). Thus, only row 1 (but not rows 2 and 3) may have "fill-in" contribution to row 4. Furthermore, since $K_{1,6}$ is non-zero (= 2), it immediately implies that there will be a "fill-in" term at location $U_{4,6}$ of row 4.

A much more efficient way to accomplish Step 3 of the symbolic phase is by creating two additional integer arrays, as defined in the following paragraphs:

ICHAINL(i) $\equiv$ Chained list for the ith row. This array will be efficiently created to identify which previous rows will have contributions to current i^{th} row. The dimension for this integer, temporary array is N.

LOCUPDATE(i) $\equiv$ Updated starting location of the ith row.

Using the data shown in Eq.(3.41), uses of the above two arrays in the symbolic phase can be described by the following step-by-step procedure:

Step 0: Initialize arrays ICHAINL $(1,2,...N) = \{0\}$ and LOCUPDATA $(1,2,...N) = \{0\}$.

Step 1: Consider row i = 1.

Step 2: Realize that the original non-zero terms occur in columns 4 and 6.

Step 3: Since the chained list ICHAINL (i =1) = 0, no other previous rows will have any contributions to row 1.

$$\text{ICHAINL (4)} = 1 \qquad (3.46)$$

$$\text{ICHAINL (1)} = 1 \qquad (3.47)$$

$$\text{LOCUPDATE (i = 1)} = 1 \qquad (3.48)$$

Eqs. (3.46-3.47) indicate that "future" row i = 4 will have to refer to row 1, and row 1 will refer to itself . Eq. (3.48) states that the updated starting location for row 1 is 1.

Step 1: Consider row i = 2.

Step 2: Realize the original non-zero term(s) only occurs in column 5.

Step 3: Since ICHAINL (i = 2) = 0, no other rows will have any contributions to

row 2.

$$ICHAINL (5) = 2 \qquad (3.49)$$

$$ICHAINL (2) = 2 \qquad (3.50)$$

$$LOCUPDATE (i = 2) = 3 \qquad (3.51)$$

Eqs.(3.49 – 3.50) indicate that "future" row i = 5 will refer to row 2, and row 2 will refer to itself. Eq.(3.51) states that the updated starting location for row 2 is 3.

Step 1: Consider row i = 3.

Step 2: The original non-zero term(s) occurs in column 5.

Step 3: Since ICHAINL (i = 3) = 0, no previous rows will have any contributions to

row 3.

The chained list for "future" row i = 5 will have to be updated in order to include row 3 on its list.

$$ICHAINL (3) = 2 \qquad (3.52)$$

$$ICHAINL (2) = 3 \qquad (3.53)$$

$$LOCUPDATE (i = 3) = 4 \qquad (3.54)$$

Thus, Eqs.(3.49, 3.53, 3.52) state that "future" row i = 5 will have to refer to row 2, row 2 will refer to row 3, and row 3 will refer to row 2. Eq.(3.54) indicates that the updated starting location for row 3 is 4.

Step 1: Consider row i = 4.

Step 2: The original non-zero term(s) occurs in column 5.

Step 3: Since ICHAINL (i = 4) = 1, and ICHAINL (1) = 1 (please refer to Eq. (3.46)), it implies row 4 will have contributions from row 1 only. The updated starting location of row 1 now will be increased by one, thus

$$LOCUPDATE\ (1) = LOCUPDATE\ (1) + 1 \tag{3.55}$$

Hence,

$$LOCUPDATE\ (1) = 1 + 1 = 2\ (please\ refer\ to\ Eq.7.49) \tag{3.56}$$

Since the updated location of non-zero term in row 1 is now at location 2 (see Eq.(3.56), the column number associated with this non-zero term is column 6 (please refer to Eq.3.41). Thus, it is obvious to see that there must be a "fill-in" term in column 6 of (current) row 4. Also, since $K_{1,6} = 2$ (or non-zero), it implies "future" row i = 6 will have to refer to row 1.

Furthermore, since the first non-zero term of row 4 occurs in column 5, it implies that "future" row 5 will also have to refer to row 4 (in addition to referring to rows 2 and 3). The chained list for "future" row 5, therefore, has to be slightly updated (so that row 4 will be included on the list) as follows:

$$ICHAINL\ (4) = 3 \tag{3.57}$$

$$ICHAINL\ (2) = 4 \tag{3.58}$$

$$LOCUPDATE\ (i = 4) = 5 \tag{3.59}$$

Notice that Eq.(3.58) will override Eq.(3.53). Thus, Eqs.(3.49, 3.58, 3.57) clearly show that symbolically factorizing "future" row i = 5 will have to refer to rows 2, then 4, and then 3, respectively.

Step 1: Consider row i = 5.

Step 2: The original non-zero term(s) occurs in column 6.

Step 3: Since

$$ICHAINL\ (i = 5) = 2 \tag{3.49, repeated}$$

$$ICHAINL\ (2) = 4 \tag{3.58, repeated}$$

$$ICHAINL\ (4) = 3 \tag{3.57, repeated}$$

It implies rows 2, then 4, and then 3 "may" have contributions (or "fill-in" effects) on row 5. However, since $K_{5,6}$ is originally a non-zero term, therefore, rows 2, 4, and 3 will NOT have any "fill-in" effects on row 5.

Step 1: There is no need to consider the last row i = N = 6 since there will be no

"fill-in" effects on the last row! It is extremely important to emphasize that

upon completion of the symbolic phase, the output array JCOLNUM (-) has

to be re-arranged to make sure that the column numbers in each row should

be in increasing order!

In this particular example (see data shown in Eq.3.41), there is one fill-in taking place in row 4. However, there is no need to perform the "ordering" operations in this case because the column numbers are already in ascending order. This observation can be verified easily by referring to Eq.(3.41). The original non-zero term of row 4 occurs in column 5 and the non-zero term (due to "fill-in" of row 4, occurs in column 6.

Thus, these column numbers are already in ascending order!

In a general situation, however, "ordering" operations may be required as can be shown in Eq.(3.60).

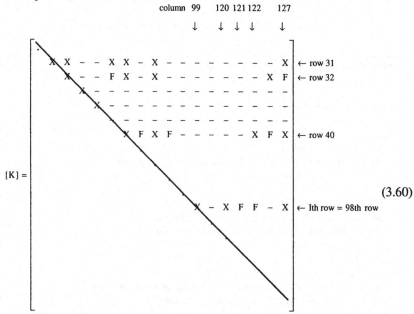

$$[K] = \tag{3.60}$$

During the factorization phase, assume that only rows 31, 32, and 40 have their contributions to the current 98th row (as shown in Eq.3.60). In this case, the original non-zero terms of row 98 occurs in column 99, 120, and 127. However, the two "fill-in" terms occur in columns 122 and then 121 (assuming current row 98 will get the contributions from rows 32, 40, and 31, respectively). Thus, the "ordering" of operations is required in this case as shown in Eq.(3.60) (since column numbers 99, 120, 127, 122, and 121 are NOT in ascending order yet!).

In subsequent paragraphs, more details (including computer codings) about symbolic factorization will be presented. To facilitate the discussion, a specific stiffness matrix data is shown in Eq.(3.61).

$$
K = \begin{matrix}
x & & x & & x & x \\
 & x & & x & \\
 & & x & & x & \\
 & & x & x & F & F \\
 & & & x & x & F \\
 & & & & x & F \\
 & & & & & x \\
\end{matrix}
\qquad (3.61)
$$

For this example, the "fill-in" effects (refer to symbols F) can be easily identified as shown in Eq.(3.61).

The symbolic factorization code is given in Table 3.2.

Table 3.2 LDL^T Sparse Symbolic Factorization

```
      subroutine symfact(n,ia,ja,iu,ju,ip,ncoef2)                    ! 001
c    .............................................
c
c    Purpose: Symbolic factorization                                 ! 002
c
c    input                                                           ! 003
c       n : order of the matrix A, n>1                               ! 004
c       ia(n+1),ja(ncoef):structure of given matrix A in compress    ! 005
c                 sparse format                                      ! 006
c
c    Output                                                          ! 007
c       ncoef2 : Number of non-zeros after factorization             ! 008
c       iu,ju  : Structure of the factorized matrix in compress      ! 009
c            sparse format                                           ! 010
c
c    working space                                                   ! 011
c       ip(n) : Chained lists of rows associated with each column    ! 012
c       IU(n) : IU is also used as a multiple switch array           ! 013
c    .............................................
      dimension ia(*),ja(*),iu(*),ju(*),ip(*)                        ! 014
      NM=N-1                                                         ! 015
      NH=N+1                                                         ! 016
      DO 10 I=1,N                                                    ! 017
      IU(I)=0                                                        ! 018
10    IP(I)=0                                                        ! 019
      JP=1                                                           ! 020
      DO 90 I=1,NM                                                   ! 021
      JPI=JP                                                         ! 022
```

```
c    if(JPI.GT.imem) THEN                                            ! 023
c    WRITE(23) ' NOT ENOUGH MEMORY '                                 ! 024
c    STOP                                                            ! 025
c    ENDIF                                                           ! 026
     JPP=N+JP-I                                                      ! 027
     MIN=NH                                                          ! 028
     IAA=IA(I)                                                       ! 029
     IAB=IA(I+1)-1                                                   ! 030
     DO 20 J=IAA,IAB                                                 ! 031
     JJ=JA(J)                                                        ! 032
     JU(JP)=JJ                                                       ! 033
     JP=JP+1                                                         ! 034
     IF(JJ.LT.MIN)MIN=JJ                                             ! 035
 20  IU(JJ)=I                                                        ! 036
 30  LAST=IP(I)                                                      ! 037
     IF(LAST.EQ.0)GO TO 60                                           ! 038
     L=LAST                                                          ! 039
 40  L=IP(L)                                                         ! 040
     LH=L+1                                                          ! 041
     IUA=IU(L)                                                       ! 042
     IUB=IU(LH)-1                                                    ! 043
     IF(LH.EQ.I)IUB=JPI-1                                            ! 044
     IU(I)=I                                                         ! 045
     DO 50 J=IUA,IUB                                                 ! 046
     JJ=JU(J)                                                        ! 047
     IF(IU(JJ).EQ.I)GO TO 50                                         ! 048
     JU(JP)=JJ                                                       ! 049
     JP=JP+1                                                         ! 050
     IU(JJ)=I                                                        ! 051
     IF(JJ.LT.MIN)MIN=JJ                                             ! 052
 50  CONTINUE                                                        ! 053
     IF(JP.EQ.JPP)GO TO 70                                           ! 054
     IF(L.NE.LAST)GO TO 40                                           ! 055
 60  IF(MIN.EQ.NH)GO TO 90                                           ! 056
 70  L=IP(MIN)                                                       ! 057
     IF(L.EQ.0)GO TO 80                                              ! 058
     IP(I)=IP(L)                                                     ! 059
     IP(L)=I                                                         ! 060
     GO TO 90                                                        ! 061
 80  IP(MIN)=I                                                       ! 062
     IP(I)=I                                                         ! 063
 90  IU(I)=JPI                                                       ! 064
     IU(N)=JP                                                        ! 065
     IU(NH)=JP                                                       ! 066
     ncoef2=iu(n+1)-1                                                ! 067
     return                                                          ! 068
     end                                                             ! 069
```

(b) Re-Ordering Column Numbers

It has been explained in Eq.(3.60) that upon exiting from the symbolic factorization phase, the coefficient (stiffness) matrix needs to be re-ordered to make sure that the column number in each row are in ascending order. This type of ordering of operation is necessary before performing the numerical factorization.

To facilitate the discussion, let's consider an unordered, rectangular matrix (shown in Figure 3.2), which can be described by Eqs.(3.62 – 3.64).

1	2	3	4	5	6	= No. columns (= NC)
		$A^{(3rd)}$		$B^{(1st)}$	$C^{(2nd)}$	1
$D^{(5th)}$			$E^{(4th)}$			2
		$F^{(6th)}$	$G^{(7th)}$			3
		$I^{(9th)}$	$J^{(8th)}$			4
	$K^{(11th)}$			$L^{(13th)}$	$M^{(12th)}$	5 = No. Rows (= NR)

Figure 3.2 Unordered, Rectangular Matrix
(Numbers in Parenthesis Represent Location Numbers)

The starting location for each row in Figure 3.2 is given as:

$$\text{ISTARTROW} \begin{pmatrix} 1 \\ 2 \\ 3 \\ 4 \\ 5 \\ 6 = NR + 1 \end{pmatrix} = \begin{Bmatrix} 1 \\ 4 \\ 6 \\ 8 \\ 11 \\ 14 \end{Bmatrix} \tag{3.62}$$

The (unordered) column numbers associated with each row is given as:

$$\text{ICOLNUM}\begin{pmatrix} 1 \\ 2 \\ 3 \\ 4 \\ 5 \\ 6 \\ 7 \\ 8 \\ 9 \\ 10 \\ 11 \\ 12 \\ 13 = \text{ISTARTROW}(6)-1 \end{pmatrix} = \begin{Bmatrix} 5 \\ 6 \\ 3 \\ 4 \\ 1 \\ 3 \\ 4 \\ 4 \\ 3 \\ 1 \\ 2 \\ 6 \\ 5 \end{Bmatrix} \qquad (3.63)$$

The "numerical" values (for the matrix shown in Figure 3.2) can be given as:

$$\text{AK}\begin{pmatrix} 1 \\ 2 \\ 3 \\ 4 \\ 5 \\ 6 \\ 7 \\ 8 \\ 9 \\ 10 \\ 11 \\ 12 \\ 13 \end{pmatrix} = \begin{Bmatrix} B \\ C \\ A \\ E \\ D \\ F \\ G \\ J \\ I \\ H \\ K \\ M \\ L \end{Bmatrix} \qquad (3.64)$$

The "unordered" matrix, shown in Figure 3.2, can be made into an "ordered" matrix by the following two-step procedure:

Step 1: If the matrix shown in Figure 3.2 is transposed, then one will obtain the following matrix (shown in Figure 3.3):

	Col. No. 1	Col. No. 2	Col. No. 3	Col. No. 4	Col. No.5
Row #1		$D^{(1st)}$		$H^{(2nd)}$	
Row #2					$K^{(3rd)}$
Row #3	$A^{(4th)}$		$F^{(5th)}$	$I^{(6th)}$	
Row #4		$E^{(7th)}$	$G^{(8th)}$	$J^{(9th)}$	
Row #5	$B^{(10th)}$				$L^{(11th)}$
Row #6	$C^{(12th)}$				$M^{(13th)}$

Figure 3.3 Transposing a Given Unordered Matrix Once

The starting locations and the associated column numbers for the matrix shown in Figure 3.3 are given as:

$$\text{ISTROW TRANSPOSE1} \begin{Bmatrix} 1 \\ 2 \\ 3 \\ 4 \\ 5 \\ 6 \\ 7 \end{Bmatrix} = \begin{Bmatrix} 1 \\ 3 \\ 4 \\ 7 \\ 10 \\ 12 \\ 14 \end{Bmatrix} \tag{3.65}$$

$$\text{ICOLN TRANSPOSE1} \begin{pmatrix} 1 \\ 2 \\ 3 \\ 4 \\ 5 \\ 6 \\ 7 \\ 8 \\ 9 \\ 10 \\ 11 \\ 12 \\ 13 \end{pmatrix} = \begin{Bmatrix} 2 \\ 4 \\ 5 \\ 1 \\ 3 \\ 4 \\ 2 \\ 3 \\ 4 \\ 1 \\ 5 \\ 1 \\ 5 \end{Bmatrix} \qquad (3.66)$$

It is important to realize from Figure 3.3 (and also from Eq.3.66) that the matrix shown in Figure 3.3 is already properly ordered. For each row of the matrix shown in Figure 3.3, increasing the location numbers also associates with increasing the column numbers.

Step 2: If the matrix shown in Figure 3.3 is transposed again (or the original matrix shown in Figure 3.2 is transposed twice), then one will obtain the matrix as shown in Figure 3.4.

1	2	3	4	5	6	
		A$^{(1st)}$		B$^{(2nd)}$	C$^{(3rd)}$	1
D$^{(4th)}$			E$^{(5th)}$			2
		F$^{(6th)}$	G$^{(7th)}$			3
H$^{(8th)}$		I$^{(9th)}$	J$^{(10th)}$			4
	K$^{(11th)}$			L$^{(12th)}$	M$^{(13th)}$	5

Figure 3.4 Transposing a Given Unordered Matrix Twice

The starting locations and the associated column numbers (for the matrix shown in Figure 3.4) are given as:

$$
\text{ISTROW TRANSPOSE2} \begin{pmatrix} 1 \\ 2 \\ 3 \\ 4 \\ 5 \\ 6 = NR + 1 \end{pmatrix} = \begin{Bmatrix} 1 \\ 4 \\ 6 \\ 8 \\ 11 \\ 14 \end{Bmatrix} \tag{3.67}
$$

$$
\text{ICOLN TRANSPOSE2} \begin{pmatrix} 1 \\ 2 \\ 3 \\ 4 \\ 5 \\ 6 \\ 7 \\ 8 \\ 9 \\ 10 \\ 11 \\ 12 \\ 13 \end{pmatrix} = \begin{Bmatrix} 3 \\ 5 \\ 6 \\ 1 \\ 4 \\ 3 \\ 4 \\ 1 \\ 3 \\ 4 \\ 2 \\ 5 \\ 6 \end{Bmatrix} \tag{3.68}
$$

Thus, one concludes that an "ordered" matrix can be obtained from the original, "unordered" matrix simply by transposing the "unordered" matrix twice! (see Table 3.3).

Table 3.3 Transposing a Sparse Matrix

```
        subroutine transa(n,m,ia,ja,iat,jat)                    ! 001
        implicit real*8(a-h,o-z)                                ! 002
        dimension ia(*),ja(*),iat(*),jat(*)                     ! 003

c       ............................................................
c
c       Purpose :                                               ! 004
c           After symbolic factorization,ja is just a merge lists   ! 005
c           WITHOUT ordering (i.e. the nonzero column numbers of a  ! 006
c           particular row are 12, 27, 14, 46, 22, 133). Upon   ! 007
c           completion, this routine will rearrange the nonzero ! 008
c           column numbers WITH ordering, to be ready for numerical ! 009
c           factorization phase (i.e. 12, 14, 22, 27, 46, 133)  ! 010
```

```
c                                                                          ! 011
c   input                                                                  ! 011
c       n : order of matrix A, n>1 also number of rows of the matrix       ! 012
c       m : number of columns of the matrix                               ! 013
c       ia(n+1),ja(ncoef): structure of matrix A in sparse format         ! 014
c
c   Output                                                                 ! 015
c       iat(n+1), jat(ncoef): structure of transposed matrix in           ! 016
c       compress sparse format                                            ! 017
c
c   Note: This algorithm can be easily modified to transpose actual       ! 018
c        "numerical values" of real matrix, instead of simply geting      ! 019
c        the transpose of the "positions" of a given matrix               ! 020
c
c   ................................................
        MH=M+1                                                            ! 021
        NH=N+1                                                            ! 022
        DO 10 I=2,MH                                                      ! 023
10 IAT(I)=0                                                               ! 024
        IAB=IA(NH)-1                                                      ! 025
C  DIR$ IVDEP                                                             ! 026
        DO 20 I=1,IAB                                                     ! 027
        J=JA(I)+2                                                         ! 028
        IF(J.LE.MH)IAT(J)=IAT(J)+1                                        ! 029
20 CONTINUE                                                               ! 030
        IAT(1)=1                                                          ! 031
        IAT(2)=1                                                          ! 032
        IF(M.EQ.1)GO TO 40                                               ! 033
        DO 30 I=3,MH                                                      ! 034
30 IAT(I)=IAT(I)+IAT(I-1)                                                 ! 035
40 DO 60 I=1,N                                                            ! 036
        IAA=IA(I)                                                         ! 037
        IAB=IA(I+1)-1                                                     ! 038
cv  IF(IAB.LT.IAA)GO TO 60                                                ! 039
C  DIR$ IVDEP                                                             ! 039
        DO 50 JP=IAA,IAB                                                  ! 040
        J=JA(JP)+1                                                        ! 041
        K=IAT(J)                                                          ! 042
        JAT(K)=I                                                          ! 043
c  ANT(K)=AN(JP)                                                          ! 044
50 IAT(J)=K+1                                                             ! 045
60 CONTINUE                                                               ! 046
        call transa2(n,m,iat,jat,ia,ja)                                   ! 047
        return                                                            ! 048
        end                                                               ! 049
```

3.5 Sparse Numerical Factorization

To facilitate the discussion, assume the factorized LDL^T of a given "dense" (say 3×3), symmetrical matrix, shown in Eq.(3.30), is sought. Equating the upper-triangular portions from both sides of Eq.(3.30), one can explicitly obtain the formulas for matrices [L] and [D] as shown in Eqs (3.69 – 3.74).

$$D(1)=K(1,1) \qquad\qquad \text{-->> solve for } D(1) \qquad\qquad (3.69)$$

$$D(1)*L(2,1)=K(1,2) \qquad\qquad \text{-->> solve for } L(2,1)=U(1,2) \quad (3.70)$$

$$D(1)*L(3,1)=K(1,3) \qquad\qquad \text{-->> solve for } L(3,1)=U(1,3) \quad (3.71)$$

$$D(1)*L(2,1)**2+D(2)=K(2,2) \qquad \text{-->> solve for } (2) \qquad\qquad (3.72)$$

$$D(1)*L(2,1)*L(3,1)+D(2)*L(3,2)=K(2,3)\text{-->>solve for } L(3,2)=U(2,3) \quad (3.73)$$

$$D(1)*L(3,1)**2+D(2)*L(3,2)**2+D(3)=K(3,3) \quad \text{-->> solve for } D(3) \quad (3.74)$$

For the general N×N dense, symmetrical matrix, the numbered values for matrices [D] and $[L]^T \equiv [U]$ can be extracted upon exiting the algorithm shown in Table 3.4 where $[D] = U_{i,i}$ and $[L]^T = U_{i,j}$ with j>i.

Table 3.4 LDL^T Dense Numerical Factorization

C	Assuming row 1 of [U] has already been factorized	!001
	do 1 i=2,n	!002
	do 2 k=1, i-1	!003
c	Computing multiplier factors	!004
	xmult=U(k,i)/U(k,k)	!005
	do 3 j=i,n	!006
	U(i,j)=U(i,j)-xmult*U(k,j)	!007
3	continue	!008
	U(k,i)=xmult	!009
2	continue	!010
1	continue	!011

Remarks about Table 3.4

Line 002: Factorization is done according to the row-by-row fashion (such as the ith row).

Line 003: The current factorized i[th] row will require information about all previously factorized k[th] rows (example: k = 1→ i-1).

Lines 006 – 008: The current, partially (or incomplete) factorized i^{th} row is actually one inside the innermost-loop 3. It should be noted that line 007 represents <u>SAXPY</u> operations [<u>S</u>ummation of a constant $\underline{A} \equiv$ xmult times a vector $\underline{x} \equiv U(k,j)$ <u>p</u>lus a vector $\underline{y} \equiv U(i,j)$].

Algorithms for symmetrical "sparse" LDL^T factorization can be obtained by making a few modifications to the symmetrical, "dense" LDL^T (see Table 3.4) as shown in Table 3.5.

Table 3.5 LDL^T Sparse Numerical Factorization

c	Assuming row 1 of [U] has already been factorized	!001
	do 1 i=2,n	!002
	do 2 k=only those previous rows which have contributions to current i^{th} row	!003
c	Computing multiplier factors	!004
	xmult=U(k,i)/U(k,k)	!005
	do 3 j=appropriated column numbers of the kth row	!006
	U(i,j)=U(i,j)-xmult*U(k,j)	!007
3	continue	!008
	U(k,i)=xmult	!009
2	continue	!010
1	continue	!011

Remarks about Table 3.5

(a) The symmetrical, "dense" and "sparse" LDL^T algorithms presented in Tables 3.4 and 3.5 look nearly identical! Only the loop index "k" (see line 003), and the loop index "j" (see line 006) have different interpretations. These differences can be easily explained by referring to the original data shown in Eq.(3.41).

(b) In Table 3.5, assume $i^{th} = 4^{th}$ row (see line 002). Then, factorizing the 4th row of Eq.(3.41) only requires the contribution of the previous $k^{th} = 1^{st}$ row (see line 003) or the "sparse" case since previous rows $k = 2^{nd}$, 3rd will have no contributions to the current 4th factorized row.

(c) In Table 3.5, the loop index "j" (see line 006), which represents the column numbers associated with "non-zero terms" in the $k^{th} = 1^{st}$ row, does <u>NOT</u> have to vary continuously from "i" to "n" (as shown on line 006 of Table 3.4) as in the dense case. In fact, the $k^{th} = 1^{st}$ row only has non-zero terms in column numbers 4 and 6. Thus, the loop index "j" (see line 006 of Table 3.5) will have values of 4, then 6, respectively.

3.6 Super (Master) Nodes (Degrees-of-Freedom)

In real life, large-scale applications, after the sparse symbolic factorization phase, one often observes that several consecutive rows have the same non-zero patterns as indicated in Eq.(3.75). The symbols F, shown in Eq.(3.75), represent non-zero terms, due to fill-in, that occurs during factorization phase.

$$
[U] = \begin{array}{ccccccccccc}
1 & 2 & 3 & 4 & 5 & 6 & 7 & 8 & 9 & 10 & \\
X & 0 & 1 & 0 & 0 & 0 & 2 & 0 & 0 & 0 & 1 \\
 & X & 3 & 0 & 0 & 4 & 5 & 6 & 0 & 0 & 2 \\
 & & X & 0 & 0 & 7 & 8 & 9 & 0 & 0 & 3 \\
 & & & X & 10 & 11 & 0 & 0 & 12 & 0 & 4 \\
 & & & & X & 13 & 0 & 0 & 14 & 0 & 5 \\
 & & & & & X & F & F & 17 & 0 & 6 \\
 & & & & & & X & 18 & 0 & 19 & 7 \\
 & & & & & & & X & 0 & 20 & 8 \\
 & & & & & & & & X & 0 & 9 \\
 & & & & & & & & & X & 10
\end{array}
\qquad (3.75)
$$

Using the sparse storage schemes discussed in Section 3.2, the non-zero terms of the matrix [U], shown in Eq.(3.75), can be described by the following two integer arrays (where n = number of degrees-of-freedom = 10, and ncoef2 = total number of non-zero terms, including fill-in terms = 20).

$$
istartr(n+1) = 1 \quad 3 \quad 7 \quad 10 \quad 13 \quad 15 \quad 18 \quad 20 \quad 21 \quad 21 \quad 21 \qquad (3.76)
$$

$$
\begin{aligned}
kindx(ncoef2) = \ & 3 \quad 7 \quad 3 \quad 6 \quad 7 \quad 8 \quad 6 \quad 7 \quad 8 \quad 5 \quad 6 \quad 9 \\
& 6 \quad 9 \quad 7 \quad 8 \quad 9 \quad 8 \quad 10 \quad 10
\end{aligned}
\qquad (3.77)
$$

It could be observed from Eq.(3.75) that rows 2 and 3, rows 4 and 5, and rows 7 and 8 have the same non-zero patterns. Therefore, the super (or master) nodes (or degrees-of-freedom) array {isupern} can be identified in Eq.(3.78).

$$
isupern(n) = 1 \quad \boxed{2 \quad 0} \quad \boxed{2 \quad 0} \quad 1 \quad \boxed{2 \quad 0} \quad 1 \quad 1 \qquad (3.78)
$$

A subroutine supnode, and the associated main program to find super-node information (see Eq.3.78), is given in Table 3.6.

Table 3.6 Algorithms to Find Super-Node Information

```
      dimension istartr(100), kindx(100), isupern(100)
c
      read(5,*) n,ncoef2
c
      read(5,*) (istartr(i),i=1,n+1)
      read(5,*) (kindx(i),i=1,ncoef2)
c
      write(6,*) 'istartr(n+1)= ',(istartr(i),i=1,n+1)
      write(6,*) 'kindx(ncoef2)= ',(kindx(i),i=1,ncoef2)
c
      call supnode(n,istartr,kindx,isupern)
c
      write(6,*) 'isupern(n)= ',(isupern(i),i=1,n)
c
      stop
      end
c++++++++++++++++++++
      subroutine supnode(n,istartr,kindx,isupern)
c     implicit real*8(a-h,o-z)
c......purposes: identify the rows which have same non-zero patterns
c......          (i.e. same column numbers lined up) for "unrolling"
c++++++++++++++++++++++++++++++++++++++++++++++++++++++++++++++++++++++
c                         1   2   3   4   5   6   7   8   9  10

c                         x   0   1   0   0   0   2   0   0   0       1
c                             x   3   0   0   4   5   6   0   0       2
c                                 x   0   0   7   8   9   0   0       3
c                                     x  10  11   0   0  12   0       4
c                                         x  13   0   0  14   0       5
c                                             x   F   F  17   0       6
c                                                 x  18   0  19       7
c                                                     x   0  20       8
c                                                         x   0       9
c                                                             x      10
cistartr(n+1)=    1   3   7  10  13  15  18  20  21  21  21
ckindx(ncoef2)=   3   7   3   6   7   8   6   7   8   5   6   9
c                 6   9   7   8   9   8  10  10
cisupern(n)=      1   2   0   2   0   1   2   0   1   1
c++++++++++++++++++++++++++++++++++++++++++++++++++++++++++++++++++++++
      dimension istartr(1),kindx(1),isupern(1)
c......initialize super node integer array
      do 1 i=1,n
 1    isupern(i)=1
c......meaning of variable "iamok" = 0 (= NOT having same non-zero patterns)
c......              = 1 (= DOES have same non-zero patterns)
c     do 2 i=2,n-1
      do 2 i=2,n
c     iamok=0
      iamok=1
c......check to see if i-th row has the same # nonzeroes as (i-1)-th row
c......also,check to see if the (i-1)-th row have a non-zero right above
c......the diagonal of the i-th row
c......if the i-th row does NOT have the same "GENERAL" non-zero patterns,
```

```
c......then, skip loop 2 immediately (no need to check any furthers !)
       kptrsi=istartr(i+1)-istartr(i)
       kptrsim1=istartr(i)-istartr(i-1)
c      if(kptrs(i).ne.kptrs(i-1)-1) go to 2
       if(kptrsi.ne.kptrsim1-1) go to 2
       ii=istartr(i-1)
       if(kindx(ii).ne.i) go to 2
c......now, check to see if the i-th row and the (i-1)-th row have the
c......same "DETAILED" non-zero patterns
c      write(6,*) 'check point 1'
       iii=istartr(i)
       jj=ii
       jjj=iii-1
c      do 3 j=1,kptrs(i)
       do 3 j=1,kptrsi
       jj=jj+1
       jjj=jjj+1
       if(kindx(jj).ne.kindx(jjj)) then
       iamok=0
c      write(6,*) 'check point 2'
       go to 222
       else
       iamok=1
c      write(6,*) 'check point 3'
       endif
  3    continue
c      if(kindx(ii).eq.n) iamok=1
c......if all"DETAILED" non-zero patterns also match, then we have "supernode"
 222   if(iamok.eq.1) then
       isupern(i)=0
c      write(6,*) 'check point 4'
       endif
  2    continue
c......finally, establish how many rows a supernode has (based on the 1's,
c......and the 0's of the array isupern(-) obtained thus far)
       icount=0
       do 11 i=2,n
c.......counting the number of zeroes encountered in array isupern(-)
       if(isupern(i).eq.0) then
       icount=icount+1
c      write(6,*) 'check point 5'
       else
       ithrow=i-1-icount
       isupern(ithrow)=icount+1
       icount=0
c      write(6,*) 'check point 6'
       endif
 11    continue
c......now, take care of the last super node
       ithrow=n-icount
       isupern(ithrow)=icount+1
       return
       end
```

3.7 Numerical Factorization with Unrolling Strategies

The "super" degrees-of-freedom (dof) information provided by Eq.(3.78) and unrolling techniques discussed in Section 2.5 can be effectively utilized during the sparse LDL^T factorization phase to enhance the performance of the algorithm as indicated in Table 3.7. The key difference between the algorithm presented in Table 3.5 (without unrolling strategies) and the one shown in Table 3.7 (with unrolling strategies) is that the loop index "k" in Table 3.7 (line 003) represents "a block of rows," or "a super row" which has the same non-zero patterns instead of a "single row" as shown on line 003 of Table 3.5.

Table 3.7 LDL^T Sparse, Unrolling Numerical Factorization

c Assuming row 1 of [U] has already been factorized	!001
do 1 i=2,n	!002
do 2 k=only those previous super rows which have contributions to current i^{th} row	!003
c Computing multiplier factors	!004
NUMSLAVES=ISUPERN(I)-1	!004a
xmult=U(k,i)/U(k,k)	!005
$xmult_1$=U(k+1,i)/U(k+1,k+1)	!005a
$xmult_2$=U(k+2,i)/U(k+2,k+2)	!005b
.	
.	
$xmult_{NS}$=U(k+NS,i)/U(k+NS,k+NS)	!005c
c NS=NUMSLAVES	!005d
do 3 j=appropriated column numbers of the super k^{th} row	!006
U(i,j)=U(i,j)-xmult*U(k,j)-$xmult_1$*U(k+1,j)-...-$xmult_{NS}$*U(k+NS,j)	!007
3 continue	!008
U(k,i)=xmult	!009
U(k+1,i)=$xmult_1$	!009a
U(k+2,i)=$xmult_2$	!009b
.	
.	
U(k+NS,i)=$xmult_{NS}$	!009c
2 continue	!010
1 continue	!011

3.8 Forward/Backward Solutions with Unrolling Strategies

The system of linear equations $[A]\bar{x} = \bar{b}$ can be solved in the following steps:

Step 1: Factorization

$$[A] = [L][D][L]^T = [U]^T [D][U] \tag{3.79}$$

$$[A] = \begin{bmatrix} 1 & 0 & 0 \\ d & 1 & 0 \\ e & f & 1 \end{bmatrix} \begin{bmatrix} a & 0 & 0 \\ 0 & b & 0 \\ 0 & 0 & c \end{bmatrix} \begin{bmatrix} 1 & d & e \\ 0 & 1 & f \\ 0 & 0 & 1 \end{bmatrix} \tag{3.80}$$

Thus

$$[L][D][L]^T \vec{x} = \vec{b} \tag{3.81}$$

Step 2: Forward solution

Define

$$[D][L]^T \vec{x} \equiv \vec{Z} \tag{3.82}$$

Hence, Eq.(3.81) becomes:

$$[L]\vec{Z} = \vec{b} \tag{3.83}$$

$$\begin{bmatrix} 1 & 0 & 0 \\ d & 1 & 0 \\ e & f & 1 \end{bmatrix} \begin{Bmatrix} Z_1 \\ Z_2 \\ Z_3 \end{Bmatrix} = \begin{Bmatrix} b_1 \\ b_2 \\ b_3 \end{Bmatrix} \tag{3.84}$$

The intermediate unknown vector $\vec{Z}$ can be solved as:

$$Z_1 = b_1 \tag{3.85}$$

$$Z_2 = b_2 - d\, Z_1 \tag{3.86}$$

$$Z_3 = b_3 - e\, Z_1 - f\, Z_2 \tag{3.87}$$

Define

$$[L]^T \vec{x} \equiv \vec{\omega} \tag{3.88}$$

Hence Eq.(3.82) becomes:

$$[D]\vec{\omega} = \vec{Z} \tag{3.89}$$

$$\begin{bmatrix} a & 0 & 0 \\ 0 & b & 0 \\ 0 & 0 & c \end{bmatrix} \begin{Bmatrix} \omega_1 \\ \omega_2 \\ \omega_3 \end{Bmatrix} = \begin{Bmatrix} Z_1 \\ Z_2 \\ Z_3 \end{Bmatrix} \qquad (3.90)$$

The intermediate vector $\bar{\omega}$ can be solved as:

$$\omega_1 = Z_1 / a \qquad (3.91)$$

$$\omega_2 = Z_2 / b \qquad (3.92)$$

$$\omega_3 = Z_3 / c \qquad (3.93)$$

Step 3: Backward solution

From Eq.(3.88), one has:

$$\begin{bmatrix} 1 & d & e \\ 0 & 1 & f \\ 0 & 0 & 1 \end{bmatrix} \begin{Bmatrix} x_1 \\ x_2 \\ x_3 \end{Bmatrix} = \begin{Bmatrix} \omega_1 \\ \omega_2 \\ \omega_3 \end{Bmatrix} \qquad (3.94)$$

The original unknown vector $\bar{x}$ can be solved as:

$$x_3 = \omega_3 \qquad (3.95)$$

$$x_2 = \omega_2 - f\, x_3 \qquad (3.96)$$

$$x_1 = \omega_1 - d\, x_2 - e\, x_3 \qquad (3.97)$$

Table 3.8 LDL^T Sparse Forward/Backward Solution

subroutine fbe(n,iu,ju,di,un,b,x)	!001
implicit real*8(a-h,o-z)	!002
dimension iu(*),ju(*),di(*),un(*),b(*),x(*)	!003
c	
c Purpose: Forward and backward substitution	!004
c This subroutine is called after the numerical factorization	!005
c	
c input	!006
c n : Order of the system, n>1	!007
c iu(n+1),ju(ncoef2),un(ncoef2),di(n): Structure of factorized	!008
c matrix which is the output of the numerical factorization	!009
c b(n) : right-hand side vector	!010
c	
c output	!011
c x(n) : vector of unknowns (displacement vector)	!012
c	
NM=N-1	!013
DO 10 I=1,N	!014
10 X(I)=B(I)	!015
DO 40 K=1,NM	!016
IUA=IU(K)	!017
IUB=IU(K+1)-1	!018
XX=X(K)	!019
DO 20 I=IUA,IUB	!020
20 X(JU(I))=X(JU(I))-UN(I)*XX	!021
30 X(K)=XX*DI(K)	!022
40 CONTINUE	!023
X(N)=X(N)*DI(N)	!024
K=NM	!025
50 IUA=IU(K)	!026
IUB=IU(K+1)-1	!027
XX=X(K)	!028
DO 60 I=IUA,IUB	!029
60 XX=XX-UN(I)*X(JU(I))	!030
X(K)=XX	!031
70 K=K-1	!032
IF(K.GT.0)GO TO 50	!033
return	!034
end	!035

The basic forward-backward sparse solution algorithms are presented in Table 3.8, in the form of FORTRAN code. An explanation of Table 3.8 is given in the following paragraphs:

Lines 013 – 024: Forward solution phase

Lines 025 – 033: Backward solution phase.

Lines 014 – 015: Initialize the intermediate vector $\vec{Z} = \vec{b}$, according to Eqs.(3.85 – 3.87).

Lines 017 – 018: Starting and ending location of the kth (= say 1st) row of the upper triangular matrix U (= L^T).

Line 019: $xx = Z_1$ (= multiplier).

Lines 020 – 021: Knowing the final, complete answer for Z_1, the partial, incomplete answers for Z_2, Z_3... are computed. Saxpy operation is used in the forward solution phase.

Line 022: $\omega_1 = Z_1 / a$, see Eq.(3.91)

Line 023: $\omega_2 = Z_2 / b$, see Eq.(3.92)

Line 024: $\omega_3 = Z_3 / c$, see Eq.(3.93)

Line 025: Backward solution begins (say $k = 2$)

Lines 026 – 027: Starting and ending location of the kth (= 2nd) row

Line 028: Initiate $\bar{x} = \begin{Bmatrix} x_3 \\ x_2 \\ x_1 \end{Bmatrix} = \begin{Bmatrix} \omega_3 \\ \omega_2 \\ \omega_1 \end{Bmatrix}$, see Eqs.(3.95 – 3.97).

Lines 030 – 031: Compute the final, complete answer for $x(k) = x_2$. Dot product operation is used in the backward solution phase.

Line 032: Decrease the counter k by 1, and continue to solve x_1.

Unrolling strategies can be incorporated into a sparse forward solution, and its "pseudo" FORTRAN code is listed in Table 3.9.

Table 3.9 Sparse Forward Solution with SAXPY Unrolling Strategies

```
      nm1=n-1                                          ! 013
      do 10 i=1,n                                      ! 014
  10 x(i)=b(i)                                         ! 015
      k=1                                              ! 016
  40 nblkrowk=isupn(k)              ! assuming = 3     ! 016a
      lastrowk=k+nblkrowk-1         ! assuming = 3     ! 016b
      iua=iu(lastrowk)              ! assuming = 75    ! 017
      iub=iu(lastrowk+1)-1          ! assuming = 79    ! 018
      numnz=iub-iua+1                                  ! 018a
  c...assuming user's unrolling level 3 is specified,
  c...and this particular supernode's size is also 3
      xx=x(k)                                          ! 019
      xx2=x(k+1)                                       ! 019a
      xx3=x(k+2)                                       ! 019b
      do 20 i=iua,iub                                  ! 020
  20 x(ju(i))=x(ju(i))-un(i)*xx3                       ! 021
                      -un(i-numnz)*xx2                 ! 021a
                      -un(i-2*numnz-1)*xx              ! 021b
```

```
c...Don't forget to take care of the          ! 021c
c...triangle region
    x(k)=xx*di(k)                              ! 022
    x(k+1)=xx2*di(k+1)                         ! 022a
    x(k+2)=xx3*di(k+2)                         ! 022b
    k=k+isupn(k)                               ! 022c
    if (k .le. nm1) go to 40                   ! 023
    x(n)=x(n)*di(n)                            ! 024
```

The "pseudo" FORTRAN algorithm for a sparse (LDL^T) forward solution with unrolling strategies is presented in Table 3.9. To facilitate the discussion in this section, Figure 3.5 should also be examined carefully.

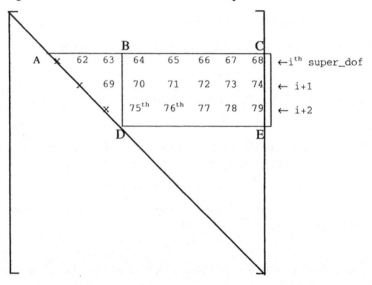

Figure 3.5 Super-Node That Has 3 Rows with the Same Non-Zero Patterns

An explantion for Table 3.9 is given in the following paragraphs.

Lines 013 – 015: Initialize the intermediate vector $\vec{Z} = \vec{b}$, according to Eqs.(3.85 – 3.87).

Lines 015 – 016a: The block size (say NBLKROWK = 3) of the $k^{th} = 1^{st}$ super-row (or super-dof)

Line 016b: The last row (of dof) number that corresponds to the $k^{th} = 1^{st}$ super- row.

Lines 017 – 018: The beginning and ending locations of the non-zero terms of the last row that corresponds to the kth super-row (see the 75th and 79th locations of Figure 3.5).

Line 018a: The number of non-zero terms (= NUMNZ) in the last row is computed.

Lines 019 – 019b: Three multipliers (= xx, xx_2, and xx_3) are calculated (assuming user's unrolling level 4 is specified, hence <u>Nunroll</u> = 3). If the block size (say NBLKROWK = 5) of the kth super-row is greater than NUMROLL(=3), then lines 019 – 019b will be augmented by

$$xx = x\,(k + 3)$$

$$xx_2 = x\,(k + 4)$$

Lines 020 – 021b: The partial, incomplete solution for the unknowns x(JU[I]) are computed. SAXPY with unrolling strategies are adopted here.

Lines 021c: The "triangular" region ABD, shown in Figure 3.5, also needs to be done here!

Lines 022 – 022b: Similar explanation as given on line 022 of Table 3.8

Line 022c: Consider the next "super" k^{th} row.

Line 024: Similar explanation as given on line 024 of Table 3.8

Remarks
   a)   The beginning and ending locations of the last row (say "i + 2"th row) of a particular super-row is easily identified (such as 75th and 79th locations, respectively), as compared to its previous $(i + 1)^{th}$ (such as 70th and 74th locations, respectively), and i^{th} rows (such as 64th and 68th locations, respectively).
   b)   Index I, I-NUMNZ, and I-2*NUMNZ-1 on lines 021, 021a, and 021b will have the values from 75th to 79th, 70th to 74th, and 64th to 68th, respectively. These same non-zero patterns are shown inside the "rectangular" region BCDE of Figure 3.5.
   c)   If the block size (= NBLKROWK) of the particular k^{th} super-row is greater than NUNROLL (say, = 3), then similar comments made about lines 019 – 019b will also be applied here.

Table 3.10 Sparse Backward Solution with DOT PRODUCT Unrolling Strategies

nsupnode=0	!001
do 1 i=1, n-1	!002
if (isupn(i).eq.0) go to 1	!003
nsupnode=nsunode+1	!004
isupnbw(nsupnode)=i	!005
1 continue	!006
ICOUNT=NSUPNODE	!007
50 k=Isupbw(ICOUNT)	!008
NBLKROWK=Isupn(K)	!009
LASTROWK=K+NBLKROWK-1	!010
IUA=IU(LASTROWK)	!011
IUB=IU(LASTROWK+1)-1	!012
NUMNZ=IUB-IUA+1	!013

```
C   Assuming use's unrolling level 3 is specified, and this particular
    $supernode's size is also 3
    xx=x(LASTROWK)                                              !014
    XX2=X(LASTROWK-1)                                           !015
    XX3=X(LASTROWK-2)                                           !016
    DO 60 i=IUA,IUB                                             !017
    XX=XX-UN(i)*X(JU(I))                                        !018
    XX2=XX2-UN(I-NUMNZ)*X(JU(I))                                !019
    XX3=XX3-UN(I-2*NUMNZ-1)*X(JU(I))                            !020
60  CONTINUE                                                    !021
C   Don't forget to take care of the triangular region         !022
    X(LASTROWK)=XX                                              !023
    X(LASTROWK-1)=XX2                                           !024
    X(LASTROWK-2)=XX3                                           !025
70  ICOUNT=ICOUNT-1                                             !026
    If(ICOUNT.GT.0) GO TO 50                                    !027
```

The "pseudo" FORTRAN algorithm for sparse (LDL^T) backward solution with unrolling strategies is presented in Table 3.10. An explanation for Table 3.10 is given in the following paragraphs:

Line 002: Assuming the super-dof array {isupn} was already computed as

$$\text{isupn} = \{1,2,3,...,10\}^T = \{1 \;\; \boxed{2 \;\; 0} \;\; \boxed{2 \;\; 0} \;\; 1 \;\; \boxed{2 \;\; 0} \;\; 1 \;\; 1\}^T \text{ (3.78 repeated)}$$

Lines 002 – 006: Upon exiting loop 1, we will have NSUPNODE = 7 (= number of super-dof) Eq.(3.98) basically gives the "beginning" row number of each super-row (or super-dof).

$$\text{isupnbw} \begin{Bmatrix} 1 \\ 2 \\ 3 \\ 4 \\ 5 \\ 6 \\ 7 = \text{NSUPNODE} \end{Bmatrix} = \begin{Bmatrix} 1 \\ 2 \\ 4 \\ 6 \\ 7 \\ 9 \\ 10 \end{Bmatrix} \tag{3.98}$$

Line 007: The counting (= ICOUNT) starts with the last super-row.

Line 008: The "actual" row number of the "beginning" row of a super-row is identified. For example, the actual row number of the beginning row of super-row 3 is equal to 4 (see Eq.3.98).

Line 009: The size of a super-row is identified.

Line 010: The "actual" row number for the last row of a super-row is calculated.

Lines 011 – 012: The "starting" and "ending" locations (= 75th, and 79th see Figure 3.5) of non-zero terms corresponding to the "last row" of a super-row

Line 013: Number of non-zero terms (= NUMNZ = 5), corresponding to the "last row" of a super-row

Lines 014 – 016: Initialize $\vec{x} = \begin{Bmatrix} \vdots \\ x_3 \\ x_2 \\ x_1 \end{Bmatrix} = \begin{Bmatrix} \vdots \\ \omega_3 \\ \omega_2 \\ \omega_1 \end{Bmatrix}$, see Eqs.(3.95 – 3.97).

Lines 017 – 021, 023 – 025: Compute the final, complete solution for x LASTROWK), x (LASTROWK-1), x (LASTROWK-2).

Line 022: The "triangular" region ABD (see Figure 3.5) needs to be considered here.

Line 026: Decrease the (super-row) counter by 1, and go backward to solve subsequent unknowns.

The "pseudo" FORTRAN codes presented in Tables 3.9 – 3.10 are only for the purpose of explaining "key ideas" for incorporating the "unroll strategies" into the sparse forward and backward solution phases. The actual, detailed FORTRAN codes for sparse forward/backward (with unrolling strategies) are given in Table 3.11.

Table 3.11 Actual FORTRAN Code for Sparse Forward & Backward (with Unrolling Strategies)

```
        implicit real*8(a-h,o-z)
c......purpose:  forward & backward sparse solution phases
c......          with saxpy, and dot product unrolling strategies
c......author:   Prof. Duc T. Nguyen's Tel = 757-683-3761 (DNguyen@odu.edu)
c......version:  february 1, 2003
c......features: different unrolling levels for forward and backward phases
c......          if checks on multiplers ( = 0 ) to skip the loop
c......notes:    (a) PLEASE DO NOT DISTRIBUTE THIS CODE TO ANY ONE WITHOUT
c                    EXPLICIT WRITTEN PERMISSION FROM Prof. D. T. Nguyen
c                (b) With unrolling level 4 (for both forward & backward)
c                    time saving is approximately 20%
c......stored at: cd ~/cee/mail/Rajan/          and also at
c......           cd ~/cee/mail/*Duc*Pub*/

        dimension x(9100),b(9100),isupn(9100),isupnode(9100),
     $            un(41000000),iu(9100),ju(41000000),di(9100)
c......purpose: create main program to test forward/backward sparse
c......          with unrolling
c......artificially created the factorized upper triangular matrix
c......assuming factorized diagonal terms di(-)=1.0 everywhere
        call pierrotime (t1)
        n=9000                          ! user input data
        nunroll=1                       ! fixed data, set by D.T. Nguyen
        iexample=4
        write(6,*) 'n, nunroll = ',n, nunroll
```

```
       inputok=0                           ! assuming input data are incorrect first
       if (iexample .le. 3  .and.  n .eq. 7) then
       inputok=1                           ! input data are correct
       endif
       if (iexample .gt. 3) then
       iblksize=n/4
       nremain=n-iblksize*4
         if (nremain .eq. 0) inputok=1         ! input data are correct
       endif
       if (inputok .eq. 1) then
       call geninput(n,ncoef1,ncoef2,di,b,iu,ju,un)
c.....artificially input/create super-node information
       call getsupern(n,isupn,isupnode,nsupnode,iexample)
c.....forward/backward sparse solution with unrolling strategies
       call pierrotime (t2)
       timeip=t2-t1
       write(6,*) 'input time timeip = ',timeip
       if (nunroll .eq. 4) then
       call fbunroll(n,x,b,isupn,isupnode,un,iu,ju,nsupnode,di,nunroll,
     $              ncoef2)
       elseif (nunroll .eq. 1) then
      call fbe(n,iu,ju,di,un,b,x)
       endif

       call pierrotime (t3)
       timefb=t3-t2
       write(6,*) 'forward & backward timefb = ',timefb
       elseif (inputok .eq. 0) then
       write(6,*) 'ERROR: sorry your input is NOT correct'
       write(6,*) 'if iexample = 1,2, or 3, then MUST have n = 7'
       write(6,*) 'if iexample = 4, then MUST have n = multiple of 4'
       endif
       stop
       end
c%%%%%%%%%%%%%%%%%%%%%%%%%%%%%%%%%%%%%%%%%%%%%%%%%%%%%%%%%%%%%%%%%%%%%%%%%%%%%
       subroutine getsupern(n,isupn,isupnode,nsupnode,iexample)
c      implicit real*8(a-h,o-z)
       dimension isupn(*),isupnode(*)
c.....purpose: user's input info. for generating super-node data
       do 6 i=1,n
       isupn(i)=0
 6     continue
c---------------------------------------------------------------------
c.....following are user's input info. to define/create/get super-node
         if (iexample .eq. 1) then
c.....example 1:  CORRECT FORWARD & CORRECT BACKWARD !
       isupn(1)=5
       isupn(6)=1
       isupn(7)=1
       write(6,*) 'example 1: isupn(1,6,7)={5,1,1}'
         elseif (iexample .eq. 2) then
c.....example 2:  CORRECT FORWARD & CORRECT BACKWARD !
       isupn(1)=2
       isupn(3)=4
       isupn(7)=1
       write(6,*) 'example 2: isupn(1,3,7)={2,4,1}'
         elseif (iexample .eq. 3) then
c.....example 3:  CORRECT FORWARD & CORRECT BACKWARD !
       isupn(1)=1
       isupn(2)=2
       isupn(4)=1
       isupn(5)=3
       write(6,*) 'example 3: isupn(1,2,4,5)={1,2,1,3}'
         elseif (iexample .eq. 4) then
c.....example 4:  CORRECT FORWARD & CORRECT BACKWARD !
       iblksize=n/4
       isupn(1)=iblksize
       isupn(1+iblksize)=iblksize
       isupn(1+2*iblksize)=iblksize
       isupn(1+3*iblksize)=iblksize
```

```
          write(6,*) 'example 4: n MUST be a multiple of nunroll=4 !!'
          endif
c------------------------------------------------------------------------
          nsupnode=0
          do 52 i=1,n
          if (isupn(i) .eq. 0) go to 52
          nsupnode=nsupnode+1
          isupnode(nsupnode)=i
 52       continue
c......output key results
          write(6,*) 'nsupnode= ',nsupnode
c         write(6,*) 'isupn(n)= ',(isupn(i),i=1,n)
c         write(6,*) 'isupnode(nsupnode)= ',(isupnode(i),i=1,nsupnode)
          return
          end
c%%%%%%%%%%%%%%%%%%%%%%%%%%%%%%%%%%%%%%%%%%%%%%%%%%%%%%%%%%%%%%%%%%%%%%%%%%
          subroutine geninput(n,ncoef1,ncoef2,di,b,iu,ju,un)
          implicit real*8(a-h,o-z)
          dimension di(*),b(*),iu(*),ju(*),un(*)
c..... purpose: artificialle create input data for the (full) coefficient matrix
c......          in sparse format, and rhs vector
c+++++++++++++++++++++++++++++++++++++++++++++++++++++++++++++++++++++++++++++++
c    <-- [L D L transpose] --> * {x} = {b}  {forward sol w}  {backward sol x}
c
c    1  3  4  5  6  7  8      x1     1          1              54,725
c
c    3  1  5  6  7  8  9      x2     4          1             -24,805
c
c    4  5  1  7  8  9  10     x3    10          1               5,984
c
c    5  6  7  1  9  10 11     x4    19          1                -979
c
c    6  7  8  9  1  11 12     x5    31          1                 121
c
c    7  8  9  10 11  1 13     x6    46          1                 -12
c
c    8  9  10 11 12 13  1     x7    64          1                   1
c+++++++++++++++++++++++++++++++++++++++++++++++++++++++++++++++++++++++++++++++
c
          ncoef1=n*(n+1)/2-n
          ncoef2=ncoef1                        ! assuming matrix is full
          icount=0
          di(n)=1.0
          b(n)=1.0                             ! initialize b(n) = 1.0
          iu(1)=1
          do 1 ir=1,n-1
          iu(ir+1)=iu(ir)+(n-ir)
          b(ir)=1.0                            ! initialize rhs vector {b} = 1.0
          di(ir)=1.0
            do 2 ic=ir+1,n
            icount=icount+1
            un(icount)=ir+ic                   ! only for n = 7, under/over-flow
for large n ??
            if (n .ne. 7) un(icount)=1.0d0/(ir+ic)!    for n .ne. 7, and n MUST be
multiple of 4
            ju(icount)=ic
 2          continue
 1        continue
          iu(n+1)=iu(n)
c......artificially created rhs vector b(-), such that FORWARD SOLUTION
c......vector {z} = {1} everywhere
          icount=0
          do 4 ir=1,n-1
            do 5 ic=ir+1,n
            icount=icount+1
            b(ic)=b(ic)+un(icount)
 5          continue
 4        continue
c......output key results
```

```
c        write(6,*) 'iu(n) = ',(iu(i),i=1,n+1)
c        write(6,*) 'di(n) = ',(di(i),i=1,n)
c        write(6,*) 'ju(ncoef2) = ',(ju(i),i=1,ncoef2)
c        write(6,*) 'un(ncoef2) = ',(un(i),i=1,ncoef2)
c        write(6,*) 'b(n) = ',(b(i),i=1,n)
         return
         end
c%%%%%%%%%%%%%%%%%%%%%%%%%%%%%%%%%%%%%%%%%%%%%%%%%%%%%%%%%%%%%%%%%%%%%%%%%
         subroutine fbunroll(n,x,b,isupn,isupnode,un,iu,ju,nsupnode,di
      $                    ,nunroll,ncoef2)
         implicit real*8(a-h,o-z)
c......purpose:  forward & backward sparse solution phases
c......          with saxpy, and dot product unrolling strategies
c......author:   Prof. Duc T. Nguyen's Tel = 757-683-3761
c......version:  february 1, 2003
         dimension x(*),b(*),isupn(*),isupnode(*),un(*),iu(*),ju(*),di(*)
         write(6,*) '**************************'
         write(6,*) 'version:  february 1, 2003'
         write(6,*) '**************************'
C+++++++++++++++++++++++++++++++++
c......forward solution phase
C+++++++++++++++++++++++++++++++++
c......initialize
         do 1 i=1,n
 1       x(i)=b(i)
c......begin with super-block row k=1
         k=1
 40      continue                          ! subsequent super-block row k
         nblkrowk=isupn(k)
         lastrowk=k+nblkrowk-1
         ntimes=nblkrowk/nunroll
         nremains=nblkrowk-ntimes*nunroll
         L=k-nunroll
         istart=iu(lastrowk)
         iend=iu(lastrowk+1)-1
         nzperow=iend-istart+1
c......FIRST: "MUST" handling complete triangular region(s) "FIRST"
c        write(6,*) 'k,lastrowk,nzperow = ',k,lastrowk,nzperow
         do 26 irow=k,lastrowk
         iua=iu(irow)
         iub=iu(irow+1)-1
c        if (iub .lt. iua) go to 26     ! ??
         istart=iua
         iend=iub-nzperow                 ! make sure nzperow has been defined
         xx=x(irow)
c        write(6,*) 'triangular region:ju(i)= ',(ju(i),i=1,ncoef2)
         do 27 i=istart,iend
         x(ju(i))=x(ju(i))-un(i)*xx
c        write(6,*) 'triangular region:i,ju(i),x(ju(i))= ',i,ju(i),x(ju(i))
 27      continue
c......diagonal scaling for final answers
         x(irow)=xx*di(irow)
c        write(6,*) 'final forward sol: irow, x(irow) = ',irow,x(irow)
 26      continue
c......SECOND: "THEN", processing rectangular region(s) "AFTERWARD"
         if (ntimes .eq. 0) go to 111
         do 14 j=1,ntimes
         L=L+nunroll
         xx=x(L)
         xx2=x(L+1)
         xx3=x(L+2)
         xx4=x(L+3)
         iua=iu(L)
         iub=iu(L+1)-1
         i1=nblkrowk-(j-1)*nunroll
         istart=iua+i1-1
         iend=iub
         if (iub .lt. iua) go to 111      ! skip loop 14
         nzperow=iend-istart+1
         const=nzperow-1
```

```
         iadd=const+(i1-1)
         iadd2=iadd+const+(i1-2)
         iadd3=iadd2+const+(i1-3)
C++++++++++++++++++++++++++++++++++++++++++++++++++++++++++
         if (xx .eq. 0.0d0 .and. xx2 .eq. 0.0d0
      $        .and. xx3 .eq. 0.0d0 .and. xx4 .eq. 0.0d0)
      $ go to 14
C++++++++++++++++++++++++++++++++++++++++++++++++++++++++++
         do 2 i=istart,iend
         x(ju(i))=x(ju(i))-un(i)*xx
      $                    -un(i+iadd)*xx2
      $                    -un(i+iadd2)*xx3
      $                    -un(i+iadd3)*xx4
c     write(6,*) 'rect regions:ju(i),x(ju(i)= ',ju(i),x(ju(i))
 2       continue
 14      continue
 111     continue
c......THIRD: processing the remaining rectangular region(s)
         L=L+nunroll
         if (nremains .eq. 0) go to 222
         go to (21,22,23),nremains
 21      iblkrowk=1
         xx=x(L)
         iua=iu(L)
         iub=iu(L+1)-1
         istart=iua+iblkrowk-1
         iend=iub
         if (iub .lt. iua) go to 222    ! ??
         nzperow=iend-istart+1
c     write(6,*) 'L,iua,iub,istart,iend,nzperow = '
c     write(6,*)  L,iua,iub,istart,iend,nzperow
         do 3 i=istart,iend
         x(ju(i))=x(ju(i))-un(i)*xx
c     write(6,*) 'remain rect regions:ju(i),x(ju(i)= ',ju(i),x(ju(i))
 3       continue
         go to 222                      ! ??
 22      iblkrowk=2
         xx=x(L)
         xx2=x(L+1)
         iua=iu(L)
         iub=iu(L+1)-1
         istart=iua+iblkrowk-1
         iend=iub
         if (iub .lt. iua) go to 222    ! ??
         nzperow=iend-istart+1
         const=nzperow-1
         iadd=const+(iblkrowk-1)
         do 4 i=istart,iend
         x(ju(i))=x(ju(i))-un(i)*xx
      $                    -un(i+iadd)*xx2
 4       continue
         go to 222                      ! ??
 23      iblkrowk=3
         xx=x(L)
         xx2=x(L+1)
         xx3=x(L+2)
         iua=iu(L)
         iub=iu(L+1)-1
         istart=iua+iblkrowk-1
         iend=iub
         if (iub .lt. iua) go to 222    ! ??
         nzperow=iend-istart+1
         const=nzperow-1
         iadd=const+(iblkrowk-1)
         iadd2=iadd+const+(iblkrowk-2)

         do 5 i=istart,iend
         x(ju(i))=x(ju(i))-un(i)*xx
      $                    -un(i+iadd)*xx2
      $                    -un(i+iadd2)*xx3
```

```
  5     continue
222     continue
c......FOURTH: diagonal scaling
 30     continue
        L=k-nunroll
        do 31 j=1,ntimes
        L=L+nunroll
        xx=x(L)
        xx2=x(L+1)
        xx3=x(L+2)
        xx4=x(L+3)
        x(L)=xx*di(L)
        x(L+1)=xx2*di(L+1)
        x(L+2)=xx3*di(L+2)
        x(L+3)=xx4*di(L+3)
c       write(6,*) 'diag scaling: L,x(L),x(L+1),x(L+2),x(L+3)= '
c       write(6,*)              L,x(L),x(L+1),x(L+2),x(L+3)
 31     continue
        L=L+nunroll
c       write(6,*) 'diagonal scaling: L,ntimes,nremains = '
c       write(6,*)              L,ntimes,nremains
        if (nremains .eq. 0) go to 44
        go to (41,42,43),nremains
 41     x(L)=x(L)*di(L)
c       write(6,*) 'remains diag scaling: L,x(L)= ',L,x(L)
        go to 44
 42     x(L)=x(L)*di(L)
        x(L+1)=x(L+1)*di(L+1)
        go to 44
 43     x(L)=x(L)*di(L)
        x(L+1)=x(L+1)*di(L+1)
        x(L+2)=x(L+2)*di(L+2)
        go to 44
 44     k=k+nblkrowk
c       write(6,*) 'k=k+nblkrowk= ',k
        if (k .lt. n) go to 40
        x(n)=x(n)*di(n)
c......output key results
        write(6,*) 'forward & diag scale sol x(n)= ',(x(i),i=1,4)
     $              ,x(n-2),x(n-1),x(n)
c       write(6,*) 'forward & diag scale sol x(n)= ',(x(i),i=1,n)
C++++++++++++++++++++++++++++++
c......backward solution phase
C++++++++++++++++++++++++++++++
        k=nsupnode                  ! consider last super-node first
333     continue                    ! consider subsequent (backward) super-nodes
        iglobeq=isupnode(k)
        nblkrowk=isupn(iglobeq)
        istarteq=iglobeq
        iendeq=istarteq+nblkrowk-1
        ntimes=nblkrowk/nunroll
        nremains=nblkrowk-ntimes*nunroll
c       write(6,*) 'backward phase: k,iglobeq,nblkrowk,istarteq = '
c       write(6,*)              k,iglobeq,nblkrowk,istarteq
c       write(6,*) 'backward phase: iendeq,ntimes,nremains = '
c       write(6,*)              iendeq,ntimes,nremains
        if (ntimes .eq. 0) then
        L=iendeq
        go to 444
        endif
        L=iendeq+nunroll
c......find number of nonzero, off-diagonal terms for the LAST ROW
c......of the current super-node
        iua=iu(iendeq)
        iub=iu(iendeq+1)-1
        nzlastrow=iub-iua+1
        if (nzlastrow .eq. 0) go to 444     ! super-node does NOT have rect. egion
c......FIRST: "MUST" take care of rectangular region(s) "FIRST"
        do 64 j=1,ntimes
        L=L-nunroll
```

```
         i1=(j-1)*nunroll
         istart=iu(L)+i1
         iend=iu(L+1)-1
         nzperow=iend-istart+1
         minus=nzperow+i1            !
         minus1=minus+nzperow+i1+1   !
         minus2=minus1+nzperow+i1+2 !
         xx=x(L)
         xx2=x(L-1)
         xx3=x(L-2)
         xx4=x(L-3)
          do 65 i=istart,iend        !
          xx=xx-un(i)*x(ju(i))
          xx2=xx2-un(i-minus)*x(ju(i))
          xx3=xx3-un(i-minus1)*x(ju(i))
          xx4=xx4-un(i-minus2)*x(ju(i))
 65       continue
         x(L)=xx
         x(L-1)=xx2
         x(L-2)=xx3
         x(L-3)=xx4
 64      continue
         L=L-nunroll                 ! ???? commented out ????
 444     continue
c......SECOND: "THEN", take care remaining rectangular region
         if (nremains .eq. 0) go to 555
         go to (71,72,73),nremains
 71      iua=iu(L)
         iub=iu(L+1)-1
         istart=iua+(nblkrowk-nremains)
         iend=iub
         nzperow=iend-istart+1
         xx=x(L)
          do 87 i=istart,iend
          xx=xx-un(i)*x(ju(i))
 87       continue
         x(L)=xx
         write(6,*) 'backward final sol: L,x(L) = ',L,x(L)
          go to 555
 72      iua=iu(L)
         iub=iu(L+1)-1
         istart=iua+(nblkrowk-nremains)
         iend=iub
         nzperow=iend-istart+1
         minus=(nblkrowk-nremains)+nzperow
         xx=x(L)
         xx2=x(L-1)
c        write(6,*) 'just before do 77 i'
         write(6,*) 'L,iua,iub,istart,iend,nzperow,minus,xx,xx2 ='
c        write(6,*)  L,iua,iub,istart,iend,nzperow,minus,xx,xx2
          do 77 i=istart,iend
          xx=xx-un(i)*x(ju(i))
          xx2=xx2-un(i-minus)*x(ju(i))
 77       continue
         x(L)=xx
         x(L-1)=xx2
c        write(6,*) 'backward final sol: L,x(L) = ',L,x(L)
c        write(6,*) 'backward final sol: L-1,x(L-1) = ',L-1,x(L-1)
          go to 555
 73      iua=iu(L)
         iub=iu(L+1)-1
         istart=iua+(nblkrowk-nremains)
         iend=iub
         nzperow=iend-istart+1
         minus=(nblkrowk-nremains)+nzperow
         minus1=minus+(nblkrowk-nremains+1)+nzperow
         xx=x(L)
         xx2=x(L-1)
         xx3=x(L-2)
          do 75 i=istart,iend
```

```
          xx=xx-un(i)*x(ju(i))
          xx2=xx2-un(i-minus)*x(ju(i))
          xx3=xx3-un(i-minus1)*x(ju(i))
 75       continue
          x(L)=xx
          x(L-1)=xx2
          x(L-2)=xx3
c         write(6,*) 'backward final sol: L,x(L) = ',L,x(L)
c         write(6,*) 'backward final sol: L-1,x(L-1) = ',L-1,x(L-1)
c         write(6,*) 'backward final sol: L-2,x(L-2) = ',L-2,x(L-2)
 555      continue
c......THIRD: "THEN", take care of complete, all triangular regions
          istart=iu(iendeq)
          iend=iu(iendeq+1)-1
          nzperow=iend-istart+1
c         write(6,*) 'istart,iend,nzperow= ',istart,iend,nzperow  ! OK here
          do 92 m=iendeq-1,istarteq,-1                            ! ??
c         do 92 m=iendeq  ,istarteq,-1                            ! ??
          iua=iu(m)
          iub=iu(m+1)-1
          istart=iua
          iend=iub-nzperow
          if (iub .lt .iua) go to 92
          xx=x(m)
            do 93 i=istart,iend
            xx=xx-un(i)*x(ju(i))
 93         continue
          x(m)=xx
c         write(6,*) 'm,x(m) = ',m,x(m)
 92       continue
c......now, move upward (or backward) to process the next super-node
c......(until hiting the 1-st super-node)
          k=k-1
c         write(6,*) 'k=k-1= ',k
          if (k .ge. 1) go to 333

c......output key results
          write(6,*) 'backward final sol x(n)= ',(x(i),i=1,4)
     $              ,x(n-2),x(n-1),x(n)
c         write(6,*) 'backward final sol x(n)= ',(x(i),i=1,n)
          return
          end
c%%%%%%%%%%%%%%%%%%%%%%%%%%%%%%%%%%%%%%%%%%%%%%%%%%%%%%%%%%%%%%%%%%%%%%%%%%%%
          subroutine fbe(n,iu,ju,di,un,b,x)
          implicit real*8(a-h,o-z)
          dimension iu(*),ju(*),di(*),un(*),b(*),x(*)
c         ....................................................................
c
c
c         Purpose: Forward and backward substitution
c             This subroutine is called after the numerical factorization
c
c         input
c             n : Order of the system, n>1
c             iu(n+1),ju(ncoef2),un(ncoef2),di(n): Structure of the factorized
c                 matrix which is the output of the numerical factorization
c             b(n) :  right-hand side vector
c         output
c             x(n) :  vector of unknowns (displacement vector)
c         ....................................................................
c
c......forward solution phase
          NM=N-1
          DO 10 I=1,N
 10       X(I)=B(I)
          DO 40 K=1,NM
          IUA=IU(K)
          IUB=IU(K+1)-1
          XX=X(K)
          DO 20 I=IUA,IUB
 20       X(JU(I))=X(JU(I))-UN(I)*XX
```

```
   30 X(K)=XX*DI(K)
   40 CONTINUE
      X(N)=X(N)*DI(N)
      write(6,*) 'forward sol = ',(x(i),i=1,4)
     $            ,x(n-2),x(n-1),x(n)
c     write(6,*) 'forward sol = ',(x(i),i=1,n)
c......backward solution phase
      K=NM
   50 IUA=IU(K)
      IUB=IU(K+1)-1
      XX=X(K)
      DO 60 I=IUA,IUB
   60 XX=XX-UN(I)*X(JU(I))
      X(K)=XX
   70 K=K-1
      IF(K.GT.0)GO TO 50
      write(6,*) 'backward sol = ',(x(i),i=1,4)
     $            ,x(n-2),x(n-1),x(n)
c     write(6,*) 'backward sol = ',(x(i),i=1,n)
      return
      end
c%%%%%%%%%%%%%%%%%%%%%%%%%%%%%%%%%%%%%%%%%%%%%%%%%%%%%%%%%%%%%%%%%%%%%%%%%
      subroutine pierrotime (time)
      real tar(2)
      real*8 time
c     ..................................................................
c     purpose :
c       This routine returns the user + system execution time
c       The argument tar returns user time in the first element and
c       system time in the second element.  The function value is the
c       sum of user and system time.  This value approximates the
c       program's elapsed time on a quiet system.
c
c       Uncomment for your corresponding platform
c
c       Note: On the SGI the resolution of etime is  1/HZ
c
c     Output
c       time: user+sytem executime time
c     ..................................................................
c     SUN -Solaris
      time=etime(tar)
c     HP - HPUX
c     time=etime_(tar)                !f90
c     time=etime_(tar)                !f77
c     COMPAQ - alpha
c     time=etime(tar)
c     CRAY
c     time=tsecnd()
c     IBM
c     time=0.01*mclock()
c     SGI origin
c     time=etime(tar)
      return
      end
```

3.9 Alternative Approach for Handling an Indefinite Matrix [3.5]

A system of sparse, symmetrical, INDEFINITE simultaneous linear equations have arisen naturally in several important engineering and science applications. Tremendous progress has been made in the past years for efficient large-scale solutions of sparse, symmetrical, definite equations. However, for a sparse indefinite system of equations, only a few efficient, robust algorithms, and software are available (especially the FORTRAN versions in the public domain). These existing indefinite sparse solvers have been discussed in recent papers [1.9, 3.2, 3.4 – 3.7].

Major difficulties involved in developing efficient sparse indefinite solvers include the need for pivoting (or 2×2 pivoting) [3.8], criteria for when and how to switch the row(s), effective strategies to predict and minimize the nonzero fill-in terms, etc.

In this work, an alternative approach is proposed for solving a system of sparse, symmetrical, indefinite equations. The key idea here is to first transform the original indefinite system into a new (modified) system of symmetrical, "definite" equations. Well-documented sparse definite solvers [1.9, 3.9 – 3.10] can be conveniently used to obtain the "intermediate solution" (in the "modified" system). This "intermediate" solution is then transformed back into the "original" space to obtain the "original" unknown vector.

To validate the formulas developed in our paper and evaluate the numerical performance of our proposed alternative approach, several NASA indefinite systems of equations have been solved (ranging from 51 to 15,637 indefinite equations) on inexpensive workstations. Our preliminary numerical results have indicated that this alterative approach may offer potential benefits in terms of accuracy, reducing incore memory requirements and even improving the computational speed over the traditional approach when the alternative approach is implemented in a parallel computer environment.

In this section, consider systems of indefinite equations in the form as shown in Eq.(3.99):

$$[A]\vec{x} = \vec{f}$$

(3.99)

where

$$[A] = \begin{bmatrix} K & a \\ a^T & 0 \end{bmatrix}$$

(3.100)

In Eq.(3.100), the symmetrically indefinite matrix [A] has the dimension n×n, and [K], [a], and [o] are sub-matrices. To simplify the discussion, it is assumed (for now) that the lower, right sub-matrix [o] has the dimension 1x1. Thus, matrix [A] has a zero on its (last) diagonal. Sub-matrix [K], shown in Eq.(3.100), is also symmetrical.

The key idea here is to transform Eq.(3.99) into a new system, such as:

$$\left[\overline{A}\right]\vec{x}^* = \vec{f} \tag{3.101}$$

where the new coefficient matrix $\left[\overline{A}\right]$ can be computed as:

$$\left[\overline{A}\right] = [A] + [\Delta A] \tag{3.102}$$

Matrix $[\Delta A]$, shown in Eq.(3.102), should be selected with the following criteria:

a.) $[\Delta A]$ should be sparse, symmetrical, and simple

b.) The resulting new matrix $[\overline{A}]$ will have better properties (such as positive definite) as compared to the original coefficient matrix [A]

Thus, one selects matrix $[\Delta A]$ as:

$$[\Delta A] = \begin{bmatrix} 0 & 0 & \cdots & \cdots & \cdots & 0 & 0 \\ & \ddots & & & & & \\ & & 0 & \cdots & \cdots & 0 & 0 \\ & & & \ddots & & \vdots & \vdots \\ & & & & \ddots & \vdots & \vdots \\ & & & & & 0 & 0 \\ & & & & & & 1 \end{bmatrix} \tag{3.103}$$

Eq.(3.103) can also be expressed as:

$$[\Delta A]_{n \times n} = \vec{d}_{n \times 1} * \vec{d}_{1 \times n}^T \tag{3.104}$$

where

$$\vec{d} = \begin{Bmatrix} 0 \\ 0 \\ 0 \\ 1 \end{Bmatrix} \tag{3.105}$$

Using Eq.(3.102), Eq.(3.99) can be re-written as:

$$\left[\overline{A} - \Delta A\right]\vec{x} = \vec{f} \tag{3.106}$$

Substituting Eq.(3.104) into Eq.(3.106), one obtains:

$$\left[\overline{A} - d_i d_i^T\right]\vec{x} = \vec{f} \tag{3.107}$$

Pre-multiplying both sides of Eq.(3.107) by $\left[\overline{A}\right]^{-1}$, one has

$$\left[I - \overline{A}^{-1} d_i d_i^T\right]\vec{x} = \vec{x}^* \tag{3.108}$$

In Eq.(3.108), [I] is an nxn identity matrix, and $\vec{x}^*$ is given as:

$$\vec{x}^* \equiv \left[\overline{A}\right]^{-1} * \vec{f} \tag{3.109}$$

Eq.(3.108) can also be expressed as:

$$\left[I - \vec{p}_i d_i^T\right]\vec{x} = \vec{x}^* \tag{3.110}$$

where

$$\vec{p}_i \equiv \left[\overline{A}\right]^{-1} * d_i \tag{3.111}$$

or

$$\left[\overline{A}\right]\vec{p}_i \equiv d_i \tag{3.112}$$

From Eq.(3.110), one obtains:

$$\vec{x}^* = \vec{x} - p_i d_i^T \vec{x} \tag{3.113}$$

Pre-multiplying Eq.(3.113) by d_i^T, one gets:

$$d_i^T \vec{x}^* = d_i^T \vec{x} - d_i^T p_i d_i^T \vec{x} \tag{3.114}$$

Since $d_i^T \vec{x}^*$ and $d_i^T \vec{x}$ are scalar quantities, hence Eq.(3.114) can be re-written as:

$$d_i^T \vec{x}^* = (1 - d_i^T p_i) d_i^T \vec{x} \tag{3.115}$$

or

$$d_i^T \vec{x} = \frac{d_i^T \vec{x}^*}{1 - d_i^T p_i} \tag{3.116}$$

From Eq.(3.113), one obtains:

$$\vec{x} = \vec{x}^* + p_i(d_i^T \vec{x}) \tag{3.117}$$

Substituting Eq.(3.116) into Eq.(3.117), one has:

$$\vec{x} = \vec{x}^* + p_i * \frac{d_i^T \vec{x}^*}{1 - d_i^T p_i} \tag{3.118}$$

Remarks on Eq.(3.118)

Both matrices [A] and $\left[\overline{A}\right]$ are assumed to be non-singular. Then, from Eqs.(3.107, 3.106, 3.102), one obtains the following relationship:

$$\left[\overline{A}\right]^{-1} * \left[\overline{A} - d_i d_i^T\right] = \left[\overline{A}\right]^{-1} * [A]$$

$$[I] - \left[\overline{A}\right]^{-1} d_i d_i^T = \text{product of two non-singular matrices } \left[\overline{A}\right]^{-1} and [A]$$

$$[I] - p_i d_i^T = \text{non-singular}$$

Thus, in a more general case, the denominator of Eq.(3.118) will also be NON-SINGULAR. The entire alternative formulation can be summarized in a convenient step-by-step procedure:

Step 0: Use Eq.(3.105) to form $\vec{d}_i$.

Then matrices $[\Delta A]$ and $\left[\overline{A}\right]$ can be computed according to Eqs.(3.104) and (3.102), respectively.

Step 1: Use Eq.(3.101) to solve for $\vec{x}.^*$

Step 2: Use Eq.(3.112) to compute $\vec{p}_i$.

Step 3: Use Eq.(3.118) to compute $\vec{x}$.

Generalized Alternative Formulation for Indefinite Systems

The alternative formulation presented in the previous section can now be generalized for cases where the original coefficient matrix [A] has multiple zero values for its diagonal terms. In this case, the sub-matrix $[\Delta A]$ will have the following form:

$$[\Delta A] = \begin{bmatrix} 0 & 0 & 0 & 0 & 0 \\ 0 & 0 & 0 & 0 & 0 \\ 0 & 0 & 0 & 0 & 0 \\ 0 & 0 & 0 & 1 & 0 \\ 0 & 0 & 0 & 0 & 1 \end{bmatrix} \begin{matrix} \\ \\ \\ \rightarrow i^{th} \text{ row} \\ \rightarrow j^{th} \text{ row} \end{matrix} \qquad (3.119)$$

Eq.(3.119) can be expressed as:

$$[\Delta A] = [\Delta A_i] + [\Delta A_j] = \begin{bmatrix} 0 & 0 & 0 & 0 & 0 \\ 0 & 0 & 0 & 0 & 0 \\ 0 & 0 & 0 & 0 & 0 \\ 0 & 0 & 0 & 1 & 0 \\ 0 & 0 & 0 & 0 & 0 \end{bmatrix} + \begin{bmatrix} 0 & 0 & 0 & 0 & 0 \\ 0 & 0 & 0 & 0 & 0 \\ 0 & 0 & 0 & 0 & 0 \\ 0 & 0 & 0 & 0 & 0 \\ 0 & 0 & 0 & 0 & 1 \end{bmatrix} \qquad (3.120)$$

or

$$[\Delta A] = \sum_{i=1}^{m} [\Delta A_i] \qquad (3.121)$$

In Eq.(3.121), m ($<<$n) represents the total number of zero diagonal terms of the original coefficient matrix $[A]$. Furthermore, Eq.(3.120) can be represented as:

$$[\Delta A]_{n \times n} = [D]_{n \times m} * [D]^T_{m \times n} \qquad (3.122)$$

where

$$[D] = \begin{bmatrix} 0 & 0 \\ 0 & 0 \\ 0 & 0 \\ 1 & 0 \\ 0 & 1 \end{bmatrix} \qquad (3.123)$$

Following exactly the same derivations given in the previous section, the "generalized" version of Eqs.(3.112) and (3.118) can be given as:

$$[\bar{A}]_{n \times n} * [P]_{n \times m} = [D]_{n \times m} \qquad (3.124)$$

$$\vec{x} = \vec{x}^* + [P]_{n \times m} * [I_{m \times m} - D^T P]^{-1} * [D]^T_{m \times n} * \vec{x}^*_{n \times 1} \qquad (3.125)$$

Remarks

(a) If $\overline{[A]}$ is a symmetrical matrix (which it is!), then $\overline{[A]}^{-1}$ is also symmetrical. One starts with the following identity:

$$[\overline{A}][\overline{A}]^{-1} = [I] = [\overline{A}]^{-1}[A] \tag{3.126}$$

Transposing both sides of Eq.(3.126), one obtains:

$$[\overline{A}^{-1}]^T[\overline{A}]^T = [I] \tag{3.127}$$

Since both matrices $\overline{[A]}$ and $[I]$ are symmetrical, Eq.(3.127) becomes:

$$[\overline{A}^{-1}]^T[\overline{A}] = [I] \tag{3.128}$$

or

$$[\overline{A}^{-1}]^T[\overline{A}] = [\overline{A}^{-1}][\overline{A}] \tag{3.129}$$

Post-multiplying both sides of Eq.(3.129) by $[\overline{A}^{-1}]$, one has:

$$[\overline{A}^{-1}]^T = [\overline{A}^{-1}] \tag{3.130}$$

Thus, $[\overline{A}^{-1}]$ is also a symmetrical matrix.

(b) Matrix product $[D]^T[P]$ is also symmetrical.

From Eq.(3.124), one has:

$$[D]^T[P] = [D]^T[\overline{A}^{-1}][D] \tag{3.131}$$

Transposing both sides of Eq.(3.131), one obtains:

$$[D^TP]^T = [D]^T[\overline{A}^{-1}]^T[D] \tag{3.132}$$

Utilizing Eq.(3.130), Eq.(3.132) becomes:

$$[D^TP]^T = [D]^T[\overline{A}^{-1}]^T[D] \tag{3.133}$$

Comparing Eq.(3.131) and Eq.(3.133), one concludes:

$$[D]^T [P] = [D^T P]^T \qquad (3.134)$$

Thus, the product of matrices $D^T P$ is also symmetrical.

(c) The order of computations of Eq.(3.125) should be done as follows:

1st Step: Compute $\vec{v} = [D]^T \vec{x}^*$ \qquad (3.135)

2nd Step: Let $[Z] \equiv I_{m \times m} - D^T P$ \qquad (3.136)

3rd Step: Let $[Z]^{-1} \vec{v} \equiv \vec{y}$ \qquad (3.137)

or $[Z]\vec{y} = \vec{v}$ \qquad (3.138)

The system of equations, shown in Eq.(3.138), in general is dense (but still is symmetrical) and can be solved for the unknown vector $\vec{y}$.

4th Step: $\vec{x} = \vec{x}^* + [P]\vec{y}$ \qquad (3.139)

d) Out-of core memory can be used to store matrix [P]. Since the matrix [P], in general, could be large (especially true when m is large), thus, a block of L columns (where L<<m<<n) for matrix [P] could be solved, and stored in the out-of-core fashion from Eq.(3.124). The specific value of L can be either specified by the user or can be automatically calculated by the computer program (based on the knowledge of available incore memory). Later on, a block of L columns can be retrieved into the core memory for the computation of $\vec{x}$ in Eq.(3.125).

e) Overhead computational costs for this alternative formulation is essentially occurred in the forward and backward solution phase to solve for matrix [P]. The number of right-hand-side columns of matrix [D] is m. In the parallel computer environments, the overhead costs can be reduced to a "single" forward and backward solution since each processor will be assigned to solve for one column of the matrix [D].

f) The computational (overhead) cost for vector $\vec{y}$ (see Eq.[3.138]) is insignificant since the dimension of matrix [Z] is small, and the right-hand-side of Eq.(3.138) is a "single" vector.

g) In practice, the value of max. A_{ii} (for $i = 1,2,..., n$) should be used in Eq.(3.119), instead of the value 1. Furthermore, the alternative formulation can be made even more stable by using Gerschgorin's theorem to determine the precise (diagonal) locations and values to be used for diagonal terms of matrix $[\Delta A]$ in Eq.(3.119). Let Z_i denote the circle in the complex plane with center a_{ii} (= diagonal terms of matrix [A] whose eigen-values are sought) and radius r_i. Thus, the eigen-values of [A] will fall into those circles:

$$Z_i = \left\{ z \in C \middle| z - a_{ii} \middle| \le r_i \right\} \tag{3.140}$$

where

$$r_i = \sum_{j=1; j \ne i}^{n} \left| a_{ij} \right| \text{ for } I = 1, 2, \ldots, n \tag{3.141}$$

Hence

$$a_{ii} - r_i \le Z_i \le a_{ii} + r_i \tag{3.142}$$

From Eq.(3.142), one can estimate the number of NEGATIVE and/or POSITIVE eigen-values of matrix [A]. For example, if $a_{ii} - r_i > 0$, then the eigen-values associated with the circle Z_i is must be positive. Adding a positive number δ to the diagonal term of [A] will shift the value of the eigen-value associated with the circle Z_i. For example, adding δ to make

$$a_{ii} - r_i + \delta_i > 0 \tag{3.143}$$

will ensure that the eigen-value associated with the circle Z_i is positive. From Eq.(3.143), one may add

$$\delta_i = r_i - a_{ii} > 0 \tag{3.144}$$

or, to be even safer

$$\delta_i = r_i - a_{ii} > \in \text{ where } \in \text{ is a small positive value} \tag{3.145}$$

to the diagonal term a_{ii} so that all eigen-values will be positive (and therefore, the new matrix $\left| A \right|$ will be positive definite).

A simple example is given in the following paragraphs to clarify the above discussion. The matrix [A] is given as:

$$[A] = \begin{bmatrix} 4 & 1 & 0 \\ 1 & 0 & 1 \\ 1 & 1 & -4 \end{bmatrix} \tag{3.146}$$

From the data shown in Eq.(3.146), one computes:

$$\left.\begin{array}{l} a_{11} = 4; r_1 = 1 \text{ and } \delta_1 = r_1 - a_{11} = -3 \\ a_{22} = 0; r_2 = 2 \text{ and } \delta_2 = r_2 - a_{22} = 2 \\ a_{33} = -4; r_3 = 2 \text{ and } \delta_3 = r_3 - a_{33} = 6 \end{array}\right\} \qquad (3.147)$$

Let $\in = 1$, it is then suggested to add positive values $\delta_2 = 2$, and $\delta_3 = 6$ to the second and third diagonal terms of $[A]$. The new matrix $\left|\overline{A}\right|$ is therefore defined as:

$$\left[\overline{A}\right] = [A] + [\Delta A]$$

$$\left[\overline{A}\right] = \begin{bmatrix} 4 & 1 & 0 \\ 1 & 0 & 1 \\ 1 & 1 & -4 \end{bmatrix} + \begin{bmatrix} 0 & 0 & 0 \\ 0 & 2 & 0 \\ 0 & 0 & 6 \end{bmatrix} \qquad (3.148)$$

$$\left[\overline{A}\right] = \begin{bmatrix} 4 & 1 & 0 \\ 1 & 2 & 1 \\ 1 & 1 & 2 \end{bmatrix}$$

The three eigen-values associated with matrices $[A]$ and $\left|\overline{A}\right|$ are $\lambda_A = \{-4.2146, -0.0556, 4.2702\}$ and $\lambda_{\overline{A}} = \{4.618, 2.382, 1.000\}$, respectively.

h) Major advantages of the generalized alternative formulation are
- Pivoting strategies are NOT required, hence the algorithm should be more robust and better solution accuracy can be expected.
- Incore memory requirements can be reduced and predicted before entering the numerical factorization phase.
- Any positive definite sparse solvers can be integrated into this alternative formulation.

i) In actual computer coding, there is no need to "push" all zero diagonal terms of $[A]$ to the bottom right of $[A]$. Furthermore, the sub-matrix $[o]$, shown in Eq.(3.100), does NOT have to be completely zero. In other words, the off-diagonal terms of this sub-matrix $[o]$ may or may not be zero (only the diagonal terms of sub-matrix $[o]$ are zero).

Numerical Applications

Based upon the proposed alternative formulation presented in previous sections, several benchmark indefinite system of equations (obtained from NASA Langley

Research Center) have been used to validate the proposed algorithms. The sizes (= Neq = Number of equations) and the original number of non-zero, off-diagonal coefficients of matrix [A] (= Ncoff) are given in Table 3.12. In the partitioned form, Eq.(3.99) can also be expressed as (please refer to Eq.3.100):

$$\begin{bmatrix} K & a \\ a^T & 0 \end{bmatrix} \begin{Bmatrix} \vec{x}_u \\ \vec{x}_\lambda \end{Bmatrix} = \begin{Bmatrix} \vec{f}_u \\ \vec{f}_\lambda \end{Bmatrix} \tag{3.149}$$

For structural mechanic applications, the vector $\vec{x}_u$ may represent the displacements, whereas the vector $\vec{x}_\lambda$ may represent the Lagrange multipliers associated with the interfaced problems. Solution accuracy of the proposed algorithms can be established by comparing the following quantities with Boeing's sparse indefinite equation solver:

1. Maximum absolute displacement (of $\vec{x}_u$, shown in Eq.[3.149])

2. Summation of absolute displacements (of $\vec{x}_u$, shown in Eq.[3.149])

3. Relative error norm (Rel Err = $\dfrac{\left\| A\vec{x} - \vec{f} \right\|}{\left\| \vec{f} \right\|}$, shown in Eq.[3.99])

The above quantities are also included in Table 3.12. It should be emphasized that CPU time comparisons are NOT yet included in this study, due to the following reasons:

(a) Boeing's sparse indefinite solver timing has been implemented earlier on the Cray-YMP supercomputer. However, the author' FORTRAN code has been recently tested on an Old Dominion University (ODU) Sun (= Norfolk) workstation since the author currently has no access to the Cray-YMP nor Boeing's source code.

(b) The re-ordering algorithms, such as Modified Minimum Degree (MMD) or Nested Disection (ND), have NOT yet been integrated into the current version of the author' FORTRAN code.

(c) Parallel processing for $[\overline{A}][P] = [D] =$ multiple RHS has not been done.

Table 3.12 Comparison of ODU and Boeing Indefinite Sparse Solvers

Neq	Ncoff	$\sum_i \left\| x_{u_i} \right\|$	$\text{Max} \left\| x_{u_j} \right\|$	Rel Err	Time (ODU-Norfolk)
51(Boeing)	218	$2.265*10^{-2}$ $(2.265*10^{-2})$	$2.000*10^{-3}$ $(1.999*10^{-3})$	$4.68*10^{-6}$ $(7.0*10^{-14})$	0.0sec N/A
247(Boeing)	2009	3.16 (3.16)	0.1525 (0.1525)	$2.63*10^{-10}$ $(4.03*10^{-10})$	0.1sec N/A
1440(Boeing)	22137	29.68 (29.68)	0.20289 (0.20289)	$4.27*10^{-11}$ $(3.26*10^{-10})$	8.7sec N/A
7767(Boeing)	76111	113.71 (113.71)	0.1610576 (0.1610576)	$5.31*10^{-8}$ $(6.00*10^{-8})$	42.7sec N/A
15367(Boeing)	286044	512.35 (512.35)	0.205696 (0.205696)	$9.22*10^{-10}$ $(4.38*10^{-11})$	5400sec N/A

Conclusions

Alternative formulations and algorithms for solving sparse system of equations have been developed. The proposed numerical algorithms have been implemented and validated through several benchmark NASA applications. Preliminary results have indicated that the proposed alternative algorithms are quite robust and are in excellent agreement with Boeing's commercial sparse indefinite solver. Detailed analysis of the proposed sparse indefinite algorithms have suggested that:

(a) The proposed algorithms are quite robust and accurate.
(b) The additional (overhead) costs for the proposed algorithms mainly occur in the forward and backward solution phases (of the associated positive definite system). These overhead costs can be easily and drastically reduced by performing the forward and backward solution phases (of multiple-right-hand-side vectors) in a parallel computer environment.
(c) In the proposed formulation, one only deals with "positive-definite" sparse systems. Thus, complex and expensive pivoting strategies are NOT required. As a consequence of these desired features, several important advantages can be realized, such as:
   - Incore memory requirements can be easily and efficiently predicted (before entering the sparse numerical factorization phase).
   - The amount of non-zero "fill-in" (and hence, the number of floating operations) can be reduced.

- Any efficient sparse "positive definite" solvers can be conveniently integrated into the proposed formulations.

Efforts are underway to incorporate various reordering algorithms (such as MMD, ND, etc.) into the proposed algorithms and implement the entire procedure in the MPI parallel computer environments. Additional results will be reported in the near future.

3.10 Unsymmetrical Matrix Equation Solver

Let's consider the following system of unsymmetrical linear equations:

$$Ax = b \qquad (3.150)$$

where the coefficient matrix A is unsymmetrical and the vectors x and b represent the unknown vector (nodal displacement) and the right-hand side (known nodal load) vector, respectively. In this section, a solver for unsymmetrical matrices, where the upper-and lower-triangular portions of the matrix are symmetrical in locations but unsymmetrical in values (see Figure 3.6), will be developed.

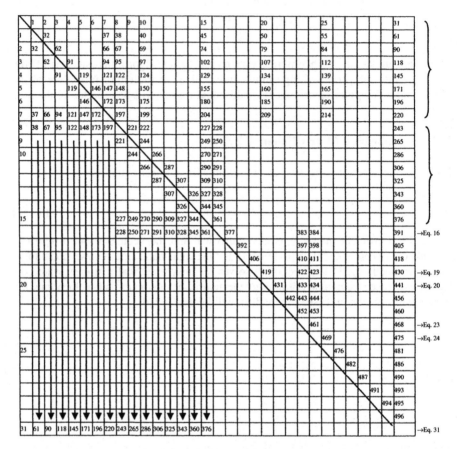

Figure 3.6 Detailed Storage Scheme for an Unsymmetrical Matrix

In order to take advantage of the algorithms discussed earlier in Sections 3.4 – 3.8 for the solution of symmetrical matrices, exploit the vector capability provided by supercomputers, take advantage of the cache provided by many workstations, and minimize the data movements into fast memory, it is necessary to arrange the data appropriately. A mixed row-wise and column-wise storage scheme is used. This storage scheme offers the advantage of applying the symbolic factorization and the super node evaluation only on one portion of the matrix, instead of the entire matrix. Compared to the symmetrical case, the re-ordering (fill-in minimization), the numerical factorization, the forward/backward substitution, and the matrix-vector multiplication sub-routines are different since the matrix is unsymmetrical in value.

Sparse Storage of the Unsymmetrical Matrix

The unsymmetrical matrix A is stored in a mixed row-oriented and column-oriented fashion. The upper portion of the matrix is stored in a sparse, row-wise NASA format as has been explained in [3.11]. The lower portion of the matrix is stored in a sparse column-wise format. Since a column-wise representation of a matrix is a row-wise representation of its transpose, and the matrix is symmetrical in locations, the arrays IA (neq + 1), JA (ncoef), which are used to determine the nonzero locations of [A], will be the same for both the upper and lower portions. AN (ncoef) will contain the coefficients of the upper portion of the matrix and a new array, AN2 (ncoef), is introduced to store the coefficient values of the lower portion of the matrix. The diagonal values will be stored in the real array AD (neq). This storage scheme allows the use of the loop unrolling technique, described in [1.9], during the factorization for both the upper-and lower-triangular portions of the matrix. Figure 3.7 shows how the coefficient matrix A is stored.

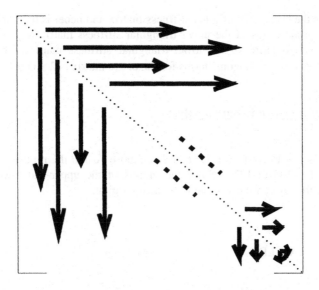

Figure 3.7 Storage Scheme for Unsymmetrical Matrix

To illustrate the usage of the adopted storage scheme, let's consider the matrix given in Eq.(3.151).

$$A = \begin{bmatrix} 11. & 0. & 0. & 1. & 0. & 2. \\ 0 & 44. & 0. & 0. & 3. & 0. \\ 0 & 0 & 66. & 0. & 4. & 0. \\ 8 & 0 & 0 & 88 & 5. & 0. \\ 0 & 10 & 11 & 12 & 110 & 7 \\ 9 & 0 & 0 & 0 & 14 & 112 \end{bmatrix} \qquad (3.151)$$

The data in Eq.(3.150) will be represented as follows:

IA (1:7 = neq + 1) = {1, 3, 4, 5, 6, 7, 7}
JA (1:6 = ncoef) = {4, 6, 5, 5, 5, 6}
AD (1:6 = neq) = {11., 44., 66., 88., 110., 112.}
AN (1:6 = ncoef) = {1., 2., 3., 4., 5., 7.}
AN2(1:6 = ncoef) = {8., 9., 10., 11., 12., 14.}

where neq is the size of the original stiffness matrix and ncoef is the number of non-zero, off diagonal terms of the upper-triangular stiffness matrix (equal to the non-zero, off diagonal terms of the lower-triangular stiffness matrix). Thus the total number of non-zero off diagonal terms for the entire matrix is 2 × ncoef.

Basic Unsymmetrical Equation Solver

One way to solve Eq.(3.150) is first to decompose A into the product of triangular matrices, either LU or LDU. Since the graphs of the upper-and lower-triangular matrices are the same, we chose the LDU factorization.

Thus

$$A = LDU \qquad (3.152)$$

where U is an upper-triangular matrix with unit diagonal, D a diagonal matrix, and L a lower-triangular matrix with unit diagonal. After factorization, the numerical values of matrix L are different from those of matrix U.

In order to better understand the general formula that we will derive for factorization of an unsymmetrical matrix, let's try to compute the factorized matrix [L], [D], and [U] from the following given 3×3 unsymmetrical matrix [A], assumed to be a full matrix, to simplify the discussion.

$$A = \begin{bmatrix} a_{11} & a_{12} & a_{13} \\ a_{21} & a_{22} & a_{23} \\ a_{31} & a_{32} & a_{33} \end{bmatrix} \tag{3.153}$$

The unsymmetrical matrix A given in Eq.(3.153) can be factorized as indicated in Eq.(3.152), or in the long form as follows:

$$\begin{bmatrix} a_{11} & a_{12} & a_{13} \\ a_{21} & a_{22} & a_{23} \\ a_{31} & a_{32} & a_{33} \end{bmatrix} = \begin{bmatrix} 1 & 0 & 0 \\ l_{21} & 1 & 0 \\ l_{31} & l_{32} & 1 \end{bmatrix} \begin{bmatrix} D_{11} & 0 & 0 \\ 0 & D_{22} & 0 \\ 0 & 0 & D_{33} \end{bmatrix} \begin{bmatrix} 1 & u_{12} & u_{13} \\ 0 & 1 & u_{23} \\ 0 & 0 & 1 \end{bmatrix} \tag{3.154}$$

The multiplication of matrices on the right-hand side of the equality gives:

$$\begin{bmatrix} a_{11} & a_{12} & a_{13} \\ a_{21} & a_{22} & a_{23} \\ a_{31} & a_{32} & a_{33} \end{bmatrix} = \begin{bmatrix} d_{11} & d_{11}u_{12} & d_{11}u_{13} \\ l_{21}d_{11} & l_{21}d_{11}u_{12}+d_{22} & l_{21}d_{11}u_{13}+d_{22}u_{23} \\ l_{31}d_{11} & l_{31}d_{11}u_{12}+l_{32}d_{22} & l_{31}d_{11}u_{13}+l_{32}d_{22}u_{23}+d_{33} \end{bmatrix} \tag{3.155}$$

where the nine unknowns (d_{11}, u_{12}, u_{13}, u_{23}, l_{21}, l_{31}, d_{22}, l_{32}, and d_{33}) from Eq.(3.154) and Eq.(3.155) can be found by simultaneously solving the following system of equations.

$$a_{11} = d_{11}$$
$$a_{12} = d_{11}u_{12}$$
$$a_{21} = l_{21}d_{11}$$
$$a_{13} = d_{11}u_{13}$$
$$a_{31} = l_{31}d_{11} \tag{3.156}$$
$$a_{22} = l_{21}d_{11}u_{12}+d_{22}$$
$$a_{23} = l_{21}d_{11}u_{13}+d_{22}u_{23}$$
$$a_{32} = l_{31}d_{11}u_{12}+l_{32}d_{22}$$
$$a_{33} = l_{31}d_{11}u_{13}+l_{32}d_{22}u_{23}+d_{33}$$

Thus from Eq.(3.156), one obtains Eq.(3.157):

$$d_{11} = a_{11}$$
$$u_{12} = \frac{a_{12}}{d_{11}}$$

$$u_{13} = \frac{a_{13}}{d_{11}}$$

$$l_{21} = \frac{a_{12}}{d_{11}}$$

$$l_{31} = \frac{a_{31}}{d_{11}}$$ (3.157)

$$d_{22} = a_{22} - l_{21}d_{11}u_{12}$$

$$u_{23} = \frac{a_{23} - l_{21}d_{11}u_{13}}{d_{22}}$$

$$l_{32} = \frac{a_{32} - l_{31}d_{11}u_{12}}{d_{22}}$$

$$d_{33} = a_{33} - (l_{31}d_{11}u_{13} + l_{32}d_{22}u_{23})$$

In solving for the unknowns in Eq.(3.157), the factorized matrices [L], [D], and [U] can be found in the following systematic pattern:

Step 1:   The 1st diagonal value of [D] can be solved for d_{11}.

Step 2:   The 1st row of the upper-triangular matrix [U] can be solved for the solution of u_{12} and u_{13}.

Step 3:   The 1st column of the lower-triangular matrix [L] can be solved for l_{21} and l_{31}.

Step 4:   The 2nd diagonal value of [D] can be solved for d_{22}.

Step 5:   The 2nd row of the upper-triangular matrix [U] can be solved for the solution of u_{23}.

Step 6:   The 2nd column of the lower-triangular matrix [L] can be solved for l_{32}.

Step 7:   The 3rd diagonal value of [D] can be solved for d_{33}.

By observing the above procedure, one can see that to factorize the term u_{ij} of the upper-triangular matrix [U], one needs to know only the factorized row i of [L] and column j of [U]. Similarly, to factorize the term l_{ij} of the lower-triangular matrix [L], one needs to know only the factorized row j of [L] and column i of [U] as shown in Figure 3.8.

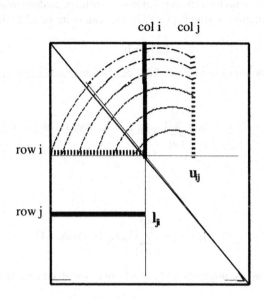

col i col j

row i

row j

u_{ij}

l_{ji}

Figure 3.8 Unsymmetrical Solver: Factorization of u_{ij} and l_{ji}

By generalizing to a matrix of dimension *neq*, the ith row elements of [U] and the ith column elements of [L] can be obtained by the formulas in Eq.(3.158) and Eq.(3.159), assuming that the rows from 1 to i-1 and column from 1 to i-1 have already been factorized:

$$u_{ij} = \frac{a_{ij} - \sum_{k=1}^{i-1} l_{ik} d_{ii} u_{kj}}{d_{ii}} \quad (j = i+1, neq) \tag{3.158}$$

$$l_{ji} = \frac{a_{ji} - \sum_{k=1}^{j-1} l_{jk} d_{jj} u_{ki}}{d_{jj}} \quad (i = j+1, neq) \tag{3.159}$$

and the diagonal values will be given by Eq.(3.160):

$$d_{ii} = a_{11} - \sum_{k=1}^{i-1} l_{ik} d_{jj} u_{ki} \tag{3.160}$$

Once the matrix is factorized, the unknown vector x is determined by the forward/ backward substitution. Using Eq.(3.152), one can write Eq.(3.150) as follows:

$$LDy = b \tag{3.161}$$

with y = Ux. The solution of Eq.(3.161) can be obtained as follows:

$$y_i^* = b_i - \sum_{k=1}^{i-1} L_{ik} y_k \ (i = 1,...,neq) \text{ with } y^* = Dy \tag{3.162}$$

and to solve

$$Ux = y \tag{3.163}$$

for x,

$$x_i = y_i - \sum_{k=i+1}^{neq} U_{ik} x_k \ (i = neq,...,1) \tag{3.164}$$

The factorization is computationally much more involved than the forward/backward substitution.

Efficient-Sparse LDU Unsymmetrical Solver

The developed sparse unsymmetrical solver is a collection of sub-routines that follow the same idea as the one given in Figure 3.8, with the sub-routines performing different tasks. Since the matrix is unsymmetrical in value, the re-ordering algorithm for the symmetric matrix is not suitable. On the other hand, by observing Figure 3.8 and the derivations in Eq.(3.152), the multipliers in the factorization of the upper portion of the matrix will be computed from the coefficients of the lower portion of the matrix and vice versa; thus, the numerical factorization will be different from the symmetrical case.

The purpose of symbolic factorization is to find the locations of all non-zero (including "fill-in" terms), off-diagonal terms of the factorized matrix [U]. Since both upper and lower portions of the matrix have the same graph, the symbolic factorization is performed only on either the upper or lower portion of the matrix. The symbolic factorization required the structure IA, JA of the matrix in an unordered representation and generates the structure IU, JU of the factorized matrix in an unordered representation. However, the numerical factorization requires IU, JU to be ordered while IA, JA can be given in an unordered representation. A symbolic transposition routine, TRANSA, which does not construct the array of non-zero of the transpose structure, will be used twice to order IU, JU, after the symbolic factorization, since we are only interested in ordering JU. One of the major goals in this phase is to predict the required computer memory for subsequent numerical factorization for either the upper or lower portion of the matrix. For an unsymmetrical case, the total memory required is twice the amount predicted by the symbolic factorization.

Ordering for Unsymmetrical Solver

Ordering algorithms, such as minimum-degree and nested dissection, have been developed for reducing fill-in during factorization of sparse symmetrical matrices. One cannot apply fill-in minimization, MMD [3.9], on the upper and lower matrices separately. Shifting rows and columns of the upper portion of the matrix will require values from the lower portion of the matrix and vice versa. Let's consider the following example:

$$
A = \begin{bmatrix}
100 & 1 & 2 & 3 & 4 \\
5 & 100 & 6 & 7 & 8 \\
9 & 10 & 100 & 11 & 12 \\
13 & 14 & 15 & 100 & 16 \\
17 & 18 & 19 & 20 & 100
\end{bmatrix} \tag{3.165}
$$

Let us assume that the application of the Modified Minimum Degree (MMD) algorithm on the graph of the matrix results in the following permutation:

$$
\text{PERM} \begin{Bmatrix} 1 \\ 2 \\ 3 \\ 4 \\ 5 \end{Bmatrix} = \begin{Bmatrix} 1 \\ 4 \\ 2 \\ 3 \\ 5 \end{Bmatrix} \tag{3.166}
$$

By switching rows and columns of the matrix given in Eq.(3.165), according to the permutation vector PERM, given in Eq.(3.166), the recorded matrix A_r becomes:

$$
A_r = \begin{bmatrix}
100 & 3 & 1 & 2 & 4 \\
13 & 100 & 14 & 15 & 16 \\
5 & 7 & 100 & 6 & 18 \\
9 & 11 & 10 & 100 & 12 \\
17 & 20 & 18 & 19 & 100
\end{bmatrix} \tag{3.167}
$$

On the other hand, if one considers only the upper portion of the matrix (as for a symmetrical case), switching rows and columns of the matrix according to the permutation vector, PERM, will result in the following reordered matrix A_s given in Eq.(3.168).

$$
A_p = \begin{bmatrix}
100 & 3 & 1 & 2 & 4 \\
3 & 100 & 7 & 11 & 16 \\
1 & 7 & 100 & 6 & 8 \\
2 & 11 & 6 & 100 & 12 \\
4 & 16 & 8 & 12 & 100
\end{bmatrix} \tag{3.168}
$$

One can see that the elements A(2,3) and A(2,4) came from the lower portion. Therefore, re-arranging the values of AN (or AN2) after the permutation vector PERM has been determined by the MMD routine will require certain elements of AN2 (or AN). The re-ordering subroutine for symmetric system has been modified to account for these changes and implemented without adding any additional working array. The portion of the skeleton Fortran code in Table 3.13 shows how to efficiently retrieve the appropriate elements from the lower (upper) portion of the matrix while constructing the re-ordered upper (lower) portion of the matrix. The permutation vector PERM and the structure IU and JU of the re-ordered matrix are assumed to be available already.

The algorithm in Table 3.13 is different for a case of a symmetrical matrix because, if only the upper portion of a symmetrical matrix is stored in memory, the numerical values in row i at the left side of the diagonal value are identical to the values in column i above the diagonal value (see Figure 3.8). Consequently, the second DO loop 231 in Table 3.13 will not be needed because all data can be retrieved from the upper portion of the matrix and one can select the appropriate pointers IJ0 and IJ00 before the inner most DO loop. On the other hand, for an unsymmetrical matrix, one should scan separately the upper and lower portion of the matrix (AN = AN2) as shown in Table 3.13.

Table 3.13 Portion of Skeleton Fortran Code for Re-ordering an Unsymmetrical Matrix

```
    DO 200 i=1, N-1
      I0=perm(i)
      DO 220 j=IU(i), IU(i+1)
      J0=perm(JU(j))
      IF(I0.LT.J0) THEN
        IJ0=I0
        IJ00=J0
        DO 230 jj=IA(IJ0), IA(IJ0+1)-1
          IF(JA(jj).NE.IJ00) GO TO 230
          UN(j)=AN(jj)
          UN2(j)=AN2(jj)
```

```
              GO TO 220
        230    CONTINUE
    ELSE
        IJ0=J0
        IJ00=I0
        DO 231 jj=IA(IJ0), IA(IJ0+1)-1
        IF(JA(jj).NE.IJ00) GO TO 231
            UN(j)=AN2(jj)
            UN2(j)=AN(jj)
            GO TO 220
        231 CONTINUE
    ENDIF
    220 CONTINUE
200 CONTINUE
```

Sparse Numerical Factorization with Loop Unrolling

By observing Figure 3.8 and the derivations in the previous section, in order to factorize an element u_{ij} of the upper-triangular matrix, one needs to know how to factorize row i of [L] and the column j of [U].

Thus, the multiplier of the upper portion of the matrix will be computed from the coefficient of the lower portion of the matrix. Table 3.14 gives the pseudo Fortran skeleton code on how the multipliers are computed and how the factorization is carried out. The diagonal matrix [D] (see Eq.3.154) is also contained by the diagonal terms of the upper- triangular matrix [U] as can be seen from line 5 of Table 3.14.

Table 3.14 Pseudo FORTRAN Skeleton Code for Sparse LDU Factorization

```
1.  c......Assuming row 1 has been factorized earlier
2.  Do 11 I=2, NEQ
3.  Do 22 K=Only those previous rows which have contributions to
               current row I
4.  c.......Compute the multipliers
5.  XMULT=U(K,I)/ U(K,K)........related to the upper triangular matrix
    XMULT2 = L(I,K)/ U(K,K)......related to the lower triangular matrix
6.  Do 33 J = appropriated column numbers of row # K
7.  U(I,J) = U(I,J) - XMULT2 * U(K,J)
    L(J,I) = L(J,I) - XMULT * L(J,K)
8.  33 Continue
9.  U(K,I) = XMULT
    L(I,K) = XMULT2
10. 22 Continue
11. 11 Continue
```

In the sparse implementation, after the symbolic factorization is completed on one portion of the matrix, the numerical factorization requires IU, JU (structure of [L] or [U]) to be ordered and the required computer memory for the factorization is known. Similar to the symmetrical case, the numerical factorization for the unsymmetrical case also requires to construct chain lists to keep track of the rows that will have contributions to the currently factorized row. Another advantage of the storage scheme that we have adopted is that the chain lists for the factorization of [L] (or [U]) will be the same as for the factorization of [U] (or [L]).

The loop unrolling strategies that have been successfully introduced earlier can also be effectively incorporated into the developed unsymmetrical sparse solver in conjunction with the master degree of freedom strategy. In the actual code implementation, "DO loops" in Eqs.(3.158 – 3.160) will be re-arranged to make use of the loop unrolling technique. The loop unrolling is applied separately for the factorization of the upper portion and for the lower portion assuming the super-nodes have already been computed (the super-nodes of the upper portion are the same as the ones for the lower portion). The skeleton FORTRAN code in Table 3.14 should be modified as shown by the pseudo, skeleton FORTRAN code in Table 3.15 for a loop unrolling level 2.

Forward and Backward Solution

The forward and backward solutions were implemented following the formula in Eqs.(3.161 – 3.163), once the factorized matrices [L], [D], and [U] are computed. In the forward solution, (Eqs.[3.161 and 3.16]), the factorized matrices [L] and [D] are used, and in the backward substitution, the upper portion of the factorized matrix [U] is used.

Table 3.15 Pseudo FORTRAN Skeleton Code for Sparse LDU Factorization with Unrolling Strategies

```
C......Assuming row 1 has been factorized earlier
       Do 11 I=2,NEQ
       Do 22 K=Only those previous "master" rows which have
                                 contributions to current row I
C......Compute the multiplier(s)
       NSLAVE DOF=MASTER (I) - 1
       XMULT = U(K,I) / U(K,K)
       XMULm = U(K+m,I) / U(K+m,K+m)
       XMULT2 = L(I,K) / U(K,K)
       XMUL2m = L(I,K+m) / U(K+m,K+m)
C......m=1,2...SLAVE DOF
       Do 33 J = appropriated column numbers of "master" row #K
          U(I,J) = U(I,J) - XMULT2 * U(K,J) - XMULT2m * U(K+m,J)
          L(J,I) = L(j,I) - XMULT * L(J,K) - XMULm * L(J,K+m)
       33 Continue
          U(K,I) = XMULT
          U(K+m,I) = XMULm
          L(I,K) = XMULT2
          L(I,K+m) = XMUL2m
       22 Continue
       11 Continue
```

Sparse Unsymmetric Matrix-Vector Multiplication

A matrix-vector multiplication sub-routine has been efficiently designed for which the unsymmetrical matrix is stored in a mixed row-wise and column-wise storage scheme. The non-zeros from the upper- and lower-triangular matrix are stored in two distinctive arrays AN and AN2 with the same structure IA and JA. Let's consider a vector *temp* (1:neq) that will contain the result of the matrix-vector multiplication. After multiplying the diagonal values by the right-hand side, the multiplication of the upper and lower portion of the matrix are efficiently implemented as shown in Table 3.16.

Table 3.16 Unsymmetrical Matrix-Vector Multiplication

```
     Do 10 i=1,n
       iaa=ia(i)
       iab=ia(i+1)-1
       Do k=iaa,iab
       kk=ja(k)
       sum=sum+an(k)*rhs(kk)
       temp(kk)=temp(kk)+an2(k)*rhs(i)
       ENDDO
       temp(i)=sum
10     Continue
```

The algorithm shown in Table 3.16 offers the advantage avoiding a conversion of a row-wise complete unordered storage that is normally used for general unsymmetric matrix into our special storage scheme (mixed-row and column-wise format).

The algorithm shown in Table 3.16 can be conveniently used to calculate the relative error norm, such as:

$$RELERR = \frac{\left\| A\vec{x} - \vec{b} \right\|}{\vec{b}} \qquad (3.169)$$

Application of the Developed Sparse Unsymmetrical Solver

Three examples are considered to evaluate the performance of the developed unsymmetrical vector sparse LDU solver (that we will refer to as UNSYNUMFA). The author has considered pivoting strategies in earlier and separate works. However, these pivoting strategies have not yet been incorporated into the current unsymmetrical sparse solver. Two applications, the HSCT (16,152 degrees of freedoms) and the SRB (54,870 degrees of freedoms) finite element models for which the static solution is known, are considered. In these first two examples, the coefficient (stiffness) matrices are known to be symmetrical. However, we still have used the developed unsymmetrical sparse solver to obtain the correct solutions! Another application, Pierrot HSCT (16,152 degrees of freedoms), is constructed by considering the structure of the HSCT FEM with the same coefficient values for the upper portion of the matrix and different values for the lower portion of the matrix to make the matrix completely unsymmetrical in value.

To check the accuracy of the results, a relative error norm is computed, as shown in Eq.(3.169), where matrix [A] is unsymmetrical. The sparse unsymmetrical matrix-vector multiplication sub-routine developed in the previous section is used to compute the product [A]{x} (where {x} is the displacement vector) that is required for error norm computation.

The numerical performance of the above three (finite element based) structural examples are presented in Tables 3.17 – 3.21.

Table 3.17 HSCT FEM: Memory Requirement for UNSYNUMFA

REORD	NCOEF	NCOEF2	Integer Memory	Real Memory	Total Memory
No Reord.	999,010	7,400,484	4,296,626	8,480,224	12,776,850
UnsynM MD	999,010	6,034,566	3,613,667	7,114,306	10,727,973

Table 3.18 HSCT FEM: Summary of Results for UNSYNUMFA1/2/8 Using UnsyMMD and Different Level of Loop Unrolling on the IBM RS6000/590 *Stretch* Machine

Loop Unrolling Level	Symfa time (sec)	Numfa time (sec)	FBE time (sec)	Total time (sec)	Max. abs. displ.	Summat abs. displ.	Relative Error Norm
1	0.480	50.010	0.310	53.350	0.477	301.291	1.34E-08
2	0.470	35.420	0.320	38.760	0.477	301.291	1.99E-08
8	0.480	28.730	0.320	32.700	0.477	301.291	1.36E-08

Table 3.19 HSCT FEM: Comparison of Results for UNSYNUMFA with No UnsyMMD and Different Level of Loop Unrolling on the IBM RS6000/590 *Stretch* Machine

Loop Unrolling Level	Symfa time (sec)	Numfa time (sec)	FBE time (sec)	Total time (sec)	Max. abs. displ.	Summat abs. displ.	Relative Error Norm
1	0.710	52.079	0.370	55.200	0.477	301.291	2.2E-09
2	0.680	35.650	0.380	38.730	0.477	301.291	2.0E-09
8	0.700	28.390	0.390	31.520	0.477	301.291	2.0E-09

Table 3.20 SRB FEM: Summary of Results for UNSYNUMFA Using UnsyMMD and Different Level of Loop Unrolling on the IBM RS6000/590 *Stretch* Machine

Loop Unrolling Level	Symfa time (sec)	Numfa time (sec)	FBE time (sec)	Total time (sec)	Max. abs. displ.	Summat abs. displ.	Relative Error Norm
1	1.93	210.500	2.820	229.560	2.061	301.291	8.1E-13
2	1.93	155.630	2.270	173.280	2.061	301.291	8.1E-13
8	1.93	133.150	1.300	150.230	2.061	301.291	8.1E-13

Table 3.21 Pierrot FEM: Summary of Results for UNSYNUMFA Using UnsyMMD and Different Level of Loop Unrolling on the IBM RS6000/590 *Stretch* Machine

Loop Unrolling Level	Symfa time (sec)	Numfa time (sec)	FBE time (sec)	Total time (sec)	Max. abs. displ.	Summat abs. displ.	Relative Error Norm
1	0.480	49.970	0.330	53.320	8.791	45.134	2.3E-07
2	0.470	35.340	0.320	38.650	8.791	45.134	1.8E-07
8	0.480	28.650	0.320	31.970	8.791	45.134	1.3E-07

3.11 Summary

In this chapter, various topics related to direct sparse equation solvers for symmetrical/unsymmetrical, positive/negative/infinite system of equations have been explained. Sparse storage schemes, impacts of sparse re-ordering algorithms on the performance of the sparse solvers, sparse symbolic/numerical factorization, and forward and backward solution phases have been discussed. Unrolling strategies, which utilize "super-row" information, have also been incorporated into the factorization and forward/backward solution phases. Several medium- to large-scale, practical applications have been used to evaluate the performance of various proposed algorithms. These sparse solvers will also be used in subsequent chapters of this textbook.

3.12 Exercises

3.1 Given the following sparse matrix:

Matrix A =

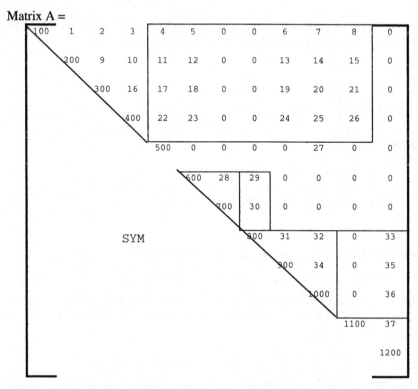

(a) Repeat the above sparse matrix [A] by using two integers and 2 real arrays
$(=\overrightarrow{IA}, \overrightarrow{JA}, \overrightarrow{AD}, \overrightarrow{AN})$, as discussed in Eqs.(3.7 – 3.10). What is the value of NCOEF1, as can be seen from Eq.(3.10)?

(b) Without any actual numerical calculation, find the number and location of non-zero terms due to "fill-in" effects during the factorization phase.

What is the value of NCOEF2, as discussed in the remarks about Eq.(3.35)?

(c) Using the results of part (b), how do you re-define the arrays $\overrightarrow{IA}$ and $\overrightarrow{JA}$ [in part (a)] to include "fill-in" terms?

3.2 Assuming the matrix given in Problem 3.1 represents the factorized matrix, find the super-row information as discussed in Eq.(3.78).

3.3 Assuming each node in Figure 3.1 has 2 dof, construct the 2 integer arrays, as discussed in Eqs.(3.37, 3.38).

3.4 Using the same data, shown in Figure 3.1 and Eqs.(3.37, 3.38), and assuming the value of an integer array IPERM (see Eq.3.39) is given as:

$$
\text{IPERM} \begin{Bmatrix} 1 \\ 2 \\ 3 \\ 4 \\ 5 \\ 6 \end{Bmatrix} = \begin{Bmatrix} 5 \\ 1 \\ 4 \\ 2 \\ 6 \\ 3 \end{Bmatrix}
$$

how will the matrix [U], shown in Eq.(3.41), be modified?

3.5 Given the indefinite matrix [A], shown in Eq.(3.146), and the right-hand side vector b = $\{5,2,-2\}^T$, using the procedures discussed in Eqs.(3.135-3.139), find the solution vector $\vec{x}$ of $[A]\vec{x} = \vec{b}$.

3.6 For the unsymmetrical matrix [A], shown in Eq.(3.165), find the factorized matrix [A] = [L] [D] [U].

4 Sparse Assembly Process

4.1 Introduction

Systems of sparse, linear, symmetrical and/or unsymmetrical equations have occurred frequently in several engineering applications such as in nonlinear thermal finite analysis using discontinuous Galerkin method, acoustic finite element analysis, etc.

A number of highly efficient sparse symmetrical equation solvers has been developed and reported in the literature. However, much less attention have been focused on the development of efficient, general symmetrical/unsymmetrical sparse assembly procedures. The objectives of this chapter, therefore, are

(a) To discuss efficient algorithms for <u>assembling</u> a general, symmetrical/ unsymmetrical, sparse system of matrix equations that arise naturally from finite element analysis.

(b) To summarize key steps involved in <u>solving</u> a general system of sparse, symmetrical/ unsymmetrical equations, and

(c) To develop a simple template, in the form of sub-routines, where all key components of the procedure can be integrated to form a complete finite element analysis code.

4.2 A Simple Finite Element Model (Symmetrical Matrices)

To facilitate the discussion in this chapter, a simple finite element model of a 2-D truss structure, with 1 bay and 1 story, is shown in Figure 4.1. This structure has pin supports at nodes 3 and 4. Thus, zero prescribed Dirichlet boundary displacements are specified at nodes 3 and 4. Each node is assumed to have 2 degree-of-freedom (dof). The applied loads and the prescribed boundary dof at the nodes are given in Figure 4.1. There are 5 (2-D) truss members in this finite element model. Young modulus (= E), and cross-sectional area (= A), for each truss member is assumed to be E =1k/in.^{2} and A = 1 in.^{2}, respectively. The base and height for this 2-D truss structure is assumed to be 12", and 9", respectively. Since there are 4 nodes, with 2 dof per node, this finite element model has a total of 8 dof (N = 8).

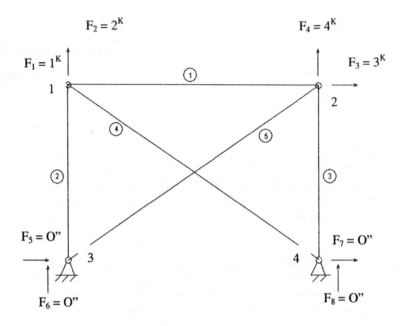

Figure 4.1 Simple 2-D Truss Finite Element Model With 1 Bay and 1 Story

$$
K =
$$

	1	2	3	4	5	6	7	8
1	x	x	x	x	0.	0.	-0.0427	+0.032
2		x	x	x	x	x	x	x
3			x	x	x	x	x	X
4				x	x	x	x	X
5					x	x		
6						x		
7							x	X
8								X

(4.1)

$$\vec{F}_{bc} = \begin{Bmatrix} F_1 \\ F_2 \\ F_3 \\ F_4 \\ F_5 \\ F_6 \\ F_7 \\ F_8 \end{Bmatrix} = \begin{Bmatrix} 1^K \\ 2^K \\ 3^K \\ 4^K \\ 0'' \\ 0'' \\ 0'' \\ 0'' \end{Bmatrix} \tag{4.2}$$

Using the "conventional" (not truly sparse) finite element assembly process discussed in an earlier chapter, the total (system) stiffness 8×8 matrix can be given in Eq.(4.1). At this stage, the exact values for various terms of the symmetrical stiffness matrix, shown in Eq.(4.1), are not important. Thus, these non-zero values of many terms are simply represented by the symbol ×. The applied joint loads, including the prescribed Dirichlet boundary conditions (= O inch), are specified in Eq.(4.2). After incorporating the Dirichlet boundary conditions, the total stiffness matrix, given by Eq.(4.1), will be modified as:

$K_{bc} =$

	1	2	3	4	5	6	7	8
1	0.0833 + 0.0427	-0.032	-0.0833	0.00				
2		0.1111 + 0.0240	0.00	0.00				
3			0.0833 + 0.0427	0.032				
4				0.1111 + 0.0240				
5					1.0			
6						1.0		
7		S	Y	M			1.0	
8								1.0

(4.3)

In Figure 4.1, there are five truss elements (with NDOFPE = 4 dof per element), thus the element-dof connectivity information can be described by the following two, integer arrays:

$$
ie \left\{ \begin{array}{c} 1 \\ 2 \\ 3 \\ 4 \\ NEL = 5 \\ 6 = NEL + 1 \end{array} \right\} = \left\{ \begin{array}{c} 1 \\ 5 \\ 9 \\ 13 \\ 17 \end{array} \right\} \begin{array}{l} \nearrow \ 1 + ndofpe \\ \\ \searrow \ 5 + ndofpe \end{array} \quad \begin{array}{l} = \text{locations for} \\ \quad \text{connectivity infomation} \end{array} \tag{4.4}
$$

$$
je \left\{ \begin{array}{c} 1 \\ 2 \\ 3 \\ 4 \\ 5 \\ 6 \\ 7 \\ 8 \\ \bullet \\ \bullet \\ \bullet \\ \bullet \\ \bullet \\ 17 \\ 18 \\ 19 \\ NEL * ndofpe = 20 \end{array} \right\} = \left\{ \begin{array}{c} 1 \\ 2 \\ 3 \\ 4 \\ 1 \\ 2 \\ 5 \\ 6 \\ \bullet \\ \bullet \\ \bullet \\ \bullet \\ 3 \\ 4 \\ 5 \\ 6 \end{array} \right\} = \begin{array}{l} \text{global dof associated} \\ \text{with each and every element} \end{array} \tag{4.5}
$$

Eqs.(4.4) and (4.5) are the sparse representation of the following (element-dof connectivity) matrix:

$$
E = \begin{array}{c} \\ 1 \\ 2 \\ 3 \\ 4 \\ 5 \end{array} \begin{array}{c} \begin{array}{cccccccc} 1 & 2 & 3 & 4 & 5 & 6 & 7 & 8 \end{array} \\ \left[\begin{array}{cccccccc} x & x & x & x & & & & \\ x & x & & & & x & x & \\ & & x & x & & & x & x \\ x & x & & & & & x & x \\ & & x & x & x & x & & \end{array} \right] \end{array} \tag{4.6}
$$

The input Dirichlet boundary conditions for the finite element model, shown in Figure 4.1, are given as:

$$\text{ibdof} \begin{Bmatrix} 1 \\ 2 \\ 3 \\ 4 = \# \text{Dirichlet boundary dof} \end{Bmatrix} = \begin{Bmatrix} 5 \\ 6 \\ 7 \\ 8 \end{Bmatrix} \qquad (4.7)$$

The dof-element connectivity matrix $[E]^T$ can be obtained by transposing the matrix Eq.(4.6):

$$[E]^T =$$

	1	2	3	4	5
1	x	x		x	
2	x	x		x	
3	x		x		x
4	x		x		x
5		x			x
6		x			x
7			x	x	
8			x	x	

(4.8)

Eq.(4.8) can be described by the following two integer arrays:

$$\text{iet} \begin{pmatrix} 1 \\ 2 \\ 3 \\ 4 \\ 5 \\ 6 \\ 7 \\ 8 \\ N+1=9 \end{pmatrix} = \begin{Bmatrix} 1 \\ 4 \\ 7 \\ 10 \\ 13 \\ 15 \\ 17 \\ 19 \\ 21 \end{Bmatrix} \qquad (4.9)$$

$$
\text{jet}
\begin{pmatrix}
1 \\
2 \\
3 \\
4 \\
5 \\
6 \\
7 \\
8 \\
9 \\
10 \\
11 \\
12 \\
13 \\
14 \\
15 \\
16 \\
17 \\
18 \\
19 \\
20 = NEL * NDOFPE
\end{pmatrix}
=
\begin{Bmatrix}
1 \\
2 \\
4 \\
1 \\
2 \\
4 \\
1 \\
3 \\
5 \\
1 \\
3 \\
5 \\
2 \\
5 \\
2 \\
5 \\
3 \\
4 \\
3 \\
4
\end{Bmatrix}
\tag{4.10}
$$

4.3 Finite Element Sparse Assembly Algorithms for Symmetrical Matrices

Having incorporated the appropriate Dirichlet boundary conditions, the total stiffness matrix (shown in Eq.[4.3]) can be represented (in the sparse format) as follows:

$$
ia
\begin{pmatrix}
1 \\
2 \\
3 \\
4 \\
5 \\
6 \\
7 \\
8 \\
9 = N+1
\end{pmatrix}
=
\begin{Bmatrix}
1 \\
4 \\
6 \\
7 \\
7 \\
7 \\
7 \\
7 \\
7
\end{Bmatrix}
\tag{4.11}
$$

$$ja\begin{pmatrix} 1 \\ 2 \\ 3 \\ 4 \\ 5 \\ 6 = ncoef1 \end{pmatrix} = \begin{Bmatrix} 2 \\ 3 \\ 4 \\ 3 \\ 4 \\ 4 \end{Bmatrix} \qquad (4.12)$$

$$ad\begin{pmatrix} 1 \\ 2 \\ 3 \\ 4 \\ 5 \\ 6 \\ 7 \\ 8 = N \end{pmatrix} = \begin{Bmatrix} 0.126 \\ 0.135 \\ 0.126 \\ 0.135 \\ 1.000 \\ 1.000 \\ 1.000 \\ 1.000 \end{Bmatrix} \qquad (4.13)$$

$$an\begin{pmatrix} 1 \\ 2 \\ 3 \\ 4 \\ 5 \\ 6 = ncoef1 \end{pmatrix} = \begin{Bmatrix} -0.032 \\ -0.083 \\ 0.000 \\ 0.000 \\ 0.000 \\ 0.032 \end{Bmatrix} \qquad (4.14)$$

4.4 Symbolic Sparse Assembly of Symmetrical Matrices

Assuming the element-dof connectivity matrix [E] (see Eq.[4.6]), the dof-element connectivity matrix $[E]^T$ (see Eq.[4.8]), and the locations of the Dirichlet boundary conditions vector {ibdof} (see Eq.[4.7]) are known. The non-zero patterns of matrices [E] and $[E]^T$ can be described by the integer arrays {ie}, {je} and {iet}, {jet} as indicated in Eqs.(4.4, 4.5), and Eqs.(4.9, 4.10), respectively.

To facilitate subsequent discussion, information about the location of Dirichlet boundary conditions (see the integer array {ibdof}, shown in Eq.[4.7]) can be "slightly modified" and "copied" into the following integer array {ia}:

$$
ia \begin{pmatrix} 1 \\ 2 \\ 3 \\ 4 \\ 5 \\ 6 \\ 7 \\ 8 = N \end{pmatrix} = \begin{cases} 0 \\ 0 \\ 0 \\ 0 \\ N = 8 \\ N = 8 \\ N = 8 \\ N = 8 \end{cases} \qquad (4.15)
$$

It should be clear from Figure 4.1 that the prescribed Dirichlet boundary conditions occurred at node 3 (or dof 5, 6) and node 4 (or dof 7, 8). This information is reflected in Eq.(4.15) since the following definitions are used for array {ia} shown in Eq.(4.15):

$$
ia(-) = \begin{cases} N, \text{at those dof (such as dof \#5,6,7,8) that correspond to Dirichlet boundary conditions} \\ 0, \text{elsewhere (such as dof \#1,2,3,4)} \end{cases}
$$

$$(4.16)$$

The purpose of a "sparse symbolic assembly" algorithm can be summarized by the following problem:

Given N (= total number dof = 8, for the example shown in Figure 4.1), the connectivity information (such as arrays {ie}, {je}, {iet}, and {jet}, as indicated in Eqs.(4.4, 4.5, 4.9, 4.10), and the Dirichlet boundary condition information (see array {ia}, shown in Eq.[4.15]) find the locations of non-zero, off-diagonal terms of the upper-triangular portion of the total stiffness matrix (shown in Eq.[4.3]). In other words, the main objective of the "sparse symbolic assembly" phase is to find two integer arrays {ia}, and {ja}, such as shown in Eqs.(4.11, 4.12), for the example illustrated in Figure 4.1. The "key ideas" for the "sparse symbolic assembly" algorithm are summarized in Table 4.1. A complete FORTRAN code (with detailed explanations) for a sparse, symmetrical symbolic assembly process is presented in Table 4.2.

Using the example shown in Figure 4.1, and executing the FORTRAN code shown in Table 4.2, then the following information can be obtained:

[a] After processing row 1 of the total stiffness matrix $[K_{bc}]$ (see Eq.[4.3]), we have:

$$
IA \begin{pmatrix} 1 \\ 2 \\ 3 \\ 4 \\ \overline{5} \\ 6 \\ 7 \\ 8 = N \end{pmatrix} = \begin{Bmatrix} 1 \\ 1 \\ 1 \\ 1 \\ \overline{N=8} \\ N=8 \\ N=8 \\ N=8 \end{Bmatrix}
$$

After processing row 1 (see Eq.[4.3])

Initialized values

$$
JA \begin{pmatrix} JP = 1 \\ JP = 2 \\ JP = 3 \end{pmatrix} = \begin{Bmatrix} 2 \\ 3 \\ 4 \end{Bmatrix}
$$

$$JP = JP + 1 = 4$$

[b] After processing row 2 of the total stiffness matrix $[K_{bc}]$ (see Eq.[4.3]), we have:

$$
IA \begin{pmatrix} 1 \\ 2 \\ 3 \\ 4 \\ \overline{5} \\ 6 \\ 7 \\ 8 = N \end{pmatrix} = \begin{Bmatrix} 1 \\ 4 \\ 2 \\ 2 \\ \overline{8=N} \\ 8 \\ 8 \\ 8 \end{Bmatrix}
$$

After processing row 2 (see Eq.4.3)

Initialized values

$$
JA \begin{pmatrix} JP = 4 \\ JP = 5 \end{pmatrix} = \begin{Bmatrix} 3 \\ 4 \end{Bmatrix}
$$

$$JP = JP + 1 = 6$$

[c] After exiting subroutine symbass, we will obtain the final output arrays {ia} and {ja} as shown in Eq.(4.11), and Eq.(4.12), respectively. The number of non-zero, off-diagonal terms of the upper-triangular matrix $[K_{bc}]$, shown in Eq.(4.3), can be computed as:

$$NCOEF1 = IA (N+1) - 1 = 7 - 1 = 6 \tag{4.17}$$

4.5 Numerical Sparse Assembly of Symmetrical Matrices

To facilitate the discussion in this section, the following variables and arrays are defined (refer to the example, shown in Figure 4.1).

Input: $\overline{IA}, \overline{JA}$ (see Eqs.[4.11, 4.12]) Descriptions of non-zero, off-diagonal terms'

locations of matrix $[K_{bc}]$ (see Eq.[4.3])

$$\overline{IDIR} = \begin{cases} 1, \text{corresponds to Dirichlet b.c. location(s)} \\ 0, \text{elsewhere} \end{cases} = \begin{Bmatrix} 0 \\ 0 \\ 0 \\ 0 \\ 1 \\ 1 \\ 1 \\ 1 \end{Bmatrix} \qquad (4.18)$$

$\overline{AE}, \overline{BE}$ = Elements stiffness matrix and element load vector (if necessary),

respectively   (4.19)

$\overline{LM}$ = Global dof numbers, associated with an element   (4.20)

$$\text{Example: } LM^1 = \begin{Bmatrix} 1 \\ 2 \\ 3 \\ 4 \end{Bmatrix}; \; LM^2 = \begin{Bmatrix} 1 \\ 2 \\ 5 \\ 6 \end{Bmatrix}$$

NN = Size of Element Matrix (= 4, for 2-D truss element, see Figure 4.1)

$\overline{B}$ = values of applied nodal loads and/or values of prescribed Dirichlet

boundary conditions   (4.21)

Remarks

[1] The 4×4 element stiffness matrix (for 2-D truss element) should be computed and stored as a 1-D array (column - by - column fashion), such as:
$$AE\,(i, j) \equiv AE\,(\text{locate}) \qquad (4.22)$$

where
$$\text{locate} = i + (j-1)*(NN = 4) \qquad (4.23)$$

$$i, j = 1 \rightarrow 4 \qquad (4.24)$$

For example: AE (3,2) ≡ AE (7)   (4.25)

[2] For the example data shown in Figure 4.1, the vector $\overline{B}$ should be initialized to:

$$
B\begin{pmatrix} 1 \\ 2 \\ 3 \\ 4 \\ 5 \\ 6 \\ 7 \\ 8=N \end{pmatrix} = \begin{Bmatrix} 1^K \\ 2^K \\ 3^K \\ 4^K \\ O'' \\ O'' \\ O'' \\ O'' \end{Bmatrix} \tag{4.26}
$$

Output: $\overline{AN}, \overline{AD}$ = Values of off-diagonal and diagonal terms of $[K_{bc}]$, respectively

(see Eqs.[4.14, 4.13])

$\overline{B}$ = Right-Hand-Side (RHS) load vector (see Eq.[4.26])

Notes: $\overline{AD}$ (N) should be initialized to $\overline{0}$ before calling this routine
N = 8 $\equiv$ NDOF $\tag{4.27}$

Working Space: $\overline{IP}$ (N), should be initialized to $\overline{0}$ before calling this routine (for

more efficient implementation)

The "key ideas" for the "sparse numerical assembly" algorithm are summarized in Table 4.3.

A complete FORTRAN code (with detailed explanation) for sparse, symmetrical numerical assembly process is presented in Table 4.4. Important remarks about Table 4.4 are given in Table 4.5.

Table 4.1 Key Ideas for Sparse Symbolic Assembly Algorithms

Input : $\overrightarrow{IE}, \overrightarrow{JE}$	Element-Node Connectivity Information (see Eqs. 4.4, 4.5 and Figure 4.1)
$\overrightarrow{IET}, \overrightarrow{JET}$	Node-Element Connectivity Information (see Eqs. 4.9, 4.10)
N	Number of Nodes (or dof) in a finite element (FE) model
$\overrightarrow{IA(N)} = \begin{cases} N, \text{correspond to Dirichlet boundary conditions (bc) location (see Eq.4.15)} \\ 0, \text{elsewhere} \end{cases}$	

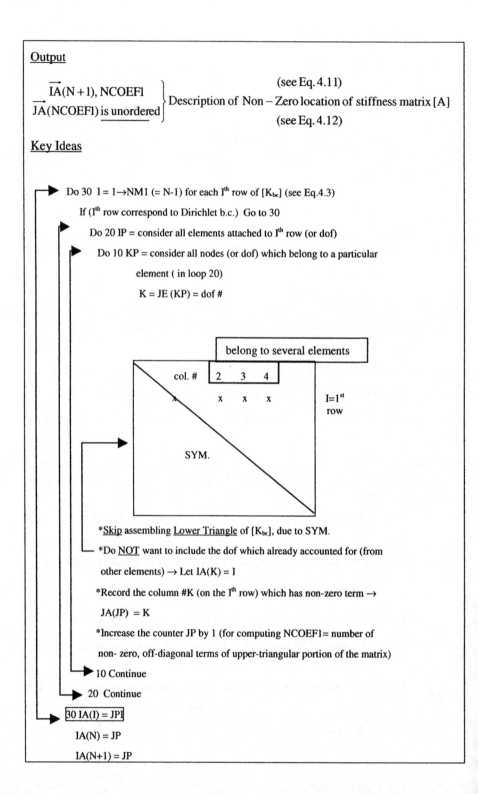

Output

$$\left.\begin{array}{l}\overrightarrow{IA}(N+1), NCOEF1 \\ \overrightarrow{JA}(NCOEF1) \text{ is } \underline{unordered}\end{array}\right\}\begin{array}{l}\text{(see Eq. 4.11)} \\ \text{Description of Non} - \text{Zero location of stiffness matrix [A]} \\ \text{(see Eq. 4.12)}\end{array}$$

Key Ideas

Do 30 I = 1→NM1 (= N-1) for each Ith row of $[K_{bc}]$ (see Eq.4.3)

If (I^{th} row correspond to Dirichlet b.c.) Go to 30

Do 20 IP = consider all elements attached to I^{th} row (or dof)

Do 10 KP = consider all nodes (or dof) which belong to a particular

element (in loop 20)

K = JE (KP) = dof #

belong to several elements

col. #   2   3   4

x   x   x   x $I = 1^{st}$
                              row

SYM.

*Skip assembling Lower Triangle of $[K_{bc}]$, due to SYM.

*Do NOT want to include the dof which already accounted for (from

other elements) → Let IA(K) = 1

*Record the column #K (on the Ith row) which has non-zero term →

JA(JP) = K

*Increase the counter JP by 1 (for computing NCOEF1= number of

non- zero, off-diagonal terms of upper-triangular portion of the matrix)

10 Continue

20 Continue

30 IA(I) = JP

IA(N) = JP

IA(N+1) = JP

Table 4.2 Complete FORTRAN Code for Sparse, Symmetrical, Symbolic Assembly Process

```
      subroutine symbass(ie,je,iet,jet,n,ia,ja)
      implicit real*8(a-h,o-z)
      dimension ie(*),je(*),iet(*),jet(*),ia(*),ja(*)


c++++++++++++++++++++++++++++++++++++++++++++++++++++++++++++++++++++++

c....PLEASE direct your questions to Dr. Nguyen (nguyen@cee.odu.edu)
c....Purposes: symmetrical, sparse symbolic assembly
c......This code is stored under file name *symb*.f, in sub-directory
c......cd ~/cee/newfem/complete*/part2.f


c++++++++++++++++++++++++++++++++++++++++++++++++++++++++++++++++++++++

c....Input: ie(nel+1)=locations (or pointers) of the first non-zero
c......      of each row (of element-dof connectivity info.)
c......      je(nel*ndofpe)=global dof column number for non-zero
c......      terms of each row (of element-dof connectivity info.)
c......      iet(ndof+1)=locations (pointers) of the first non-zero
c......      of each row (of dof-element connectivity info.)
c......      jet(nel*ndofpe)=locations (pointers) of the first non-
zero
c......      of each row (of dof-element connectivity info.)
c......      ia(ndof)= ndof in the positions correspond to Dirichlet
b.c.
c......                0  elsewhere
c....Output:ia(ndof+1)=starting locations of the first non-zero
c......      off-diagonal terms for each row of structural stiffness
c......      matrix
c......      ja(ncoeff)=column numbers (unordered) correspond to
c......      each nonzero, off-diagonal term of each row of
structural
c......      stiffness matrix


c++++++++++++++++++++++++++++++++++++++++++++++++++++++++++++++++++++++

      write(6,*) '                                                  '
      write(6,*) '***************************************************'
      write(6,*) '***************************************************'
      write(6,*) '                                                  '


c++++++++++++++++++++++++++++++++++++++++++++++++++++++++++++++++++++++

      jp=1                        !001
      nml=n-1                     !002
      do 30 i=1,nml               !003 last row (= eq) will be skipped
      jpi=jp                      !004 delayed counter for ia(-) array
      if ( ia(i) .eq. n ) go to 30 ! 005 skip row which correspond to
                                   ! Dirichlet b.c.
      ieta=iet(i)                 !006 begin index (to find how many elements)
      ietb=iet(i+1)-1             !007 end index (to find how many elements)
        do 20 ip=ieta,ietb ! 008 loop covering ALL elements attached
```

```
to
                             ! row i
          j=jet(ip)              !009 actual "element number" attached to row
i
          iea=ie(j)              !010 begin index (to find how many nodes
                                 !   attached to element j)


            ieb=ie(j+1)-1  !011 end index (to find how many nodes
attached
                             !   to element j)
             do 10 kp=iea,ieb
                                 !012 loop covering ALL nodes attached to
                                 !   element j
             k=je(kp)            !013 actual "node, or column number" attached
                                 !   to element j
          if (k .le. i) go to 10     !014 skip, if it involves with
                                          !LOWER triangular portion
          if ( ia(k) .ge. i ) go to 10  !015 skip, if same node
                                          !already been accounted by
                                          !earlier elements
          ja(jp)=k                 !016 record "column number" associated
                                 !   with non-zero off-diag. term
          jp=jp+1                  !017 increase "counter" for column
                                 !   number array ja(-)
          ia(k)=i                  !018 record node (or column number) k
                                 !   already contributed to row i
   10          continue           !019
   20          continue           !020
   30     ia(i)=jpi               !021  record "starting location" of
non-
                                 !   zero off-diag.
c                                !   terms associated with row i
          ia(n)=jp                !022  record "starting location" of
non-
                                 !   zero term of
                                 !   LAST ROW
          ia(n+1)=jp              !023  record "starting location"
                                 !   of non-zero term of
                                 !   LAST ROW + 1
c
          ncoef1=ia(n+1)-1
c         write(6,*) 'ia(-) array = ',(ia(i),i=1,n+1)
c         write(6,*) 'ja(-) array = ',(ja(i),i=1,ncoef1)
c
          return
          end
```

Table 4.3 Key Ideas for Sparse Numerical Assembly Algorithms

Note: Before calling this routine, initialize $\overline{IP} \begin{pmatrix} 1 \\ 2 \\ 3 \\ \cdot \\ Ndof \end{pmatrix} = \overline{0}$ (4.28)

Do 40 L = 1, NN(= Element local row dof#)
- Get I = global row dof # (correspond to L) = LM(L)
- If (I is Dirichlet b.c.) Go To 40 ($\Rightarrow$ skip)
- Assemble diagonal terms {AD} of $[K_{bc}]$ from [AE]
- Assemble the Ith row of {B}, _if we have load acting on element_
- KK = 0 $\Rightarrow$ Indicator that entire row 1 of [AE] has no contribution to $[K_{bc}]$

   For example, entire row I of [AE] is in lower triangle of [A]

Do 20 LL=1, NN (= element local column dof #)
- Find the location K of $[AE]_{L,LL}$ in the 1-D array of [AE]
- If (LL=L) Go To 20 $\Rightarrow$ skip, since diagonal term of [AE] already taken care before loop 20
- Get J = LM(LL) = global column dof #
- If (J is Dirichlet b.c.) Go To 10 $\Rightarrow$ skip, but modify the RHS vector B first
- If (J.LT.I) Go To 20 $\Rightarrow$ skip lower triangle
- IP(J) = K$\Rightarrow$ record col. # J(associate with row #I) & the correspond Kth location of [AE] = $[AE]_{L,LL}$
- KK = 1 $\Rightarrow$ Indicator, that row L of [AE] does have contribution to row 1 of $[K_{bc}]$ Go To 20

10   B(I) = B(I) –B(J)* AE(K) $\Rightarrow$ modify RHS B, for Dirich b. c. only

20   Continue

   If (KK = 0) Go To 40 $\Rightarrow$ if row L of [AE] has no contribution

       Do 30 J = IA(I) , IA(I+1) –1 $\Rightarrow$ pointer to col.# of row of $[K_{bc}]$

       Icol. # = JA(J)

       K = IP(Icol.#)

       If (K = 0) Go To 30 $\Rightarrow$ skip (the current element has nothing to do

                    with Icol# of $[K_{bc}]$

       AN(J) = AN(J) + AE(K) $\Rightarrow$ contribution of element stiffness (= AE) to

                    total stiffness (= AN)

       IP(Icol. #) = 0 $\Rightarrow$ initialize again, before considering next row
30   Continue
40 Continue

Table 4.4 Complete FORTRAN Code for Sparse, Symmetrical Numerical Assembly Process

```
C%%%%%%%%%%%%%%%%%%%%%%%%%%%%%%%%%%%%%%%%%%%%%%%%%%%%%%%%%%%%%%%%%%%%%%%%%

c       Table 4.4  Complete FORTRAN Code for Sparse, Symmetrical
c                  Numerical Assembly Process

        subroutine numass(ia,ja,idir,ae,be,lm,ndofpe,an,ad,b,ip)
        implicit real*8(a-h,o-z)
        dimension ia(*),ja(*),idir(*),ae(*),be(*),lm(*),an(*)
        dimension ad(*),b(*),ip(*)

C+++++++++++++++++++++++++++++++++++++++++++++++++++++++++++++++++++++++

c......PLEASE direct your questions to Dr. Nguyen (nguyen@cee.odu.edu)
c......Purposes: symmetrical, sparse numerical assembly
c......           This code is stored under file name *symb*.f, in sub-
directory
c......               ~

C+++++++++++++++++++++++++++++++++++++++++++++++++++++++++++++++++++++++

c......Input: ia(ndof+1)=starting locations of the first non-zero
c......        off-diagonal terms for each row of structural stiffness
c......        matrix
c......         ja(ncoeff)=column numbers (unordered) correspond to
c......        each nonzero, off-diagonal term of each row of structural
c......        stiffness matrix
c......        idir(ndof)= 1 in the positions correspond to Dirichlet b.c.
c......               0 elsewhere
c......         ae(ndofpe**2),be(ndofpe)= element (stiffness) matrix,
c......         and element (load) vector
c......        lm(ndofpe)= global dof associated with a finite element
c......         ndofpe= number of dof per element
c......        b(ndof)= before using this routine, values of b(-) should
c......        be initialized to:
c......        Ci, values of prescribed Dirichlet bc at proper locations
c......        or  values of applied nodal loads
c
c......Output: an(ncoeff1)= values of nonzero, off-diagonal terms of
c......         structural stiffness matrix
c......        ad(ndof)= values off diagonal terms of structural stiffness
c......        matrix
c......         b(ndof)= right-hand-side (load) vector of system of linear
c......         equations
c......Temporary Arrays:
c......        ip(ndof)= initialized to 0
c......               then IP(-) is used and reset to 0

C%%%%%%%%%%%%%%%%%%%%%%%%%%%%%%%%%%%%%%%%%%%%%%%%%%%%%%%%%%%%%%%%%%%%%%%%%

        do 40 L=1,ndofpe              !001 local "row" dof
        i=lm(L)                       !002 global"row" dof
        if ( idir(i) .ne. 0 ) go to 401 !003 skip, if DIRICHLET b.c.
        k=L-ndofpe                    !004 to find location of element k-
diag
        ad(i)=ad(i)+ae(k+L*ndofpe)    !005 assemble K-diag
        b(i)=b(i)+be(L)               !006 assemble element rhs load vector
        kk=0                          !007 flag, to skip contribution of
                                      !    entire
c                                     !    global row i if all global col #
c                                     !    j < i, or if entire row i belongs
c                                     !    to LOWER triangle
            do 20 LL=1,ndofpe         !008 local "column" dof
            k=k+ndofpe                !009 find location of element
```

```
stiffness k
          if (LL .eq. L) go to 20            !010 skip, diag. term already taken
care
          j=lm(LL)                           !011 global column dof
          if ( idir(j) .ne. 0 ) go to 10     !012 skip, if DIRICHLET b.c.
          if (j .lt. i) go to 20             !013 skip, if LOWER portion
          ip(j)=k                            !014 record global column # j
(associated
                                             !    with
c                                            !    global row #i) correspond to k-th
c                                            !    term
                                             !    of element stiffness k
          kk=1                               !015 FLAG, indicate row L of [k] do
have
c                                            !    contribution to global row I of
[K]
          go to 20                              !016
  10      b(i)=b(i)-b(j)*ae(k)               !017 modify rhs load vector due to
                                             !    DIRICHLET b.c.
  20      continue                           !018
          if (kk .eq. 0) go to 40            !019 skip indicator (see line 007)
          iaa=ia(i)                          !020 start index
          iab=ia(i+1)-1                      !021 end index
          do 30 j=iaa,iab                    !022 loop covering all col numbers
                                             !    associated
c                                            !    with global row i
          k=ip( ja(j) )                      !023 ip ( col # ) already defined on
line
                                             !    014
c                                            !    or initialized to ZERO initially
          if (k .eq. 0) go to 30             !024 skip, because eventhough row L of
                                             !    [k] do
c                                            !    have contribution to global row I
of
c                                            !    [K},
                                             !    some terms of row L (of [k])
which
c                                            !    associated
should                                       !    with DIRICHLET b.c. columns
                                             !    be SKIPPED
          an(j)=an(j)+ae(k)                  !025 assemble [K] from [k]
          ip( ja(j) )=0                      !026 reset to ZERO for col # j before
                                             !    considering
c                                            !    the next row L
  30      continue                           !027
          go to 40
 401      ad(i)=1.0                          !    reset the diagonal of K(i,i)=1.0,
                                             !    due to b.c.
  40      continue                           !028
c......print debugging results
c        ndof=12
c        ncoeff1=ia(ndof+1)-1
c        write(6,*) 'at the end of routine numass'
c        write(6,*) 'ia(-) array = ',(ia(i),i=1,ndof+1)
c        write(6,*) 'ja(-) array = ',(ja(i),i=1,ncoeff1)
c        write(6,*) 'ad(-) array = ',(ad(i),i=1,ndof)
c        write(6,*) 'an(i) array = ',(an(i),i=1,ncoeff1)
         return
         end

c%%%%%%%%%%%%%%%%%%%%%%%%%%%%%%%%%%%%%%%%%%%%%%%%%%%%%%
```

Table 4.5 Important Remarks about Table 4.4

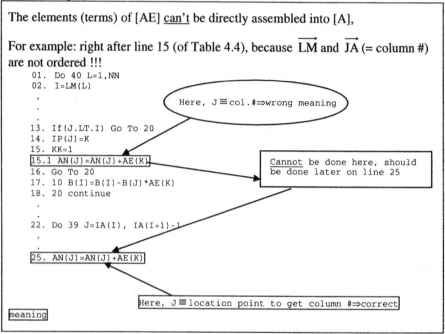

The elements (terms) of [AE] <u>can't</u> be directly assembled into [A],

For example: right after line 15 (of Table 4.4), because $\overrightarrow{LM}$ and $\overrightarrow{JA}$ (= column #) are not ordered !!!

```
01.  Do 40 L=1,NN
02.  I=LM(L)
 .
 .
 .
13.  If(J.LT.I) Go To 20
14.  IP(J)=K
15.  KK=1
15.1 AN(J)=AN(J)+AE(K)
16.  Go To 20
17.  10 B(I)=B(I)-B(J)*AE(K)
18.  20 continue
 .
 .
22.  Do 39 J=IA(I), IA(I+1)-1
 .
 .
25.  AN(J)=AN(J)+AE(K)
```

Here, J≡col.#⇒wrong meaning

Cannot be done here, should be done later on line 25

Here, J≡location point to get column #⇒correct meaning

4.6 Step-by-Step Algorithms for Symmetrical Sparse Assembly

The complete FORTRAN code, which consists of the main program and its sub-routines, for finite element sparse assembly is given in Tables 4.6, 4.7, 4.2, and 4.4. Outlines for this entire code are summarized in the following step-by-step algorithms:

Step 1: [a] Input joint coordinates

　　　　(see sub-routine jointc, Table 4.6)

　　　　[b] Input joint loads

　　　　(see sub-routine jointload, Table 4.6)

　　　　[c] Input element connectivity (see sub-routine elco01, Table 4.6)

　　　　for 2-D truss element, there are 4 dofpe, hence ndofpe = 4

　　　　[d] Input Dirichlet boundary conditions (see sub-routine boundary, Table 4.6)

Step 2: "Symbolic" sparse assembly

　　　　(a) Call transa2 ($nel, ndof,$ $\overrightarrow{ie}, \overrightarrow{je}, \overrightarrow{iet}, \overrightarrow{jet}$), see Table 4.7

　　　　(b) Call symbass ($\overrightarrow{ie}, \overrightarrow{je}, \overrightarrow{iet}, \overrightarrow{jet}, ndof, \overrightarrow{ia}, \overrightarrow{ja})$, see Table 4.2

The dimensions for output arrays: iet(ndof+1), jet(nel*ndofpe)

Step 3: "Numerical" Sparse Assembly

 Do 4 iel = 1, nel

 [a] Get the global dof (and stored in array $\overrightarrow{lm}$) associated with element # iel

 [b] Get element information (see call elinfo, Table 4.6), such as joint coordinates, material properties, etc.

 [c] Get element stiffness matrix $\overrightarrow{ae}$ (and element equivalent joint load, $\overrightarrow{be}$, if applicable) in global coordinate reference (see call elstif, Table 4.6)

 [d] Call numass ($\overrightarrow{ia}, \overrightarrow{ja}, \overrightarrow{iboundc}, \overrightarrow{ae}, \overrightarrow{be}, \overrightarrow{lm}, \overrightarrow{ndofpe}, \overrightarrow{an}, \overrightarrow{ad}, \overrightarrow{b}, \overrightarrow{ip}$), see Table 4.4

 where

$$ip \ (ndof+1) = \text{temporary working array} = \overrightarrow{O} \ (\text{initially}) \qquad (4.29)$$

$$iboundc \begin{pmatrix} 1 \\ 2 \\ 3 \\ 4 \\ 5 \\ 6 \\ 7 \\ 8 \end{pmatrix} = \begin{cases} 0 \\ 0 \\ 0 \\ 0 \\ 1 = \text{Dirichlet b.c.} \\ 1 = \text{Dirichlet b.c.} \\ 1 = \text{Dirichlet b.c.} \\ 1 = \text{Dirichlet b.c.} \end{cases} \qquad (4.30)$$

$$\overrightarrow{b} \text{ should be initialized as } b \begin{pmatrix} 1 \\ 2 \\ 3 \\ 4 \\ 5 \\ 6 \\ 7 \\ 8 \end{pmatrix} = \begin{cases} 1^K \\ 2^K \\ 3^K \\ 4^K \\ 0'' \\ 0'' \\ 0'' \\ 0'' \end{cases} \qquad (4.31)$$

 4   Continue

Remarks

(1) After completing the "sparse" assembly (including the imposed Dirichlet boundary conditions), the matrix $[K_{bc}]$ and right-hand-side (load) vector $\vec{F}_{bc}$ can be "directly" generated and stored by the following arrays:

$\{ia\}$, see Eq.(4.11)

$\{ja\}$, see Eq.(4.12)

$\{ad\}$, see Eq.(4.13)

$\{an\}$, see Eq.(4.14)

$\{B\}$, see Eq.(4.26)

(2) If non-zero (say = 1.00″) prescribed displacements are specified at each Dirichlet (support) boundary DOF, then the modified RHS load vector $\{B\}$ (see Eq.[4.26]) will become:

$$\vec{F}_{bc} = \begin{Bmatrix} F_1 \\ F_2 \\ F_3 \\ F_4 \\ F_5 \\ F_6 \\ F_7 \\ F_8 \end{Bmatrix} = \begin{Bmatrix} 1.0107^K \\ 2.1031^K \\ 3.0747^K \\ 4.1671^K \\ 1.00'' \\ 1.00'' \\ 1.00'' \\ 1.00'' \end{Bmatrix} \begin{matrix} \rightarrow 1^K - (K_{15}=0)(Z_5=1') - (K_{16}=0)(Z_6=1') \\ - (K_{17}=-0.0427)(Z_7=1'') \\ - (K_{18}=0.032)(Z_8=1'') \\ \\ \\ \\ \\ \end{matrix} \qquad (4.32)$$

(3) The following example (2 Bay-1 Story) can be used to further validate the above step-by-step sparse assembly procedure.

$E = 1K/in.^2$

$A = 1in.^2$

Height $= 9''$

Base $= 12''$

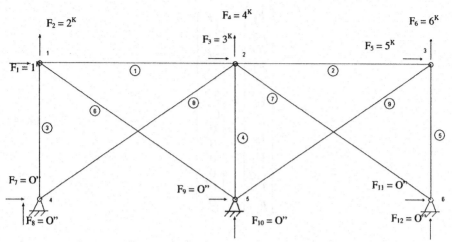

Figure 4.2 2-Bay, 1-Story Truss Structure

A 2-D truss structure with the applied loads, Dirichlet support boundary conditions, element connectivities, material properties, and the geometrical dimensions are all given in Figure 4.2. The non-zero terms for the total stiffness matrix (before imposing the Dirichlet boundary conditions) are indicated by Eq.(4.33). The applied joint load vector (including the prescribed boundary displacements) is given by Eq.(4.34). The total (= system) stiffness matrix, after imposing boundary conditions, is shown in Eq.(4.35).

		1	2	3	4	5	6	7	8	9	10	11	12
	1	x	x	x	x			x	x	x	x		
	2		x	x	x			x	x	x	x		
	3			x	x	x	x	x	x	x	x	x	x
	4				x	x	x	x	x	x	x	x	x
[K] =	5					x	x			x	x	x	x
	6						x			x	x	x	x
	7							x	x				
	8								x				
	9		S	Y	M					x	x		
	10									x	x		
	11											x	x
	12												X

(4.33)

$$F_{bc} = \begin{Bmatrix} F_1 \\ F_2 \\ F_3 \\ F_4 \\ F_5 \\ F_6 \\ F_7 \\ F_8 \\ F_9 \\ F_{10} \\ F_{11} \\ F_{12} \end{Bmatrix} = \begin{Bmatrix} 1^K \\ 2^K \\ 3^K \\ 4^K \\ 5^K \\ 6^K \\ 0'' \\ 0'' \\ 0'' \\ 0'' \\ 0'' \\ 0'' \end{Bmatrix} \tag{4.34}$$

$[K_{bc}] =$

	1	2	3	4	5	6	7	8	9	10	11	12
1	x	1st	2nd	3rd								
2		x	4th	5th								
3			X	6th	7th	8th						
4				x	9th	10th						
5					x	11th						
6						X						
7							1.0					
8								1.0				
9									1.0			
10										1.0		
11											1.0	
12												1.0

(4.35)

(4) Different finite element types can also be easily incorporated into the presented sparse assembly algorithms. As an example, the 1-Bay, 1-Story truss structure with five truss elements (shown in Figure 4.1) is now added with two triangular elements as illustrated in Figure 4.3.

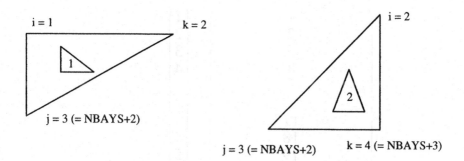

Figure 4.3 Two Triangular Elements Added to Figure 4.1

The "artificial" or "faked" element stiffness matrix for the triangular finite element can be given as:

$$
\left[k^{(e)} \right]_{6\times6} = \left[k_{i,j} \right] = \begin{array}{c} \\ 1 \\ 2 \\ 3 \\ 4 \\ 5 \\ 6 \end{array}
\begin{array}{cccccc}
1 & 2 & 3 & 4 & 5 & 6 \\
\left[\begin{array}{cccccc}
2 & 3 & 4 & 5 & 6 & 7 \\
3 & 4 & 5 & 6 & 7 & 8 \\
4 & 5 & 6 & 7 & 8 & 9 \\
5 & 6 & 7 & 8 & 9 & 10 \\
6 & 7 & 8 & 9 & 10 & 11 \\
7 & 8 & 9 & 10 & 11 & 12
\end{array}\right]
\end{array}
\tag{4.36}
$$

where i, j = 1,2,...6 and $[k_{i,j}] = i + j$ (4.37)

The input element connectivity arrays for this mixed finite element model (for NBAYS = 1 = NSTORY) becomes:

$$
ie \left\{
\begin{array}{c}
1 \\
2 \\
3 \\
4 \\
5 = NELTYPE(1) \\
6 \\
7 = NELTYPE(1) + NELTYPE(2) \\
8 = \left[\sum_i NELTYPE(i) \right] + 1
\end{array}
\right\} = \left\{
\begin{array}{c}
1 \\
5 \\
9 \\
13 \\
17 \\
21 \\
27 \\
33
\end{array}
\right\}
\tag{4.38}
$$

$$
je \begin{pmatrix}
1 \\
2 \\
3 \\
4 \\
\\
17 \\
18 \\
19 \\
20 = \text{NELTYPE(1)} * 4 \\
\\
21 \\
22 \\
23 \\
24 \\
25 \\
26 \\
\\
27 \\
28 \\
29 \\
30 \\
31 \\
32 = \text{ncoef1}
\end{pmatrix}
=
\begin{Bmatrix}
1 \\
2 \\
3 \\
4 \\
\\
3 \\
4 \\
5 \\
6 \\
\\
1 \\
2 \\
5 \\
6 \\
3 \\
4 \\
\\
3 \\
4 \\
5 \\
6 \\
7 \\
8
\end{Bmatrix}
\qquad (4.39)
$$

since NELTYPE(1) = 5 truss elements with ndofpe = 4 (= number of dof per element)

NELTYPE(2) = 2 triangular elements with ndofpe = 6

$$\text{ncoef1} = ie(8)-1 = 33-1 = 32 = \sum_{i=1}^{NELTYPES} NDOFPE_i *NELTYPE(i)$$

where NLETYPE(1), and NELTYPE(2) represents the number of finite element type 1 (= 2-D truss element), and type 2 (= 2-D triangular element), respectively.

The general flowchart for symmetrical sparse assembly process for mixed finite element model is outlined in Figure 4.4.

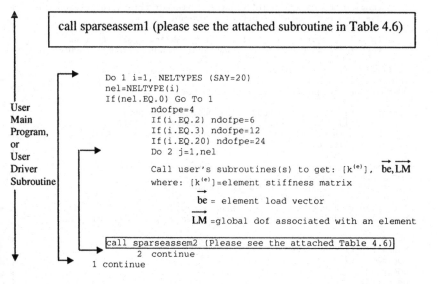

Figure 4.4 Outlines for Symmetrical Sparse Assembly Process with Mixed Finite

Element Types

Using the computer program, shown in Tables (4.6, 4.7, 4.2, and 4.4), the input and output data (arrays IA, JA, AD, AN, and B) are shown in Table 4.8 for the mixed finite element model shown in Figures 4.1 and 4.3.

Table 4.6 Inputs and Outputs for Mixed (2-D Truss and Triangular) Finite Element Sparse Assembly Process

```
c       Table 4.6  Main Program and Its Subroutines For Symmetrical
c                      Finite Element Sparse Assembly
c
c......version: June 1, 2002 ---->> part1.f
c*******************************************************************
c......PLEASE do "NOT" distribute this code to any persons, or any
c       organizations without explicit written agreement from
c       Dr. Duc T. NGUYEN (Tel= 757-683-3761,  dnguyen@odu.edu)
c       This file is stored under ~/cee/*commer*/part1.f
c*******************************************************************
c......PURPOSES:
c......(a)   Simple demonstration to users on EFFICIENT SPARSE ASSEMBLY
c......(b)   Simple 2-D "artificial" truss, 2-d triangular elements
c......      /material/load data is "automatically" generated
c......      Small examples (5 "real" truss & 2 "faked" triangular elements
c       has been successfully/correctly assembled.
c......(c)   For "testing" users
c----->      Part 1 = "source codes" of main program/"several" routines
c----->      Part 2 = "object codes" of routines: TRANSA2, SYMBASS, NUMASS
c
C++++++++++++++++++++++++++++++++++++++++++++++++++++++++++++++++++++
        implicit real*8(a-h,o-z)
        common/junk1/height,base,ireal,nbays,nstory
c
        dimension b(1000), ibdof(1000), ia(1000), ir(1000000),
      $            elk(24,24), be(24), ae(576), lm(24), ad(1000),
      $            x(1000), y(1000), z(1000), ja(1000000), an(1000000),
      $            neltype(20)
c
C++++++++++++++++++++++++++++++++++++++++++++++++++++++++++++++++++++
c......input general information
        nbays=1        ! User just needs to change this input
        nstory=1       ! User just needs to change this input
        bcdispl=0.0    ! User inputs zero prescribed displ. at each bc dof
c       bcdispl=1.0    !             prescribed displ. at each bc dof
c
        height=9.0
        base=12.0
        ndofpemax=24
        ndofpn=2
        neltypes=20
        mtoti=1000000
c
        nboundc=(nbays+1)*ndofpn
        nel1=nstory*(4*nbays+1)
        nodes=(nbays+1)*(nstory+1)
        ndof=nodes*ndofpn
c
        do 202 i=1,neltypes
 202    neltype(i)=0
        neltype(1)=nel1                ! # of 2-d truss elements
        neltype(2)=2                   ! # of 2-d triangular elements
c
        write(6,*) 'Date:  June 1, 2002'
        write(6,*) '   '
        write(6,*) 'nbays, nstory, # real truss/faked triangular el. = '
        write(6,*) nbays, nstory, neltype(1), neltype(2)
c
        call jointc(x,y,z,nodes)
        call jointload(ndof,b)
C++++++++++++++++++++++++++++++++++++++++++++++++++++++++++++++++++++
        nelcum=0        ! total # elements (all element types)
        ndofpenel=0
c
        do 203 i=1,neltypes
```

```
          if ( neltype(i) .ne. 0 ) then
            if (i .eq. 1) ndofpe=4
            if (i .eq. 2) ndofpe=6
          nelcum=nelcum+neltype(i)
          ndofpenel=ndofpenel+ndofpe*neltype(i)
          endif
 203    continue
c
        iebeg=1
        jebeg=iebeg+(nelcum+1)
        ibcbeg=jebeg+(ndofpenel)
        ietbeg=ibcbeg+ndof
        jetbeg=ietbeg+(ndof+1)
        ipbeg=jetbeg+(ndofpenel)
        last=ipbeg+ndof
        if (mtoti .lt. last) then
         write(6,*) 'error: not enough memory for mtoti !!!'
        endif
        call elco01(ir(jebeg),nel1,ir(iebeg),neltypes,neltype)
C+++++++++++++++++++++++++++++++++++++++++++++++++++++++++++++++
c......input (or "artificially" generated) connectivity information
c......{ie} = locations for the connectivity information
c......       = 1, 1+1*ndofpe, 1+2*ndofpe, 1+3*ndofpe, etc...
c......       = 1, 5,         9,          13,        17, etc...
c
        write(6,*) 'ie= ', (ir(iebeg-1+i),i=1,nelcum+1)
c......{je} = global dof associated with EACH and every element
c......       = 13, 14, 1, 2,      13, 14, 7, 8,      etc...
        write(6,*) 'je = ', (ir(jebeg-1+i),i=1,ndofpenel)
c......initialize the boundary condition flag, and rhs load vector
        do 1 i=1,ndof
        ir(ibcbeg-1+i)=0
 1      continue
c......input (or "artificially" generated) boundary dof
c......{ibdof} = the global dof associated with DIRICLET boundary conditions
c......          = 13, 14, 15, 16
        call boundary(nodes,nbays,ibdof)

        write(6,*) ' ibdof = ', (ibdof(i),i=1,nboundc)
c......modify the boundary condition flag
        do 2 i=1,nboundc
        j=ibdof(i)
        ir(ibcbeg-1+j)=1 ! will be used later in sparse numerical assembly
        b(j)=bcdispl ! prescribed dirichlet bc
 2      continue
C+++++++++++++++++++++++++++++++++++++++++++++++++++++++++++++++++
        call sparseassem1(nelcum,ndof,ndofpenel,mtoti,ir,iflag,
      $ ae,be,lm,ia,ja,ad,an,b,ndofpe)
C+++++++++++++++++++++++++++++++++++++++++++++++++++++++++++++++++
        do 106 i=1,neltypes
          if ( neltype(i) .eq. 0 ) go to 106
            if (i .eq. 1) then
            nel=neltype(i)
            ndofpe=4
            elseif (i .eq. 2) then
            nel=neltype(i)
            ndofpe=6
            endif
c......loop for all elements
        do 4 iel=1,nel
c......get the global dof associated with each element
            if (i .eq. 1) then
        iend=iel*ndofpe
        istart=iend-ndofpe
        do 76 j=1,ndofpe
c76     lm(j)=je(istart+j)
        lm(j)=ir(jebeg-1+istart+j)
 76     continue
c
        call elinfo(iel,x,y,z,ir(jebeg),youngm,areaa,xl,cx,cy,cz,lm)
```

```
        m=ndofpemax
        call elstif(iel,youngm,areaa,xl,cx,cy,elk,ndofpe,m,nel)
        if(iel .eq. 1 .or. iel .eq. nel) then
        write(6,*) ' elem # ', iel, '*** passed elstiff ***'
        endif
c
        go to 678
            elseif (i .eq. 2) then
c......     get lm(-), elk(-,-), and be(-) = {0} for 2-d triangular element
type
c-------------------------------------------------------
c......"artificial" 2-d triangular elements structural data
c......  hard coded element connectivities, elk(-,-), and be(-) !
        do 86 j=1,ndofpe
c86     lm(j)=je(istart+j)
        locate=jebeg-1+neltype(1)*4+(iel-1)*ndofpe+j
        lm(j)=ir(locate)
c
        be(j)=0.
c
            do 88 k=1,ndofpe
            elk(j,k)=j+k
 88         continue
c
 86     continue
c-------------------------------------------------------
            go to 678
            endif
c......convert element stiffness from 2-D array into 1-D array
 678    continue
        do 78  j=1,ndofpe
        do 77 jj=1,ndofpe
        locate=jj+(j-1)*ndofpe
        ae(locate)=elk(jj,j)
 77     continue
 78     continue
c......sparse "numerical" assembly
        call sparseassem2(nelcum,ndof,ndofpenel,mtoti,ir,iflag,
     $ ae,be,lm,ia,ja,ad,an,b,ndofpe)
        if(iel .eq. 1 .or. iel .eq. nel) then
        write(6,*) ' elem # ', iel, '*** passed numass ***'
        endif
c......
 4      continue
 106    continue    ! different element type loop
c......
        ncoef1=ia(ndof+1)-ia(1)
        write(6,*) 'ndof,ncoef1 = ',ndof,ncoef1
        write(6,*) 'ia(ndof+1)= ',(ia(i),i=1,ndof+1)
        write(6,*) 'ja(ncoef1)= ',(ja(i),i=1,ncoef1)
        write(6,*) 'ad(ndof)= ',(ad(i),i=1,ndof)
        write(6,*) 'an(ncoef1)= ',(an(i),i=1,ncoef1)
        write(6,*) 'b(ndof)= ',(b(i),i=1,ndof)
        stop
        end
c%%%%%%%%%%%%%%%%%%%%%%%%%%%%%%%%%%%%%%%%%%%%%%%%%%%%%%%%%%%%
        subroutine jointc(x,y,z,nodes)
        implicit real*8(a-h,o-z)
c******purposes: input joint coordinates
        dimension x(1),y(1),z(1)
        common/junk1/height,base,ireal,nbays,nstory
c-------------------------------------------------------
c......"artificial" structural datas
        joint=0
        do 11 i=1,nstory+1
        ycor=(nstory+1-i)*height
        xcor=-base
        do 12 j=1,nbays+1
        joint=joint+1
        z(joint)=0.
```

```
        xcor=xcor+base
        x(joint)=xcor
        y(joint)=ycor
 12     continue
 11     continue
c......
        write(6,*) 'joint coordinates'
        do 21 i=1,nodes
        write(6,*) i,x(i),y(i),z(i)
 21     continue
        return
        end
c%%%%%%%%%%%%%%%%%%%%%%%%%%%%%%%%%%%%%%%%%%%%%%%%%%%%%%%%%%%%%%%%%%
        subroutine elco01(je,nel,ie,neltypes,neltype)
        implicit real*8(a-h,o-z)
c******purpose: input element connectivity
c******explain:
        dimension je(*),ie(*),neltype(*)
        common/junk1/height,base,ireal,nbays,nstory
c
        do 102 ii=1,neltypes
        if ( neltype(ii) .ne. 0 ) then
          if (ii .eq. 1) then
            ndofpe=4
c----------------------------------------------------------
c......"artificial" 2-d truss elements structural data
c......(first) horizontal members
        memhor=nbays*nstory
        ielnum=0
        istart=-nbays
        do 11 i=1,nstory
        istart=istart+(nbays+1)
        do 12 j=1,nbays
        ielnum=ielnum+1
C++++++++++++++++++++++++++++++++++
        nodei=istart+j-1
        nodej=istart+j
          ielstart=(ielnum-1)*4
          je(ielstart+1)=nodei*2-1
          je(ielstart+2)=nodei*2
          je(ielstart+3)=nodej*2-1
          je(ielstart+4)=nodej*2
C++++++++++++++++++++++++++++++++++
 12     continue
 11     continue
c......(second) vertical members
        memver=nstory*(nbays+1)
        ielnum=0
        do 21 i=1,nbays+1
        do 22 j=1,nstory
        ielnum=ielnum+1
        iel=ielnum+memhor
        nodei=i+(nbays+1)*(j-1)
        nodej=nodei+(nbays+1)
C++++++++++++++++++++++++++++++++++
          ielstart=(iel-1)*4
          je(ielstart+1)=nodei*2-1
          je(ielstart+2)=nodei*2
          je(ielstart+3)=nodej*2-1
          je(ielstart+4)=nodej*2
C++++++++++++++++++++++++++++++++++
 22     continue
 21     continue
c......(third) diagonal members
        memhv=memhor+memver
        ielnum=0
        incr=nbays*nstory
        do 31 i=1,nbays
        do 32 j=1,nstory
        ielnum=ielnum+1
```

```
          iel=ielnum+memhv
c......diagonal members of type "\\\\\\\\"
          nodei=i+(nbays+1)*(j-1)
          nodej=nodei+(nbays+2)
C+++++++++++++++++++++++++++++++++++
          ielstart=(iel-1)*4
          je(ielstart+1)=nodei*2-1
          je(ielstart+2)=nodei*2
          je(ielstart+3)=nodej*2-1
          je(ielstart+4)=nodej*2
C+++++++++++++++++++++++++++++++++++
c......diagonal members of type "/////////"
c         nodeij(iel+incr,1)=nodeij(iel,1)+1
c         nodeij(iel+incr,2)=nodeij(iel,2)-1
          ielstart=(iel-1)*4
          iwhat=je(ielstart+2)/2
          jwhat=je(ielstart+4)/2
          nodei=iwhat+1
          nodej=jwhat-1
c
          ielstart=(iel+incr-1)*4
C+++++++++++++++++++++++++++++++++++
          je(ielstart+1)=nodei*2-1
          je(ielstart+2)=nodei*2
          je(ielstart+3)=nodej*2-1
          je(ielstart+4)=nodej*2
C+++++++++++++++++++++++++++++++++++
 32       continue
 31       continue
c......
          write(6,*) 'element connectivity nodei&j'
          do 61 i=1,nel
          istart=(i-1)*4
          write(6,*) i,je(istart+1),je(istart+2),je(istart+3),je(istart+4)
 61       continue
C+++++++
          ie(1)=1
          do 71 i=1,nel
          ie(i+1)=ie(i)+ndofpe
 71       continue
          go to 102
c-------------------------------------------------------
c......"artificial" 2-d triangular elements structural data
c...... hard coded element connectivities !
          elseif (ii .eq. 2) then
          ndofpe=6
          ie( neltype(1)+1 )=4*neltype(1)+1
          ie( neltype(1)+2 )=4*neltype(1)+7
          ie( neltype(1)+3 )=4*neltype(1)+13
          je( 4*neltype(1)+1 )=1
          je( 4*neltype(1)+2 )=2
          je( 4*neltype(1)+3 )=(nbays+2)*2-1
          je( 4*neltype(1)+4 )=(nbays+2)*2
          je( 4*neltype(1)+5 )=3
          je( 4*neltype(1)+6 )=4
          je( 4*neltype(1)+7 )=3
          je( 4*neltype(1)+8 )=4
          je( 4*neltype(1)+9 )=(nbays+2)*2-1
          je( 4*neltype(1)+10)=(nbays+2)*2
          je( 4*neltype(1)+11)=(nbays+3)*2-1
          je( 4*neltype(1)+12)=(nbays+3)*2
c-------------------------------------------------------
          endif
          endif
 102      continue
c
 999      return
          end
c%%%%%%%%%%%%%%%%%%%%%%%%%%%%%%%%%%%%%%%%%%%%%%%%%%%%%
          subroutine boundary(nodes,nbays,ibdof)
```

```
          dimension ibdof(*)
          lastbn=nodes
          ifirstbn=lastbn-nbays
          icount=0
          do 35 nodebc=ifirstbn,lastbn
          icount=icount+1
          ibdof(icount)=nodebc*2-1
c
          icount=icount+1
          ibdof(icount)=nodebc*2
 35       continue
          return
          end
c%%%%%%%%%%%%%%%%%%%%%%%%%%%%%%%%%%%%%%%%%%%%%%%%%%%%%%%%%%%
          subroutine elstif(ithel,youngm,areaa,xl,cx,cy,elk,ndofpe,m,nel)
          implicit real*8(a-h,o-z)
c******purpose: to generate element stiffness matrix
c******explain:
          dimension elk(m,m)
c-------------------------------------------
          c=youngm*areaa/xl
c......element stiffness matrix for 2-D "truss"
          elk(1,1)=c*cx*cx
          elk(2,2)=c*cy*cy
          elk(1,2)=c*cx*cy
          elk(3,3)=elk(1,1)
          elk(2,3)=-elk(1,2)
          elk(1,3)=-elk(1,1)
          elk(4,4)=elk(2,2)
          elk(3,4)=elk(1,2)
          elk(2,4)=-elk(2,2)
          elk(1,4)=-elk(1,2)
c......
          do 1 i=1,ndofpe
          do 2 j=i,ndofpe
          elk(j,i)=elk(i,j)
 2        continue
 1        continue
c......
c          if(ithel .eq. 1 .or. ithel .eq. nel) then
c          write(6,*) 'el. stiff [k] for member ',ithel
c          write(6,*) 'youngm,areaa,xl,c,cx,cy,cz= '
c          write(6,*) youngm,areaa,xl,c,cx,cy,cz
c          do 33 i=1,ndofpe
c          write(6,34) (elk(i,j),j=1,ndofpe)
c34        format(2x,6e13.6)
c33        continue
c          endif
c------------------------
          return
          end
c%%%%%%%%%%%%%%%%%%%%%%%%%%%%%%%%%%%%%%%%%%%%%%
          subroutine elinfo(ithel,x,y,z,je,youngm,areaa,xl,cx,cy,cz,lm)
          implicit real*8(a-h,o-z)
c******purpose: to obtain all necessary information for element stiffness
c******          generation
c******explain:
          dimension x(1),y(1),z(1),je(*),lm(*)
c-----------------------------------------------------------------------
          youngm=1.00
          areaa=1.00
c......
c          nodei=nodeij(ithel,1)
c          nodej=nodeij(ithel,2)
          ielstart=(ithel-1)*4
          nodei=je(ielstart+2)/2
          nodej=je(ielstart+4)/2
c
          lm(1)=je(ielstart+1)
          lm(2)=je(ielstart+2)
```

```
        lm(3)=je(ielstart+3)
        lm(4)=je(ielstart+4)
c......
        deltax=x(nodej)-x(nodei)
        deltay=y(nodej)-y(nodei)
        deltaz=z(nodej)-z(nodei)
        xl=deltax**2+deltay**2+deltaz**2
        xl=dsqrt(xl)
        cx=deltax/xl
        cy=deltay/xl
        cz=deltaz/xl
c......
c       write(6,*)'youngm,areaa,xl,cx,cy,cz,ithel = ',youngm
c     $,areaa,xl,cx,cy,cz,ithel
        return
        end
c%%%%%%%%%%%%%%%%%%%%%%%%%%%%%%%%%%%%%%%%%%
        subroutine jointload(ndof,b)
        implicit real*8(a-h,o-z)
        dimension b(*)
c......generate "artificial" horizontal force in the x-direction
c......at dof #1 of 2-D truss structure
        do 1 i=1,ndof
        b(i)=i
  1     continue
        return
        end
c@@@@@@@@@@@@@@@@@@@@@@@@@@@@@@@@@@@@@@@@@@@@@@@@@@@@@@@@@@@@@@@@@@@@
c  NO  SOURCE  CODE  FROM  HERE  DOWN  !
c@@@@@@@@@@@@@@@@@@@@@@@@@@@@@@@@@@@@@@@@@@@@@@@@@@@@@@@@@@@@@@@@@@@@
        subroutine sparseassem1(nel,ndof,ndofpenel,mtoti,ir,iflag,
     $  ae,be,lm,ia,ja,ad,an,b,ndofpe)
        implicit real*8(a-h,o-z)
c++++++++++++++++++++++++++++++++++++++++++++++++++++++++++++++++++++++++
c......Purposes:
c...... Symbolic sparse assembly (for symmetrical matrix)
c       according to ODU sparse formats
c......Author(s), Dated Version:
c       Prof. Duc T. NGUYEN, June 1'2002
c......Input Parameters:
c       nel = total (cumulative) number of finite elements (including
c             all different element types
c       ndof = total # degree-of-freedom (dof), including all Dirichlet b.c.
c       ndofpenel = summation of (ndofpe type1 * nel type1 +
c                                 ndofpe type2 * nel type2 + etc...)
c       mtoti = available INTEGER words of memory
c       ir(mtoti) = ir(iebeg=1,..., jebeg-1) = same as ie(-) array
c                 = ir(jebeg ,..., ibcbeg-1) = same as je(-) array
c                 = ir(ibcbeg,..., ietbeg-1) = Dirichlet b.c. = either 0 or 1
c
c       iflag(40) = unused for now
c       ae(ndofpemax**2) = unused for now
c       be(ndofpemax) = unused for now
c       lm(ndofpemax) = unused for now
c       b(ndof) = unused for now
c       ndofpe = # dof per element (correspond to a specific finite element
type
c
c......Output Parameters:
c       ir(mtoti) = ir(ietbeg,..., jetbeg-1) = same as ie(-) transpose  array
c                 = ir(jetbeg,..., ipbeg-1)  = same as je(-) transpose  array
c                 = ir(ipbeg,..., last-1)    = same as ip(-) array
c       ia(ndof+1) = starting location of FIRST non-zero, off-diag. term
c                    for each row of the assembled sparse (symmetrical) matrix
c       ja(ncoef1) = column numbers associated with non-zero terms on each row
c                    where ncoef1 = ia(ndof+1)-1 of the assembled sparse
c (sym.)
c                    matrix
c       ad(ndof) = unused for now
c       an(ncoef1) = unused for now
```

```
c          b(ndof) = unused for now
c......Temporary (working) Parameters:
c          ir(mtoti) = ir(iebeg=1,..., jebeg-1, ..., mtoti)
c
C++++++++++++++++++++++++++++++++++++++++++++++++++++++++++++++++++++++++++++++
        dimension ir(*), iflag(*),ae(*),be(*),lm(*),
     $             ia(*),ja(*),ad(*),an(*),b(*)
c
        iebeg=1
        jebeg=iebeg+(nel+1)
        ibcbeg=jebeg+(ndofpenel)
        ietbeg=ibcbeg+ndof
        jetbeg=ietbeg+(ndof+1)
        ipbeg=jetbeg+(ndofpenel)
        last=ipbeg+ndof
        if (mtoti .lt. last) then
         write(6,*) 'error: not enough memory for mtoti !!!'
        endif
c
c......copy boundary flag for later usage in Duc's sparse symbolic assembly
        do 3 i=1,ndof
        if ( ir(ibcbeg-1+i) .eq. 0 ) ia(i)=0
        if ( ir(ibcbeg-1+i) .eq. 1 ) ia(i)=ndof
 3      continue
c
c        write(6,*) 'just before call transa2: ie(-)= '
c        write(6,*) (ir(iebeg-1+i),i=1,nel+1)
c        write(6,*) 'just before call transa2: je(-)= '
c        write(6,*) (ir(jebeg-1+i),i=1,ndofpenel)
        call transa2(nel,ndof,ir(iebeg),ir(jebeg),ir(ietbeg),ir(jetbeg))
c        write(6,*) 'just after call transa2: iet(-)= '
c        write(6,*) (ir(ietbeg-1+i),i=1,ndof+1)
c        write(6,*) 'just after call transa2: jet(-)= '
c        write(6,*) (ir(jetbeg-1+i),i=1,ndofpenel)
        write(6,*) '*** passed transa2 ***'
c......sparse "symbolic" assembly
        call symbass(ir(iebeg),ir(jebeg),ir(ietbeg),ir(jetbeg)
     $              ,ndof,ia,ja)
        ncoef1=ia(ndof+1)-1
        write(6,*) '*** passed symbass ***'
c        write(6,*) 'after symbass: ia = ',(ia(i),i=1,ndof+1)
c        write(6,*) 'after symbass: ja = ',(ja(i),i=1,ncoef1)
c        write(6,*) 'after symbass: ncoef1 = ',ncoef1
c......initialize before numerical sparse assembling
        do 95 j=1,ncoef1
 95     an(j)=0.
c......initialize
        do 6 i=1,ndof
        ad(i)=0.
        ir(ipbeg-1+i)=0
 6      continue
c
        return
        end
c@@@@@@@@@@@@@@@@@@@@@@@@@@@@@@@@@@@@@@@@@@@@@@@@@@@@@@@@@@@@@@@@@@@@@@
        subroutine sparseassem2(nel,ndof,ndofpenel,mtoti,ir,iflag,
     $  ae,be,lm,ia,ja,ad,an,b,ndofpe)
        implicit real*8(a-h,o-z)
C++++++++++++++++++++++++++++++++++++++++++++++++++++++++++++++++++++++++++++++
c......Purposes:
c...... Numerical sparse assembly (for symmetrical matrix)
c        according to ODU sparse formats
c......Author(s), Dated Version:
c        Prof. Duc T. NGUYEN, June 1'2002
c......Input Parameters:
c        nel = total (cumulative) number of finite elements (including
c              all different element types
c        ndof = total # degree-of-freedom (dof), including all Dirichlet b.c.
c        ndofpenel = summation of (ndofpe type1 * nel type1 +
c                                  ndofpe type2 * nel type2 + etc...)
```

```
c        mtoti = available INTEGER words of memory
c        ir(mtoti) = ir(iebeg=1,..., jebeg-1) = same as ie(-) array
c                  = ir(jebeg ,..., ibcbeg-1) = same as je(-) array
c                  = ir(ibcbeg,..., ietbeg-1) = Dirichlet b.c. = either 0 or 1
c
c                  = ir(ietbeg,..., jetbeg-1) = same as ie(-) transpose  array
c                  = ir(jetbeg,..., ipbeg-1) = same as je(-) transpose  array
c                  = ir(ipbeg,..., last-1)   = same as ip(-) array
c        iflag(40) = unused for now
c        ae(ndofpemax**2) = element stiffness matrix, stored as 1-d array
c                          (column-wise)
c                          where ndofpemax = maximum #dof per element (for all
c                          el. types)
c        be(ndofpemax) = equivalent joint load vector
c                          (i.e: distributed, pressure load acting on an element)
c        lm(ndofpemax) = global dof # associated with each element
c        b(ndof) = rhs vector," contains the applied joint loads "AND" known
value
c                  of DIRICHLET b.c.
c        ndofpe = # dof per element (correspond to a specific finite element
type
c        ia(ndof+1) = starting location of FIRST non-zero, off-diag. term
c                      for each row of the assembled sparse (symmetrical) matrix
c        ja(ncoef1) = column numbers associated with non-zero terms on each row
c                      where ncoef1 = ia(ndof+1)-1 of the assembled sparse
(sym.)
c                      matrix
c
c......Output Parameters:
c        ad(ndof) = values of diagonal terms of the assembled sparse (sym)
matrix
c        an(ncoef1) = values of off-diagonal, non-zero terms of the sparse
(sym)
c                      matrix
c        b(ndof) = "modified" rhs vector to include "effects" of known
DIRICHLET
c                  b.c.
c                  of DIRICHLET b.c.
c......Temporary (working) Parameters:
c        ir(mtoti) = ir(iebeg=1,..., jebeg-1, ..., mtoti)
c
C+++++++++++++++++++++++++++++++++++++++++++++++++++++++++++++++++++++++++++++
c
         dimension ir(*), iflag(*),ae(*),be(*),lm(*),
      $            ia(*),ja(*),ad(*),an(*),b(*)
c
         iebeg=1
         jebeg=iebeg+(nel+1)
         ibcbeg=jebeg+(ndofpenel)
         ietbeg=ibcbeg+ndof
         jetbeg=ietbeg+(ndof+1)
         ipbeg=jetbeg+(ndofpenel)
         last=ipbeg+ndof
         if (mtoti .lt. last) then
           write(6,*) 'error: not enough memory for mtoti !!!'
         endif
c
         call numass(ia,ja,ir(ibcbeg),ae,be,lm,ndofpe,an,ad,b,ir(ipbeg))
c
         return
         end
c@@@@@@@@@@@@@@@@@@@@@@@@@@@@@@@@@@@@@@@@@@@@@@@@@@@@@@@@@@@@@@@@@@@@@
c@@@@@@@@@@@@@@@@@@@@@@@@@@@@@@@@@@@@@@@@@@@@@@@@@@@@@@@@@@@@@@@@@@@@@
```

Table 4.7 Transposing a Sparse Matrix

```
c       Table 4.7  Transposing a Sparse Matrix

c@@@@@@@@@@@@@@@@@@@@@@@@@@@@@@@@@@@@@@@@@@@@@@@@@@@@@@@@@@@@@@@
c..... file = part2.o  (only object codes) ...............
        subroutine transa2(n,m,ia,ja,iat,jat)
        dimension ia(*),ja(*),iat(*),jat(*)
c
        MH=M+1
        NH=N+1
        DO 10 I=2,MH
   10 IAT(I)=0
        IAB=IA(NH)-1
        DO 20 I=1,IAB
        J=JA(I)+2
        IF(J.LE.MH)IAT(J)=IAT(J)+1
   20 CONTINUE
        IAT(1)=1
        IAT(2)=1
        IF(M.EQ.1)GO TO 40
        DO 30 I=3,MH
   30 IAT(I)=IAT(I)+IAT(I-1)
   40 DO 60 I=1,N
        IAA=IA(I)
        IAB=IA(I+1)-1
        IF(IAB.LT.IAA)GO TO 60
        DO 50 JP=IAA,IAB
        J=JA(JP)+1
        K=IAT(J)
        JAT(K)=I
c       ANT(K)=AN(JP)
   50 IAT(J)=K+1
   60 CONTINUE
        return
        end
```

Table 4.8 Inputs and Outputs for Mixed (2-D Truss and Triangular) Sparse Assembly Process

Date: June 1, 2002

nbays, nstory, # real truss/faked triangular el. =
1 1 5 2

joint coordinates
1   0.   9.0000000000000 0.
2   12.0000000000000 9.0000000000000 0.
3   0.   0.   0.
4   12.0000000000000   0.   0.

element connectivity node i&j

1   1 2 3 4
2   1 2 5 6
3   3 4 7 8
4   1 2 7 8
5   3 4 5 6

```
ie =   1 5 9 13 17 21 27 33
je =   1 2 3 4 1 2 5 6 3 4 7  8 1 2 7 8 3 4 5 6 1 2 5 6
   3 4 3 4 5 6 7 8
ibdof =    5 6 7 8
* * * passed transa2 * * *

* * * * * * * * * * * * * * * * * * * * * * * * *
* * * * * * * * * * * * * * * * * * * * * * * * *

* * * passed symbass * * *
elem #   1* * * passed elstiff * * *
elem #   1* * * passed numass * * *
elem #   5* * * passed elstiff * * *
elem #   5* * * passed numass * * *
elem #   1* * * passed numass * * *
elem #   2* * * passed numass * * *
 ndof , ncoef1 =    8 6
 ia(ndof +1) =    1 4 6 7 7 7 7 7
 ja(ncoef1) =    2 3 4 3 4 4
 ad(ndof) =    2.1260000000000 4.1351111111111 2.1260000000000
16.1351111111111 1.0000000000000 1.0000000000000 1.0000000000000
1.0000000000000
an(ncoef1) =    2.9680000000000 5.9166666666667 7.0000000000000
7.0000000000000 8.0000000000000 14.0320000000000
b(ndof) =    1.0000000000000 2.0000000000000 3.0000000000000
4.0000000000000 0. 0. 0. 0.


real              0.0
user              0.0
sys               0.0
```

4.7 A Simple Finite Element Model (Unsymmetrical Matrices)

In this section, three different algorithms for general and efficient sparse assembly of a system of finite-element-based, unsymmetrical equations are presented. These three unsymmetrical sparse assembly algorithms will be incorporated into different unsymmetrical sparse solvers such as the ones developed by I. Duff's group [3.2, 3.6, 3.7], by D. Nguyen's team [1.9], and by SGI library subroutines [3.1]. METIS re-ordering algorithms [3.3], for minimizing the fill-in terms, symbolic factorization, numerical factorizations, master degree-of-freedom, are also discussed. Simple examples are used to facilitate the discussions, and medium-to large-scale examples are used to evaluate the numerical performance of different algorithms.

To facilitate the discussions, a simple/academic finite element model that consists of four rectangular elements, with 1 degree-of-freedom (dof) per node, and its loading condition (R_i), is shown in Figure 4.5. There are 4 dof per element (ndofpe = 4).

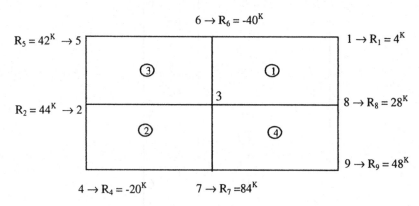

Figure 4.5 a Simple, Unconstrained Finite Element Model

The global dof, associated with each of the eth rectangular element (see Figure 4.5), are given by the following "element-dof" connectivity matrix [E], as follows:

$$[E] = \begin{array}{c|ccccccccc} & 1 & 2 & 3 & 4 & 5 & 6 & 7 & 8 & 9 \\ \hline 1 & x & & x & & & x & & x & \\ 2 & & x & x & x & & & x & & \\ 3 & & x & x & & x & x & & & \\ 4 & & & x & & & & x & x & x \end{array} \qquad (4.40)$$

In Eq.(4.40), the number of rows and columns will correspond to the number of (finite) elements, (NEL = 4) and number of dof, (NDOF = 9) respectively. Using the sparse storage scheme, matrix [E] in Eq.(4.40) can be described by the following two integer arrays:

$$
IE\begin{pmatrix} 1 \\ 2 \\ 3 \\ 4 \\ 5 = NEL+1 \end{pmatrix} = \begin{Bmatrix} 1 \\ 5 \\ 9 \\ 13 \\ 17 \end{Bmatrix} \tag{4.41}
$$

$$
JE\begin{pmatrix} 1 \\ 2 \\ 3 \\ 4 \\ 5 \\ 6 \\ 7 \\ 8 \\ 9 \\ 10 \\ 11 \\ 12 \\ 13 \\ 14 \\ 15 \\ 16 = NEL*ndofpe \end{pmatrix} = \begin{Bmatrix} 3 \\ 8 \\ 1 \\ 6 \\ 7 \\ 3 \\ 2 \\ 4 \\ 5 \\ 2 \\ 3 \\ 6 \\ 7 \\ 9 \\ 8 \\ 3 \end{Bmatrix} \tag{4.42}
$$

Thus, Eq.(4.41) gives the <u>starting location</u> of the first non-zero term for each row of matrix [E] while Eq.(4.42) contains the global dof associated with each eth (= 1, 2, 3, 4) rectangular element. The transpose of the matrix [E] can be given as:

	1	2	3	4
1	x			
2		x	x	
3	x	x	x	x
4		x		
5			x	
6	x		x	
7		x		x
8	x			x
9				x

$$[E]^T =$$

(4.43)

which can be described by the following two integer arrays:

$$
IET \begin{pmatrix} 1 \\ 2 \\ 3 \\ 4 \\ 5 \\ 6 \\ 7 \\ 8 \\ 9 \\ 10 = NDOF+1 \end{pmatrix} = \begin{Bmatrix} 1 \\ 2 \\ 4 \\ 8 \\ 9 \\ 10 \\ 12 \\ 14 \\ 16 \\ 17 \end{Bmatrix}
$$

(4.44)

$$
\text{JET}\begin{Bmatrix} 1 \\ & 2 \\ & 3 \\ 4 \\ 5 \\ 6 \\ 7 \\ & 8 \\ 9 \\ & 10 \\ & 11 \\ 12 \\ 13 \\ & 14 \\ & 15 \\ 16 \end{Bmatrix} = \begin{Bmatrix} 1 \\ 2 \\ 3 \\ 1 \\ 2 \\ 3 \\ 4 \\ 2 \\ 3 \\ 1 \\ 3 \\ 2 \\ 4 \\ 1 \\ 4 \\ 4 \end{Bmatrix}
\qquad (4.45)
$$

The "exact" numerical values for the 4×4 (rectangular) element stiffness matrix [ELK] are "unimportant" at this stage of the discussion, and therefore is assumed to be given by the following formula:

$$
\text{ELK}(i, j) = (i + j) * \text{iel} \qquad (4.46)
$$

and for $i \neq j$, then

$$
\text{ELK}(j, i) = -\text{ELK}(i, j) \qquad (4.47)
$$

In Eqs.(4.46 and 4.47), $i, j = 1, 2, 3, 4$ and

$$
\text{iel} = \text{rectangular element number} (= 1, 2, 3 \text{ or } 4)
$$

Thus, according to Eqs.(4.46, 4.47), element stiffness matrix for rectangular elements ② is given as:

$$
[\text{ELK}] = \begin{bmatrix} 2 & 3 & 4 & 5 \\ -3 & 4 & 5 & 6 \\ -4 & -5 & 6 & 7 \\ -5 & -6 & -7 & 8 \end{bmatrix} * (\text{iel} = 2) = \begin{bmatrix} 4 & 6 & 8 & 10 \\ -6 & 8 & 10 & 12 \\ -8 & -10 & 12 & 14 \\ -10 & -12 & -14 & 16 \end{bmatrix} \qquad (4.48)
$$

Hence, the "simulated" elements stiffness matrix, given by Eq.(4.48), is unsymmetrical in value, but it is still symmetrical in location!

Following the usual finite element procedures, the system of linear, unsymmetrical equations can be assembled as:

$$[K]\{Z\} = \{R\} \qquad (4.49)$$

where

$$[K]=$$

	1	2	3	4	5	6	7	8	9
1	6		-4			7		-5	
2		24	5	14	-9	18	-8		
3	4	-5	60	12	-12	26	-26	-25	-24
4		-14	-12	16	F	F	-10	F	F
5		9	12		6	15	F	F	F
6	-7	-18	-26		-15	32	F	-6	F
7		8	26	10			12	16	12
8	5		25			6	-16	28	-20
9			24				-12	20	16

(4.50)

$$\{R\} = \begin{Bmatrix} 4 \\ 44 \\ 10 \\ -20 \\ 42 \\ -40 \\ 84 \\ 28 \\ 48 \end{Bmatrix} \qquad (4.51)$$

The hand-calculator solution for Eqs.(4.49 – 4.51) is:

$$\{Z\}^T = \{1.0, 1.0, 1.0, 1.0, 1.0, 1.0, 1.0, 1.0, 1.0\} \qquad (4.52)$$

which have been independently confirmed by the developed computer program, shown in Table 4.9.

Remarks

For the upper-triangluar portion of Eq.(4.50), there are nine fill-in terms, denoted by the symbols F (fill-in terms are NOT shown for the lower-triangular portion, although they also exist!).

4.8 Re-Ordering Algorithms

In order to minimize the number of fill-in terms during the symbolic and numerical factorization phases, re-ordering algorithms such as MMD, ND, and METIS [3.4, 4.1, 3.3] can be used. In this work, METIS re-ordering algorithm is selected due to its low memory and low number of fill-in requirements.

For the example shown in Figure 4.5, one needs to prepare the following two adjacency (integer) input arrays for METIS

$$\text{IADJ} = \begin{cases} 1, & 4, & 9, & 17, \\ 20, & 23, & 28, & 33, \\ 38, & 41 \end{cases} \tag{4.53}$$

$$\text{JADJ} = \boxed{3, 8, 6}, \ \boxed{7, 3, 4, 5, 6}, \ \boxed{8, 1, 6, 7, 2, 4, 5, 9}, \ \boxed{7, 3, 2}, \tag{4.54}$$

$$\boxed{2, 3, 6}, \ \boxed{3, 8, 1, 5, 2}, \ \boxed{3, 2, 4, 9, 8}, \qquad \boxed{3, 1, 6, 7, 9},$$

$$\boxed{7, 8, 3}$$

$$\text{NODE(or DOF)} = \begin{cases} 1, & 2, & 3, & 4, \\ 5, & 6, & 7, & 8, \\ 9 \end{cases} \tag{4.55}$$

Equation (4.55) represents the node (or dof) number of the finite element model. Eq.(4.54) represents the node (or dof) numbers connected to the ith node, and Eq.(4.53) contains the starting locations associated with the array JADJ.

The output of METIS, for example, will be given by the following two integer arrays:

$$\text{IPERM} \begin{bmatrix} 1 \\ 2 \\ 3 \\ 4 \\ 5 \\ 6 \\ 7 \\ 8 \\ 9 \end{bmatrix} = \begin{Bmatrix} 5 \\ 1 \\ 6 \\ 9 \\ 4 \\ 7 \\ 8 \\ 3 \\ 2 \end{Bmatrix} \tag{4.56}$$

or IPERM (old numbering dof system) = {new numbering dof system} (4.57)

and

$$\text{INVP} \begin{pmatrix} 1 \\ 2 \\ 3 \\ 4 \\ 5 \\ 6 \\ 7 \\ 8 \\ 9 \end{pmatrix} = \begin{Bmatrix} 2 \\ 9 \\ 8 \\ 5 \\ 1 \\ 3 \\ 6 \\ 7 \\ 4 \end{Bmatrix} \qquad (4.58)$$

or INVP(new numbering dof system)={old numbering dof system} (4.59)

Using the array IPERM(-), shown in Eq.(4.56), the system matrix [K], shown in Eq.(4.50), can be re-arranged as:

$[K_{metis}]=$

	1	2	3	4	5	6	7	8	9
1	24			-9		5	18	-8	14
2		16	20			24		-12	
3		-20	28		5	25	6	-16	
4	9			6		12	15	F	F
5			-5		6	-4	7	F	
6	-5	-24	-25	-12	4	60	26	-26	12
7	-18		-6	-15	-7	-26	32	F	F
8	8	12	16			26		12	10
9	-14					-12		-10	16

(4.60)

Similarly, the system right-hand-side load vector {R}, shown in Eq.(4.51), can be re-arranged as:

$$[R_{metis}] = \begin{Bmatrix} 44 \\ 48 \\ 28 \\ 42 \\ 4 \\ 10 \\ -40 \\ 84 \\ -20 \end{Bmatrix} \tag{4.61}$$

Remarks

For the upper–triangular portion of Eq.(4.60), there will be five fill-in terms, denoted by the symbol F, upon completing the numerical factorization phase.

Using Eq.(4.56) for the array IPERM(-), Eqs.(4.40–4.45) will become Eqs.(4.62–4.67):

		1	2	3	4	5	6	7	8	9
$[E_{metis}]=$	1			x		x	x	X		
	2	x					x		x	x
	3	x			x		x	X		
	4		x	x			x		x	

$$\tag{4.62}$$

$$IE_{metis} \begin{Bmatrix} 1 \\ 2 \\ 3 \\ 4 \\ 5 \end{Bmatrix} = \begin{Bmatrix} 1 \\ 5 \\ 9 \\ 13 \\ 17 \end{Bmatrix} \tag{4.63}$$

$$
JE_{metis} = \begin{pmatrix} 1 \\ 2 \\ 3 \\ 4 \\ 5 \\ 6 \\ 7 \\ 8 \\ 9 \\ 10 \\ 11 \\ 12 \\ 13 \\ 14 \\ 15 \\ 16 \end{pmatrix} = \begin{Bmatrix} 6 \\ 3 \\ 5 \\ 7 \\ 8 \\ 6 \\ 1 \\ 9 \\ 4 \\ 1 \\ 6 \\ 7 \\ 8 \\ 2 \\ 3 \\ 6 \end{Bmatrix} \qquad (4.64)
$$

$[E^T{}_{metis}] =$

	1	2	3	4
1		x	X	
2				x
3	x			x
4			X	
5	x			
6	x	x	X	x
7	x		X	
8		x		x
9		x		

$$(4.65)$$

$$
\mathbf{IET}_{\text{metis}}
\begin{pmatrix}
1 \\ 2 \\ 3 \\ 4 \\ 5 \\ 6 \\ 7 \\ 8 \\ 9 \\ 10
\end{pmatrix}
=
\begin{Bmatrix}
1 \\ 3 \\ 4 \\ 6 \\ 7 \\ 8 \\ 12 \\ 14 \\ 16 \\ 17
\end{Bmatrix}
\tag{4.66}
$$

$$
[\mathbf{JET}_{\text{metis}}] =
\begin{pmatrix}
1 & & \\
2 & & \\
 & 3 & \\
4 & & \\
5 & & \\
 & 6 & \\
7 & & \\
 & 8 & \\
 & 9 & \\
 & 10 & \\
 & 11 & \\
12 & & \\
13 & & \\
 & 14 & \\
 & 15 & \\
16 & &
\end{pmatrix}
=
\begin{Bmatrix}
2 & \\
3 & \\
 & 4 \\
1 & \\
4 & \\
 & 3 \\
1 & \\
 & 1 \\
 & 2 \\
 & 3 \\
 & 4 \\
1 & \\
3 & \\
 & 2 \\
 & 4 \\
2 &
\end{Bmatrix}
\tag{4.67}
$$

4.9 Imposing Dirichlet Boundary Conditions

Assuming the following Dirichlet boundary conditions are imposed at dof 2, 4, 5 (see Figure 4.5):

$$\text{ibc} \begin{pmatrix} 2 \\ 4 \\ 5 \end{pmatrix} = \begin{bmatrix} \text{ndof} \\ \text{ndof} \\ \text{ndof} \end{bmatrix} = \begin{Bmatrix} 9 \\ 9 \\ 9 \end{Bmatrix} \tag{4.68}$$

and the corresponding right-hand-side load (and prescribed Dirichlet boundary conditions) vector $\{R\}$ becomes:

$$R \begin{pmatrix} 2 \\ 4 \\ 5 \end{pmatrix} = \begin{Bmatrix} 0.2'' \\ 0.4'' \\ 0.5'' \end{Bmatrix} \tag{4.69}$$

Then Eqs.(4.60 – 4.61) will become:

$[K_{\text{metis, bc}}] =$

	1	2	3	4	5	6	7	8	9
1	1								
2		16	20			24		-12	
3		-20	28		5	25	6	-16	
4				1					
5			-5		6	-4	7		
6		-24	-25		4	60	26	-26	
7			-6		-7	-26	32		
8		12	16			26		12	
9									1

$$\tag{4.70}$$

$$\left\{R_{\text{metis,bc}}\right\} = \left\{ \begin{array}{c} 0.2'' \\ 48^K \\ 28^K \\ 0.5'' \\ 4^K \\ 12.2^K \\ -28.9^K \\ 78.4^K \\ 0.4'' \end{array} \right\} \tag{4.71}$$

It should be mentioned that fill-in terms associated with Eq.(4.70) are NOT shown here.

The hand-calculator solution of

$$[K_{\text{metis, bc}}] * \{Z_{\text{metis, bc}}\} = \{R_{\text{metis, bc}}\} \tag{4.72}$$

can be given as:

$$\{Z_{\text{metis, bc}}\}^T = \{\underline{0.2}, 0.975, 0.996, \underline{0.5}, 1.575, 1.265, 0.656, 1.490, \underline{0.4}\} \tag{4.73}$$

which have been independently confirmed by the developed computer program shown in Table 4.9.

Table 4.9 Unsymmetrical Sparse Assembly

```
c    Table 4.9   Unsymmetrical Sparse Assembly
c
c%%%%%%%%%%%%%%%%%%%%%%%%%%%%%%%%%%%%%%%%%%%%%%%%%
     subroutine part1unsym
     implicit real*8(a-h,o-z)
c++++++++++++++++++++++++++++++++++++++++++++++++++++++++++++++++++++++++
c    purpose: d.t. nguyen's general, unsymmetrical sparse assembly
c    version: November 12, 2002
c    author:  Prof. Duc T. Nguyen  (Email= dnguyen@odu.edu)
c    notes:   these files (part1unsym.f, part2unsym.f) are stored
c             under sub-directory ~/cee/newfem/*complete*sparse*/
c
c             Please do "NOT" distribute this source code to anyone
c             without explicit written agreement by Prof. D.T. Nguyen
c++++++++++++++++++++++++++++++++++++++++++++++++++++++++++++++++++++++++
     dimension ie(1000),je(4000),iet(1000),jet(4000)
     dimension ia(1000),ja(4000),iakeep(1000)
     dimension iakeept(1000),jat(4000),ibc(1000)
     dimension elk(4,4),ae(16),be(4),lm(4)
     dimension b(1000),an(4000),ip(1000),ant(4000)
     dimension iakepp(1000),jj(4000),aa(4000)
     dimension index(1000),itempo1(1000),tempo1(1000)
```

```
      dimension iperm(1000),invp(1000)
      dimension ikeep(1000,5),iw(1000,8),w(1000) ! required by ma28
      dimension ad(1000),an2(4000),x(1000),iut(1000),jut(4000) ! Duc-Pierrot
      dimension idir(1000),isupd(1000),iu(1000),ju(4000)  ! Duc-Pierrot
      dimension di(1000),un(4000),iup(1000),un2(4000),di2(1000) ! Duc-Pierrot

c
      write(6,*) ' ************************* '
      write(6,*) ' version: November 12, 2002 '
      write(6,*) ' ************************* '
c
      read(5,*) nel,ndof,ndofpe,method,nbc,metis,iunsolver
c......input finite element mesh (connectivities)
      call femesh(nel,ndofpe,ie,je)         !    from Willie
      write(6,*) 'nel,ndof,ndofpe,method,nbc,metis,iunsolver= '
      write(6,*) nel,ndof,ndofpe,method,nbc,metis,iunsolver
      write(6,*) 'method = 1 = sorting column # by sparse TRANSPOSE'
      write(6,*) '         2 = sorting column # by SORTING'
      write(6,*) 'iunsolver  =  1 = Duff''s unsymmetrical sparse algo'
      write(6,*) '           =  2 = Duc-Pierrot''s unsym sparse solver'
      write(6,*) '           =  3 = SGI library unsym sparse solver'
      write(6,*) 'ie= ',(ie(i),i=1,nel+1)
      write(6,*) 'je= ',(je(i),i=1,nel*ndofpe)
c......input "system" rhs load vector(s)
      call rhsload(ndof,b)                  !    from Willie
c......applying boundary conditions
      call boundaryc(ndof,ibc,nbc,b)        !    from Willie
c
      call transa2(nel,ndof,ie,je,iet,jet)
      write(6,*) 'iet= ',(iet(i),i=1,ndof+1)
      write(6,*) 'jet= ',(jet(i),i=1,nel*ndofpe)
c++++++++++++++++++++++++++++++++++++++++++++++++++++++++++++++++++++
c......use METIS or not ??
      do 11 i=1,ndof
      iperm(i)=i
      invp(i)=i
 11   continue
      if (metis .eq. 1) then
      call adjacency(ie,je,iet,jet,ndof,ia,ja,iakeep)
      write(6,*) 'iakeep(-)= ',(iakeep(i),i=1,ndof+1)
      jcount=iakeep(ndof+1)-1
      write(6,*) 'ja(-)= ',(ja(i),i=1,jcount)
      call metisreord(ndof,iakeep,ja,iperm,invp)
      write(6,*) 'iperm(-)= ',(iperm(i),i=1,ndof)
c.......arrays je(-), and iet(-), jet(-) need be re-defined (due to IPERM)
      icount=nel*ndofpe
      do 12 j=1,icount
      itempo1(j)=je(j)
 12   continue
c
      call newje(itempo1,iperm,icount,je)
      write(6,*) 'ie= ',(ie(i),i=1,nel+1)
      write(6,*) 'je(-)= ',(je(i),i=1,nel*ndofpe)
      call transa2(nel,ndof,ie,je,iet,jet)
      write(6,*) 'iet= ',(iet(i),i=1,ndof+1)
      write(6,*) 'jet= ',(jet(i),i=1,nel*ndofpe)
c.......rhs load vector {b}, & Dirichlet bc  need be re-defined (due to IPERM)
      do 16 j=1,ndof
      iakepp(j)=ibc(j)
 16   tempo1(j)=b(j)
```

```
         do 17 j=1,ndof
         new=iperm(j)
         b(new)=tempo1(j)
         ibc(new)=iakepp(j)
17       continue
         endif
c++++++++++++++++++++++++++++++++++++++++++++++++++++++++++++++++++
c......nguyen's unsymmetric symbolic sparse assembly
         do 42 i=1,ndof+1
         ia(i)=0
42       iakeep(i)=0
         do 43 i=1,4000   ! at this moment, dimension of ja(-) is unknown ?
43       ja(i)=0
         if (iunsolver .ne. 2) then
         call symbass(ie,je,iet,jet,ndof,ia,ja,iakeep,ibc)
c
         ncoef1=iakeep(ndof+1)-iakeep(1)
         write(6,*) 'iakeep(ndof+1)= ',(iakeep(i),i=1,ndof+1)
         write(6,*) 'ja(ncoef1)= ',(ja(i),i=1,ncoef1)
c......nguyen's unsymmetric numerical sparse assembly
         call unsymasem(ibc,ncoef1,an,ip,ndof,nel,elk,be,ndofpe,
      $  lm,je,ae,iakeep,ja,b,nbc)
c        write(6,*) 'rhs load = ',(b(i),i=1,ndof)
c
         elseif (iunsolver .eq. 2) then
c......unsym sparse assembly by Duc-Pierrot's formats
         do 77 i=1,ndof
         ia(i)=ibc(i)
77       continue
         call symbassym(ie,je,iet,jet,ndof,ia,ja)
         write(6,*) 'duc-pierrot ia(-)= ',(ia(i),i=1,ndof+1)
         ncoef1=ia(ndof+1)-1
         write(6,*) 'duc-pierrot ja(-)= ',(ja(i),i=1,ncoef1)
c%%%%%%%%%%%%%%%%%%%%%%%%%%%%%%%%%%%%%%%%%%%%%%%%%%
c
         do 93 i=1,ncoef1
         an2(i)=0.
93       an(i)=0.
c
         do 92 i=1,ndof
92       ip(i)=0
c
c        icount=0
         do 94 i=1,nel
         call elunsym(i,elk,ndofpe,be,lm,je,ae)    ! from Willie
         write(6,*) 'el# = ',i,' UNSYM. row-wise elk(4,4)= '
         write(6,8) ((elk(ir,jc),jc=1,4),ir=1,4)
8        format(2x,4(e13.6,1x))
         write(6,*) 'el# = ',i,' UNSYM. col-wise elk(4,4)= '
         do 88 jc=1,4
         istart=(jc-1)*ndofpe+1
         iend=jc*ndofpe
         write(6,8) (ae(ir),ir=istart,iend)
88       continue
c
         call numassuns(ia,ja,ibc,ae,be,lm,ndofpe,an,ad,b,ip,an2,itempo1)
c
94       continue
c        call transagain(ndof,ndof,ia,ja,iakeep,jat,an,ant)
c        call transagain(ndof,ndof,ia,ja,iakeep,jat,an2,aa)
```

```
      write(6,*) 'before symfact ia(-)= ',(ia(i),i=1,ndof+1)
      write(6,*) 'before symfact ja(-)= ',(ja(i),i=1,ncoef1)
      call symfact(ndof,ia,ja,iu,ju,ip,ncoef2)
      write(6,*) 'duc-pierrot iu(-)= ',(iu(i),i=1,ndof+1)
      write(6,*) 'duc-pierrot ju(-)= ',(ju(i),i=1,ncoef2)
      call transa(ndof,ndof,iu,ju,iut,jut) ! TWICE, but no real numbers
      write(6,*) 'after transa TWICE iu(-)= ',(iu(i),i=1,ndof+1)
      write(6,*) 'after transa TWICE ju(-)= ',(ju(i),i=1,ncoef2)
c     call supnode(n,iu,ju,isupern)
c%%%%%%%%%%%%%%%%%%%%%%%%%%%%%%%%%%%%%%%%%%%%%%%%%
c
      write(6,*) 'an(ncoef1) = ',(an(i),i=1,ncoef1)
      write(6,*) 'an2(ncoef1) = ',(an2(i),i=1,ncoef1)
      write(6,*) 'ad(ndof)= ',(ad(i),i=1,ndof)
      endif
c......ordering column numbers (of each row) in ascending orders
      if (method .eq. 2) then
c.......using "sorting" ideas !
         if (iunsolver .ne. 2) then
      call sorting(ndof,iu,itempo1,ju,tempo1,index,b,an) ! here an=upper+lower
         elseif (iunsolver .eq. 2) then
      call sortingan2(ndof,iu,itempo1,ju,tempo1,index,    ! here an=upper
     $ b,an,an2,x)                                        !  an2=lower
         endif
c.......using "sparse transposing" ideas !
      elseif (method .eq. 1) then
         if (iunsolver .ne. 2) then
      call transa2m(ndof,ndof,iakeep,ja,iakepp,jat,an,ant)! here an=upper+lower
         elseif (iunsolver .eq. 2) then
      call transa2m(ndof,ndof,iakeep,ja,iakepp,jat,an2,aa)! here an2=lower
         endif
      endif
      write(6,*) 'an(ncoef1) = ',(an(i),i=1,ncoef1)
      write(6,*) 'an2(ncoef1) = ',(an2(i),i=1,ncoef1)
      write(6,*) 'ad(ndof)= ',(ad(i),i=1,ndof)
c++++++++++++++++++++++++++++++++++++++++++++++++++++++++++++++++++++++++++++
      if (iunsolver .eq. 1) then
c......get unsym. matrix [A] in MA28's (row,column) formats
      call rownuma28(iakeep,ndof,iakepp)
      nz=iakeep(ndof+1)-1
      licn=4*nz
      lirn=licn
      u=0.
      write(6,*) 'ndof,nz,licn,lirn,u = ',ndof,nz,licn,lirn,u
      write(6,*) 'irn(-)= ',(iakepp(i),i=1,nz)
      write(6,*) 'icn(-)= ',(ja(i),i=1,nz)
      write(6,*) ' b(-)= ',( b(i),i=1,ndof)
c......no need tonumbers in increasing order before using ma28 ?
c     call transa2m(ndof,ndof,iakeep,ja,itempo1,jat,an,ant)
c     call transa2m(ndof,ndof,itempo1,jat,iakeep,ja,ant,an)
c     write(6,*) 'irn(-)= ',(iakepp(i),i=1,nz)
c     write(6,*) 'icn(-)= ',(ja(i),i=1,nz)
c......LU factorization
      write(6,*) ' an(-)= ',( an(i),i=1,nz)
      call ma28ad(ndof,nz,an,licn,iakepp,lirn,ja,u,ikeep,iw,
     $w,iflag)
      write(6,*) 'passed ma28ad'
c......forward-backward solution phases
      nrhs=1
      mtype=1
```

```
        do 62 i=1,nrhs
        call ma28cd(ndof,an,licn,ja,ikeep,b,w,mtype)
        write(6,*) 'passed ma28cd'
 62     continue
c......print solutions
        write(6,*) 'unknown= ',(b(i),i=1,ndof)
        elseif (iunsolver .eq. 2) then
c......get unsym. matrix [A] in Duc-Pierrot's ODU-NASA unsym. formats
        do 80 i=1,ndof
        isupd(i)=1
 80     continue
        write(6,*) 'just before call unsynumfa1'
        call unsynumfa1(ndof,ia,ja,ad,an,iu,ju,di,un,ip,iup,
     $           isupd,an2,un2,di2)
        write(6,*) 'just before call unsyfbe'
        call unsyfbe(ndof,iu,ju,di,un,b,x,un2)
        write(6,*) 'just after call unsyfbe'
        write(6,*) 'unknown x(-)= ',(x(i),i=1,ndof)
c
        elseif (iunsolver .eq. 3) then
c......get unsym. matrix [A] in SGI's column-wise formats
c      call SGI library routine for unsym. sparse solver,
c      based on input arrays previously generated:
c      iakepp(-), jat(-), ant(-)
        write(6,*) 'SGI: iakepp(-)= ',(iakepp(i),i=1,ndof+1)
        write(6,*) 'SGI: jat(-)= ',(jat(i),i=1,ncoef1)
        write(6,*) 'SGI: ant(-)= ',(ant(i),i=1,ncoef1)
        endif
c+++++++++++++++++++++++++++++++++++++++++++++++++++++++++++++++++++++++
c
c     stop
        return
        end
c%%%%%%%%%%%%%%%%%%%%%%%%%%%%%%%%%%%%%%%%%%%%%%%%%%%%%%%
        subroutine supnode(n,istartr,kindx,isupern)
c      implicit real*8(a-h,o-z)
c......purposes: identify the rows which have same non-zero patterns
c......       (i.e. same column numbers lined up) for "unrolling"
        dimension istartr(1),kindx(1),isupern(1)
c......initialize super node integer array
        do 1 i=1,n
 1      isupern(i)=1
c......meaning of variable "iamok" = 0 (= NOT having same non-zero patterns)
c......               = 1 (= DOES have same non-zero patterns)
c      do 2 i=2,n-1
        do 2 i=2,n
c      iamok=0
        iamok=1
c......check to see if i-th row has the same # nonzeroes as (i-1)-th row
c......also,check to see if the (i-1)-th row have a non-zero right above
c......    the diagonal of the i-th row
c......if the i-th row does NOT have the same "GENERAL" non-zero patterns,
c......then, skip loop 2 immediately (no need to check any furthers !)
        kptrsi=istartr(i+1)-istartr(i)
        kptrsim1=istartr(i)-istartr(i-1)
c      if(kptrs(i).ne.kptrs(i-1)-1) go to 2
        if(kptrsi.ne.kptrsim1-1) go to 2
        ii=istartr(i-1)
        if(kindx(ii).ne.i) go to 2
c......now, check to see if the i-th row and the (i-1)-th row have the
```

```
c......same "DETAILED" non-zero patterns
c    write(6,*) 'check point 1'
     iii=istartr(i)
     jj=ii
     jjj=iii-1
c    do 3 j=1,kptrs(i)
     do 3 j=1,kptrsi
     jj=jj+1
     jjj=jjj+1
     if(kindx(jj).ne.kindx(jjj)) then
     iamok=0
c    write(6,*) 'check point 2'
     go to 222
     else
     iamok=1
c    write(6,*) 'check point 3'
     endif
 3   continue
c    if(kindx(ii).eq.n) iamok=1
c......if all"DETAILED" non-zero patterns also match, then we have "supernode"
 222   if(iamok.eq.1) then
     isupern(i)=0
c    write(6,*) 'check point 4'
     endif
 2   continue
c......finally, establish how many rows a supernode has (based on the 1's,
c......and the 0's of the array isupern(-) obtained thus far)
     icount=0
     do 11 i=2,n
c.......counting the number of zeroes encountered in array isupern(-)
     if(isupern(i).eq.0) then
     icount=icount+1
c    write(6,*) 'check point 5'
     else
     ithrow=i-1-icount
     isupern(ithrow)=icount+1
     icount=0
c    write(6,*) 'check point 6'
     endif
 11    continue
c......now, take care of the last super node
     ithrow=n-icount
     isupern(ithrow)=icount+1
     return
     end
C%%%%%%%%%%%%%%%%%%%%%%%%%%%%%%%%%%%%%%%%%%%%%%
     subroutine symfact(n,ia,ja,iu,ju,ip,ncoef2)
c    implicit real*8(a-h,o-z)
     dimension ia(*),ja(*),iu(*),ju(*),ip(*)
C......Purposes: Symbolic factorization
c    input: ia(n+1),ja(ncoef1)  structure of given matrix A in RR(U)U.
c......    ia(n+1)          starting location of 1-st non-zero term
c    n              order of matrix A and of matrix U.
c......    ncoef1 = ia(n+1)-1
c    output: iu(n+1),ju(ncoef2)  structure of resulting matrix U in RR(U)U.
c    working space: ip(n)        of dimension N.  Chained lists of rows
c                              associated with each column.
c                              The array IU is also used as the multiple
c                              switch array.
     NM=N-1
```

```
      NH=N+1
      DO 10 I=1,N
      IU(I)=0
   10 IP(I)=0
      JP=1
      DO 90 I=1,NM
      JPI=JP
      JPP=N+JP-I
      MIN=NH
      IAA=IA(I)
      IAB=IA(I+1)-1
c     write(6,*) 'SYMFACT: i,nm,jpi,jp,n,jpp,min,iaa,iab = '
c     write(6,*) i,nm,jpi,jp,n,jpp,min,iaa,iab
      IF(IAB.LT.IAA)GO TO 30
      DO 20 J=IAA,IAB
      JJ=JA(J)
      JU(JP)=JJ
c     write(6,*) 'SYMFACT:insideloop20: j,jj,jp,ju(jp),min = '
c     write(6,*) j,jj,jp,ju(jp),min
      JP=JP+1
c     write(6,*) 'SYMFACT:insideloop20: j,jj,jp,ju(jp),min = '
c     write(6,*) j,jj,jp,ju(jp),min
c     IF(JJ.LT.MIN)MIN=JJ
      IF(JJ.LT.MIN) then
      min=jj
c     write(6,*) 'SYMFACT:insideloop20: min = '
c     write(6,*) min
      endif
c  20 IU(JJ)=I
      IU(JJ)=I
c     write(6,*) 'SYMFACT:insideloop20: jj,iu(jj),i = '
c     write(6,*) jj,iu(jj),i
   20 continue
   30 LAST=IP(I)
c     write(6,*) 'SYMFACT: i,ip(i),last = '
c     write(6,*) i,ip(i),last
      IF(LAST.EQ.0)GO TO 60
      L=LAST
   40 L=IP(L)
      LH=L+1
      IUA=IU(L)
      IUB=IU(LH)-1
c     write(6,*) 'SYMFACT: last,l,ip(l),lh,iu(l),iua,iu(lh),iub = '
c     write(6,*) last,l,ip(l),lh,iu(l),iua,iu(lh),iub
c     IF(LH.EQ.I)IUB=JPI-1
      IF(LH.EQ.I) then
      iub=jpi-1
c     write(6,*) 'SYMFACT: lh,i,jpi,iub = '
c     write(6,*) lh,i,jpi,iub
      endif
      IU(I)=I
c     write(6,*) 'SYMFACT: i,iu(i) = '
c     write(6,*) i,iu(i)
      DO 50 J=IUA,IUB
      JJ=JU(J)
c     write(6,*) 'SYMFACT:insideloop50: j,iua,iub,ju(j),jj,iu(jj),i = '
c     write(6,*) j,iua,iub,ju(j),jj,iu(jj),i
      IF(IU(JJ).EQ.I)GO TO 50
      JU(JP)=JJ
c     write(6,*) 'SYMFACT:insideloop50: jp,ju(jp)= '
```

```
c    write(6,*) jp,ju(jp)
     JP=JP+1
     IU(JJ)=I
c    write(6,*) 'SYMFACT:insideloop50: jp,jj,iu(jj),i= '
c    write(6,*) jp,jj,iu(jj),i
c    IF(JJ.LT.MIN)MIN=JJ
     IF(JJ.LT.MIN) then
c    write(6,*) 'SYMFACT:insideloop50: jj,min = ',jj,min
     min=jj
c    write(6,*) 'SYMFACT:insideloop50: jj,min = ',jj,min
     endif
 50  CONTINUE
c    write(6,*) 'SYMFACT: jp,jpp,l,last,min,nh = '
c    write(6,*) jp,jpp,l,last,min,nh
     IF(JP.EQ.JPP)GO TO 70
     IF(L.NE.LAST)GO TO 40
 60  IF(MIN.EQ.NH)GO TO 90
 70  L=IP(MIN)
c    write(6,*) 'SYMFACT: min,ip(min),l = '
c    write(6,*) min,ip(min),l
     IF(L.EQ.0)GO TO 80
     IP(I)=IP(L)
c    write(6,*) 'SYMFACT: l,ip(l),i,ip(i) = '
c    write(6,*) l,ip(l),i,ip(i)
     IP(L)=I
c    write(6,*) 'SYMFACT: l,ip(l),i,ip(i) = '
c    write(6,*) l,ip(l),i,ip(i)
     GO TO 90
 80  IP(MIN)=I
     IP(I)=I
c    write(6,*) 'SYMFACT: min,ip(min),i,ip(i) = '
c    write(6,*) min,ip(min),i,ip(i)
 90  IU(I)=JPI
     IU(N)=JP
     IU(NH)=JP
c    write(6,*) 'SYMFACT: i,iu(i),jpi,n,iu(n),jp,nh,iu(nh) = '
c    write(6,*) i,iu(i),jpi,n,iu(n),jp,nh,iu(nh)
     ncoef2=iu(n+1)-1
     write(6,*) 'inside symfact: ncoef2= ', ncoef2
     return
     end
c%%%%%%%%%%%%%%%%%%%%%%%%%%%%%%%%%%%%%%%%%%%%
     subroutine rownuma28(ia,ndof,irow)
     dimension ia(*),irow(*)
c
     do 1 i=1,ndof
     istart=ia(i)
     iend=ia(i+1)-1
     do 2 j=istart,iend
     irow(j)=i
 2   continue
 1   continue
     return
     end
c%%%%%%%%%%%%%%%%%%%%%%%%%%%%%%%%%%%%%%%%%%%%
     subroutine elunsym(iel,elk,ndofpe,be,lm,je,ae)    ! from Willie
     implicit real*8(a-h,o-z)
     dimension elk(ndofpe,ndofpe)
     dimension be(*),lm(*),je(*),ae(*)
c......purpose: generate "FAKED" unsym. element stiffness
```

```
      do 1 i=1,ndofpe
      do 2 j=i,ndofpe
      elk(i,j)=(i+j)*iel
      if (i .ne. j) then
      elk(j,i)=-elk(i,j)
      endif
2     continue
1     continue
c
      jcount=0
      do 13 k=1,ndofpe
      be(k)=0.
      icount=icount+1
      lm(k)=je(icount)
13    continue
c
      do 28 ic=1,ndofpe
      do 29 ir=1,ndofpe
      jcount=jcount+1
      ae(jcount)=elk(ir,ic)
29      continue
28    continue
      return
      end
c%%%%%%%%%%%%%%%%%%%%%%%%%%%%%%%%%%%%%%%%%%%%%%%%%%%%%%
      subroutine femesh(nel,ndofpe,ie,je)
      dimension ie(*),je(*)
      read(5,*) (ie(i),i=1,nel+1)
      read(5,*) (je(i),i=1,nel*ndofpe)
      return
      end
c%%%%%%%%%%%%%%%%%%%%%%%%%%%%%%%%%%%%%%%%%
      subroutine rhsload(ndof,b)
      implicit real*8(a-h,o-z)
      dimension b(*)
c--------------------
c     do 22 i=1,ndof
c22    b(i)=i
c--------------------
c......such that unknown displ. vector = {1.0, 1.0, ..., 1.0}
      b(1)=4.0
      b(2)=44.0
      b(3)=10.0
      b(4)=-20.0
      b(5)=42.0
      b(6)=-40.0
      b(7)=84.0
      b(8)=28.0
      b(9)=48.0
c--------------------
      return
      end
c%%%%%%%%%%%%%%%%%%%%%%%%%%%%%%%%%%%%%%%
      subroutine boundaryc(ndof,ibc,nbc,b)
      implicit real*8(a-h,o-z)
      dimension ibc(*),b(*)
c......assuming no boundary conditions ---->> answers are correct
      do 1 i=1,ndof
1     ibc(i)=0
c......applied boundary conditions ---->> answers are "ALSO" correct !!
```

```
      if (nbc .ne. 0) then
      ibc(2)=ndof
      ibc(4)=ndof
      ibc(5)=ndof
       b(2)=0.2
       b(4)=0.4
       b(5)=0.5
      endif
      return
      end
c%%%%%%%%%%%%%%%%%%%%%%%%%%%%%%%%%%%%%%%%%%%%%%
      subroutine unsymasem(ibc,ncoef1,an,ip,ndof,nel,elk,be,ndofpe,
     $ lm,je,ae,iakeep,ja,b,nbc)
      implicit real*8(a-h,o-z)
      dimension ibc(*),an(*),ip(*),elk(ndofpe,ndofpe),be(*),
     $      lm(*),je(*),ae(*),iakeep(*),ja(*),b(*)
      if (nbc .ne. 0) then
c......applied boundary conditions ---->> answers are "ALSO" correct !!
c    ibc(2)=1    ! should be consistent, such as ibc(2)=ndof
c    ibc(4)=1    ! hence, we should NOT need these 3 stmts
c    ibc(5)=1
      endif
c
      do 43 i=1,ncoef1
43    an(i)=0.
c
      do 42 i=1,ndof
42    ip(i)=0
c
      icount=0
      do 12 i=1,nel
      call elunsym(i,elk,ndofpe,be,lm,je,ae)    ! from Willie
c
      call unnumass(iakeep,ja,ibc,ae,be,lm,ndofpe,an,b,ip)
c
      write(6,*) 'el# = ',i,' UNSYM. elk(4,4)= '
      write(6,8) ((elk(ir,jc),jc=1,4),ir=1,4)
8     format(2x,4(e13.6,1x))
12    continue
      return
      end
c%%%%%%%%%%%%%%%%%%%%%%%%%%%%%%%%%%%%%%%%%%%%%%
      subroutine sorting(ndof,iakeep,itempo1,ja,tempo1,index,b,an)
      implicit real*8(a-h,o-z)
      dimension iakeep(*),itempo1(*),ja(*),tempo1(*),index(*),b(*)
      dimension an(*)
       write(6,*) '----------------------------'
       write(6,*) 'use method = 2 = sorting'
       write(6,*) '----------------------------'
      do 34 irow=1,ndof
       jbeg=iakeep(irow)
       jend=iakeep(irow+1)-1
       ncol=iakeep(irow+1)-iakeep(irow)
        do 35 icol=1,ncol
        itempo1(icol)=ja(jbeg+icol-1)
        tempo1(icol)=an(jbeg+icol-1)
35      continue
      call indexi(ncol,itempo1,index)
      call sorti(ncol,itempo1,index,ja(jbeg))
      call sortr(ncol, tempo1,index,an(jbeg))
```

```
 34   continue
c......output row-by-row
      do 48 irow=1,ndof
      jbeg=iakeep(irow)
      jend=iakeep(irow+1)-1
      write(6,*) 'irow,jbeg,jend = ',irow,jbeg,jend
      write(6,*) 'an row-wise = ',(an(j),j=jbeg,jend)
      write(6,*) 'rhs load = ',b(irow)
 48   continue
      return
      end
c%%%%%%%%%%%%%%%%%%%%%%%%%%%%%%%%%%%%%%%%%%%
      subroutine sortingan2(ndof,iakeep,itempo1,ja,tempo1,index,
    $ b,an,an2,x)
      implicit real*8(a-h,o-z)
      dimension iakeep(*),itempo1(*),ja(*),tempo1(*),index(*),b(*)
      dimension an(*),an2(*),x(*)
      write(6,*) '----------------------------'
      write(6,*) 'use method = 2 = sorting'
      write(6,*) '----------------------------'
      do 34 irow=1,ndof
      jbeg=iakeep(irow)
      jend=iakeep(irow+1)-1
      ncol=iakeep(irow+1)-iakeep(irow)
        do 35 icol=1,ncol
        itempo1(icol)=ja(jbeg+icol-1)
        tempo1(icol)= an(jbeg+icol-1)
          x(icol)=an2(jbeg+icol-1)
 35     continue
      call indexi(ncol,itempo1,index)
      call sorti(ncol,itempo1,index,ja(jbeg))
      call sortr(ncol, tempo1,index,an(jbeg))
      call sortr(ncol, x,index,an2(jbeg))
 34   continue
c......output row-by-row
      do 48 irow=1,ndof
      jbeg=iakeep(irow)
      jend=iakeep(irow+1)-1
      write(6,*) 'irow,jbeg,jend = ',irow,jbeg,jend
      write(6,*) 'an row-wise = ',(an(j),j=jbeg,jend)
      write(6,*) 'an2 col-wise = ',(an2(j),j=jbeg,jend)
      write(6,*) 'rhs load = ',b(irow)
 48   continue
      return
      end
c%%%%%%%%%%%%%%%%%%%%%%%%%%%%%%%%%%%%%%%%%%%
c%%%%%%%%%%%%%%%%%%%%%%%%%%%%%%%%%%%%%%%%%%%
      subroutine transpose(ndof,iakeep,ja,iakeept,jat,an,ant,b,aa)
      implicit real*8(a-h,o-z)
      dimension iakeep(*),ja(*),iakeept(*),jat(*),an(*),ant(*),b(*)
      dimension aa(*)
      write(6,*) '----------------------------'
      write(6,*) 'use method = 1 = transa'
      write(6,*) '----------------------------'
c
      call transa2m(ndof,ndof,iakeep,ja,iakeept,jat,an,ant)
      call transa2m(ndof,ndof,iakeept,jat,iakepp,jj,ant,aa)
c......output row-by-row
      do 32 irow=1,ndof
        jbeg=iakeep(irow)
```

```
      jend=iakeep(irow+1)-1
      write(6,*) 'irow,jbeg,jend = ',irow,jbeg,jend
      write(6,*) 'an row-wise = ',(aa(j),j=jbeg,jend)
      write(6,*) 'rhs load = ',b(irow)
 32   continue
      return
      end
c%%%%%%%%%%%%%%%%%%%%%%%%%%%%%%%%%%%%%%%%%%%%
      subroutine newje(jeold,iperm,icount,je)
      dimension jeold(*),iperm(*),je(*)
c
      do 1 i=1,icount
      iold=jeold(i)
      new=iperm(iold)
      je(i)=new
 1    continue
c
      return
      end
c%%%%%%%%%%%%%%%%%%%%%%%%%%%%%%%%%%%%%%%%%%%%
      subroutine transa(n,m,ia,ja,iat,jat)
c     implicit real*8(a-h,o-z)
      dimension ia(*),ja(*),iat(*),jat(*)
c++++++++++++++++++++++++++++++++++++++++++++++++++++++++++++++++++++++
c......PLEASE direct your questions to Dr. Nguyen (nguyen@cee.odu.edu)
c......purposes: After symbolic factorization,ja is just a merge lists
c......      WITHOUT ordering (i.e. the nonzero column numbers of a
c......      particular row are 12, 27, 14, 46, 22, 133). Upon
c......      completion, this routine will rearrange the nonzero column
c......      numbers WITH ordering, to be ready for numerical factorization
c......      phase (i.e. 12, 14, 22, 27, 46, 133)
c     input: ia,ja,an    given matrix in RR(C)U.
c        n        number of rows of the matrix.
c        m        number of columns of the matrix.
c     output: iat,jat,ant  transposed matrix in RR(C)O.
c++++++++++++++++++++++++++++++++++++++++++++++++++++++++++++++++++++++
c     dimension ia(6), ja(7), iat(6), jat(7)
c     dimension ia(neq+1), ja(ncoef1), iat(neq+1), jat(ncoef1)
c     neq=5
c     ncoef1=7
c     nrows=5
c     ncols=5
c......    x 2 0 4 5  ---> input unordered col # 2,5,4
c......    x 3 4 0  ---> input unordered col # 4,3
c......        x 0 5  ---> input col # 5
c......          x 5  ---> input col # 5
c......          x
c     ia(1)=1
c     ia(2)=4
c     ia(3)=6
c     ia(4)=7
c     ia(5)=8
c     ia(6)=8
c
c     ja(1)=2
c     ja(2)=5
c     ja(3)=4
c     ja(4)=4
c     ja(5)=3
c     ja(6)=5
```

```
c      ja(7)=5
c
c      call transa(nrows,ncols,ia,ja,iat,jat)
c      write(6,*) 'after reordered, ia(-)= ',(ia(i),i=1,neq+1)
c      write(6,*) 'after reordered, ja(-)= ',(ja(i),i=1,ncoef1)
c      stop
c      end
c%%%%%%%%%%%%%%%%%%%%%%%%%%%%%%%%%%%%%%%%%%%%%%%%%%%%%%%
c
       MH=M+1
       NH=N+1
       DO 10 I=2,MH
    10 IAT(I)=0
       IAB=IA(NH)-1
       DO 20 I=1,IAB
       J=JA(I)+2
       IF(J.LE.MH)IAT(J)=IAT(J)+1
    20 CONTINUE
       IAT(1)=1
       IAT(2)=1
       IF(M.EQ.1)GO TO 40
       DO 30 I=3,MH
    30 IAT(I)=IAT(I)+IAT(I-1)
    40 DO 60 I=1,N
       IAA=IA(I)
       IAB=IA(I+1)-1
       IF(IAB.LT.IAA)GO TO 60
       DO 50 JP=IAA,IAB
       J=JA(JP)+1
       K=IAT(J)
       JAT(K)=I
c      ANT(K)=AN(JP)
    50 IAT(J)=K+1
    60 CONTINUE
       call transa2(n,m,iat,jat,ia,ja)
       return
       end
c%%%%%%%%%%%%%%%%%%%%%%%%%%%%%%%%%%%%%%%%%%%%%%%%%%%%%%%
       subroutine transagain(n,m,ia,ja,iat,jat,an,ant)
       implicit real*8(a-h,o-z)
       dimension ia(*),ja(*),iat(*),jat(*),an(*),ant(*)
c
       MH=M+1
       NH=N+1
       DO 10 I=2,MH
    10 IAT(I)=0
       IAB=IA(NH)-1
       DO 20 I=1,IAB
       J=JA(I)+2
       IF(J.LE.MH)IAT(J)=IAT(J)+1
    20 CONTINUE
       IAT(1)=1
       IAT(2)=1
       IF(M.EQ.1)GO TO 40
       DO 30 I=3,MH
    30 IAT(I)=IAT(I)+IAT(I-1)
    40 DO 60 I=1,N
       IAA=IA(I)
       IAB=IA(I+1)-1
       IF(IAB.LT.IAA)GO TO 60
```

```
      DO 50 JP=IAA,IAB
      J=JA(JP)+1
      K=IAT(J)
      JAT(K)=I
      ANT(K)=AN(JP)
 50 IAT(J)=K+1
 60 CONTINUE
      call transa2m(n,m,iat,jat,ia,ja,ant,an)
      return
      end
c%%%%%%%%%%%%%%%%%%%%%%%%%%%%%%%%%%%%%%%%%%
c
c     Table 4.9  Unsymmetrical Sparse Assembly (continued)
c
c@@@@@@@@@@@@@@@@@@@@@@@@@@@@@@@@@@@@@@@@@@@@@@@@@@
c++++++++++++++++++++++++++++++++++++++++++++++++++++++++++++++++
c     purpose: d.t. nguyen's general, unsymmetrical sparse assembly
c     version: November 12, 2001
c     author:  Prof. Duc T. Nguyen  (Email= dnguyen@odu.edu)
c     notes:   these files (part1unsym.f, part2unsym.f) are stored
c              under sub-directory ~/cee/newfem/*complete*sparse*/
c
c              Please do "NOT" distribute this source code to anyone
c              without explicit written agreement by Prof. D.T. Nguyen
c++++++++++++++++++++++++++++++++++++++++++++++++++++++++++++++++
c@@@@@@@@@@@@@@@@@@@@@@@@@@@@@@@@@@@@@@@@@@@@@@@@@@
      subroutine transa2(n,m,ia,ja,iat,jat)
      dimension ia(*),ja(*),iat(*),jat(*)
c
      MH=M+1
      NH=N+1
      DO 10 I=2,MH
 10 IAT(I)=0
      IAB=IA(NH)-1
      DO 20 I=1,IAB
      J=JA(I)+2
      IF(J.LE.MH)IAT(J)=IAT(J)+1
 20 CONTINUE
      IAT(1)=1
      IAT(2)=1
      IF(M.EQ.1)GO TO 40
      DO 30 I=3,MH
 30 IAT(I)=IAT(I)+IAT(I-1)
 40 DO 60 I=1,N
      IAA=IA(I)
      IAB=IA(I+1)-1
      IF(IAB.LT.IAA)GO TO 60
      DO 50 JP=IAA,IAB
      J=JA(JP)+1
      K=IAT(J)
      JAT(K)=I
c     ANT(K)=AN(JP)
 50 IAT(J)=K+1
 60 CONTINUE
      return
      end
c%%%%%%%%%%%%%%
      subroutine transa2m(n,m,ia,ja,iat,jat,an,ant)
      implicit real*8(a-h,o-z)
      dimension ia(*),ja(*),iat(*),jat(*)
```

```
      dimension an(*),ant(*)
c
      MH=M+1
      NH=N+1
      DO 10 I=2,MH
   10 IAT(I)=0
      IAB=IA(NH)-1
      DO 20 I=1,IAB
      J=JA(I)+2
      IF(J.LE.MH)IAT(J)=IAT(J)+1
   20 CONTINUE
      IAT(1)=1
      IAT(2)=1
      IF(M.EQ.1)GO TO 40
      DO 30 I=3,MH
   30 IAT(I)=IAT(I)+IAT(I-1)
   40 DO 60 I=1,N
      IAA=IA(I)
      IAB=IA(I+1)-1
      IF(IAB.LT.IAA)GO TO 60
      DO 50 JP=IAA,IAB
      J=JA(JP)+1
      K=IAT(J)
      JAT(K)=I
      ANT(K)=AN(JP)
   50 IAT(J)=K+1
   60 CONTINUE
      return
      end
c%%%%%%%%%%%%%%%%%%%%%%%%%%%%%%%%%%%%%%%%%%%%%%%%%%%%%%%%%%%%%
      subroutine symbass(ie,je,iet,jet,n,ia,ja,iakeep,ibc)
      implicit real*8(a-h,o-z)
c++++++++++++++++++++++++++++++++++++++++++++++++++++++++++++++++++
c    purpose: nguyen's general, unsym. symbolic sparse assembly
c    note:    these files (part1unsym.f, part2unsym.f) are stored
c             under sub-directory ~/cee/newfem/*complete*/
c++++++++++++++++++++++++++++++++++++++++++++++++++++++++++++++++++
      dimension ie(*),je(*),iet(*),jet(*),ia(*),ja(*)
      dimension iakeep(*),ibc(*)
c++++++++++++++++++++++++++++++++++++++++++++++++++++++++++++++++++
c......PLEASE direct your questions to Dr. Nguyen (dnguyen@odu.edu)
c......Purposes: UN-symmetrical, sparse symbolic assembly
c......          This code is stored under file name *symb*.f, in sub-directory
c++++++++++++++++++++++++++++++++++++++++++++++++++++++++++++++++++
c......Input: ie(nel+1)=locations (or pointers) of the first non-zero
c......       of each row (of element-dof connectivity info.)
c......       je(nel*ndofpe)=global dof column number for non-zero
c......       terms of each row (of element-dof connectivity info.)
c......       iet(ndof+1)=locations (pointers) of the first non-zero
c......       of each row (of dof-element connectivity info.)
c......       jet(nel*ndofpe)=locations (pointers) of the first non-zero
c......       of each row (of dof-element connectivity info.)
c......       ia(ndof)= ndof in the positions correspond to Dirichlet b.c.
c......             0 elsewhere
c......Output:ia(ndof+1)=starting locations of the first non-zero
c......       off-diagonal terms for each row of structural stiffness
c......       matrix
c......       ja(ncoeff)=column numbers (unordered) correspond to
c......       each nonzero, off-diagonal term of each row of structural
c......       stiffness matrix
```

```
c++++++++++++++++++++++++++++++++++++++++++++++++++++++++++++++++++++++
      write(6,*) '                         '
      write(6,*) '                         '
c++++++++++++++++++++++++++++++++++++++++++++++++++++++++++++++++++++++
      jp=1                        !001
      nm1=n-1                     !002
      do 30 i=1,n                 !003 last row (=eq) will NOT be skipped
      jpi=jp                      !004 delayed counter for ia(-) array
c     if ( ia(i) .eq. n ) then    !005 skip row which correspond to Dirichlet b.c.
      if ( ibc(i) .eq. n ) then
      ja(jp)=i
      jp=jp+1
      ia(i)=i
      iakeep(i)=jpi
      go to 30
      endif
      ieta=iet(i)                 !006 begin index (to find how many elements)
      ietb=iet(i+1)-1             !007 end index (to find how many elements)
        do 20 ip=ieta,ietb        !008 loop covering ALL elements attached to row i
        j=jet(ip)                 !009 actual "element number" attached to row i
        iea=ie(j)                 !010 begin index (to find how many nodes attached to element j)
        ieb=ie(j+1)-1             !011 end index (to find how many nodes attached to element j)
        do 10 kp=iea,ieb          !012 loop covering ALL nodes attached to element j
        k=je(kp)                  !013 actual "node, or column number" attached to element j
c       if (k .le. i) go to 10    !014 skip, if it involves with LOWER triangular portion
        if ( ia(k) .ge. i ) go to 10 !015 skip, if same node already been accounted by earlier elements
        if ( ibc(k) .eq. n ) go to 10 !  skip this column (correspond to bc)
        ja(jp)=k                  !016 record "column number" associated with non-zero off-diag. term
c       write(6,*) 'jp,ja(jp)= ',jp,ja(jp)
        jp=jp+1                   !017 increase "counter" for column number array ja(-)
        ia(k)=i                   !018 record node (or column number) k already contributed to row i
c       write(6,*) 'k,ia(k)= ',k,ia(k)
 10     continue                  !019
 20     continue                  !020
c30     ia(i)=jpi                 !021 record "starting location" of non-zero off-diag.
 30     iakeep(i)=jpi             !021 record "starting location" of non-zero off-diag.
c                                 !   terms associated with row i
c     ia(n)=jp                    !022 record "starting location" of non-zero term of LAST ROW
c     ia(n+1)=jp                  !023 record "starting location" of non-zero term of LAST ROW + 1
      iakeep(n+1)=jp              !023 record "starting location" of non-zero term of LAST ROW + 1
c
c     ncoef1=ia(n+1)-1
c     write(6,*) 'ia(-) array = ',(ia(i),i=1,n+1)
c     write(6,*) 'ja(-) array = ',(ja(i),i=1,ncoef1)
c
      return
      end
c%%%%%%%%%%%%%%%%%%%%%%%%%%%%%%%%%%%%%%%%%%%%%%%%%%%%%%%
      subroutine numass(ia,ja,idir,ae,be,lm,ndofpe,an,ad,b,ip)
      implicit real*8(a-h,o-z)
      dimension ia(*),ja(*),idir(*),ae(*),be(*),lm(*),an(*)
      dimension ad(*),b(*),ip(*)
c++++++++++++++++++++++++++++++++++++++++++++++++++++++++++++++++++++++
c......PLEASE direct your questions to Dr. Nguyen (nguyen@cee.odu.edu)
c......Purposes: symmetrical, sparse numerical assembly
c......     This code is stored under file name *symb*.f, in sub-directory
c......     ~
c++++++++++++++++++++++++++++++++++++++++++++++++++++++++++++++++++++++
c......Input: ia(ndof+1)=starting locations of the first non-zero
c......     off-diagonal terms for each row of structural stiffness
```

```
c......    matrix
c......       ja(ncoeff)=column numbers (unordered) correspond to
c......       each nonzero, off-diagonal term of each row of structural
c......       stiffness matrix
c......       idir(ndof)= 1 in the positions correspond to Dirichlet b.c.
c......          0 elsewhere
c......       ae(ndofpe**2),be(ndofpe)= element (stiffness) matrix,
c......       and element (load) vector
c......       lm(ndofpe)= global dof associated with a finite element
c......       ndofpe= number of dof per element
c......       b(ndof)= before using this routine, values of b(-) should
c......       be initialized to:
c......       Ci, values of prescribed Dirichlet bc at proper locations
c......       or values of applied nodal loads
c
c......Output: an(ncoeff1)= values of nonzero, off-diagonal terms of
c......       structural stiffness matrix
c......       ad(ndof)= values off diagonal terms of structural stiffness
c......       matrix
c......       b(ndof)= right-hand-side (load) vector of system of linear
c......       equations
c......Temporary Arrays:
c......       ip(ndof)= initialized to 0
c......          then IP(-) is used and reset to 0
c%%%%%%%%%%%%%%%%%%%%%%%%%%%%%%%%%%%%%%%%%%%%%%%%%%%
       do 40 L=1,ndofpe          !001 local "row" dof
       i=lm(L)                   !002 global"row" dof
       if ( idir(i) .ne. 0 ) go to 401   !003 skip, if DIRICHLET b.c.
       k=L-ndofpe                !004 to find location of element k-diag
       ad(i)=ad(i)+ae(k+L*ndofpe)     !005 assemble K-diag
       b(i)=b(i)+be(L)                !006 assemble element rhs load vector
       kk=0                      !007 flag, to skip contribution of entire
c                               !  global row i if all global col #
c                               !  j < i, or if entire row i belongs
c                               !  to LOWER triangle
       do 20 LL=1,ndofpe          !008 local "column" dof
       k=k+ndofpe                 !009 find location of element stiffness k
       if (LL .eq. L) go to 20     !010 skip, diag. term already taken care
       j=lm(LL)                   !011 global column dof
       if ( idir(j) .ne. 0 ) go to 10   !012 skip, if DIRICHLET b.c.
       if (j .lt. i) go to 20      !013 skip, if LOWER portion
       ip(j)=k                    !014 record global column # j (associated with
c                               !  global row # i) correspond to k-th term
c                               !  of element stiffness k
       kk=1                       !015 FLAG, indicate row L of [k] do have
c                               !  contribution to global row I of [K]
       go to 20                   !016
10     b(i)=b(i)-b(j)*ae(k)           !017 modify rhs load vector due to DIRICHLET b.c.
20     continue                   !018
       if (kk .eq. 0) go to 40     !019 skip indicator (see line 007)
       iaa=ia(i)                  !020 start index
       iab=ia(i+1)-1              !021 end index
       do 30 j=iaa,iab            !022 loop covering all col numbers associated
c                               !  with global row i
       k=ip( ja(j) )              !023 ip ( col # ) already defined on line 014
c                               !    or initialized to ZERO initially
       if (k .eq. 0) go to 30      !024 skip, because eventhough row L of [k] do
c                               !  have contribution to global row I of [K],
c                               !  some terms of row L (of [k]) which associated
c                               !  with DIRICHLET b.c. columns should be SKIPPED
```

```
        an(j)=an(j)+ae(k)           !025 assemble [K] from [k]
        ip( ja(j) )=0               !026 reset to ZERO for col # j before considering
c                               !  the next row L
 30     continue                    !027
        go to 40
401     ad(i)=1.0                   !  reset the diagonal of K(i,i)=1.0, due to b.c.
 40     continue                    !028
c......print debugging results
c       ndof=12
c       ncoeff1=ia(ndof+1)-1
c       write(6,*) 'at the end of routine numass'
c       write(6,*) 'ia(-) array = ',(ia(i),i=1,ndof+1)
c       write(6,*) 'ja(-) array = ',(ja(i),i=1,ncoeff1)
c       write(6,*) 'ad(-) array = ',(ad(i),i=1,ndof)
c       write(6,*) 'an(i) array = ',(an(i),i=1,ncoeff1)
        return
        end
c%%%%%%%%%%%%%%%%%%%%%%%%%%%%%%%%%%%%%%%%%%%%%%%%%%%%%%%%%%%%%
        subroutine unnumass(ia,ja,idir,ae,be,lm,ndofpe,an,b,ip)
        implicit real*8(a-h,o-z)
        dimension ia(*),ja(*),idir(*),ae(*),be(*),lm(*),an(*)
c       dimension ad(*),b(*),ip(*)
        dimension b(*),ip(*)
c++++++++++++++++++++++++++++++++++++++++++++++++++++++++++++++++++++++++
c......PLEASE direct your questions to Dr. Nguyen (nguyen@cee.odu.edu)
c......Purposes: UN-symmetrical, sparse numerical assembly
c......      This code is stored under file name *symb*.f, in sub-directory
c......          ~
c++++++++++++++++++++++++++++++++++++++++++++++++++++++++++++++++++++++++
c......Input: ia(ndof+1)=starting locations of the first non-zero
c......      off-diagonal terms for each row of structural stiffness
c......      matrix
c......      ja(ncoeff)=column numbers (unordered) correspond to
c......      each nonzero, off-diagonal term of each row of structural
c......      stiffness matrix
c......      idir(ndof)= 1 in the positions correspond to Dirichlet b.c.
c......          0 elsewhere
c......      ae(ndofpe**2),be(ndofpe)= element (stiffness) matrix,
c......      and element (load) vector
c......      lm(ndofpe)= global dof associated with a finite element
c......      ndofpe= number of dof per element
c......      b(ndof)= before using this routine, values of b(-) should
c......      be initialized to:
c......      Ci, values of prescribed Dirichlet bc at proper locations
c......      or  values of applied nodal loads
c
c......Output: an(ncoeff1)= values of nonzero, off-diagonal terms of
c......      structural stiffness matrix
c......      ad(ndof)= values off diagonal terms of structural stiffness
c......      matrix
c......      b(ndof)= right-hand-side (load) vector of system of linear
c......      equations
c......Temporary Arrays:
c......      ip(ndof)= then initialized to 0
c......          then IP(-) is used and reset to 0
c%%%%%%%%%%%%%%%%%%%%%%%%%%%%%%%%%%%%%%%%%%%%%%%%%%%%%%%%%%%%%
        do 40 L=1,ndofpe           !001 local "row" dof
        i=lm(L)                    !002 global"row" dof
        if ( idir(i) .ne. 0 ) go to 401   !003 skip, if DIRICHLET b.c.
        k=L-ndofpe                 !004 to find location of element k-diag
```

```
c     ad(i)=ad(i)+ae(k+L*ndofpe)        !005 assemble K-diag will be done later
      b(i)=b(i)+be(L)                    !006 assemble element rhs load vector
c     kk=0                               !007 flag, to skip contribution of entire
c     write(6,*) 'L,i,k = ',L,i,k
c                                !   global row i if all global col #
c                                !   j < i, or if entire row i belongs
c                                !   to LOWER triangle
      do 20 LL=1,ndofpe                  !008 local "column" dof
      k=k+ndofpe                         !009 find location of element stiffness k
c     if (LL .eq. L) go to 20            !010 skip, diag. term already taken care
      j=lm(LL)                           !011 global column dof
      if ( idir(j) .ne. 0 ) go to 10     !012 skip, if DIRICHLET b.c.
c     if (j .lt. i) go to 20             !013 skip, if LOWER portion
      ip(j)=k                            !014 record global column # j (associated with
c                                !   global row # i) correspond to k-th term
c                                !   of element stiffness k
c     kk=1                               !015 FLAG, indicate row L of [k] do have
c     write(6,*) 'LL,k,j,ip(j) = '
c     write(6,*) LL,k,j,ip(j)
c                                !   contribution to global row I of [K]
      go to 20                           !016
10    b(i)=b(i)-b(j)*ae(k)               !017 modify rhs load vector due to DIRICHLET b.c.
c     write(6,*) 'i,j,k,b(j),b(i),ae(k) = '
c     write(6,*) i,j,k,b(j),b(i),ae(k)
20    continue                           !018
c     if (kk .eq. 0) go to 40            !019 skip indicator (see line 007)
      iaa=ia(i)                          !020 start index
      iab=ia(i+1)-1                      !021 end index
      do 30 j=iaa,iab                    !022 loop covering all col numbers associated
c                                !   with global row i
      k=ip( ja(j) )                      !023 ip ( col # ) already defined on line 014
c     write(6,*) 'j,ja(j),k = '
c     write(6,*) j,ja(j),k
c                                !         or initialized to ZERO initially
      if (k .eq. 0) go to 30             !024 skip, because eventhough row L of [k] do
c                                !   have contribution to global row I of [K],
c                                !   some terms of row L (of [k]) which associated
c                                !   with DIRICHLET b.c. columns should be SKIPPED
      an(j)=an(j)+ae(k)                  !025 assemble [K] from [k]
      ip( ja(j) )=0                      !026 reset to ZERO for col # j before considering
c     write(6,*) 'j,k,ae(k),an(j),ja(j), = '
c     write(6,*) j,k,ae(k),an(j),ja(j)
c                                !   the next row L
30    continue                           !027
      go to 40
c......find the location of diagonal term,
c......which corresponds to Dirichlet b.c.
401   locate=ia(i)
      an(locate)=1.0               !   reset diagonal term K(i,i)=1.0, due to b.c.
40    continue                           !028
c......print debugging results
      ncoeff1=ia(ndof+1)-1
c     write(6,*) 'at the end of routine unnumass'
c     write(6,*) 'ia(-) array = ',(ia(i),i=1,ndof+1)
c     write(6,*) 'ja(-) array = ',(ja(i),i=1,ncoeff1)
c     write(6,*) 'an(i) array = ',(an(i),i=1,ncoeff1)
c     write(6,*) 'rhs b = ',(b(i),i=1,ndof)
      return
      end
c%%%%%%%%%%%%%%%%%%%%%%%%%%%%%%%%%%%%%%%%%%%%%%%%%%%%%%%%%%%%%%%%
```

```
      SUBROUTINE indexi(n,irr,indx)
      INTEGER n,indx(n),M,NSTACK,irr(n)
      PARAMETER (M=7,NSTACK=50)
      INTEGER i,indxt,ir,itemp,j,jstack,k,l,istack(NSTACK)
      do 11 j=1,n
        indx(j)=j
11    continue
      jstack=0
      l=1
      ir=n
1     if(ir-l.lt.M)then
        do 13 j=l+1,ir
          indxt=indx(j)
          iii=irr(indxt)
          do 12 i=j-1,1,-1
            if(irr(indx(i)).le.iii)goto 2
            indx(i+1)=indx(i)
12        continue
          i=0
2         indx(i+1)=indxt
13      continue
        if(jstack.eq.0)return
        ir=istack(jstack)
        l=istack(jstack-1)
        jstack=jstack-2
      else
        k=(l+ir)/2
        itemp=indx(k)
        indx(k)=indx(l+1)
        indx(l+1)=itemp
        if(irr(indx(l+1)).gt.irr(indx(ir)))then
          itemp=indx(l+1)
          indx(l+1)=indx(ir)
          indx(ir)=itemp
        endif
        if(irr(indx(l)).gt.irr(indx(ir)))then
          itemp=indx(l)
          indx(l)=indx(ir)
          indx(ir)=itemp
        endif
        if(irr(indx(l+1)).gt.irr(indx(l)))then
          itemp=indx(l+1)
          indx(l+1)=indx(l)
          indx(l)=itemp
        endif
        i=l+1
        j=ir
        indxt=indx(l)
        iii=irr(indxt)
3       continue
          i=i+1
        if(irr(indx(i)).lt.iii)goto 3
4       continue
          j=j-1
        if(irr(indx(j)).gt.iii)goto 4
        if(j.lt.i)goto 5
        itemp=indx(i)
        indx(i)=indx(j)
        indx(j)=itemp
        goto 3
```

```
5     indx(l)=indx(j)
      indx(j)=indxt
      jstack=jstack+2
      if(jstack.gt.NSTACK)pause 'NSTACK too small in indexx'
      if(ir-i+1.ge.j-l)then
        istack(jstack)=ir
        istack(jstack-1)=i
        ir=j-1
      else
        istack(jstack)=j-1
        istack(jstack-1)=l
        l=i
      endif
      endif
      goto 1
      END
c%%%%%%%%%%%%%%%%%%%%%%%%%%%%%%%%%%%%%%%%%%%%%%%%%%%%%%%%%
      subroutine sorti(n,in,index,iout)
      dimension in(*),index(*),iout(*)
      do 1 i=1,n
      locate=index(i)
      iout(i)=in(locate)
1     continue
      return
      end
c%%%%%%%%%%%%%%%%%%%%%%%%%%%%%%%%%%%%%%%%%%%%%%%%%%%%%%%%%
      subroutine sortr(n,ain,index,out)
      implicit real*8(a-h,o-z)
      dimension ain(*),index(*),out(*)
      do 1 i=1,n
      locate=index(i)
      out(i)=ain(locate)
1     continue
      return
      end
c%%%%%%%%%%%%%%%%%%%%%%%%%%%%%%%%%%%%%%%%%%%%%%%%%%%%%%%%%
      subroutine adjacency(ie,je,iet,jet,n,ia,ja,iakeep)
c++++++++++++++++++++++++++++++++++++++++++++++++++++++++++++++++++
c    purpose: obtaining adjacency arrays for METIS/MMD etc... reordering
c    note:    could be applied to either "node" number, or "dof" number !
c             stored under sub-directory ~/cee/newfem/*complete*/adjacency.f
c++++++++++++++++++++++++++++++++++++++++++++++++++++++++++++++++++
      dimension ie(*),je(*),iet(*),jet(*),ia(*),ja(*)
      dimension iakeep(*)
c++++++++++++++++++++++++++++++++++++++++++++++++++++++++++++++++++
c......Input: ie(nel+1)=locations (or pointers) of the first non-zero
c......    of each row (including LOWER & UPPER TRIANGULAR, excluding DIAGONAL TERMS,
c......    of element-dof [or node] connectivity info.)
c......    je(nel*ndofpe)=global dof column number for non-zero
c......    terms of each row (of element-dof [or node] connectivity info.)
c
c......    iet(ndof+1)=locations (pointers) of the first non-zero
c......    of each row (of dof [or node] -element connectivity info.)
c......    jet(nel*ndofpe)=locations (pointers) of the first non-zero
c......    of each row (of dof-element connectivity info.)
c
c......notes: ndof may represent total # dof, or total # nodes
c
c         this subroutine is "almost identical" to Duc T. Nguyen's routine
c         for UNSYM. SPARSE "SYMBOLIC ASSEMBLY", where we just simply SKIPING the
```

```
c        diagonal term (since METIS/MMD/ND does NOT count DIAGONAL terms
c        in the 2 adjacency arrays). Also, this routine will NOT require to
c        consider Dirichlet boundary conditions !
c
c......Output:ia(ndof+1)=starting locations of the first non-zero
c......     terms for each row of structural stiffness matrix
c......     ja(ncoeff)=column numbers (unordered) correspond to
c......     each nonzero term of each row of structural
c......     stiffness matrix
c+++++++++++++++++++++++++++++++++++++++++++++++++++++++++++++++++++++++
      jp=1                     !001
      do 30 i=1,n              !003 last row (=eq) will NOT be skipped
      jpi=jp                   !004 delayed counter for ia(-) array
      ieta=iet(i)              !006 begin index (to find how many elements)
      ietb=iet(i+1)-1          !007 end index (to find how many elements)
        do 20 ip=ieta,ietb     !008 loop covering ALL elements attached to row i
        j=jet(ip)              !009 actual "element number" attached to row i
        iea=ie(j)              !010 begin index (to find how many nodes attached to element j)
        ieb=ie(j+1)-1          !011 end index (to find how many nodes attached to element j)
          do 10 kp=iea,ieb     !012 loop covering ALL nodes attached to element j
          k=je(kp)             !013 actual "node, or column number" attached to element j
c-------------------------------------------------------------------------------
          if (k .eq. i) go to 10   !014 SKIP, if it involves with DIAGONAL terms
c-------------------------------------------------------------------------------
          if ( ia(k) .ge. i ) go to 10 !015 skip, if same node already been accounted by earlier elements
          ja(jp)=k             !016 record "column number" associated with non-zero off-diag. term
c     write(6,*) 'jp,ja(jp)= ',jp,ja(jp)
          jp=jp+1              !017 increase "counter" for column number array ja(-)
          ia(k)=i             !018 record node (or column number) k already contributed to row i
c     write(6,*) 'k,ia(k)= ',k,ia(k)
  10      continue             !019
  20    continue               !020
  30  iakeep(i)=jpi            !021 record "starting location" of non-zero terms associated with row i
      iakeep(n+1)=jp           !023 record "starting location" of non-zero term of LAST ROW + 1
c
      ncoef1=iakeep(n+1)-1
      write(6,*) 'iakeep(-) array = ',(iakeep(i),i=1,n+1)
      write(6,*) 'ja(-) array = ',(ja(i),i=1,ncoef1)
c
      return
      end
c%%%%%%%%%%%%%%%%%%%%%%%%%%%%%%%%%%%%%%%%%%%%%%%%%%%%%
      subroutine symbassym(ie,je,iet,jet,n,ia,ja)
      dimension ie(*),je(*),iet(*),jet(*),ia(*),ja(*)
c+++++++++++++++++++++++++++++++++++++++++++++++++++++++++++++++++++++++
c......PLEASE direct your questions to Dr. Nguyen (dnguyen@odu.edu)
c......Purposes: symmetrical, sparse symbolic assembly
c......     This code is stored under file name *symb*.f, in sub-directory
c......
c+++++++++++++++++++++++++++++++++++++++++++++++++++++++++++++++++++++++
c......Input: ie(nel+1)=locations (or pointers) of the first non-zero
c......     of each row (of element-dof connectivity info.)
c......     je(nel*ndofpe)=global dof column number for non-zero
c......     terms of each row (of element-dof connectivity info.)
c......     iet(ndof+1)=locations (pointers) of the first non-zero
c......     of each row (of dof-element connectivity info.)
c......     jet(nel*ndofpe)=locations (pointers) of the first non-zero
c......     of each row (of dof-element connectivity info.)
c......     ia(ndof)= ndof in the positions correspond to Dirichlet b.c.
c......             0 elsewhere
```

```
c......Output:ia(ndof+1)=starting locations of the first non-zero
c......    off-diagonal terms for each row of structural stiffness
c......    matrix
c......    ja(ncoeff)=column numbers (unordered) correspond to
c......    each nonzero, off-diagonal term of each row of structural
c......    stiffness matrix
c+++++++++++++++++++++++++++++++++++++++++++++++++++++++++++++++++++++++++
      jp=1                    !001
      nml=n-1                 !002
      do 30 i=1,nml           !003 last row (= eq) will be skipped
      jpi=jp                  !004 delayed counter for ia(-) array
      if ( ia(i) .eq. n ) go to 30    !005 skip row which correspond to Dirichlet b.c.
      ieta=iet(i)             !006 begin index (to find how many elements)
      ietb=iet(i+1)-1         !007 end index (to find how many elements)
        do 20 ip=ieta,ietb    !008 loop covering ALL elements attached to row i
      j=jet(ip)               !009 actual "element number" attached to row i
      iea=ie(j)               !010 begin index (to find how many nodes attached to element j)
      ieb=ie(j+1)-1           !011 end index (to find how many nodes attached to element j)
        do 10 kp=iea,ieb      !012 loop covering ALL nodes attached to element j
      k=je(kp)                !013 actual "node, or column number" attached to element j
      if (k .le. i) go to 10  !014 skip, if it involves with LOWER triangular portion
      if ( ia(k) .ge. i ) go to 10  !015 skip, if same node already been accounted by earlier elements
      ja(jp)=k                !016 record "column number" associated with non-zero off-diag. term
      jp=jp+1                 !017 increase "counter" for column number array ja(-)
      ia(k)=i                 !018 record node (or column number) k already contributed to row i
10    continue                !019
20    continue                !020
30    ia(i)=jpi               !021 record "starting location" of non-zero off-diag.
!     terms associated with row i
      ia(n)=jp                !022 record "starting location" of non-zero term of LAST ROW
      ia(n+1)=jp              !023 record "starting location" of non-zero term of LAST ROW + 1
c
      ncoefl=ia(n+1)-1
c     write(6,*) 'ia(-) array = ',(ia(i),i=1,n+1)
c     write(6,*) 'ja(-) array = ',(ja(i),i=1,ncoefl)
c
      return
      end
c%%%%%%%%%%%%%%%%%%%%%%%%%%%%%%%%%%%%%%%%%%%%%%%%%%%%%%%%
      subroutine numassuns(ia,ja,idir,ae,be,lm,ndofpe,an,ad,b,ip
     $ ,an2,itempol)
      implicit real*8(a-h,o-z)
      dimension ia(*),ja(*),idir(*),ae(*),be(*),lm(*),an(*),an2(*)
      dimension ad(*),b(*),ip(*),itempol(*)
c+++++++++++++++++++++++++++++++++++++++++++++++++++++++++++++++++++++++++
c......PLEASE direct your questions to Dr. Nguyen (nguyen@cee.odu.edu)
c......Purposes: UNsymmetrical, sparse numerical assembly
c......      This code is stored under file name *symb*.f, in sub-directory
c......      ~
c+++++++++++++++++++++++++++++++++++++++++++++++++++++++++++++++++++++++++
c......Input: ia(ndof+1)=starting locations of the first non-zero
c......    off-diagonal terms for each row of structural stiffness
c......    matrix
c......    ja(ncoeff)=column numbers (unordered) correspond to
c......    each nonzero, off-diagonal term of each row of structural
c......    stiffness matrix
c......    idir(ndof)= 1 in the positions correspond to Dirichlet b.c.
c......         0 elsewhere
c......    ae(ndofpe**2),be(ndofpe)= element (stiffness) matrix,
c......    and element (load) vector
```

```
c......      lm(ndofpe)= global dof associated with a finite element
c......      ndofpe= number of dof per element
c......      b(ndof)= before using this routine, values of b(-) should
c......      be initialized to:
c......      Ci, values of prescribed Dirichlet bc at proper locations
c......      or  values of applied nodal loads
c
c......Output: an(ncoeff1)= values of nonzero, off-diagonal terms of
c......      structural stiffness matrix (upper triangular portion)
c......      an2(ncoeff1)= values of nonzero, off-diagonal terms of
c......      structural stiffness matrix (lower triangular portion)
c......      ad(ndof)= values off diagonal terms of structural stiffness
c......      matrix
c......      b(ndof)= right-hand-side (load) vector of system of linear
c......      equations
c......Temporary Arrays:
c......      ip(ndof)= initialized to 0
c......          then IP(-) is used and reset to 0
c%%%%%%%%%%%%%%%%%%%%%%%%%%%%%%%%%%%%%%%%%%%%%%%%%%
        do 40 L=1,ndofpe          !001 local "row" dof
        i=lm(L)                   !002 global"row" dof
        if ( idir(i) .ne. 0 ) go to 401      !003 skip, if DIRICHLET b.c.
        k=L-ndofpe                !004 to find location of element k-diag
c       k2=(L-1)*ndofpe+L
        k2=(L-1)*ndofpe
        write(6,*) 'local row #L,k,k2= ',L,k,k2
        ad(i)=ad(i)+ae(k+L*ndofpe)      !005 assemble K-diag
        b(i)=b(i)+be(L)           !006 assemble element rhs load vector
        kk=0                      !007 flag, to skip contribution of entire
!       global row i if all global col #
!       j < i, or if entire row i belongs
!       to LOWER triangle
        do 20 LL=1,ndofpe         !008 local "column" dof
        k=k+ndofpe                !009 find location of element stiffness k
        k2=k2+1                   !  find location of el stiff (lower)
        write(6,*) 'local col #LL,k,k2= ',LL,k,k2
        if (LL .eq. L) go to 20   !010 skip, diag. term already taken care
c       k2=k2+1                   !  find location of el stiff (lower)
        j=lm(LL)                  !011 global column dof
        if ( idir(j) .ne. 0 ) go to 10   !012 skip, if DIRICHLET b.c.
        if (j .lt. i) go to 20    !013 skip, if LOWER portion
        ip(j)=k                   !014 record global column # j (associated with
        itempo1(j)=k2
!  global row # i) correspond to k-th term
!  of element stiffness k
        kk=1                      !015 FLAG, indicate row L of [k] do have
!  contribution to global row I of [K]
        go to 20                  !016
10      b(i)=b(i)-b(j)*ae(k)      !017 modify rhs load vector due to DIRICHLET b.c.
20      continue                  !018
        if (kk .eq. 0) go to 40   !019 skip indicator (see line 007)
        iaa=ia(i)                 !020 start index
        iab=ia(i+1)-1             !021 end index
        do 30 j=iaa,iab           !022 loop covering all col numbers associated
!  with global row i
        k=ip( ja(j) )             !023 ip ( col # ) already defined on line 014
        k2=itempo1( ja(j) )
!         or initialized to ZERO initially
        if (k .eq. 0) go to 30    !024 skip, because eventhough row L of [k] do
c                                 !  have contribution to global row I of [K},
```

```
c                        !   some terms of row L (of [k]) which associated
c                        !   with DIRICHLET b.c. columns should be SKIPPED
      write(6,*) 'j,k,k2= ',j,k,k2
      write(6,*) 'an(j),ae(k),an2(j),ae(k2)= '
      write(6,*)  an(j),ae(k),an2(j),ae(k2)
      an(j)=an(j)+ae(k)          !025 assemble [K] from [k]
      an2(j)=an2(j)+ae(k2)       !025 assemble [K] from [k]
      write(6,*) 'an(j),ae(k),an2(j),ae(k2)= '
      write(6,*)  an(j),ae(k),an2(j),ae(k2)
      ip( ja(j) )=0              !026 reset to ZERO for col # j before considering
!   the next row L
30    continue                  !027
      go to 40
401   ad(i)=1.0                 !   reset the diagonal of K(i,i)=1.0, due to b.c.
40    continue                  !028
c......print debugging results
c     ndof=12
c     ncoeff1=ia(ndof+1)-1
c     write(6,*) 'at the end of routine numass'
c     write(6,*) 'ia(-) array = ',(ia(i),i=1,ndof+1)
c     write(6,*) 'ja(-) array = ',(ja(i),i=1,ncoeff1)
c     write(6,*) 'ad(-) array = ',(ad(i),i=1,ndof)
c     write(6,*) 'an(i) array = ',(an(i),i=1,ncoeff1)
      return
      end
c%%%%%%%%%%%%%%%%%%%%%%%%%%%%%%%%%%%%%%%%%%%%%%%%%%%%%%%%%
c%%%%%%%%%%%%%%%%%%%%%%%%%%%%%%%%%%%%%%%%%%%%%%%%%%%%%%%%%
```

4.10 Unsymmetrial Sparse Equations Data Formats

In this section, the system stiffness matrix given by Eq.(4.70) is used as an illustrative example for I. Duff's [3.2, 3.6, 3.7, 4.2, 4.3], Runesha/Nguyen's [3.11], and SGI's [3.1] input data format requirements.

[A] I. Duff's Unsymmetrical Sparse Matrix Storage Scheme[3.2, 3.6, 3.7, 4.2, 4.3]

The matrix equation, shown in Eq.(4.70), can be described by the following storage schemes:

$$
\text{IAROW}_{\text{MA28}} \begin{Bmatrix} 1 \\ 2 \\ 3 \\ 4 \\ 5 \\ 6 \\ 7 \\ 8 \\ 9 \\ 10 \\ 11 \\ 12 \\ 13 \\ 14 \\ 15 \\ 16 \\ 17 \\ 18 \\ 19 \\ 20 \\ 21 \\ 22 \\ 23 \\ 24 \\ 25 \\ 26 \\ 27 \\ 28 \\ 29 \\ 30 \\ 31 \end{Bmatrix} = \begin{Bmatrix} 1 \\ 2 \\ 2 \\ 2 \\ 2 \\ 3 \\ 3 \\ 3 \\ 3 \\ 3 \\ 3 \\ 4 \\ 5 \\ 5 \\ 5 \\ 5 \\ 6 \\ 6 \\ 6 \\ 6 \\ 6 \\ 6 \\ 7 \\ 7 \\ 7 \\ 7 \\ 8 \\ 8 \\ 8 \\ 8 \\ 9 \end{Bmatrix}
\tag{4.74}
$$

$$\{JACOL_{ma28} (1,2,3,...31)\}^T = \{\boxed{1}, \quad \boxed{2,3,6,8}, \quad \boxed{2,3,5,6,7,8}, \quad \boxed{4}, \quad (4.75)$$
$$\boxed{3,5,6,7}, \quad \boxed{2,3,5,6,7,8}, \quad \boxed{3,5,6,7},$$
$$\boxed{2,3,6,8}, \quad \boxed{9}\}$$

$$\{AN_{ma28}(1,2,3,...,31)\}^T = \{\boxed{1}, \boxed{16., 20., 24., -12.}, \boxed{-20., 28., 5., 25., 6., -16.} \quad (4.76)$$
$$\boxed{1.0}, \boxed{-5., 6., -4., 7.}, \boxed{-24., -25., 4., 60., 26., -26.}$$
$$\boxed{-6., -7., -26., 32.}, \boxed{12., 16., 26., 12.}, \boxed{1.}\}$$

Remarks

It is required that column number, shown in Eq.(4.76), be ordered by increasing number.

[B] Nguyen-Runesha's Unsymmetrical Sparse Matrix Storage Scheme[3.11]

The matrix equation, shown in Eq.(4.70), can be described by the following mixed row-wise (for the upper – triangular portion), and column-wise (for the lower – triangular portion) storage schemes. Here, we assume the matrix Eq.(4.70) is still symmetrical in non-zero locations.

$$IA_{N-R} \begin{pmatrix} 1 \\ 2 \\ 3 \\ 4 \\ 5 \\ 6 \\ 7 \\ 8 \\ 9 \\ 10 = NDOF+1 \end{pmatrix} = \begin{Bmatrix} 1 \\ 1 \\ 4 \\ 8 \\ 8 \\ 9 \\ 10 \\ 12 \\ 12 \\ 12 \end{Bmatrix} \qquad (4.77)$$

$$
JA_{N-R} \begin{Bmatrix} 1 \\ 2 \\ 3 \\ 4 \\ 5 \\ 6 \\ 7 \\ 8 \\ 9 \\ 10 \\ 11 \end{Bmatrix} = \begin{Bmatrix} 3 \\ 6 \\ 8 \\ & 5 \\ & 6 \\ & 7 \\ & 8 \\ 6 \\ 7 \\ & 7 \\ & 8 \end{Bmatrix} \tag{4.78}
$$

$$
AN_{N-R} \begin{Bmatrix} 1 \\ 2 \\ 3 \\ 4 \\ 5 \\ 6 \\ 7 \\ 8 \\ 9 \\ 10 \\ 11 \end{Bmatrix} = \begin{Bmatrix} 20. \\ 24. \\ -12. \\ & 5. \\ & 25. \\ & 6. \\ & -16. \\ -4. \\ 7. \\ & 26. \\ & -26. \end{Bmatrix} \tag{4.79}
$$

$$
AN2_{N-R}
\begin{pmatrix}
1 \\ 2 \\ 3 \\ 4 \\ 5 \\ 6 \\ 7 \\ 8 \\ 9 \\ 10 \\ 11
\end{pmatrix}
=
\begin{Bmatrix}
-20. \\
-24. \\
12. \\
& -5. \\
& -25. \\
& -6. \\
& 16. \\
4. \\
-7. \\
& -26. \\
& 26.
\end{Bmatrix}
\tag{4.80}
$$

$$
AD
\begin{pmatrix}
1 \\ 2 \\ 3 \\ 4 \\ 5 \\ 6 \\ 7 \\ 8 \\ 9
\end{pmatrix}
=
\begin{Bmatrix}
1. \\
16. \\
28. \\
1. \\
6. \\
60. \\
32. \\
12. \\
1.
\end{Bmatrix}
\tag{4.81}
$$

Remarks

The column numbers, shown in Eq.(4.78), need **NOT** be ordered.

[C] SGI's Unsymmetrical Sparse Matrix Storage Scheme[3.1]

The matrix equation, shown in Eq.(4.70), can be described by the following column – wise storage schemes:

$$
\text{IA}_{SGI}
\begin{pmatrix}
1 \\
2 \\
3 \\
4 \\
5 \\
6 \\
7 \\
8 \\
9 \\
10 = \text{NDOF} + 1
\end{pmatrix}
=
\begin{Bmatrix}
1 \\
2 \\
6 \\
12 \\
13 \\
17 \\
23 \\
27 \\
31 \\
32
\end{Bmatrix}
\tag{4.82}
$$

$$
\{JA_{SGI}\,(1,2,3,4,...,31)\}^T = \{\boxed{1}, \boxed{2,3,6,8}, \boxed{2,3,5,6,7,8}, \boxed{4}, \boxed{3,5,6,7}, \tag{4.83}
$$
$$
\boxed{2,3,4,5,6,7,8}, \boxed{3,5,6,7}, \boxed{2,3,6,8}, \boxed{9}\}
$$

$$
\{AN_{SGI}\,(1,2,3,4,...,31)\}^T = \{\boxed{1.0}, \boxed{16.0, -20.0, -24.0, 12.0}, \tag{4.84}
$$
$$
\boxed{20., 28., -5., -25., -6., 16.}, \boxed{1.0}, \boxed{5., 6., 4., -7.},
$$
$$
\boxed{24., 25., -4., 60., -26., 26.}, \boxed{6., 7., 26., 32.},
$$
$$
\boxed{-12., -16., -26., 12.}, \boxed{1.0}\,\}
$$

Remarks

The column numbers, given by Eq.(4.83), need NOT be ordered.

4.11 Symbolic Sparse Assembly of Unsymmetrical Matrices

It should be very helpful to understand the symbolic sparse assembly for "symmetrical" matrices first, as shown in Table 4.1, since only minor changes in the symmetrical case are required to make it work in the unsymmetrical case.

The following minor changes in Table 4.1 are required for handling the unsymmetrical case:

(a) Do 30 I = 1, N (last row will NOT be skipped)
(b) Introduce a new integer array IAKEEP (N + 1) that plays the role of array IA(-), for example: IAKEEP(I) = JPI (see sub-routine symbass in Table 4.9).

(c) Remove the IF statement (in Table 4.1) that skips the lower–triangle portion. As a consequence of this, the original array IA(-) will contain some additional unwanted terms.

(d) The output from "unsymmetrical" sparse assembly will be stored by IAKEEP(-) and JA(-), instead of IA(-) and JA(-) as in the symmetrical case!

A complete FORTRAN code for <u>unsymmetrical</u> symbolic sparse assembly is shown in a sub-routine symbass of Table 4.9.

4.12 Numerical Sparse Assembly of Unsymmetrical Matrices

The numerical sparse assembly for an "unsymmetrical" case is quite similar to the "symmetrical" case. The only minor change required is both upper– and lower–triangular portions of the element stiffness matrices, see Eq.(4.46 – 4.48), are used during the unsymmetrical assembly process.

A complete FORTRAN code for <u>unsymmetrical</u> numerical sparse assembly is shown in subroutine unnumass of Table 4.9.

4.13 Step-by-Step Algorithms for Unsymmetrical Sparse Assembly and Unsymmetrical Sparse Equation Solver

<u>Step 1</u>: Input General Control Information such as NEL, NDOF, NDOFPE, METHOD, NBC, METIS, IUNSOLVER

where

NEL = total number of elements in the finite element model

NDOF = total number of degree-of-freedom (dof) in the finite element model

   (including the Dirchlet boundary conditions)

NDOFPE = number of dof per element

METHOD = { 1, if sorting column numbers by transposing a matrix twice

        2, if sorting column numbers by using numerical recipe's sub-routines}

NBC = number of boundary (Dirichlet) conditions

METIS = { 1, if METIS re-ordering algorithm is used

      0, if METIS re-ordering algorithm is <u>NOT</u> used}

IUNSOLVER = { 1, if MA28 unsymmetrical sparse solver is used   [see 4.3]

          2, if Nguyen-Runesha's sparse server is used   [see 3.11]

          3, if SGI sparse solver is used}   [see 3.1]

Step 2: Input Element-DOF Connectivity Information call femesh (nel, ndofpe, $\vec{ie}$, $\vec{je}$) where arrays ie(-) and je(-) have already been explained in Eqs.(4.41 – 4.42).

Step 3: Input System Right-Hand-Side (RHS) Load Vector Call rhsload (ndof, $\vec{b}$) where the array b(-) plays the same role as array {R} in Eq.(4.51)

$$
b\begin{pmatrix} 1 \\ 2 \\ 3 \\ 4 \\ 5 \\ 6 \\ 7 \\ 8 \\ 9 \end{pmatrix} = \begin{Bmatrix} 4 \\ 44 \\ 10 \\ -20 \\ 42 \\ -40 \\ 84 \\ 28 \\ 48 \end{Bmatrix} \tag{4.85}
$$

Step 4: Input Dirichlet Boundary Conditions Call boundaryc (ndof, $\vec{ibc}$, nbc, $\vec{b}$) where the integer array ibc (NDOF) contains the boundary flags such as the one shown in Eq.(4.68)

$$
ibc\begin{pmatrix} 2 \\ 4 \\ 5 \end{pmatrix} = \begin{Bmatrix} ndof \\ ndof \\ ndof \end{Bmatrix} \tag{4.86}
$$

Here nbc = number of boundary conditions = 3

The values of Eq.(4.85) should be modified to incorporate Dirichlet boundary conditions such as:

$$b \begin{pmatrix} 1 \\ 2 \\ 3 \\ 4 \\ 5 \\ 6 \\ 7 \\ 8 \\ 9 \end{pmatrix} = \begin{Bmatrix} 4^K \\ 0.2'' \\ 10^K \\ 0.4'' \\ 0.5'' \\ -40^K \\ 84^K \\ 28^K \\ 81^K \end{Bmatrix} \qquad (4.87)$$

Step 5:   Find the transpose of matrix [E], and express the answers in the form of two integer arrays IET (-) and JET (-), as explained in Eqs.(4.44 – 4.45)

call transa2 (nel, ndof, $\vec{ie}$, $\vec{je}$, $\vec{iet}$, $\vec{jet}$)

Step 6:   Compute the two adjacency arrays IADJ (-) and JADJ (-) as explained in Eqs.(4.53 – 4.54).

call adjacency ($\vec{ie}$, $\vec{je}$, $\vec{iet}$, $\vec{jet}$, ndof, $\vec{ia}$, $\vec{ja}$, $\overrightarrow{iakeep}$)

where ia (NDOF+1) is a temporary working array, and iakeep (-) and ja (-) arrays play the same roles as IADJ (-) and JADJ (-).

Step 7:   Use METIS [3.3] re-ordering algorithms to minimize fill-in terms call metisreord (ndof, $\overrightarrow{iakeep}$, $\vec{ja}$, $\overrightarrow{iperm}$, $\overrightarrow{invp}$) where the output integer arrays IPERM (-) and INVP (-) have already been explained in Eqs.(4.56 – 4.59).

Step 8:   Re-compute the input connectivity array je(-), shown in Step 2, using IPERM (-) information

call newje ($\overrightarrow{itempol}$, $\overrightarrow{iperm}$, icount, $\vec{je}$)

where itermpol (NDOF) is a temporary, integer array, and icount = nel*ndofpe is the dimension for array je (-).

Step 9:   Re-do Step 5

Step 10: Only apply for MA28, or SGI unsymmetrical solvers

[A] Perform the <u>unsymmetrical</u> sparse <u>symbolic</u> <u>assembly</u>

call symbass ($\overrightarrow{ie}$, $\overrightarrow{je}$, $\overrightarrow{iet}$, $\overrightarrow{jet}$, ndof, $\overrightarrow{ia}$, $\overrightarrow{ja}$, $\overrightarrow{iakeep}$, $\overrightarrow{ibc}$)

The input arrays iakeep (NDOF+1), and ja (ncoef1) where

ncoef1 = iakeep (ndof + 1) – iakeep (1) will play a similar role as Eqs.(4.74 – 4.75), or Eqs.(4.82 – 4.83) for MA28, or SGI formats, respectively.

[B] Perform the <u>unsymmetrical</u> sparse <u>numerical</u> <u>assembly</u>

call unsymasem($\overline{\overline{ibc}}$, ncoef1, $\overline{an}$, $\overline{ip}$, ndof, nel, $\overline{el_k}$, $\overline{be}$, ndofpe, $\overline{lm}$, $\overline{je}$, $\overline{ae}$, $\overrightarrow{iakeep}$, $\overline{ja}$, $\overline{b}$, nbc)

where

ibc (ndof) = containing flags for (Dirichlet) boundary condition code (= ndof or 0)
ncoef1 = iakeep (ndof + 1) –1 = number of non-zero terms for the coefficient matrix (including upper + lower + diagonal terms for iunsovlver $\neq$ 2, and including only upper terms for iunsolver = 2)

an (ncoef1) = numerical values of non-zero terms of the coefficient matrix

ip (ndof) = integer, temporary working arrays

ndof = total number of degree-of-freedom (dof)

nel = total number of elements

elk (ndofpe, ndofpe) = element stiffness matrix

be (ndofpe) = element (nodal) load vector

ndofpe = number of dof per element

lm (ndofpe) = global dof associated with an element

je (ndofpe*nel) = connectivity information (= global dof associated with each and every element)

ae (ndofpe*ndofpe) = element stiffness matrix (stored in 1-D array, column-wise)

iakeep (ndof + 1) = starting location of the 1st non-zero term for each row of the coefficient matrix (including upper + diagonal + lower terms)

ja(ncoef1) = column numbers (unordered) associated with non-zero terms for each row of the coefficient matrix

b (ndof) = right-hand-side nodal load vector

nbc = number of boundary conditions

Step 11: Only apply for Nguyen-Runesha's unsymmetrical sparse solver [3.11]

[A] Perform ONLY the "upper" triangular sparse <u>symbolic</u> <u>assembly</u>, since we assume the unsymmetrical matrix is still symmetrical in non-zero locations!

$$\text{call symbassym} \left(\vec{ie}, \vec{je}, \vec{iet}, \vec{jet}, \text{ndof}, \vec{ia}, \vec{ja} \right)$$

where the output arrays ia(-) and ja(-) play similar roles as indicated in Eqs.(4.77 – 4.78), which IPERM (-) information has already been assumed to be incorporated.

[B] Obtain the "faked" element stiffness matrix, ELK (_,_), and element load vector, be(-)

$$\text{call elunsym} \left(i, \vec{elk}, \text{ndofpe}, \vec{be}, \vec{lm}, \vec{je}, \vec{ae} \right)$$

[C] Perform the "upper and lower" triangular sparse <u>numerical</u> <u>assembly</u>

$$\text{call numassuns} \left(\vec{ia}, \vec{ja}, \vec{ibc}, \vec{ae}, \vec{be}, \vec{lm}, \text{ndofpe}, \vec{an}, \vec{ad}, \vec{b}, \vec{ip}, \vec{an2}, \vec{itempol} \right)$$

[D] Perform <u>symbolic factorization</u>

$$\text{call symfact} \left(\text{ndof}, \vec{ia}, \vec{ja}, \vec{iu}, \vec{ju}, \vec{ip}, \text{ncoef2} \right)$$

[E] Transpose twice the arrays iu(-) and ju(-) so that the column numbers associated with each non-zero term for each row will be ordered (in increasing number)

$$\text{call transa} \left(\text{ndof}, \text{ndof}, \vec{iu}, \vec{ju}, \vec{iut}, \vec{jut} \right)$$

[F] Only apply if "unrolling" strategies for factorization will be used.

$$\text{call supnode} \left(\text{ndof}, \vec{iu}, \vec{ju}, \vec{isupern} \right)$$

<u>Step 12</u>: Only apply if (method = 2) ordering column numbers (of each row) by using "Numerical Recipe's Sorting" routines

[A] if MA28 or SGI (iunsolver = 1, or 3) unsymmetrical solver is used, then

$$\text{call sorting} \left(\text{ndof}, \vec{iu}, \vec{itempll}, \vec{ju}, \vec{tempol}, \text{index}, \vec{b}, \vec{an}, \vec{an2}, \vec{x} \right)$$

[B] If Nguyen-Runesha's (iunsolver = 2) unsymmetrical sparse solver is used:

$$\text{call transa2m} \left(\text{ndof}, \text{ndof}, \vec{iakeep}, \vec{ja}, \vec{iakepp}, \vec{jat}, \vec{an2}, \vec{aa} \right)$$

<u>Step 14</u>: Unsymmetrical Equation Solver's Solutions

[A] if MA28 (iunsolver = 1) solver is used:

(A.1) Converting the coefficient matrix into MA28's formats

$$\text{call rownuma28} \left(\vec{iakepp}, \text{ndof}, \vec{iakepp} \right)$$

(A.2) Performing LU factorization

$$\text{call ma28ad} \left(\text{ndof}, \text{nz}, \vec{an}, \text{licn}, \vec{iakepp}, \text{lirn}, \vec{ja}, u, \vec{ikeep}, iw, w, \text{iflag} \right)$$

(A.3) Performing Forward/Backward Solution

$$\text{call ma28cd} \left(\text{ndof}, \vec{an}, \text{licn}, \vec{ja}, \vec{ikeep}, \vec{b}, w, \text{mtype} \right)$$

[B] If Nguyen-Runesha's (iunsolver = 2) solver is used:

(B.1) Performing LU factorization (without unrolling strategies)

call unsynumfa1 $(ndof, \vec{ia}, \vec{ja}, \vec{ad}, \vec{an}, \vec{iu}, \vec{ju}, \vec{di}, \vec{un}, \vec{ip}, \vec{iup}, i \sup d, \overrightarrow{an2}, \overrightarrow{un2}, \overrightarrow{di2})$

(B.2) Performing forward/backward solutions

call unsyfbe $\left(ndof, \vec{iu}, \vec{ju}, \vec{di}, \vec{un}, \vec{b}, \vec{x}, \overrightarrow{un2}\right)$

[C] if SGI's Unsymmetrical Sparse Solver (iunsolver = 3) is used:
    [C.1] Converting the coefficient matrix into SGI formats
    [C.2] Performing SGI unsymmetrical sparse factorization
    [C.3] Performing SGI unsymmetrical sparse forward/backward solution.

4.14 A Numerical Example

In this section, a small-scale (academic) example is provided for the following objectives:
    (a) To validate the developed computer program and the associated sub-routines (for option to do sparse assembly process only)
    (b) To help the user to know how to prepare the input date file(s)

[A] Descriptions of the Illustrated Example:

Please refer to Section 4.7.

[B] Typical Input Date File (= test. dat)

Please refer to Section 4.13.

4, 9, 4, 1, 1, 1, 3........Input general control information   (see Step 1)

1, 5, 9, 13, 17........input or generate ie(-) array   (see Step 2, and Eq.[4.41])

3. 8, 1, 6, 7, 3, 2, 4, 5, 2, 3, 6, 7, 9, 8, 3........input or generate je(-) array..............
                                                         (see Step 2, and Eq.[4.42])

Remarks

Different options/capabilities provided by the developed software can be specified by changing a few input values in the general control information (see Table 4.9 and Section 4.13).

4.15 Summary

In this chapter, numerical strategies and the computer software implementation for both symmetrical and unsymmetrical sparse assembly algorithms and solvers have been described. Detailed step-by-step procedures, in the form of the developed "template," have been presented. A simple numerical example is given to validate the proposed algorithms and their developed computer software.

Finally, detailed input data file has been explained so that the users will have sufficient background and experience to incorporate the developed procedures into his/her large-scale application codes.

4.16 Exercises

4.1

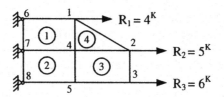

For the above finite element model, which consists of three rectangular and one triangular elements, assuming each node has only one degree-of-freedom (dof), and the prescribed Dirichlet boundary conditions at nodes 6 – 8 are given as:

$$ibc \begin{pmatrix} 6 \\ 7 \\ 8 \end{pmatrix} = \begin{Bmatrix} 1'' \\ 2'' \\ -3'' \end{Bmatrix}$$

assuming the symmetrical element stiffness matrix for the typical rectangular, or triangular elements, are given as:

$$[k^{(e)}] = k_{i,j} = \begin{cases} (i+j)*4, \text{for } i = j \\ i+j, \text{for } i \neq j \end{cases}$$

(a) Find the integer arrays $\{ie\}$, $\{je\}$, $\{iet\}$, and $\{jet\}$
(b) Compute the arrays $\{ia\}$, $\{ja\}$, $\{ad\}$, $\{an\}$, and $\{b\}$, which represent the total stiffness matrix and load vector, respectively (before imposing the Dirichlet boundary conditions)
(c) Re-do part (b), after imposing boundary conditions

4.2 Re-do problem 4.1 for the case of <u>unsymmetrical</u> element stiffness matrix,

where

$$[k^{(e)}] = k_{i,j} = \begin{cases} (i+j)*4, \text{for } i = j \\ i+j, \text{for } j > i \\ -(i+j), \text{for } j < i \end{cases}$$

and Nguyen-Runesha's unsymmetrical sparse storage scheme is used (see Section 4.10).

4.3 Re-do problem 4.2, but use SGI unsymmetrical sparse storage scheme (see

Section 4.10).

4.4 Re-do problem 4.2, but use I. Duff's unsymmetrical sparse storage scheme (see Section 4.10)

5 Generalized Eigen-Solvers

5.1 Introduction

Eigen-value problems arise naturally in many engineering applications such as free structural vibrations, structural dynamics, earthquake engineering, control-structure interactions, etc.

Solution algorithms for both "standard" and "generalized" eigen-value problems are well documented in the literature [5.1 – 5.10]. The focus of this chapter is to explain some efficient algorithms, which take full advantages of sparse technologies, to find the solution of the following <u>generalized</u> eigen-problem:

$$[K]\phi_i = \lambda_i [M]\phi_i \qquad (5.1)$$

where

$[K] \equiv n{\times}n$ stiffness matrix

$[M] \equiv n{\times}n$ mass matrix

$\phi_i \equiv$ eigen-vectors

$\lambda_i \equiv$ eigen-values

If the mass matrix [M] is equal to an identity matrix, then the generalized eigen-equation (5.1) will become the following <u>standard</u> eigen-problem:

$$[K]\phi_i = \lambda_i \phi_i \qquad (5.2)$$

5.2 A Simple Generalized Eigen-Example

Considering the following numerical data:

$$[K] = \begin{bmatrix} 5 & -2 \\ -2 & 2 \end{bmatrix} \qquad (5.3)$$

$$[M] = \begin{bmatrix} \dfrac{5}{4} & 0 \\ 0 & \dfrac{1}{5} \end{bmatrix} \qquad (5.4)$$

From Eq.(5.1), one has:

$$K\phi - \lambda M\phi = 0 \qquad (5.5)$$

$$[K - \lambda M]\phi = 0 \qquad (5.6)$$

$$\begin{bmatrix} 5 - \dfrac{5\lambda}{4} & -2 \\ -2 & 2 - \dfrac{\lambda}{5} \end{bmatrix} \{\phi\} = \{0\} \qquad (5.7)$$

The homogeneous system of Eq.(5.7) will have non-trivial solution if we set:

$$\det \begin{vmatrix} 5 - \dfrac{5\lambda}{4} & -2 \\ -2 & 2 - \dfrac{\lambda}{5} \end{vmatrix} = 0 \qquad (5.8)$$

The corresponding two roots (= eigen-values) of the determinant Eq.(5.8) are given as:

$$\lambda = \lambda_1 = 2 \qquad (5.9)$$

$$\lambda = \lambda_2 = 12 \qquad (5.10)$$

Substituting Eq.(5.9) into Eq.(5.6), one obtains:

$$\begin{bmatrix} 2.5 & -2 \\ -2 & 1.6 \end{bmatrix} \begin{Bmatrix} \phi_1^{(1)} \\ \phi_2^{(1)} \end{Bmatrix} = \begin{Bmatrix} 0 \\ 0 \end{Bmatrix} \qquad (5.11)$$

Observing the Eq.(5.11), one recognizes that the two equations are __NOT__ independent, and therefore, there is only one (= n−1= 2−1) independent equation. Since there are two unknowns (= $\phi_1^{(1)} and \phi_2^{(1)}$) and only one independent equation, there is an infinite number of solutions to Eq.(5.11). A solution can be found by selecting, for example

$$\phi_2^{(1)} = 1 \qquad (5.12)$$

Then the other unknown $\phi_1^{(1)}$ can be found from Eq.(5.11) as:

$$\phi_1^{(1)} = \frac{4}{5} \qquad (5.13)$$

Similarly, substituting Eq.(5.10) into Eq.(5.6), and selecting

$$\phi_2^{(2)} = 2 \qquad (5.14)$$

Then, $\phi_1^{(2)}$ can be solved as:

$$\phi_1^{(2)} = -\frac{2}{5} \qquad (5.15)$$

Thus, the eigen-matrix $[\Phi]$ can be formed as:

$$[\Phi] = \begin{bmatrix} \phi_1^{(1)} & \phi_1^{(2)} \\ \phi_2^{(1)} & \phi_2^{(2)} \end{bmatrix} = \begin{bmatrix} \dfrac{4}{5} & -\dfrac{2}{5} \\ 1 & 2 \end{bmatrix} \tag{5.16}$$

Following are two properties of the eigen-matrix $[\Phi]$, for the appropriated choice of normalized eigen-vectors:

$$[\Phi]^T [M][\Phi] = [I] = \text{Identity matrix} \tag{5.17}$$

$$[\Phi]^T [K][\Phi] = \begin{bmatrix} \lambda_1 & 0 \\ 0 & \lambda_2 \end{bmatrix} \tag{5.18}$$

5.3 Inverse and Forward Iteration Procedures

An eigen-value and its associated eigen-vector of Eq.(5.1) can be found by the following "inverse iteration" procedure, which is shown in Table 5.1:

Table 5.1 Inverse Iteration Procedure

Step 1: Guess eigen-pair solution

$$\lambda = 1 \tag{5.19}$$

$$\{\phi\} = \{x_1\} \tag{5.20}$$

Step 2: Substitute Eqs.(5.19-5.20) into Eq.(5.1), hence

$$[K]\{x_1\} \neq (1)[M]\{x_1\} \tag{5.21}$$

The new, better guess $\{x_2\}$ can be found from:

$$[K]\{x_2\} = [M]\{x_1\} \tag{5.22}$$

Step 3: If convergence is not yet achieved, for example

$$\left\| \vec{x_2} - \vec{x_1} \right\| > \varepsilon \tag{5.23}$$

then $\{x_1\}$ and $\{x_2\}$ in Eq.(5.22) will be replaced by $\{x_2\}$ and $\{x_3\}$, respectively.

The "forward iteration" procedure is very similar to the inverse iteration. The only difference is the new, better guess $\{x_2\}$ can be found from the following (instead of Eq.[5.22]):

$$[K]\{x_1\} = [M]\{x_2\} \tag{5.24}$$

It is, however, preferable to solve for a new, better vector $\{x_2\}$ from Eq.(5.22) rather than from Eq.(5.24), since the former is involved with stiffness matrix, which is usually positive definite, and therefore, the solution for $\{x_2\}$ can be found more easily!

The inverse iteration procedures can often be incorporated with the usage of an orthonormality condition, as illustrated in Table 5.2.

The following assumptions are made:
  (a) The stiffness matrix [K] in Eq.(5.1) is positive definite.
  (b) The mass matrix [M] in Eq.(5.1) can be a diagonal mass, with or without zeros on the diagonal. Matrix [M] can also be sparse and/or banded.
  (c) If the stiffness matrix [K] is only positive semi-definite, a "shift" should be used prior to the iterations.

Table 5.2 Inverse Iterations with Orthornormality Conditions

Step 1: Guess eigen-pair solution

$$\lambda = 1$$

$$\{\phi\} = \{x_1\}$$

Step 2: For k = 1, 2, 3... until converge, solve for $\overline{x}_{k+1}$ from the following equation

$$[K]\{\overline{x}_{k+1}\} = [M]\{x_k\} \tag{5.25}$$

In order to satisfy the mass orthonormality condition

$$\{x_{k+1}^T\} * [M] * \{x_{k+1}\} = 1 \tag{5.26}$$

We need to orthonormalize the vector $\{\overline{x}_{k+1}\}$ obtained from solving

Eq.(5.25), as following:

$$\{x_{k+1}\} = \frac{\{\overline{x}_{k+1}\}}{(\overline{x}_{k+1}^T M \overline{x}_{k+1})^{1/2}} \tag{5.27}$$

The following example is used to illustrate the details involved in the algorithms shown in Table 5.2.

Example: The stiffness and mass matrices [K] and [M] are given as:

$$[K] = \begin{bmatrix} 2 & -1 & 0 & 0 \\ -1 & 2 & -1 & 0 \\ 0 & -1 & 2 & -1 \\ 0 & 0 & -1 & 1 \end{bmatrix} \tag{5.28}$$

$$[M] = \begin{bmatrix} 0 & & & \\ & 2 & & \\ & & 0 & \\ & & & 1 \end{bmatrix} \tag{5.29}$$

Initial guesses:

$$\{x_1\}^T = \{1, \quad 1, \quad 1, \quad 1\} \text{ and } \lambda = 1 \tag{5.30}$$

Improved guessed vector $\{\bar{x}_2\}$ is solved from:

$$[K]\{\bar{x}_2\} = [M]\{x_1\} \tag{5.31}$$

Thus

$$\{\bar{x}_2\}^T = \{3, \quad 6, \quad 7, \quad 8\} \tag{5.32}$$

Compute:

$$\{\bar{x}_2\}^T [M]\{\bar{x}_2\} = 136 \tag{5.33}$$

Impose the mass orthonormality condition to obtain the new vector $\{x_2\}$:

$$\{x_2\} = \frac{1}{\sqrt{136}} * \begin{Bmatrix} 3 \\ 6 \\ 7 \\ 8 \end{Bmatrix} \tag{5.34}$$

The above steps are repeated again, to obtain:

$$[K]\{\bar{x}_3\} = [M]\{x_2\} \tag{5.35}$$

Solve for

$$\{\bar{x}_3\} = \left(\frac{1}{\sqrt{136}}\right) * \begin{Bmatrix} 20 \\ 40 \\ 48 \\ 56 \end{Bmatrix} \tag{5.36}$$

$$\{\bar{x}_3\}^T M\{\bar{x}_3\} = \frac{6336}{136} \tag{5.37}$$

$$\{x_3\} = \left(\frac{1}{\sqrt{6336}}\right) * \begin{Bmatrix} 20 \\ 40 \\ 48 \\ 56 \end{Bmatrix} = \begin{Bmatrix} 0.251 \\ 0.503 \\ 0.603 \\ 0.704 \end{Bmatrix} \tag{5.38}$$

The corresponding "exact" eigen-vector is:

$$\phi^{(1)} = \begin{Bmatrix} 0.250 \\ 0.500 \\ 0.606 \\ 0.707 \end{Bmatrix} \tag{5.39}$$

$$[K]\{\phi^{(1)}\} = \lambda_1 [M]\{\phi^{(1)}\} \tag{5.40}$$

It should be emphasized here that it is difficult to assure the convergence to a specific (and arbitrary selected) eigen-value and the corresponding eigen-vector.

$$\begin{bmatrix} 2 & -1 & 0 & 0 \\ -1 & 2 & -1 & 0 \\ 0 & -1 & 2 & -1 \\ 0 & 0 & -1 & 1 \end{bmatrix} \begin{Bmatrix} 0.250 \\ 0.500 \\ 0.602 \\ 0.707 \end{Bmatrix} = \lambda_1 \begin{bmatrix} 0 & & & \\ & 2 & & \\ & & 0 & \\ & & & 1 \end{bmatrix} \begin{Bmatrix} 0.250 \\ 0.500 \\ 0.602 \\ 0.707 \end{Bmatrix} \tag{5.41}$$

From Eq.(5.41), one has:

$$\lambda_1 = 0.148 \tag{5.42}$$

5.4 Shifted Eigen-Problems

The following example will demonstrate "shifting" procedures can be used to obtain eigen-solutions in the case where the stiffness matrix [K] is singular.

Example: The stiffness and mass matrices [K] and [M] are given as:

$$[K] = \begin{bmatrix} 3 & -3 \\ -3 & 3 \end{bmatrix}; [M] = \begin{bmatrix} 2 & 1 \\ 1 & 2 \end{bmatrix} \tag{5.43}$$

The "shifted" stiffness matrix $[\hat{K}]$ can be computed as:

$$[\hat{K}] = [K] - \rho[M] \tag{5.44}$$

where ρ is the shifted value.

The "new" eigen-problem is defined as:

$$[\hat{K}]\{\psi\} = \mu[M]\{\psi\} \tag{5.45A}$$

Substituting Eq.(5.44) into Eq.(5.45) to get:

$$(K - \rho M)\psi = \mu M \psi \tag{5.45B}$$

or

$$[K]\{\psi\} = (\rho + \mu)[M]\{\psi\} \tag{5.46}$$

Comparing the "new" eigen-problem of Eq.(5.46) with the "original" eigen-problem shown in Eq.(5.1), one has:

$$\lambda = \rho + \mu \tag{5.47}$$

and

$$\{\phi\} = \{\psi\} \tag{5.48}$$

Thus, the eigen-vectors of Eq.(5.1) and Eq.(5.46) are the same, and the eigen-values of Eqs.(5.1, 5.46) are different only by the shifted value ρ.

For the numerical data shown in Eq.(5.43), one has:

$$\begin{bmatrix} 3 & -3 \\ -3 & 3 \end{bmatrix}\{\phi\} = \lambda \begin{bmatrix} 2 & 1 \\ 1 & 2 \end{bmatrix}\{\phi\} \tag{5.49}$$

or

$$\begin{bmatrix} 3-2\lambda & -3-\lambda \\ -3-\lambda & 3-2\lambda \end{bmatrix}\{\phi\} = \begin{Bmatrix} 0 \\ 0 \end{Bmatrix} \tag{5.50}$$

For non-trival solution, we require the determinant of the coefficient matrix $(= K - \lambda M)$ in Eq.(5.50) to be vanished, thus:

$$\det|K - \lambda M| = 0 = 3\lambda^2 - 18\lambda = 0 \tag{5.51}$$

or

$$\lambda = \lambda_1 = 0 \tag{5.52}$$

$$\lambda = \lambda_2 = 6 \tag{5.53}$$

Substituting Eq.(5.52) into Eq.(5.50), one obtains the first eigen-vector:

$$\begin{Bmatrix} \phi_1^{(1)} \\ \phi_2^{(1)} \end{Bmatrix} = \begin{Bmatrix} 1/\sqrt{6} \\ 1/\sqrt{6} \end{Bmatrix} \tag{5.54}$$

Substituting Eq.(5.53) into Eq.(5.50), one obtains the second eigen-vector:

$$\begin{Bmatrix} \phi_1^{(2)} \\ \phi_2^{(2)} \end{Bmatrix} = \begin{Bmatrix} 1/\sqrt{2} \\ -1/\sqrt{2} \end{Bmatrix} \tag{5.55}$$

Now, using the shifted value $\rho = -2$ into Eq.(5.45B), one obtains:

$$\begin{bmatrix} 7 & -1 \\ -1 & 7 \end{bmatrix}\{\psi\} = \mu \begin{bmatrix} 2 & 1 \\ 1 & 2 \end{bmatrix}\{\psi\} \tag{5.56}$$

The two eigen-values from Eq.(5.56) can be obtained as:

$$\mu = \mu_1 = 2 \text{, hence } \lambda_1 = \rho + \mu_1 = -2 + 2 = 0 \tag{5.57}$$

$$\mu = \mu_2 = 8 \text{, hence } \lambda_2 = \rho + \mu_2 = -2 + 8 = 6 \tag{5.58}$$

Substituting the eigen-values μ from Eqs.(5.57 – 5.58) into Eq.(5.56), one obtains the following eigen-vectors:

$$\left\{\psi^{(1)}\right\} = \left\{\begin{matrix} 1/\sqrt{6} \\ 1/\sqrt{6} \end{matrix}\right\} \tag{5.59}$$

$$\left\{\psi^{(2)}\right\} = \left\{\begin{matrix} 1/\sqrt{2} \\ -1/\sqrt{2} \end{matrix}\right\} \tag{5.60}$$

5.5 Transformation Methods

The eigen-vectors $\left\{\phi^{(i)}\right\}$ obtained from solving the generalized eigen-equation (5.1) have the following properties:

$$[\Phi]^T [K][\Phi] = \begin{bmatrix} \searcher & \lambda & \\ & & \searcher \end{bmatrix} \tag{5.61}$$

$$[\Phi]^T [M][\Phi] = \begin{bmatrix} \searcher & I & \\ & & \searcher \end{bmatrix} \tag{5.62}$$

Where the eigen-matrix $[\Phi]$ is defined as:

$$[\Phi] = \begin{bmatrix} \phi^{(1)} & \phi^{(2)} & \ldots\ldots & \phi^{(n)} \end{bmatrix} \tag{5.63}$$

The basic idea in the transformation methods is to reduce (or transform) the matrices [K] and [M] into diagonal forms by using successive pre- and post- multiplication by matrices $[P_k]^T$, and $[P_k]$, respectively. These ideas are based on the properties given in Eqs.(5.61–5.62). The main ideas behind the transformation method can be summarized in Table 5.3.

Table 5.3 Transformation Methods

$$\text{Let } K_1 = K \text{ and } M_1 = M \tag{5.64}$$

Then

$$K_2 = P_1^T K_1 P_1 \tag{5.65}$$

$$K_3 = P_2^T K_2 P_2 = P_2^T (P_1^T K_1 P_1) P_2 \tag{5.66}$$

$$\vdots$$

$$K_{k+1} = P_k^T K_k P_k = (P_n^T P_{n-1}^T \cdots P_1^T) K_1 (P_1 P_2 \cdots P_n) \tag{5.67}$$

Similarly:

$$M_2 = P_1^T M_1 P_1 \tag{5.68}$$

$$M_3 = P_2^T M_2 P_2 = P_2^T (P_1^T M_1 P_1) P_2 \tag{5.69}$$

$$\vdots$$

$$M_{k+1} = P_k^T M_k P_k = (P_n^T P_{n-1}^T \cdots P_1^T) M_1 (P_1 P_2 \cdots P_n) \tag{5.70}$$

Comparing Eqs.(5.67, 5.70) and Eqs.(5.61, 5.62), one clearly sees that if matrices P_k are properly selected, then

$$[K_{k+1}] \rightarrow [\lambda] \text{ and } [M_{k+1}] \rightarrow [\, I \,] \text{ as } k \rightarrow \infty$$

in which case; with l being the last iteration, then the eigen-marix $[\Phi]$ can be symbolically represented as:

$$\Phi = P_1 P_2 \cdots P_l \tag{5.71}$$

In practice, it is <u>NOT</u> necessary that $[K_{k+1}]$ converges to the diagonal eigen-values matrix $[\lambda]$, and $[M_{k+1}]$ converges to the identity matrix $[I]$. Rather, it is only required that both $[K_{k+1}]$ and $[M_{k+1}]$ will converge to diagonal matrices, then

$$[\lambda] = \frac{\text{diagonal } k_r^{(l+1)}}{\text{diagonal } M_r^{(l+1)}} \tag{5.72}$$

and

$$[\Phi] = (P_1 \quad P_2 \quad \cdots \quad P_l) * \frac{1}{\sqrt{\text{diagonal } M_r^{(l+1)}}} \tag{5.73}$$

[A] <u>Jacobi Method for the Standard Eigen-Problems</u>

In this section, the following standard eigen-problem

$$[K]\{\phi\} = \lambda\{\phi\} \tag{5.74}$$

will be solved by the Jacobi method. Eq.(5.74) is a special case of the generalized eigen-equation (5.1) where [M] is set to be the identity matrix.

From Eqs.(5.70, 5.62), and realizing [M] = [I], one has:

$$M_{k+1} = P_k^T[I]P_k = [I]$$

or

$$[P_k]^T[P_k] = [I] \tag{5.75}$$

Thus, the transformation methods (such as Jacobi method) require the selected matrix [P_k] to have the properties as indicated in Eq.(5.75). The matrix [P_k], therefore, should have the following form:

$$
P_k =
\begin{bmatrix}
1 & & & \vdots & & & & & \vdots & & & \\
 & 1 & & \vdots & & & & & \vdots & & & \\
 & & 1 & \vdots & & & & & \vdots & & & \\
 & & & \cos\theta & & & & & -\sin\theta & & & \rightarrow i^{th}\text{row} \\
 & & & \vdots & 1 & & & & \vdots & & & \\
 & & & \vdots & & 1 & & & \vdots & & & \\
 & & & \vdots & & & 1 & & \vdots & & & \\
 & & & \vdots & & & & 1 & \vdots & & & \\
\cdots & \cdots & \cdots & \sin\theta & \cdots & \cdots & \cdots & \cdots & \cos\theta & \cdots & \cdots & \rightarrow j^{th}\text{row} \\
 & & & \vdots & & & & & \vdots & 1 & & \\
 & & & \vdots & & & & & \vdots & & 1 \\
\end{bmatrix} \tag{5.76}
$$

$$i^{th}\text{column} \qquad\qquad j^{th}\text{column}$$

Where θ is selected such that the element (i, j) of [K_{k+1}] at $(k+1)^{th}$ iteration will become zero (since the main objective of the Jacobi method is to transform the original matrix [K] into a diagonal matrix where all the off-diagonal terms $K_{i,j}$ will become zeros).

Considering the following triple products:

$$K_{k+1} = P_k^T K_k P_K, \text{ or} \tag{5.77}$$

$$
\begin{bmatrix}
a & b & c & d & e & f \\
b & \bar{g} & h & i & \bar{j} & k \\
c & h & l & m & n & o \\
d & i & m & p & q & r \\
e & \bar{j} & n & q & \bar{s} & t \\
f & k & o & r & t & u
\end{bmatrix} =
$$

$$
\begin{bmatrix}
1 & \cdot & \cdot & \cdot & \cdot & \cdot \\
\cdot & \cos\theta & & \sin\theta & & \cdot \\
\cdot & & 1 & & & \cdot \\
\cdot & & & 1 & & \cdot \\
\cdot & -\sin\theta & & \cos\theta & & \cdot \\
\cdot & & & & & 1
\end{bmatrix}
\begin{bmatrix}
a & b & c & d & e & f \\
b & g & h & i & j & k \\
c & h & l & m & n & o \\
d & i & m & p & q & r \\
e & j & n & q & s & t \\
f & k & o & r & t & u
\end{bmatrix}
\begin{bmatrix}
1 & \cdot & \cdot & \cdot & \cdot & \cdot \\
\cdot & \cos\theta & & -\sin\theta & & \cdot \\
\cdot & & 1 & & & \cdot \\
\cdot & & & 1 & & \cdot \\
\cdot & \sin\theta & & \cos\theta & & \cdot \\
\cdot & & & & & 1
\end{bmatrix}
$$

(5.78)

Assuming that $i = 2^{th}$ and $j = 5^{th}$ locations, and observing Eq.(5.78), one clearly sees that $[K_{k+1}]$ and $[K_k]$ are only different at locations $(2, 2)$, $(2, 5)$, $(5, 2)$, and $(5, 5)$, and all remaining terms of matrices $[K_{k+1}]$ and $[K_k]$ are unchanged. Thus, at the $(k+1)^{th}$ iteration, one only needs to consider the following portions of the triple product:

$$
K_{k+1} = P_k^T K_k P_k = \begin{bmatrix} \cos\theta & \sin\theta \\ -\sin\theta & \cos\theta \end{bmatrix} \begin{bmatrix} K_{ii}^{(k)} & K_{ij}^{(k)} \\ K_{ji}^{(k)} & K_{jj}^{(k)} \end{bmatrix} \begin{bmatrix} \cos\theta & -\sin\theta \\ \sin\theta & \cos\theta \end{bmatrix}
$$

(5.79)

where

$$
[K_{k+1}] = \begin{bmatrix} K_{ii}^{(k+1)} & K_{ij}^{(k+1)} \\ K_{ji}^{(k+1)} & K_{jj}^{(k+1)} \end{bmatrix}
$$

The objective to select the matrix P_k (at the kth iteration) is to make sure that in the next iteration (or the $[k+1]^{th}$ iteration), the off-diagonal terms of $[K_{k+1}]$ will be driven to zeros. Hence, setting the off-diagonal terms $K_{ij}^{(k+1)} = K_{ji}^{(k+1)} = 0$, and equating with the corresponding right-hand-side terms of Eq.(5.78), one can solve for the unknown θ, from either of the following formulas:

$$
\tan(2\theta) = \frac{2K_{ij}^{(k)}}{K_{ii}^{(k)} - K_{jj}^{(k)}}, \text{ for } K_{ii}^{(k)} \neq K_{jj}^{(k)}
$$

(5.80)

$$
\text{or } \theta = \frac{\pi}{4}, \text{ for } K_{ii}^{(k)} = K_{jj}^{(k)}
$$

(5.81)

If "l" is the last iteration, then

$$[K_{\ell+1}] \approx \begin{bmatrix} \searrow \\ \qquad \lambda \\ \qquad \qquad \searrow \end{bmatrix} = \text{diagonal eigen-values matrix} \qquad (5.82)$$

convergence to a tolerance ξ has been achieved, provided that the following conditions are met:

(a) Diagonal terms do <u>NOT</u> change much, or

$$\frac{\left| K_{ii}^{(\ell+1)} - K_{ii}^{(\ell)} \right|}{K_{ii}^{(\ell+1)}} \leq 10^{-\xi} \quad ; \quad i = 1, 2, \cdots, n \qquad (5.83)$$

(b) Off-diagonal terms approach zeros, or

$$\left[\frac{K_{ij}^{(\ell+1) \, 2}}{K_{ii}^{(\ell+1)} K_{jj}^{(\ell+1)}} \right]^{\frac{1}{2}} \leq 10^{-\xi} \quad ; \quad i < j \qquad (5.84)$$

The step-by-step Jacobi method for the solution of the standard eigen-value problem can be given as follows:

<u>Step 1</u>: At the sth sweep, a threshold for a near zero tolerance value is defined as 10^{-2s}.

<u>Step 2</u>: For all $K_{i,j}$ terms (with i<j, and i, j = 1, 2, ..., n = size of matrix [K]), compute the coupling factor according to the left-hand side of Eq.(5.84), and apply the transformation (see Eq.[5.79]) if the factor is larger than the current threshold.

<u>Step 3</u>: Use Eq.(5.83) to check for convergence. If convergence is achieved, then the process is stopped; otherwise, return to Step 1 for another sweep.

<u>Example</u>

$$\text{Given } [K] = \begin{bmatrix} 5 & -4 & 1 & 0 \\ -4 & 6 & -4 & 1 \\ 1 & -4 & 6 & -4 \\ 0 & 1 & -4 & 5 \end{bmatrix} \equiv [K_1]$$

For sweep 1 we have as a threshold 10^{-2}. Therefore we obtain the following results. For i = 1, j = 2, applying Eq.(5.80), one gets:

$$\cos\theta = 0.7497 \; ; \; \sin\theta = 0.6618$$

and thus, from Eq (5.76), one has:

$$P_1 = \begin{bmatrix} 0.7497 & -0.6618 & 0 & 0 \\ 0.6618 & 0.7497 & 0 & 0 \\ 0 & 0 & 1 & 0 \\ 0 & 0 & 0 & 1 \end{bmatrix}$$

$$P_1^T K P_1 = \begin{bmatrix} 1.169 & 0 & -1.898 & 0.6618 \\ 0 & 9.531 & -3.661 & 0.7497 \\ -1.898 & -3.661 & 6 & -4 \\ 0.6618 & 0.7497 & -4 & 5 \end{bmatrix} = [K_2]$$

For i = 1, j = 3:

$$\cos\theta = 0.9398 \; ; \; \sin\theta = 0.3416$$

$$P_2 = \begin{bmatrix} 0.9398 & 0 & -0.3416 & 0 \\ 0 & 1 & 0 & 0 \\ 0.3416 & 0 & 0.9398 & 0 \\ 0 & 0 & 0 & 1 \end{bmatrix}$$

$$P_2^T P_1^T K P_1 P_2 = \begin{bmatrix} 0.7792 & -1.2506 & 0 & -0.7444 \\ -1.2506 & 9.5314 & -3.4402 & 0.7497 \\ 0 & -3.4402 & 6.6891 & -3.9853 \\ -0.7444 & 0.7497 & -3.9853 & 5. \end{bmatrix} = [K_3]$$

$$P_1 P_2 = \begin{bmatrix} 0.7046 & -0.6618 & -0.2561 & 0 \\ 0.6220 & 0.7497 & -0.2261 & 0 \\ 0.3416 & 0 & 0.9398 & 0 \\ 0 & 0 & 0 & 1 \end{bmatrix}$$

For i = 1, j = 4:

$$\cos\theta = 0.9857 \; ; \; \sin\theta = 0.1687$$

$$P_3 = \begin{bmatrix} 0.9857 & 0 & 0 & -0.1687 \\ 0 & 1 & 0 & 0 \\ 0 & 0 & 1 & 0 \\ 0.1687 & 0 & 0 & 0.9857 \end{bmatrix}$$

$$P_3^T P_2^T P_1^T K P_1 P_2 P_3 = \begin{bmatrix} 0.6518 & -1.106 & -0.6725 & 0 \\ -1.106 & 9.531 & -3.440 & 0.9499 \\ -0.6725 & -3.440 & 6.690 & -3.928 \\ 0 & 0.9499 & -3.928 & 5.127 \end{bmatrix} = [K_4]$$

$$P_1 P_2 P_3 = \begin{bmatrix} 0.6945 & -0.6618 & -0.2561 & -0.1189 \\ 0.6131 & 0.7497 & -0.2261 & -0.1050 \\ 0.3367 & 0 & 0.9398 & -0.0576 \\ 0.1687 & 0 & 0 & 0.9857 \end{bmatrix}$$

For i = 2, j = 3:

$$\cos\theta = 0.8312 ;\ \sin\theta = -0.5560$$

$$P_4 = \begin{bmatrix} 1 & 0 & 0 & 0 \\ 0 & 0.8312 & 0.5560 & 0 \\ 0 & -0.5560 & 0.8312 & 0 \\ 0 & 0 & 0 & 1 \end{bmatrix}$$

$$P_4^T P_3^T P_2^T P_1^T K P_1 P_2 P_3 P_4 = \begin{bmatrix} 0.6518 & 0.5453 & -1.174 & 0 \\ -0.5453 & 11.83 & 0 & 2.974 \\ -1.174 & 0 & 4.388 & -2.737 \\ 0 & 2.974 & -2.737 & 5.127 \end{bmatrix} = [K_5]$$

$$P_1 P_2 P_3 P_4 = \begin{bmatrix} 0.6945 & -0.4077 & -0.5808 & -0.1189 \\ 0.6131 & 0.7488 & 0.2289 & -0.1050 \\ 0.3367 & -0.5226 & 0.7812 & -0.0576 \\ 0.1682 & 0 & 0 & 0.9857 \end{bmatrix}$$

To complete the first sweep, we zero element (3,4), using

$$\cos\theta = 0.7335 ;\ \sin\theta = -0.6797$$

$$P_6 = \begin{bmatrix} 1 & 0 & 0 & 0 \\ 0 & 1 & 0 & 0 \\ 0 & 0 & 0.7335 & 0.6797 \\ 0 & 0 & -0.6797 & 0.7335 \end{bmatrix}$$

and hence the approximation obtained for Λ and Φ are

$$\Lambda = P_6^T \dots P_1^T K P_1 \dots P_6$$

i.e.,

$$\Lambda = \begin{bmatrix} 0.6518 & -0.5098 & 0.9926 & -0.6560 \\ 0.5098 & 12.96 & -0.7124 & -0.6601 \\ -0.9926 & -0.7124 & 6.7596 & 0 \\ -0.6560 & -0.6602 & 0 & 1.6272 \end{bmatrix}$$

and

$$\Phi \doteq P_1 \dots P_6$$

i.e.,

$$\Phi \doteq \begin{bmatrix} 0.6945 & -0.4233 & -0.4488 & -0.3702 \\ 0.6131 & 0.6628 & 0.4152 & -0.1113 \\ 0.3367 & -0.5090 & 0.4835 & 0.6275 \\ 0.1687 & 0.3498 & -0.6264 & 0.6759 \end{bmatrix}$$

And after the third sweep we have

$$\Lambda \doteq \begin{bmatrix} 0.1459 & & & \\ & 13.09 & & \\ & & 6.854 & \\ & & & 1.910 \end{bmatrix}$$

$$\Phi \doteq \begin{bmatrix} 0.3717 & -0.3717 & -0.6015 & -0.6015 \\ 0.6015 & 0.6015 & 0.3717 & -0.3717 \\ 0.6015 & -0.6015 & 0.3717 & 0.3717 \\ 0.3717 & 0.3717 & -0.6015 & 0.6015 \end{bmatrix}$$

The approximation for Λ is diagonal to the precision given and we can use

$$\lambda_1 \doteq 0.1459; \qquad \phi_1 \doteq \begin{bmatrix} 0.3717 \\ 0.6015 \\ 0.6015 \\ 0.3717 \end{bmatrix}$$

$$\lambda_2 \doteq 1.910; \qquad \phi_2 \doteq \begin{bmatrix} -0.6015 \\ -0.3717 \\ 0.3717 \\ 0.6015 \end{bmatrix}$$

$$\lambda_3 \doteq 6.854; \qquad \phi_3 \doteq \begin{bmatrix} -0.6015 \\ 0.3717 \\ 0.3717 \\ -0.6015 \end{bmatrix}$$

$$\lambda_4 \doteq 13.09; \qquad \phi_4 \doteq \begin{bmatrix} -0.3717 \\ 0.6015 \\ -0.6015 \\ 0.3717 \end{bmatrix}$$

[B] Generalized Jacobi Method for Generalized Eigen-Problems

For the generalized eigen-problem as described by Eq.(5.1), we wish to diagonalize both the stiffness and mass matrices [K], and [M], respectively. Thus, the transformation matrix $[P_k]$ will be selected as:

$$P_k = \begin{bmatrix} 1 & & & & & & & & \\ & 1 & & & & & & & \\ & & 1 & & & & & & \\ & & & 1 & & \theta_2 & & & \\ & & & & 1 & & & & \\ & & & & & 1 & & & \\ & & \theta_1 & & & 1 & & \\ & & & & & & 1 & \\ & & & & & & & 1 \end{bmatrix} \begin{matrix} \\ \\ \\ i^{\text{th}} \text{ row} \\ \\ \\ j^{\text{th}} \text{ row} \\ \\ \\ \end{matrix} \qquad (5.85)$$

The new, updated matrices [K] and [M] in the next iteration can be computed from the current iteration as follows:

$$[K_{k+1}] = P_k^T K_k P_k \qquad (5.86)$$

or

$$
\begin{bmatrix} K_{ii}^{(k+1)} & K_{ij}^{(k+1)} \\ K_{ji}^{(k+1)} & K_{jj}^{(k+1)} \end{bmatrix} = \begin{bmatrix} 1 & \theta_1 \\ \theta_2 & 1 \end{bmatrix} \begin{bmatrix} K_{ii}^{(k)} & K_{ij}^{(k)} \\ K_{ji}^{(k)} & K_{jj}^{(k)} \end{bmatrix} \begin{bmatrix} 1 & \theta_2 \\ \theta_1 & 1 \end{bmatrix} \tag{5.87}
$$

$$
= \begin{bmatrix} K_{ii}^{(k)} + \theta_1 K_{ji}^{(k)} & K_{ij}^{(k)} + \theta_1 K_{jj}^{(k)} \\ K_{ji}^{(k)} + \theta_2 K_{ii}^{(k)} & K_{jj}^{(k)} + \theta_2 K_{ij}^{(k)} \end{bmatrix} \begin{bmatrix} 1 & \theta_2 \\ \theta_1 & 1 \end{bmatrix} \tag{5.88}
$$

or

$$
\begin{bmatrix} K_{ii}^{(k+1)} & 0 \\ 0 & K_{jj}^{(k+1)} \end{bmatrix} = \begin{bmatrix} \times & \theta_2\{K_{ii}^{(k)} + \theta_1 K_{ji}^{(k)}\} + \{K_{ij}^{(k)} + \theta_1 K_{jj}^{(k)}\} \\ \times & \times \end{bmatrix} \tag{5.89}
$$

The off-diagonal terms on both sides of Eq.(5.89) are equated to each other, thus

$$
K_{ij}^{(k+1)} = 0 = \theta_2 K_{ii}^{(k)} + \{1 + \theta_1\theta_2\}K_{ij}^{(k)} + \theta_1 K_{jj}^{(k)} \tag{5.90}
$$

Similarly, one has:

$$
[M_{k+1}] = P_k^T M_k P_k \tag{5.91}
$$

$$
M_{ij}^{(k+1)} = 0 = \theta_2 M_{ii}^{(k)} + \{1 + \theta_1\theta_2\}M_{ij}^{(k)} + \theta_1 M_{jj}^{(k)} \tag{5.92}
$$

The two unknowns θ_1 and θ_2 can be found by solving Eqs.(5.90,5.92) simultaneously [5.5]:

$$
\theta_1 = \frac{-G_1^{(k)}}{G_4^{(k)}} \quad \text{and} \quad \theta_2 = \frac{G_2^{(k)}}{G_4^{(k)}}, \tag{5.93}
$$

where

$$
G_1^{(k)} \equiv K_{ii}^{(k)} M_{ij}^{(k)} - M_{ii}^{(k)} K_{ij}^{(k)} \tag{5.94}
$$

$$
G_2^{(k)} \equiv K_{jj}^{(k)} M_{ij}^{(k)} - M_{jj}^{(k)} K_{ij}^{(k)} \tag{5.95}
$$

$$
G_3^{(k)} \equiv K_{ii}^{(k)} M_{jj}^{(k)} - K_{jj}^{(k)} M_{ii}^{(k)} \tag{5.96}
$$

$$
G_4^{(k)} = \frac{G_3^{(k)}}{2} + \text{sign}\{G_3^{(k)}\} * \sqrt{\left\{\frac{G_3^{(k)}}{2}\right\}^2 + G_1^{(k)} * G_2^{(k)}} \tag{5.97}
$$

Remarks

(a) Eq.(5.93) for θ_1 and θ_2 has been developed for the case [M] is a positive definite, full or banded matrix, and it can be proved that $G_4^{(k)}$ is always non-zero.

(b) The generalized Jacobi procedure can also be adopted for the case [M] is a diagonal matrix, with or without zero diagonal elements.

(c) Assuming "l" is the last iteration, then convergence is achieved if

$$\frac{\left| \lambda_i^{(l+1)} - \lambda_i^{(l)} \right|}{\lambda_i^{(l+1)}} \leq 10^{-\xi}, \text{ for } i = 1, 2, ..., n \tag{5.98}$$

where

$$\lambda_i^{(l)} \equiv \frac{K_{ii}^{(l)}}{M_{ii}^{(l)}} \text{ and } \lambda_i^{(l+1)} \equiv \frac{K_{ii}^{(l+1)}}{M_{ii}^{(l+1)}} \tag{5.99}$$

and

$$\left[\frac{\left\{ K_{ij}^{(l+1)} \right\}^2}{K_{ii}^{(l+1)} K_{jj}^{(l+1)}} \right]^{1/2} \leq 10^{-\xi} \tag{5.100}$$

$$\left[\frac{\left\{ M_{ij}^{(l+1)} \right\}^2}{M_{ii}^{(l+1)} M_{jj}^{(l+1)}} \right]^{1/2} \leq 10^{-\xi} \tag{5.101}$$

for all (i, j) with i<j

5.6 Sub-Space Iteration Method[5.5]

Sub-space iteration method is an effective algorithm to find the "few" lowest eigen-pairs of a fairly large generalized eigen-problem. Both inverse iteration and the generalized Jacobi iteration methods have been incorporated into the sub-space iteration algorithm. The main steps involved in this method can be described as follows:

Assuming the first "m" eigen-pair solution for Eq.(5.1) is sought. One computes:
$$L= \text{Minimum } (2 * m, m + 8) \tag{5.102}$$

and $L \leq n$ (= size of matrix [K])

"Guess" the first L eigen-vectors matrix $[X_1]_{n \times L}$

for k = 1, 2, 3, ... until convergence is achieved.

- Solve $\overline{X}_{k+1}$ from:

$$[K]\left[\overline{X}_{k+1}\right]=[M][X_k] \qquad (5.103)$$

- Find the "reduced" stiffness and mass matrices from:

$$\left[K_{k+1}^{R}\right]_{L\times L}=\left[\overline{X}_{k+1}^{T}\right]_{L\times n}[K]_{n\times n}\left[\overline{X}_{k+1}\right]_{n\times L} \qquad (5.104)$$

$$\left[M_{k+1}^{R}\right]_{L\times L}=\left[\overline{X}_{k+1}^{T}\right][M]\left[\overline{X}_{k+1}\right] \qquad (5.105)$$

- Solve the "reduced" eigen-problem

$$\left[K_{k+1}^{R}\right]_{L\times L}\left[Q_{k+1}\right]_{L\times L}=\left[\wedge_{k+1}\right]_{L\times L}\left[M_{k+1}^{R}\right]_{L\times L}\left[Q_{k+1}\right]_{L\times L} \qquad (5.106)$$

- Find the improved eigen-vectors

$$[X_{k+1}]_{n\times L}=\left[\overline{X}_{k+1}\right]_{n\times L}[Q_{k+1}]_{L\times L} \qquad (5.107)$$

Then
$$\left[\wedge_{k+1}\right]\to[\lambda]\equiv eigen-values \text{ and } [X_{k+1}]\to[\Phi]\equiv \text{eigen-vectors (see Eq.5.1)}$$
as $k\to\infty$

Remarks

(a) Inverse iteration method is employed in Eq.(5.103)
(b) Generalized Jacobi iteration method can be used to solve the reduced eigen-equation (5.106).
(c) The initial guess eigen-vector $[X_1]$ should contain independent columns. The simplest guess for $[X_1]$ is:

$$[X_1]_{n\times L}=\left[e^{(1)} \quad e^{(2)} \quad \cdots \quad e^{(L)}\right] \qquad (5.108)$$

where

$$e^{(i)}=\begin{Bmatrix} 0 \\ 0 \\ \vdots \\ 1 \\ 0 \\ 0 \\ 0 \end{Bmatrix}_{n\times 1} \to i^{th}\text{row} = \text{unit vector} \qquad (5.109)$$

Example

The numerical data for the stiffness and mass matrix [K] and [M] are given as:

$$[K] = \begin{bmatrix} 2 & -1 & 0 \\ -1 & 4 & -1 \\ 0 & -1 & 2 \end{bmatrix} \quad \text{and} \quad [M] = \begin{bmatrix} \frac{1}{2} & & \\ & 1 & \\ & & \frac{1}{2} \end{bmatrix} \qquad (5.110)$$

Assuming the initial guess for eigen-vectors is given by

$$[X_1] = \begin{bmatrix} 0 & 2 \\ 1 & 1 \\ 2 & 0 \end{bmatrix} \qquad (5.111)$$

$[\overline{X}_2]$ can be solved from Eq.(5.103):

$$[\overline{X}_2] = \left(\frac{1}{4}\right) \begin{bmatrix} 1 & 3 \\ 2 & 2 \\ 3 & 1 \end{bmatrix} \qquad (5.112)$$

Reduced stiffness and mass matrices can be found from Eqs.(5.104 – 5.105):

$$[K_2^R] = \overline{X}_2^T K \overline{X}_2 = \left(\frac{1}{4}\right) \begin{bmatrix} 5 & 3 \\ 3 & 5 \end{bmatrix} \qquad (5.113)$$

$$[M_2^R] = \overline{X}_2^T M \overline{X}_2 = \left(\frac{1}{16}\right) \begin{bmatrix} 9 & 7 \\ 7 & 9 \end{bmatrix} \qquad (5.114)$$

Solution of the reduced eigen-problem:

$$\left[K_2^R \right] [Q_2] = [\Lambda_2] \left[M_2^R \right] [Q_2]$$

can be solved "iteratively" (by using Generalized Jacobi method), or can also be solved "directly," and "exactly" since the system is involved with only 2×2 matrices.

Thus

$$[\Lambda_2] = \begin{bmatrix} 2 & 0 \\ 0 & 4 \end{bmatrix} = \text{eigen-values of the reduced system} \qquad (5.115)$$

$$[Q_2] = \begin{bmatrix} \frac{1}{\sqrt{2}} & 2 \\ \frac{1}{\sqrt{2}} & -2 \end{bmatrix} = \text{eigen-vectors of the reduced system} \qquad (5.116)$$

The new, improved eigen-vectors can be obtained from Eq.(5.107):

$$[X_2] = [\bar{X}_2][Q_2] = \begin{bmatrix} \dfrac{1}{\sqrt{2}} & -1 \\ \dfrac{1}{\sqrt{2}} & 0 \\ \dfrac{1}{\sqrt{2}} & 1 \end{bmatrix} \tag{5.117}$$

In this particular example, we obtained the "exact" eigen-values (see Eq.[5.115]), and "exact" eigen-vectors (see Eq.[5.117]) in one iteration (of the sub-space iteration loop)! This quick convergence is due to the fact that the starting iteration vectors $[X_1]$ (see Eq.[5.111]) span the sub-space defined as the "exact" eigen-vectors $\phi^{(1)}$ and $\phi^{(2)}$.

Proof (for the above paragraph)

The first two "exact" eigen-vectors from a 3×3 system given by Eq.(5.110) are:

$$\phi^{(1)} = \begin{Bmatrix} 1/\sqrt{2} \\ 1/\sqrt{2} \\ 1/\sqrt{2} \end{Bmatrix} \text{ and } \phi^{(2)} = \begin{Bmatrix} -1 \\ 0 \\ 1 \end{Bmatrix} \tag{5.118}$$

$$\phi^{(1)} = \begin{Bmatrix} 1/\sqrt{2} \\ 1/\sqrt{2} \\ 1/\sqrt{2} \end{Bmatrix} = a_1 \begin{Bmatrix} 0 \\ 1 \\ 2 \end{Bmatrix} + a_2 \begin{Bmatrix} 2 \\ 1 \\ 0 \end{Bmatrix} \tag{5.119}$$

1st column of $[X_1]$ 2nd column of $[X_1]$

where $a_1 = 1/(2\sqrt{2}) = a_2$

and

$$\phi^{(2)} = \begin{Bmatrix} -1 \\ 0 \\ 1 \end{Bmatrix} = a_3 \begin{Bmatrix} 0 \\ 1 \\ 2 \end{Bmatrix} + a_4 \begin{Bmatrix} 2 \\ 1 \\ 0 \end{Bmatrix} \tag{5.120}$$

where $a_3 = \dfrac{1}{2}; a_4 = \dfrac{-1}{2}$

In other words, the exact first two eigen-vectors $\phi^{(1)}$ and $\phi^{(2)}$ can be expressed as linear combinations of the columns of the starting iteration vectors $[X_1]$. Thus, sub-space iteration algorithm converges in just one iteration.

5.7 Lanczos Eigen-Solution Algorithms

For a large-scale system, if one has interest in obtaining the "first few" eigen-pair solutions for the generalized eigen-equation (5.1), then sub-space iteration algorithms can be employed. However, if the number of requested eigen-pair solutions increase, then the computational cost involved in sub-space iteration algorithms will increase very quickly. Thus, in this case, Lanczos eigen-solution algorithms are highly recommended (due to their highly computational efficiency).

5.7.1 Derivation of Lanczos Algorithms

The main steps involved in the Lanczos eigen-solution algorithms can be summarized in Table 5.4.

Table 5.4 Basic Lanczos Algorithms for Eigen-Solution of Generalized Eigen-
Equation $K\phi = \lambda M\phi$

Step 1: Input data
Given the stiffness matrix $[K]_{n \times n}$, and mass matrix $[M]_{n \times n}$
Let $m \equiv$ the number of requested eigen-values, and the corresponding eigen-vectors
$$\text{Let } r \equiv (2 \text{ or } 3) * m \qquad (5.121)$$
Step 2: Factorized the sparse stiffness matrix
$$[K] = [L] [D] [L]^T \qquad (5.122)$$
Step 3: Select an arbitrary vector $\{x\}$, and normalize this vector with respect to the mass matrix $[M]$, by the following operations
$$b = \sqrt{x^T M x} \qquad (5.123)$$
$$\{x^{(1)}\} = \text{first Lancozos vector} = \frac{\{x\}}{b} \qquad (5.124)$$
$$\text{Let } b_1 = 0 \qquad (5.125)$$
Step 4: Compute subsequent Lanczos vectors $\{x^{(i)}\}$
where $i = 2, 3, 4, \ldots, r$
(a) Performing forward and backward solution to solve for vector $\{ \bar{x}^{(i)} \}$:

$$[K]\{\bar{x}^{(i)}\} = [M]\{x^{(i-1)}\}$$ (5.126)

(b) Compute the scalars

$$a_{i-1} = \{\bar{x}^{(i)}\}^T \underbrace{[M]\{x^{(i-1)}\}}$$ (5.127)

already calculated in step 4(a)

(c) <u>Orthogonalize</u> the Lanczos vector with respect to the mass matrix [M]

$$\{\tilde{x}^{(i)}\} = \{\bar{x}^{(i)}\} - a_{i-1}\{x^{(i-1)}\} - b_{i-1}\{x^{(i-2)}\}$$ (5.128)

(d) <u>Normalize</u> the Lanczos vector with respect to [M]

$$b_i = \sqrt{\tilde{x}^{(i)^T} M \tilde{x}^{(i)}}$$ (5.129)

Get subsequent Lanczos vectors

$$\{x^{(i)}\} = \frac{\{\tilde{x}^{(i)}\}}{b_i}$$ (5.130)

<u>Step 5</u>: Form the symmetrical, tri-diagonal matrix $[T]_{r \times r}$

$$[T_r] = \begin{bmatrix} a_1 & b_2 & 0 & 0 & 0 & \cdots & \cdots & 0 \\ b_2 & a_2 & b_3 & 0 & 0 & \cdots & \cdots & 0 \\ 0 & b_3 & a_3 & b_4 & 0 & \cdots & \cdots & 0 \\ & & \ddots & \ddots & \ddots & & & \vdots \\ & & & \ddots & \ddots & \ddots & & \vdots \\ & & & & \ddots & \ddots & \ddots & \vdots \\ \cdots & \cdots & \cdots & \cdots & \cdots & b_{r-1} & a_{r-1} & b_r \\ \cdots & \cdots & \cdots & \cdots & \cdots & & b_r & a_r \end{bmatrix}$$ (5.131)

<u>Step 6</u>: Compute the standard eigen-solution of the reduced, tri-diagonal system

$$[T_r]\{\phi^*\} = \mu\{\phi^*\}$$ (5.132)

where $\lambda = \dfrac{1}{\mu}$ (5.133)

<u>Step 7</u>: Recover the original eigen-vectors ϕ

$$\phi = [X]\phi^*$$ (5.134)

where $[X] = [\bar{x}^{(1)}, \quad \bar{x}^{(2)}, \quad \ldots\ldots, \quad \bar{x}^{(r)}] =$ Lanczos vectors (5.135)

Few Remarks about Larízos Algorithms in Table 5.4

(1) The relationship between the original, generalized eigen-equation $K\phi = \lambda M\phi$, and the reduced, standard, tri-diagonal eigen-equation $T_r\phi^* = \mu\phi^*$ can be established as follows:

Pre-multiplying both sides of Eq.(5.1) by MK^{-1}, one has

$$MK^{-1}(K\phi) = \lambda MK^{-1}(M\phi) \qquad (5.136)$$

or

$$\frac{1}{\lambda}M\phi = MK^{-1}M\phi \qquad (5.137)$$

Substitute Eq.(5.134) into Eq.(5.137), to get:

$$\left(\frac{1}{\lambda}\right)M(X\phi^*) = MK^{-1}M(X\phi^*) \qquad (5.138)$$

Pre-multiply both sides of Eq.(5.138) by X^T to get:

$$\left(\frac{1}{\lambda}\right)X^T MX\phi^* = X^T MK^{-1}MX\phi^* \qquad (5.139)$$

Since the Lanczos vectors [X] is orthonormalize with respect to the mass [M] matrix, hence

$$[X]^T[M][X] = [I] \qquad (5.140)$$

Using Eq.(5.140), Eq.(5.139) becomes:

$$(X^T MK^{-1}MX)\phi^* = \left(\frac{1}{\lambda}\right)\phi^* \qquad (5.141)$$

Comparing Eq.(5.141) with Eq.(5.132), one concludes:

$$[T_r] \equiv [X]^T[M]\left[K^{-1}\right][M][X] \qquad (5.142)$$

and

$$\mu = \frac{1}{\lambda}, \text{ or } \lambda = \frac{1}{\mu} \qquad (5.133, \text{ repeated})$$

(2) In practice, we never form the tri-diagonal matrix $[T_r]$ according to Eq(5.142). Rather, $[T_r]$ is formed by Eq.(5.131). To prove that Eq.(5.131) can be obtained by the transformation of Eq.(5.142), one starts with Eq.(5.126):

$$\bar{x}^{(i)} = K^{-1}Mx^{(i-1)} \qquad (5.143)$$

Also, from Eq.(5.128), one obtains:

$$\bar{x}^{(i)} = a_{i-1}x^{(i-1)} + b_{i-1}x^{(i-2)} + \tilde{x}^{(i)} \tag{5.144}$$

Using Eq.(5.130), Eq.(5.144) becomes:

$$\bar{x}^{(i)} = a_{i-1}x^{(i-1)} + b_{i-1}x^{(i-2)} + b_i x^{(i)} \tag{5.145}$$

Comparing Eqs. (5.143, 5.145), one obtains:

$$\bar{x}^{(i)} = K^{-1}Mx^{(i-1)} = a_{i-1}x^{(i-1)} + b_{i-1}x^{(i-2)} + b_i x^{(i)} \tag{5.146}$$

Using the above relation for i = 2, 3, 4, ..., r (and recalled $b_1 = 0$), one obtains:

$$\bar{x}^{(2)} = a_1 x^{(1)} + b_2 x^{(2)} \tag{5.147}$$

$$\bar{x}^{(3)} = a_2 x^{(2)} + b_2 x^{(1)} + b_3 x^{(3)} \tag{5.148}$$

$$\bar{x}^{(4)} = a_3 x^{(3)} + b_3 x^{(2)} + b_4 x^{(4)} \tag{5.149}$$

$$\vdots$$

$$\bar{x}^{(r)} = a_{r-1}x^{(r-1)} + b_{r-1}x^{(r-2)} + b_r x^{(r)} \tag{5.150}$$

Eqs.(5.147 – 5.150) can also be represented as:

$$\left[\bar{x}^{(i)}\right]_{n\times r} = \left[x^{(1)}, \; x^{(2)}, \; ..., \; x^{(r)}\right]_{n\times r}
\begin{bmatrix}
a_1 & b_2 & 0 & 0 & 0 & \cdots & \cdots & 0 \\
b_2 & a_2 & b_3 & 0 & 0 & \cdots & \cdots & 0 \\
0 & b_3 & a_3 & b_4 & 0 & \cdots & \cdots & 0 \\
& & \ddots & \ddots & \ddots & & & \vdots \\
& & & \ddots & \ddots & \ddots & & \vdots \\
& & & & \ddots & \ddots & \ddots & \vdots \\
\cdots & \cdots & \cdots & \cdots & \cdots & b_{r-1} & a_{r-1} & b_r \\
\cdots & \cdots & \cdots & \cdots & \cdots & & b_r & a_r
\end{bmatrix}_{r\times r}$$

$$+ b_r
\begin{bmatrix}
0 & \cdots & \cdots & \cdots & 0 & x_1^{(r)} \\
0 & \cdots & \cdots & \cdots & 0 & x_2^{(r)} \\
\vdots & & & & \vdots & \vdots \\
\vdots & & & & \vdots & \vdots \\
\vdots & & & & \vdots & \vdots \\
0 & \cdots & \cdots & \cdots & 0 & x_n^{(r)}
\end{bmatrix}_{n\times r} \tag{5.151}$$

Thus, Eq.(5.146) can be expressed as:

$$K^{-1}MX^{(j)} = X^{(j)}[T_r] + b_j \left\{x^{(j+1)}\right\}_{n\times 1} \left\{e^{(j)}\right\}_{1\times n}^T \tag{5.152}$$

where

$$\left\{e^{(j)}\right\}_{1\times n}^T = \{0, \; 0, \; \cdots, \; 0, \; 1\} \tag{5.153}$$

at the jth location

Pre-multiplying both sides of Eq.(5.152) by $X^{(j)}$ M, and using the M-orthonormality of Lanczos vectors $X^{(j)}$, one has:

$$X^{(j)^T}MK^{-1}MX^{(j)} = \underbrace{X^{(j)^T}MX^{(j)}}_{[Identity]}[T_r] + b_j\underbrace{X^{(j)^T}Mx^{(j+1)}}_{[Zero]}e^{(j)^T} \qquad (5.154)$$

$$X^{(j)^T}MK^{-1}MX^{(j)} = [T_r] \qquad \text{(5.142, repeated)}$$

Eqs.(5.150 – 5.152) have clearly indicated the tri-diagonal matrix $[T_r]$ can also be assembled from Eq.(5.131), rather than from Eq.(5.142).

(3) When the size of $[T_r]$ becomes n×n, $\bar{x}^{(n)}$ calculated from Eq.(5.128) becomes{0} since the complete space is spanned by $[X]_{n \times n}$ and no vector M-orthogonal to all vectors in $[X]_{n \times n}$ exists!

(4) Either QL or Jacobi algorithms can be used to solve the standard, tri-diagonal eigen-problem shown in Eq.(5.132). However, QL algorithms seem to perform better in this step!

(5) For a large-scale finite element model (where $n \geq 10^6$ dof), the memory requirements to store (2 or 3) *(m = # requested eigen-pair) Lanczos vectors could be very significant.

(6) Handling an Indefinite Matrix

For some applications, such as linearized buckling analysis, or when a diagonal (lumped) mass matrix has zero diagonal terms, then the generalized eigen-equation can be expressed as:

$$([A] - \lambda_i[B])\phi_i = 0 \qquad (5.154A)$$

where

[A] = real, symmetric, positive definite, linear tangent stiffness matrix

[B] = real, symmetric (does NOT have to be positive definite), linearized geometric stiffness matrix

To avoid numerical difficulties, (such as the square root operations defined in Eqs.[5.123, 5.129]) associated with Lanczos algorithm for Eq.(5.154A), it is fairly common [5.11] to solve the following modified eigen-equation that deals with two positive definite matrices:

$$([\hat{A}] - \gamma_i[A])\phi_i = 0 \qquad (5.154B)$$

where

$$[\hat{A}] \equiv A + B \tag{5.154C}$$

$$\gamma_i \equiv \left(\frac{-\lambda_i - 1}{-\lambda_i} \right) \tag{5.154D}$$

The equivalence between Eq.(5.154A) and Eq.(5.154B) can be easily established by substituting Eqs.(5.154C and 5.154D) into Eq.(5.154B):

$$\left\{ [A + B] - \left(\frac{-\lambda_i - 1}{-\lambda_i} \right) [A] \right\} \phi_i = 0 \tag{5.154E}$$

Multiplying both sides of the above equation by $(-\lambda_i)$, one obtains:

$$\{ -A\lambda_i - B\lambda_i + \lambda_i A + A \} \phi_i = 0 \tag{5.154F}$$

The above equation is exactly identical to Eq.(5.154A)!

5.7.2 Lanczos Eigen-Solution Error Analysis

From Eq.(5.137), one has:

$$MK^{-1}M\phi = \frac{1}{\lambda} M\phi \tag{5.137, repeated}$$

Assuming [M] is a positive definite matrix, then [M] can be decomposed according to the Cholesky factorization:

$$[M] = [L][L]^T \equiv [U]^T[U] \tag{5.155}$$

where $[L]^T \equiv [U] =$ upper-triangular matrix. If lumped mass formulation is used, then [M] is a diagonal matrix, which may have some zero diagonal elements. In this case, one first needs to perform "static condensation" to remove those massless degree-of-freedoms.

Substituting Eq.(5.155) into Eq.(5.137), one gets:

$$(LL^T)K^{-1}(LL^T)\phi = \frac{1}{\lambda}(LL^T)\phi \tag{5.156}$$

Define

$$\psi \equiv L^T \phi \tag{5.157}$$

Then Eq.(5.156) becomes:

$$LL^T K^{-1} L(\psi) = \frac{1}{\lambda} L(\psi) \tag{5.158}$$

Using the definition given by Eq.(5.133), the above equation becomes:

$$LL^T K^{-1} L\psi = \mu L\psi \tag{5.159}$$

Pre-multiplying Eq.(5.159) by L^{-1}, one obtains:

$$(L^T K^{-1} L)\psi = (\mu)\psi \tag{5.160}$$

Eq.(5.160) can be recognized as the standard eigen-problem where the coefficient matrix is $L^T K^{-1} L$.

The "residual" vector obtained from the standard eigen-problem (see Eq.[5.160]) is given as:

$$\left\| r^{(i)} \right\| = \left\| L^T K^{-1} L\psi^{(i)} - \mu_i \psi^{(i)} \right\| \tag{5.161}$$

Substituting Eq.(5.157) into Eq.(5.161), one gets:

$$\left\| r^{(i)} \right\| = \left\| L^T K^{-1} LL^T \phi^{(i)} - \mu_i L^T \phi^{(i)} \right\| \tag{5.162}$$

Using Eq.(5.134), Eq.(5.162) becomes:

$$\left\| r^{(i)} \right\| = \left\| L^T K^{-1} LL^T X \phi^{*(i)} - \mu_i L^T X \phi^{*(i)} \right\| \tag{5.163}$$

$$\left\| r^{(i)} \right\| = \left\| L^T (K^{-1} LL^T X - \mu_i X)\phi^{*(i)} \right\| \tag{5.164}$$

Pre-multiplying both sides of Eq.(5.132) by [X], one obtains:

$$[X][T_r]\phi^{*(i)} = \mu_i [X]\phi^{*(i)} \tag{5.165}$$

Comparing both sides of Eq.(5.165), one concludes:

$$\mu_i [X] = [X][T_r] \tag{5.166}$$

Substituting Eq.(5.166) into Eq.(5.164), one obtains:

$$\left\| r^{(i)} \right\| = \left\| L^T (K^{-1} LL^T X - XT_r)\phi^{*(i)} \right\| \tag{5.167}$$

or

$$\left\| r^{(i)} \right\| = \left\| L^T (K^{-1} MX - XT_r)\phi^{*(i)} \right\| \tag{5.168}$$

Using Eq.(5.152), Eq.(5.168) can be re-written as:

$$\left\| r^{(i)} \right\| = \left\| L^T (b_j x^{(j+1)} e^{(j)})^T \phi^{*(i)} \right\|$$

The Lanczos vectors $[X] \equiv \left[\{x^{(1)}\} \ \{x^{(2)}\} \ ..., \ \{x^{(r)}\} \right]$ are orthonormalized with respect to the mass matrix [M], hence

$$X^T MX = [I] \tag{5.169}$$

Substituting Eq.(5.155) into Eq.(5.169), one has:

$$X^T (LL^T)X = [I]$$

$$or \ (X^T L)(L^T X) = [I]$$

$$or \ (L^T X)^T (L^T X) = [I] \tag{5.170}$$

From Eq.(5.170), one concludes:

$$\left\| L^T X \right\| = 1 \tag{5.171}$$

Using Eq.(5.171), Eq.(5.168) can be re-written as:

$$\left\| r^{(i)} \right\| = \left\| b_j e^{(j)^T} \phi^{*(i)} \right\| \tag{5.171A}$$

Referring to the definition given in Eq.(5.153), Eq.(5.171A) becomes:

$$\left\| r^{(i)} \right\| = \left\| b_j \phi^{*j,i} \right\| \tag{5.172}$$

where $\phi^{*j,i}$ is the jth element of the vector $\phi^{*(i)}$ of Eq.(5.132).

It can be proved that [5.5]

$$\left\| r^{(i)} \right\|_2 \geq \min \left| \lambda_i - \bar{\lambda} \right| \tag{5.173}$$

where $\bar{\lambda}$ is the calculated eigen-value, and λ_i is the exact i^{th} eigen-value.

Therefore

$$\min \left| \lambda_i - \bar{\lambda} \right| \leq \left| b_j \phi^{*j,i} \right| \tag{5.174}$$

Since we do not know the exact eigen-values λ_i, the error bound equation(5.174) will not tell us if the calculated eigen-value $\bar{\lambda}$ is close to which exact eigen-value. A Sturm sequence check, therefore, needs be applied for this purpose.

Derivation of Equation(5.173)

For the "standard" eigen-problem, [M] = [I], and the generalized eigen-equation (5.1) becomes:

$$K\phi = \lambda[M = I]\phi = \lambda\phi \tag{5.175}$$

Let $\bar{\lambda}$ and $\bar{\phi}$ be the computed eigen-value, and the corresponding eigen-vector of Eq.(5.175). The residual (or error) vector can be expressed as:

$$r = K\bar{\phi} - \bar{\lambda}\bar{\phi} \tag{5.176}$$

Since [M] = [I], Eqs.(5.61 – 5.62) become:

$$\Phi^T K \Phi = [\Lambda] = \begin{bmatrix} \lambda_1 & & & \\ & \lambda_2 & & \\ & & \ddots & \\ & & & \lambda_n \end{bmatrix} \tag{5.177}$$

$$\Phi^T \Phi = [I] \Rightarrow \Phi^T = \Phi^{-1} \tag{5.178}$$

where

$$[\Phi] = \left[\left\{ \phi^{(1)} \right\} \quad \left\{ \phi^{(2)} \right\} \quad \dots, \quad \left\{ \phi^{(n)} \right\} \right] \tag{5.179}$$

Pre-multiply and post-multiply both sides of Eq.(5.177) by Φ, and Φ^T, respectively (and also using Eq.[5.178]), then Eq.(5.177) becomes:

$$K = \Phi[\Lambda]\Phi^T \tag{5.180}$$

Substituting Eq.(5.180) into Eq.(5.176), one obtains:

$$r = \Phi \Lambda \Phi^T \bar{\phi} - \bar{\lambda}\, \bar{\phi} \tag{5.181}$$

or

$$r = \Phi \Lambda \Phi^T \bar{\phi} - \bar{\lambda}[\Phi \Phi^T]\bar{\phi}$$

or

$$r = \Phi(\Lambda - \bar{\lambda}I)\Phi^T \bar{\phi} \tag{5.182}$$

Pre-multiply both sides of Eq.(5.182) by Φ^T to obtain:

$$\Phi^T r = (\Lambda - \bar{\lambda}I)\Phi^T \bar{\phi}$$

or

$$(\Lambda - \bar{\lambda}I)^{-1}\Phi^T r = \Phi^T \bar{\phi} \tag{5.183}$$

Pre-multiply both sides of Eq.(5.183) by Φ to obtain:

$$\Phi(\Lambda - \bar{\lambda}I)^{-1}\Phi^T r = \bar{\phi} \tag{5.184}$$

Taking the norm on both sides of Eq.(5.184), one has:

$$\left\| \bar{\phi} \right\|_2 = \left\| \Phi(\Lambda - \bar{\lambda}I)^{-1}\Phi^T r \right\|_2 \tag{5.185}$$

Thus

$$\left\| \bar{\phi} \right\|_2 \le \left\| \Phi \right\| * \left\| (\Lambda - \bar{\lambda}I)^{-1} \right\|_2 * \left\| \Phi^T \right\|_2 * \left\| r \right\|_2 \tag{5.186}$$

The calculated eigen-vector $\bar{\phi}$ can be normalized, so that

$$\left\| \bar{\phi} \right\|_2 = 1$$

Hence, Eq.(5.186) becomes:

$$1 \le \|\Phi\| * \left\|(\Lambda - \bar{\lambda} I)^{-1}\right\|_2 * \left\|\Phi^T\right\|_2 * \|r\|_2 \tag{5.187}$$

Also, since

$$\Phi\Phi^T = I$$

Hence Eq.(5.187) becomes:

$$1 \le \left\|(\Lambda - \bar{\lambda} I)^{-1}\right\|_2 * \|r\|_2 \tag{5.188}$$

Also, since

$$\left\|(\Lambda - \bar{\lambda} I)^{-1}\right\|_2 = \max \frac{1}{\left|\lambda_i - \bar{\lambda}\right|} \tag{5.189}$$

Therefore

$$\min\left|\lambda_i - \bar{\lambda}\right| \le \|r\|_2 \tag{5.190}$$

Example: Consider the standard eigen-problem $K\phi = \lambda\phi$, where

$$[K] = \begin{bmatrix} 2 & -1 \\ -1 & 2 \end{bmatrix}$$

The eigen-values of the above matrix are roots of the following characteristic equation:

$$\det \begin{vmatrix} 2-\lambda & -1 \\ -1 & 2-\lambda \end{vmatrix} = 0 = 4 + \lambda^2 - 4\lambda - 1$$

or

$$\lambda^2 - 4\lambda + 3 = 0$$

hence

$$\lambda = \frac{4 \pm \sqrt{16-12}}{2} = \frac{4 \pm \sqrt{4}}{2}$$

Thus

$$\lambda = \lambda_1 = 1$$

$$\lambda = \lambda_2 = 3$$

The two "normalized" eigen-vectors $\phi^{(1)}$ and $\phi,^{(2)}$ which correspond to the two eigen-values $\lambda_1 = 1$ and $\lambda_2 = 3$, respectively, can be computed as:

$$\phi^{(1)} = \frac{1}{\sqrt{2}} \begin{Bmatrix} 1 \\ 1 \end{Bmatrix} = \begin{Bmatrix} \sqrt{2}/2 \\ \sqrt{2}/2 \end{Bmatrix}$$

$$\phi^{(2)} = \frac{1}{\sqrt{2}}\begin{Bmatrix}-1\\1\end{Bmatrix} = \begin{Bmatrix}-\sqrt{2}\big/2\\ \sqrt{2}\big/2\end{Bmatrix}$$

assuming the calculated $\overline{\lambda}_1 = 1.1$ and $\overline{\phi}^{(1)} = \begin{Bmatrix}0.6\\0.8\end{Bmatrix}$ are approximations to λ_1 and $\phi^{(1)}$.

The residual vector $r^{(1)}$ can be computed as:

$$r^{(1)} = K\overline{\phi}^{(1)} - \overline{\lambda}_1\overline{\phi}^{(1)}$$

$$r^{(1)} = \begin{bmatrix}2 & -1\\-1 & 2\end{bmatrix}\begin{Bmatrix}0.6\\0.8\end{Bmatrix} - (1.1)\begin{Bmatrix}0.6\\0.8\end{Bmatrix}$$

or

$$r^{(1)} = \begin{Bmatrix}0.4\\1.0\end{Bmatrix} - \begin{Bmatrix}0.66\\0.88\end{Bmatrix} = \begin{Bmatrix}-0.26\\0.12\end{Bmatrix}$$

Hence

$$\left\|r^{(1)}\right\|_2 = 0.286$$

From the error-bound Eq.(5.190), one has:

$$\left\|\lambda_1 - \overline{\lambda}_1\right\| \leq 0.286$$

$$\left\|1 - 1.1\right\| \leq 0.286$$

$$0.1 \leq 0.286, \text{ which is true!}$$

For the "generalized" eigen-problem, $K\phi = \lambda M\phi$, assuming the eigen-pair $\overline{\lambda}$ and $\overline{\phi}$ have already been approximately calculated, the residual can be computed as:

$$r_M = K\overline{\phi} - \overline{\lambda}M\overline{\phi} \tag{5.191}$$

In order to see the similarity between this case and the "standard" eigen-problem, we first assume [M] is a positive definite matrix, hence $M = L L^T$. Therefore, Eq.(5.191) becomes:

$$r_M = K\overline{\phi} - \overline{\lambda}LL^T\overline{\phi} \tag{5.192}$$

Pre-multiplying both sides of Eq.(5.192) with L^{-1}, one gets:

$$L^{-1}r_M = L^{-1}K\overline{\phi} - \overline{\lambda}L^T\overline{\phi} \tag{5.193}$$

Let

$$L^T\overline{\phi} \equiv \tilde{\phi} \tag{5.194}$$

Hence

$$\bar{\phi} = L^{-T} \tilde{\phi} \qquad (5.195)$$

Substituting Eq.(5.195) into Eq.(5.193), one obtains:

$$L^{-1} r_M = L^{-1} K L^{-T} \tilde{\phi} - \bar{\lambda}\, \tilde{\phi} \qquad (5.196)$$

or

$$r = \tilde{K}\tilde{\phi} - \bar{\lambda}\bar{\phi} \qquad (5.197)$$

where

$$r \equiv L^{-1} r_m \qquad (5.198)$$

$$\tilde{K} \equiv L^{-1} K L^{-T} \qquad (5.199)$$

Eq.(5.197) plays a similar role as Eq.(5.176).

During the above process (Eq.[5.191]→Eq.[5.199]), factorization of [M] is required. If this is <u>not</u> possible (for the case of diagonal mass matrix with some zero diagonal terms), or for the case where factorization of [K] is already available, we may consider a "different generalized" eigen-equation, such as:

$$M\phi = \frac{1}{\lambda} K\phi \qquad (5.200)$$

In this case, the residual vector can be computed as:

$$r_k = M\bar{\phi} - \frac{1}{\bar{\lambda}} K\bar{\phi} \qquad (5.201)$$

Assuming [K] is positive definite, hence

$$[K] = L_k L_k^T \qquad (5.202)$$

Eq.(5.201) becomes:

$$r_k = M\bar{\phi} - \frac{1}{\bar{\lambda}} L_k L_k^T \bar{\phi} \qquad (5.203)$$

or

$$L_k^{-1} r_k = L_k^{-1} M\bar{\phi} - \frac{1}{\bar{\lambda}} L_k^{-1} L_k L_k^T \bar{\phi} \qquad (5.204)$$

$$L_k^{-1} r_k = L_k^{-1} M\bar{\phi} - \frac{1}{\bar{\lambda}} L_k^T \bar{\phi} \qquad (5.205)$$

Let

$$L_k^T \bar{\phi} \equiv \tilde{\phi}_k \qquad (5.206)$$

Hence

$$\overline{\phi} = L_k^{-T} \tilde{\phi}_k \tag{5.207}$$

Substituting Eq.(5.207) into Eq.(5.205), one obtains:

$$L_k^{-1} r_k = L_k^{-1} M L_k^{-T} \tilde{\phi}_k - \frac{1}{\overline{\lambda}} L_k^{T} L_k^{-T} \tilde{\phi}_k \tag{5.208}$$

or

$$L_k^{-1} r_k = L_k^{-1} M L_k^{-T} \tilde{\phi}_k - \frac{1}{\overline{\lambda}} \tilde{\phi}_k \tag{5.209}$$

or

$$r = \tilde{M} \tilde{\phi}_k - \frac{1}{\overline{\lambda}} \tilde{\phi}_k \tag{5.210}$$

where

$$r \equiv L_k^{-1} r_k \tag{5.211}$$

$$\tilde{M} \equiv L_k^{-1} M L_k^{-T} \tag{5.212}$$

Thus, using Eq.(5.210), instead of using Eq.(5.197), one can establish the bounds on $\frac{1}{\overline{\lambda}}$, and hence the bounds on $\overline{\lambda}$, without factorizing the mass matrix [M].

Remarks

While the above error bounds can be computed, it is much simpler if we compute the error as:

$$\text{error} = \frac{\left\| K\overline{\phi} - \lambda M \overline{\phi} \right\|_2}{\left\| K\overline{\phi} \right\|_2} \tag{5.213}$$

5.7.3 Sturm Sequence Check

The Sturm sequence check can be briefly described as follows:

Step 1: Given the stiffness matrix $[K]_{n \times n}$ and the shifted value ρ

Step 2: Compute the stiffness matrix $[\hat{K}]_{n \times n}$ as follows:

$$\hat{K} = K - \rho I \tag{5.214}$$

Step 3: Factorize $[\hat{K}]$ as follows:

$$\hat{K} = [L][D][L]^T \tag{5.215}$$

where [L] and [D] are the lower triangular, and diagonal matrices, respectively.

<u>Step 4</u>: Count the number of "NEGATIVE" elements in [D], (say $= m \le n$). Then, we conclude that there are "m" eigen-values of [K] whose values are less than ρ.

<u>Example</u>: Using the data from the previous example, assuming we have no ideas that the computed eigen-value $\bar{\lambda}$ is an approximation of λ_2, hence

$$\left\| \lambda_i - \bar{\lambda} \right\| \le \left\| \bar{r} \right\| = 0.286 \tag{5.216}$$

or

$$\left\| \lambda_i - 1.1 \right\| \le 0.286 \tag{5.217}$$

Thus

$$1.1 - 0.286 \le \lambda_i \le 1.1 + 0.286 \tag{5.218}$$

$$0.814 \le \lambda_i \le 1.386 \tag{5.219}$$

Let us use the lower bound 0.8, and upper bound 1.4:

(a) Consider lower bound case:

$$\hat{K} = K - 0.8I = \begin{bmatrix} 1.2 & -1 \\ -1 & 1.2 \end{bmatrix} = [L][D][L]^T \tag{5.220}$$

$$= \begin{bmatrix} 1 & 0 \\ -\dfrac{5}{6} & 1 \end{bmatrix} \begin{bmatrix} \dfrac{6}{5} & 0 \\ 0 & \dfrac{11}{30} \end{bmatrix} \begin{bmatrix} 1 & -\dfrac{5}{6} \\ 0 & 1 \end{bmatrix} \tag{5.221}$$

Conclusion: there is no eigen-values (of $K\phi = \lambda\phi$) that are less than 0.8 (because there is NO negative terms in [D]).

(b) Consider an upper-bound case:

$$\hat{K} = K - 1.4I = \begin{bmatrix} 0.6 & -1 \\ -1 & 0.6 \end{bmatrix} = [L][D][L]^T \tag{5.222}$$

$$= \begin{bmatrix} 1 & 0 \\ -\dfrac{5}{3} & 1 \end{bmatrix} \begin{bmatrix} 0.6 & 0 \\ 0 & -\dfrac{16}{15} \end{bmatrix} \begin{bmatrix} 1 & -\dfrac{5}{3} \\ 0 & 1 \end{bmatrix} \tag{5.223}$$

Conclusion: Since there is one negative term in [D], hence, there must be one eigen-value (of $K\phi = \lambda\phi$) that is less than 1.4. Thus $\lambda_1 \leq 1.4$ and therefore, $\bar{\lambda} = 1.1$ must be an approximation of λ_1 !

Consider the generalized eigen-problem, as indicated by

$$[K]\,\phi_i = \lambda_i\,[M]\,\phi_i \qquad\qquad (5.1, \text{repeated})$$

The corresponding "shifted" eigen-problem can be given by

$$[\hat{K}]\,\psi_i = \mu_i\,[M]\,\psi_i \qquad\qquad (5.45A, \text{repeated})$$

where

$$[\hat{K}] = [K] - \rho[M] \qquad\qquad (5.44, \text{repeated})$$

$$\lambda_i = \rho + \mu_i \qquad\qquad (5.47, \text{repeated})$$

and

$$\rho = \text{shifted value}$$

It should be recalled here that Eqs.(5.1 and 5.45A) will have the same eigen-vectors, and their corresponding eigen-values are related by Eq.(5.47).

If one factorizes the above matrix $[\hat{K}] = [L]\,[D]\,[L]^T$, as explained in Chapter 3, then the number of "negative" terms appearing in the diagonal matrix [D] will be equal to the number of eigen-values that have values less than the shifted value ρ [see 5.5].

The above Sturm Sequence Properties can be used to find the number of eigen-values between the specified lower and upper limits as illustrated by the following example.

Given the following stiffness and mass matrices:

$$[K] = \begin{bmatrix} 5 & -4 & 1 & 0 \\ -4 & 6 & -4 & 1 \\ 1 & -4 & 6 & -4 \\ 0 & 1 & -4 & 5 \end{bmatrix} \quad \text{and} \quad [M] = [I]$$

Find the number of eigen-values between

$$(\lambda_L = 1) \langle \lambda_i \langle (\lambda_u = 7)$$

Solution

Step 1

Find $\hat{K}_1 = K - (\rho = \lambda_u = 7) * [M = I]$

$$\text{or } [\hat{K}_1] = \begin{bmatrix} -2 & -4 & 1 & 0 \\ -4 & -1 & -4 & 1 \\ 1 & -4 & -1 & -4 \\ 0 & 1 & -4 & -2 \end{bmatrix}$$

Step 2

Factorize $\hat{K}_1 = L_1 D_1 L_1^{T}$ (refer to Chapter 3)

For this case, the diagonal matrix [D_1] will have "3 negative terms." Thus, according to the Sturm Squence Properties, there are 3 eigen-values that have values less than $\lambda_u = 7$.

Step 3

Find $\hat{K}_2 = K - (\rho = \lambda_L = 1) * [M = I]$

$$\text{or } [\hat{K}_2] = \begin{bmatrix} 4 & -4 & 1 & 0 \\ -4 & 5 & -4 & 1 \\ 1 & -4 & 5 & -4 \\ 0 & 1 & -4 & 4 \end{bmatrix}$$

Step 4

Factorize $\hat{K}_2 = L_2 D_2 L_2^{T}$

For this case, the diagonal matrix [D_2] will have "1 negative term." Thus, there is 1 eigen-value that has the value less than $\lambda_L = 1$. Hence, there are m = 2 eigen-values between the specified ranges $[\lambda_L = 1 \quad , \quad \lambda_u = 7]$. This conclusion can be easily verified since this same example has already been solved in Section 5.5 (using the Jacobi method), and the 4 eigen-values found are $\lambda_i = (0.1459, 1.910, 6.854, 13.09)$.

Step 5

Use the Lanczos eigen-algorithms, discussed in Section 5.7, to solve for m = 2 lowest eigen-values of the following shifted eigen-equation:

$$\hat{K}_2 \, \phi_i = \mu \, M \, \phi_i$$

Remarks

(a) The "METIS" [3.3] re-ordering algorithms (for minimizing fill-in terms during the factorization phase), and the symbolic factorization phase (see Chapter 3), which are needed before performing the task mentioned in Step 2 (for factorizating $\hat{K}_1$), can be re-used in $\boxed{\text{Step 4}}$ (for factorizating $\hat{K}_2$). This is possible because both matrices $\hat{K}_1$ and $\hat{K}_2$ have the same non-zero locations.

(b) In practical computer implementation, $\hat{K}_2$ should overwrite $\hat{K}_1$ in order to save computer memory requirements.

5.7.4 Proving the Lanczos Vectors Are M-Orthogonal

The 2nd Lanczos vector $x^{(2)}$ expressed in Eq.(5.130) and the vector $\tilde{x}^{(2)}$ given by Eq.(5.128) are parallel to each other since these two vectors are different only by a scalar factor. Thus, it is sufficient to prove $\tilde{x}^{(2)}$ to be M-orthogonal to the 1st Lanczos vector $x^{(1)}$.

For i = 2, pre-multiplying both sides of Eq.(5.128) by $x^{(1)^T}M$, one obtains:

$$x^{(1)^T}M\tilde{x}^{(2)} = x^{(1)^T}M\bar{x}^{(2)} - a_1 x^{(1)^T}Mx^{(1)} - (b_1 = 0)*x^{(0)} \qquad (5.224)$$

Substituting a_1 from Eq.(5.127) into Eq.(5.224), one gets:

$$x^{(1)^T}M\tilde{x}^{(2)} = x^{(1)^T}M\bar{x}^{(2)} - \bar{x}^{(2)^T}Mx^{(1)} * x^{(1)^T}Mx^{(1)} \qquad (5.225)$$

Since the 1st Lanczos vector has already been normalized with respect to the mass [M] matrix (see Eqs.[5.123 – 5.124]), hence the scalar $x^{(1)^T}Mx^{(1)} = 1$, and the above (scalar) equation (5.225) becomes:

$$x^{(1)^T}M\tilde{x}^{(2)} = x^{(1)^T}M\bar{x}^{(2)} - (\bar{x}^{(2)^T}Mx^{(1)})^T = 0$$

Thus, $x^{(1)}$ is M-orthogonal to $\tilde{x}^{(2)}$, or $x^{(2)}$.

Next, we will prove that $x^{(3)}$ (or $\tilde{x}^{(3)}$) is M-orthogonal to $x^{(1)}$, and $x^{(2)}$.

For i = 3, pre-multiplying both sides of Eq.(5.128) by $x^{(2)^T}M$, one obtains:

$$x^{(2)^T}M\tilde{x}^{(3)} = x^{(2)^T}M\bar{x}^{(3)} - a_2 x^{(2)^T}Mx^{(2)} - b_2 x^{(2)^T}Mx^{(1)}$$

or

$$x^{(2)^T}M\tilde{x}^{(3)} = x^{(2)^T}M\bar{x}^{(3)} - a_2 \qquad (5.226)$$

Substituting Eq.(5.127) into Eq.(5.226), one has:

$$x^{(2)^T}M\tilde{x}^{(3)} = x^{(2)^T}M\bar{x}^{(3)} - \bar{x}^{(3)^T}Mx^{(2)} = 0$$

Thus, $\tilde{x}^{(3)}$ (or $x^{(3)}$) is M-orthogonal to $x^{(2)}$.

The following section will prove that $\tilde{x}^{(3)}$ is also M-orthogonal to $x^{(1)}$.

For i = 3, Eq.(5.128) becomes:

$$\tilde{x}^{(3)} = \bar{x}^{(3)} - a_2 x^{(2)} - b_2 x^{(1)} \tag{5.227}$$

Substituting $\bar{x}^{(3)}$ from Eq.(5.126) and a_2 from Eq.(5.127) into Eq.(5.227), one obtains:

$$\tilde{x}^{(3)} = K^{-1}Mx^{(2)} - \left\{\bar{x}^{(3)}\right\}^T Mx^{(2)} * x^{(2)} - b_2 x^{(1)}$$

or

$$\tilde{x}^{(3)} = K^{-1}Mx^{(2)} - \left\{K^{-1}Mx^{(2)}\right\}^T Mx^{(2)} * x^{(2)} - b_2 x^{(1)}$$

Pre-multiplying both sides of the above equation by $x^{(1)^T}M$, one obtains:

$$x^{(1)^T}M\tilde{x}^{(3)} = x^{(1)^T}MK^{-1}Mx^{(2)} - x^{(1)^T}M\left\{K^{-1}Mx^{(2)}\right\}^T Mx^{(2)} * x^{(2)} - b_2 x^{(1)^T}Mx^{(1)}$$

or

$$x^{(1)^T}M\tilde{x}^{(3)} = x^{(1)^T}MK^{-1}Mx^{(2)} - x^{(1)^T}M\left\{x^{(2)^T}MK^{-T}\right\}Mx^{(2)} * x^{(2)}$$
$$- b_2 x^{(1)^T}Mx^{(1)} \tag{5.228}$$

Let scalar $\equiv x^{(2)^T}MK^{-T}Mx^{(2)}$

or scalar $\equiv \{x^{(2)^T}MK^{-T}Mx^{(2)}\}^T = x^{(2)^T}MK^{-1}Mx^{(2)}$

Comparing the above two expressions for scalar, one concludes:

$$K^{-T} = K^{-1}$$

Thus K^{-1} is also symmetrical.

Eq.(5.228) can be re-written as:

$$x^{(1)^T}M\tilde{x}^{(3)} = x^{(1)^T}MK^{-1}Mx^{(2)} - \underbrace{x^{(1)^T}Mx^{(2)}}_{zero} * scalar - b_2$$

or

$$x^{(1)^T}M\tilde{x}^{(3)} = x^{(2)^T}MK^{-1}Mx^{(1)} - b_2 \tag{5.229}$$

From Eq.(5.126), one gets:

$$\bar{x}^{(2)} = K^{-1}Mx^{(1)}$$

Substituting the above expression into Eq.(5.229), one obtains:

$$x^{(1)^T}M\tilde{x}^{(3)} = x^{(2)^T}M\bar{x}^{(2)} - b_2 \tag{5.230}$$

From Eq.(5.128), one gets:

$$\overline{x}^{(2)} = \widetilde{x}^{(2)} + a_1 x^{(1)} + (b_1 = 0) * x^{(0)}$$

Thus, Eq.(5.230) can be written as:

$$x^{(1)^T} M \widetilde{x}^{(3)} = x^{(2)^T} M\{\widetilde{x}^{(2)} + a_1 x^{(1)}\} - b_2$$

$$= x^{(2)^T} M \widetilde{x}^{(2)} + a_1 x^{(2)^T} M x^{(1)} - b_2$$

$$x^{(1)^T} M \widetilde{x}^{(3)} = x^{(2)^T} M\{\widetilde{x}^{(2)}\} - b_2$$

or, using Eq.(5.130), the above equation becomes:

$$x^{(1)^T} M \widetilde{x}^{(3)} = x^{(2)^T} M\{x^{(2)} * b_2\} - b_2$$

or

$$x^{(1)^T} M \widetilde{x}^{(3)} = [x^{(2)^T} M x^{(2)}] * b_2 - b_2 = 0$$

Hence, the vector $\widetilde{x}^{(3)}$ (or $x^{(3)}$) is M-orthogonal to vector $x^{(1)}$.

Remarks

It can be proved that if vector $x^{(1)}$ is M-orthogonal to vectors $x^{(2)}, x^{(3)}, x^{(4)}, ..., x^{(j)}$, then $x^{(1)}$ will also be M-orthogonal to the vector $x^{(j+1)}$.

5.7.5 "Classical" Gram-Schmidt Re-Orthogonalization

Given a set of vectors $a^{(1)}, a^{(2)}, ..., a^{(k)}$, the Gram-Schmidt re-orthogonalization procedures will generate a sequence of vectors $q^{(1)}, q^{(2)}, ..., q^{(k)}$, such that:
  (a) $q^{(1)}$ is different from $a^{(1)}$ by only a scaled factor, so that $q^T q = 1$
  (b) $q^{(1)}$ and $q^{(2)}$ are orthonormal in the plane formed by $a^{(1)}$ and $a^{(2)}$
  (c) similarly, $q^{(1)}, q^{(2)}$, and $q^{(3)}$ are orthonormal in the space spanned by $a^{(1)}, a^{(2)}$, and $a^{(3)}$, etc.

Without a scaled factor, $q^{(k)}$ can be generated by the following equation:

$$q^{(k)} = a^{(k)} - (q^{(1)^T} a^{(k)}) q^{(1)} - (q^{(2)^T} a^{(k)}) q^{(2)} - ... - (q^{(k-1)^T} a^{(k)}) q^{(k-1)} \quad (5.231)$$

The Gram-Schmidt process can be visualized and explained as shown in Figure 5.1.

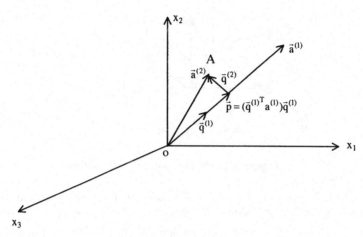

Figure 5.1 Projection of $\vec{a}^{(2)}$ Onto a Line

For k = 2, Eq.(5.231) yields:

$$q^{(2)} = a^{(2)} - (q^{(1)^T} a^{(2)}) q^{(1)} \tag{5.232}$$

The closest distance from point A (= the tip of vector $a^{(2)}$) to vector $q^{(1)}$ (or $a^{(1)}$) is the perpendicular distance AB. Point B is the projection of $\vec{a}^{(2)}$ onto unit vector $\vec{q}^{(1)}$.

Thus, $\overrightarrow{OB} \equiv \vec{p} = (q^{(1)^T} a^{(2)}) * \vec{q}^{(1)}$

From the right triangle OBA, one clearly sees that the right-hand side of Eq.(5.232) represents the vector $\overrightarrow{BA}$ or $\vec{q}^{(2)}$, which is perpendicular to $\vec{q}^{(1)}$.

The vector $\vec{q}^{(2)}$, therefore, can be considered as the "error," the component of vector $a^{(2)}$ that is <u>not</u> in the direction of $q^{(1)}$. This "error" vector is perpendicular to $q^{(1)}$.
For k = 3, Eq.(5.231) yields:

$$q^{(3)} = a^{(3)} - (q^{(1)^T} a^{(3)}) q^{(1)} - (q^{(2)^T} a^{(3)}) q^{(2)} \tag{5.233}$$

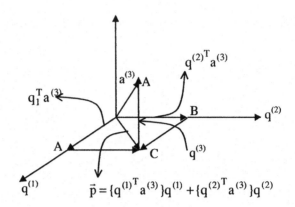

Figure 5.2 Projection of $\vec{a}^{(3)}$ Into a Plane

The 2nd and 3rd terms of the right-hand side of Eq.(5.233) implied that the components of vector $a^{(3)}$ along the direction of $q^{(1)}$ and $q^{(2)}$ are removed, respectively. Hence, the remaining component $q^{(3)} = (a^{(3)} - \vec{p})$ of $a^{(3)}$ is perpendicular to both $q^{(1)}$ and $q^{(2)}$ as shown in Figure 5.2.

Thus, in the "classical" Gram-Schmidt process, the current vector $q^{(k)}$ is orthonormalized with all previous $q^{(1)}, q^{(2)}, ..., q^{(k-1)}$ vectors.

FORTRAN coding for the "classical" Gram-Schmidt is listed in Table 5.5.

In theory, Lanczos vectors are [M] orthogonal to each others. However, in actual numerical computation, the orthogonal properties among Lanczos vectors are lost due to round-off errors. Ths "classical" Gram-Schmidt orthogonalization procedure can be used to re-establish the orthogonal condition as described in Table 5.5.

Table 5.5 Classical Gram-Schmidt Orthogonalization

Step 1: Given the mass matrix [M], and an arbitrary number of vectors $a^{(k)}$. The following process will orthogonalize (with respect to [M]) each successive vector $\vec{q}$ against all of the previous ones.

Step 2: For k = 1 to n

(a) $q^{(k)} = a^{(k)}$

(b)   for j = 1 to k-1

(c)   Compute the scalar $r_{jk} = q^{(j)^T}[M]a^{(k)}$

(d)   Compute $q^{(k)} = q^{(k)} - r_{jk}q^{(j)}$

(e)   End

(f) Compute the scalar $r_{kk} = \left\| q^{(k)^T} M q^{(k)} \right\|$

(g) Compute $q^{(k)} = q^{(k)} \Big/ r_{kk}$

(h) End

Example

Given the matrix $[A] = \begin{bmatrix} a^{(1)} & a^{(2)} & a^{(3)} \end{bmatrix} = \begin{bmatrix} 1 & -1 & 1 \\ 1 & -0.5 & 0.25 \\ 1 & 0 & 0 \\ 1 & 0.5 & 0.25 \\ 1 & 1 & 1 \end{bmatrix}$

Normalize the above $a^{(i)}$ vectors, thus

$$[A] = \begin{bmatrix} 0.4472 & -0.6325 & 0.6860 \\ 0.4472 & -0.3162 & 0.1715 \\ 0.4472 & 0 & 0 \\ 0.4472 & 0.3162 & 0.3162 \\ 0.4472 & 0.6325 & 0.6325 \end{bmatrix}$$

To simplify the analysis, assuming $[M]_{5 \times 5} = [I]$ = identity matrix

$$r_{11} = \left\| q^{(1)} \right\| = \left\| a^{(1)} \right\| = \cdots$$

$$q^{(1)} = \frac{q^{(1)}}{r_{11}} = \cdots$$

$$q^{(2)} = a^{(2)}$$

$$r_{12} = q^{(1)^T} a^{(2)} = \cdots$$

$$q^{(2)} = q^{(2)} + r_{12}q^{(1)} = \cdots$$

$$r_{22} = \left\| q^{(2)} \right\| = \cdots$$

$$q^{(2)} = \frac{q^{(2)}}{r_{22}} = \cdots$$

$$q^{(3)} = a^{(3)}$$

$$r_{13} = q^{(1)^T} a^{(3)} = \cdots$$

$$q^{(3)} = q^{(3)} + r_{13}q^{(1)} = \cdots$$

$$r_{23} = q^{(2)^T} a^{(3)} = \cdots$$

$$q^{(3)} = q^{(3)} + r_{23}q^{(2)} = \cdots$$

$$r_{33} = \left\| q^{(3)} \right\| = \cdots$$

$$q^{(3)} = \frac{q^{(3)}}{r_{33}} = \cdots$$

The complete FORTRAN coding for the classical Gramschmidt algorithm is listed in the following box:

```
c%%%%%%%%%%%%%%%%%%%%%%%%%%%%%%%%%%%%%%%%%%%%%%%%%%%%%%%
c...... Clasical Gramschmidt algorithm
c...... In conjunction with Table 5.5 (of D.T. Nguyen's book)
c
c......"EDUCATIONAL" version March 22, 2002    !
c......written by:    Prof. Duc T. Nguyen
c......stored at:     cd ~/cee/*odu*clas*/classic*.f
c......Remarks: get the same correct answers as the "modified" Gramschmidt
c......         given on Heath's book, page 99
cjcol =    1
caa(-,-) =     1.0000000000000    1.0000000000000    1.0000000000000
c    1.0000000000000    1.0000000000000
cjcol =    2
caa(-,-) =    -1.0000000000000   -0.50000000000000   0.    0.50000000000000
c    1.0000000000000
cjcol =    3
caa(-,-) =     1.0000000000000    0.25000000000000   0.    0.25000000000000
c    1.0000000000000
cjcol =    1
cqq(-,-) =     0.44721359549996    0.44721359549996    0.44721359549996
c    0.44721359549996    0.44721359549996
cjcol =    2
cqq(-,-) =    -0.63245553203368   -0.31622776601684    1.5700924586838D-17
c    0.31622776601684    0.63245553203368
cjcol =    3
cqq(-,-) =     0.53452248382485   -0.26726124191242   -0.53452248382485
c   -0.26726124191242    0.53452248382485
```

```
C++++++++++++++++++++++++++++++++++++++++++++++++
      implicit real*8(a-h,o-z)
      dimension aa(100,100),qq(100,100),ak(100),qj(100),qk(100)
c
c......input data
c
      read(5,*) n,nvectors
      do 1 jcol=1,nvectors
      read(5,*) (aa(irow,jcol),irow=1,n)
      write(6,*) 'jcol = ',jcol
      write(6,*) 'aa(-,-) = ',(aa(irow,jcol),irow=1,n)
 1    continue
c
c......clasical Gramschmidt algorithms
c
      do 11 k=1,nvectors
c
      do 12 i=1,n
      ak(i)=aa(i,k)
      qk(i)=ak(i)
 12   continue
c
      do 13 j=1,k-1
c
      do 14 i=1,n
      qj(i)=qq(i,j)
 14   continue
c
      call dotprod(qj,ak,n,rjk)
      do 15 i=1,n
      qk(i)=qk(i)-rjk*qj(i)
 15   continue
 13   continue
c
      call dotprod(qk,qk,n,rkk)
      rkk=dsqrt(rkk)
c
      do 16 i=1,n
      qk(i)=qk(i)/rkk
      qq(i,k)=qk(i)
 16   continue
 11   continue
c......output results
      do 21 jcol=1,nvectors
      write(6,*) 'jcol = ',jcol
      write(6,*) 'qq(-,-) = ',(qq(irow,jcol),irow=1,n)
 21   continue
c......verify to make sure all vectors qj(-) are orthogonalization
      do 31 k=1,nvectors
c
      do 32 i=1,n
      qk(i)=qq(i,k)
 32   continue
c
      do 33 j=1,k-1
c
      do 34 i=1,n
      qj(i)=qq(i,j)
 34   continue
c
      call dotprod(qk,qj,n,scalar)
      scalar=dabs(scalar)
      if (scalar .gt. 0.000001) then
      write(6,*) 'vector qk(-) NOT orthogonal to vector qj(-) !!'
      write(6,*) 'scalar = should be zero !! = ',scalar
      write(6,*) 'vectors k = ',k,' vector j = ',j
      endif
 33   continue
 31   continue
      stop
```

```
        end
c%%%%%%%%%%%%%%%%%%%%%%%%%%%%%%%%%%%%%%%%%%%%%%%%%%%%
        subroutine dotprod(a,b,n,scalar)
        implicit real*8(a-h,o-z)
        dimension a(*),b(*)
c
        scalar=0.
        do 1 i=1,n
        scalar=scalar+a(i)*b(i)
  1     continue
        return
        end
```

5.7.6 Detailed Step-by-Step Lanczos Algorithms

The Lanczos algorithms outlined in Table 5.4 are not detailed enough (such as the re-orthogonalization process) for computer implementation. Thus, more detailed step-by-step Lanczos algorithms are described in Table 5.6.

Table 5.6 Detailed Lanczos Algorithms for Eigen-Solution of Generalized Eigen-

Equation $K\phi = \lambda M\phi$

Step 1: Input data

Given the stiffness matrix $[K]_{n \times n}$, and mass matrix $[M]_{n \times n}$

Let $m \equiv$ the number of requested eigen-values, and the corresponding eigen-vectors

Let $r \equiv (2 \text{ or } 3) * m$ (5.234)

Step 2: Factorized the sparse stiffness matrix

$$[K] = [L] [D] [L]^T$$ (5.235)

Step 3: Select an arbitrary vector $\{x\}$, and normalize this vector with respect to the mass matrix [M], by the following operations

$$b = \sqrt{x^T M x}$$ (5.236)

$$\{x^{(1)}\} = \text{first Lancozos vector} = \frac{\{x\}}{b}$$ (5.237)

$$\text{Let } b_1 = 0$$ (5.238)

Step 4: Compute subsequent Lanczos vectors $\{x^{(i)}\}$

where $i = 2, 3, 4, \ldots, r$

(a) Performing forward and backward solution to solve for vector $\{\bar{x}^{(i)}\}$:

$$[K]\{\bar{x}^{(i)}\} = [M]\{x^{(i-1)}\}$$ (5.239)

(b) Compute the scalars

$$a_{i-1} = \underbrace{\left\{\overline{x}^{(i)}\right\}^T [M]\left\{x^{(i-1)}\right\}}_{}$$ (5.240)

already calculated in step 4(a)

(c) <u>Orthogonalize</u> the Lanczos vector with respect to the mass matrix [M]

$$\left\{\tilde{x}^{(i)}\right\} = \left\{\overline{x}^{(i)}\right\} - a_{i-1}\left\{x^{(i-1)}\right\} - b_{i-1}\left\{x^{(i-2)}\right\}$$ (5.241)

(d) Normalize the Lanczos vector with respect to [M]

$$b_i = \sqrt{\tilde{x}^{(i)^T} M \tilde{x}^{(i)}}$$ (5.242)

Get subsequent Lanczos vectors

$$\left\{x^{(i)}\right\} = \frac{\left\{\tilde{x}^{(i)}\right\}}{b_i}$$ (5.243)

(e) Re-orthogonalization, or not ??

For any new Lanczos vector $x^{(i)}$, compute and check the orthogonal conditions between the new, and all previous Lanczos vectors:

$$E_j = x^{(j)^T} M x^{(i)}, \text{ where } j = 1, 2, ..., i\text{-}1$$ (5.244)

If $E_j > \sqrt{\varepsilon_{mp}}$, then (ε_{mp} = machine/computer precision). $x^{(i)}$ should be M-orthogonal to $x^{(j)}$, by employing the GramSchmidt re-rothogonalization process. Algorithms shown in Table 5.5 can be used. However, the do loop index "k" (in step 2) should be set to k = i (since we only want to re-orthogonalize the current Lanczos vector $x^{(i)}$ with appropriated previous Lanczos vectors. We do not, however, want to reorthogonalize all Lanczos vectors with each other. Even though the "modified GramSchmidt" process shown in Table 5.5, the latter is a more appropriate choice for GramSchmidt process!)

<u>Step 5</u>: Form the symmetrical, tri-diagonal, expanded matrix [T] $_{rxr}$

$$[T_r] = \begin{bmatrix} a_1 & b_2 & 0 & 0 & 0 & \cdots & \cdots & 0 \\ b_2 & a_2 & b_3 & 0 & 0 & \cdots & \cdots & 0 \\ 0 & b_3 & a_3 & b_4 & 0 & \cdots & \cdots & 0 \\ & & \ddots & \ddots & \ddots & & & \vdots \\ & & & \ddots & \ddots & \ddots & & \vdots \\ & & & & \ddots & \ddots & \ddots & \vdots \\ \cdots & \cdots & \cdots & \cdots & \cdots & b_{r-1} & a_{r-1} & b_r \\ \cdots & \cdots & \cdots & \cdots & \cdots & & b_r & a_r \end{bmatrix}$$ (5.245)

Step 6: "Occasional" compute the standard eigen-solution of the reduced, tri-diagonal system

$$[T_r]\{\phi^*\} = \mu\{\phi^*\} \tag{5.246}$$

$$\text{where } \lambda = \frac{1}{\mu} \tag{5.247}$$

Remarks

(1) If "m" lowest eigen-pairs are requested, one usually provides approximately "2*m" starting vectors. Thus, the first time we want to solve for eigen-solution from the tri-diagonal, standard eigen-problem is when i = 2m (say m = 100 requested lowest eigen-pair solution, hence i = 200).

(2) Check for convergence

Using Eq.(5.174), one computes:

$$\text{Error}(j) \approx \frac{b_{i+1}\phi_i^{*(j)}}{(\lambda_j)_{\text{computed}}} \tag{5.248}$$

(3) Assuming 80 (out of the m = 100 requested eigen-pair) have already been converged, we define:

NEVNC = $\underline{N}$umber of $\underline{E}$igen-$\underline{V}$alues $\underline{N}$ot $\underline{C}$onverged

Thus

NEVNC = m−80 = 100−80 = 20

(4) The next time to solve the tri-diagonal eigen-system will be approximately at (2*NEVNC) iterations later (assuming i + 2*NEVNC ≤ r, where r = maximum number of Lanczos steps).

(5) Go back to Step 4(a), until all requested "m" eigen-values are converged.

(6) If we wish to obtain, 100 eigen-values greater than certain threshold value (say $\rho \geq 5$), then a shift should be applied, such as:

$$\hat{K} = K - \rho M$$

(7) Recover the original eigen-vectors

$$\phi = [X]\phi^*$$

5.7.7 Educational Software for Lanczos Algorithms

Based upon the detailed step-by-step Lanczos algorithms presented in Table 5.6, an educational (non-sparse) version of Lanczos eigen-solver has been coded in FORTRAN and is listed in Table 5.7.

Many statements have been inserted in Table 5.7. Together with Table 5.6, the readers should be able to read and understand the presented Lanczos eigen-solver without too much difficulty.

A far more efficient "sparse" Lanczos eigen-solver can be easily constructed simply by replacing the numerical recipe's dense solver [5.12] embedded inside Table 5.7 with the sparse solver as described in Chapter 3 of this book.

Table 5.7 Educational Lanczos Eigen-Solver (Version 3)

```
c    Table 5.7: Educational Lanczos Eigen-Solver (Version 3)

     implicit real*8(a-h,o-z)
c$$$$$$$$$$$$$$$$$$$$$$$$$$$$$$$$$$$$$$$$$$$$$$$$$$$$$$$$$$$$$$$$$$$$$$$$$$$$$$$$$$$
c......educational version of D.T. Nguyen's Lanczos code
c......DTN's original version= Feb. 1, 2000 ---->> Sept. 17'2001
c......                       (JQ debuged few errors/improvements)
c++++++This code is stored under the file name education_lanczos_3.f
c++++++in sub-directory ~/cee/educational*2001/
c                 education_lanczos_3.f
c
c    notes:
c
c    (a) The main program is followed by all subroutines (in alphabet
c        order)
c    (b) JQ has made the following changes in DTNguyen's educational Lanczos
c        [01] The order of computation for sub-matrices [T] should be:
c             T(1,1), T(1,2), T(2,1), then T((2,2), T(2,3), T(3,2) etc...
c             instead of
c             T(1,1), then T(1,2), T(2,1), T(2,2), then T(2,3), T(3,2), T(3,3)
c        [02] Error norm check should use the original [K]
c        [03] Suppose after say 20 Lanczos steps, it is time to solve
c             the reduced, tri-diagonal [T] eigen-values. Just before this
c             step, we need to save/copy the original [T] (upto this point),
c             since matrix [T] will be destroyed/overwritten/updated during
c             this eigen-solution, and therefore if not all evalues converged
c             yet, then we kneed to proceed for more Lanczos steps, by
c             continuing to build up (or expand) the previous, original [T],
c             and "NOT" by expanding the previous "destroyed/overwritten"
c             [T]
c        [04] JQ used the same "approx. error estimation" for evalues as
c             the one JQ did in the SPARSEPACK, or in ooc sparse Lanczos,
c             which is different than the formula used by DTNguyen's
c             educational Lanczos version
c        [05] JQ used routine TQ2, instead of using DTNguyen's Jacobi
c             to solve the reduced [T] evalues problem, because:
c             TQ2 is more efficient to solve the "STANDARD" evalues problem
c             JACOBI requires "positive" definite matrix
c        [06] JQ added the option for finding n-evalues above
c             the threshold value (this option should be useful, especially
c             for electromagnetic problems, where users do NOT want to
```

```
c          find a lot of "near zero" evalues, which are useless !
c
c          The value of ISHIFT need be activated ( = user's specified input )
c          for this option. This option is also useful for "singular", or
c          "near singular" stiffness matrices.
c      [07] Each time that we generate a "new" Lanczos vector, we have also
c          had generated the associated constants "a" and "b", which are
c          used to construct the [T] matrix. After "re-orthogonalize"
c          the current Lanczos vector with all previous Lanczos vectors, JQ
c          used the "new" constant "b" in constructing the [T] matrix
c      [08] in re-ortho. routine, JQ is faster because of vector*vector operations
c          in re-ortho. routine, DTN is slower because of matrix*vector operations
c  (c) Upon completing all changes mentioned in item (b), DTNguyen has
c          already "successfully" tested different options in this educational
c          version of Lanczos (but not 100% thoroughly tested)
c......Latest test = July 20, 2004
c
c......Could also obtained CORRECT Lanczos solution for Subby's tiny CST
c      structural problem:
c
c      [K] = 1.53*D+06   -308378   -245739
c                        1.1596D+06     0
c                                   1.53D+06
c
c      [Diag. M] = 0.026   0      0
c                    0    0.052   0
c                    0     0     0.052
c
c      ABUQUS's eigen-values = Subby's code = (2.02D+07, 2.84D+07, 6.2D+07)
c
c      [09] In this educational Lanczos (2-nd) version, if the problem size n_dof
c          is too small (n_dof less than certain threshold value, say n_dof < 20 or 30),
c          we will implement Lanczos procedures (not JACOBI procedures) to find "ALL"
c          eigenvalues !
c
c      [10] Yes, we can request "ALL" eigen-values (from 1 to n_eigen-values) by Lanczos
procedures.
c
c      [11] Duc T. Nguyen provided additional option, where user defines the the "full"
c          stiffness and mass matrices.
c
c      [12] This educational Lanczos (3-rd) version will "re-orthogonalize"
c          the current Lanczos vector with "ALL previous Lanczos vectors". "Selective"
c          re-orthogonalized wrt only those "few previous Lanczos vectors" (which failed
c          the re-orthogonal checks) will NOT be robust, and might give WRONG solutions
c          to some cases. However, in this current implementation (including the ooc_
c          sparse_lanczos) we do NOT want to re-orthogonal every Lanczos vector with all
c          the other vectors either (since it will be highly inefficient). Thus, in this
c          version (and including the ooc_sparse_lanczos), we do NOT care about the array
c          lossorth(-)
c
c      [13] Duc's implementation of "estimated" convergence check formular will take
c          more Lanczos iterations to converge, as compared to JQ's formular
```

```
c$$$$$$$$$$$$$$$$$$$$$$$$$$$$$$$$$$$$$$$$$$$$$$$$$$$$$$$$$$$$$$$$$$$$$$$$$$$$$$$$$$$
      real*8 xk(1000,1000), xm(1000,1000), xx(1000), tempo1(1000),
     $ tempo2(1000), b(1000), x(1000,1000), a(1000), xj(1000),
     $ xitutar(1000), z(1000,1000), evalamda(1000),
     $ phi(1000,1000), omega(1000), xkorig(1000,1000)
      real*8 xmi(1000,1000), tridiam(1000,1000)
      real*8 W(1000),w1(1000),w2(1000),w4(1000),eps1,eps2
      integer lossorth(1000),indx(1000)
c
      np=1000    ! only for dimension purposes
      nsweeps=35
c......read master control information
      write(6,*) 'Re-vised Version, or latest test Date: 07-20-2004'
      write(6,*)'LanFlag,ishift,irepeat,n,neig,lumpk,lumpm,ntimes
     $        ,jacobonly,eps1,eps2'
      write(6,*) 'LanFlag = 0 = DTNguyen^s (original) slower version '
      write(6,*) 'because matrix vector operations in reorthog routine'
      write(6,*) 'LanFlag = 1 = JQ^s faster version '
      write(6,*) 'because vector vector operations in reorthog routine'
c##################################################################################
      open(unit=7,file="K.INFO",status="old",form="formatted")
      read(7,*) LanFlag,ishift,irepeat,n,neig,lumpk,lumpm,
     1 ntimes,jacobonly,eps1,eps2,iauto_data
      write(6,*) 'LanFlag,ishift,irepeat,n,neig,lumpk,lumpm,
     1 ntimes,jacobonly,eps1,eps2,iauto_data = '
      write(6,*)  LanFlag,ishift,irepeat,n,neig,lumpk,lumpm,
     1 ntimes,jacobonly,eps1,eps2,iauto_data
      if ( n .gt. 999 ) then
       write(*,*)'Error: n cannot be larger than 999 !'
        stop
      end if
c++++++ n      = actual size of stiffness matrix
c++++++ neig    = number of requested eigen-pair solutions
c++++++ lumpk   = 1 (if diagonal stiffness is used), else general
c               stiffness is used
c++++++ lumpm   = 1 (if diagonal mass is used), else general mass is used
c++++++ ntimes  = factor to compute lanmax (lanmax = ntimes * neig).
c++++++          = 3 or 4 is a recommended value for ntimes
c++++++            lanmax = max. number of Lanczos vectors generated
c++++++ jacobonly = 1 (if "ALL" eigen-values/vectors are computed using
CJACOBI method)
c++++++          = 0 (if a "few" lowest eigen-values/vectors are computed
Cby Lanczos method)
c++++++ eps1, eps2 = small positive error tolerance used for convergence
c              check, or
c++++++            orthogonalized check
c...... iauto_data = 1 (if [K] & [M] data are automatically generated)
c...... iauto_data = 0 (if [K] & [M] data are specified by the user's input file K.INFO)
c##################################################################################
c......Option for user's input data file (=K.INFO) to define [K] & [M] matrices
      if (iauto_data .eq. 0) then   !
       read(7,*) ((xk(i,j),j=1,n),i=1,n)   ! read STIFFNESS matrix, row-wise
       read(7,*) ((xm(i,j),j=1,n),i=1,n)   ! read MASS matrix, row-wise
```

```
c
      do 318 i=1,n
      do 319 j=1,n
      xk(i,j)=xk(i,j)-ishift*xm(i,j)
319   continue
318   continue
c
      do 321 i=1,n
      do 322 j=1,n
      xkorig(i,j)=xk(i,j)
322   continue
321   continue
c......step I: automatically generate stiffness and mass matrices
      else                    !
      do 1 i=1,n
      do 2 j=1,n
      xk(i,j)=0.0d0
2     xm(i,j)=0.0d0
1     continue
c
      if (lumpm .eq. 1) then
      do 6 i=1,n
c     xm(i,i)=1.0
      xm(i,i)=1.0d0
6     continue
      else
      do 7 i=1,n
        do 8 j=i,n
c       xm(i,j)=(-1.0)**(i+j)/(i+j)
c       if (i .eq. j) xm(i,i)=i
        xm(i,j) = 0.000001/(i+j)
        if ( i .eq. j ) xm(i,j) = 1.0D0
        xm(j,i)=xm(i,j)
8       continue
7     continue
      endif
      jjj = ishift/1    !1000
      write(*,*)'JJJ ,irepeat= ',jjj,irepeat

      if (lumpk .eq. 1) then
      do 3 i=1,n
      xk(i,i)=10.0d0 * i  - ishift * xm(i,i)
c     xk(i,i)=-i
      xkorig(i,i)=xk(i,i)
3     continue
      else
      do 4 i=1,n
        do 5 j=i,n
c       xk(i,j)=(i+j)*(-1.0)**(i+j)
c       if (i .eq. j) xk(i,i)=(i+i)*i*1000.0
        xk(i,j) = 0.001*(i+j)  - ishift * xm(i,j)
        if ( i .eq. j ) then
        xk(i,j) = 1.0D0 * i - ishift * xm(i,j)
```

```
         if ( i .gt.jjj .and. i .le. jjj+irepeat )
   1     xk(i,j) = 1.0D0 * (1+jjj) - ishift * xm(i,j)
         end if
         xk(j,i)=xk(i,j)
         xkorig(i,j)=xk(i,j)
         xkorig(j,i)=xk(j,i)
   5     continue
   4   continue
       endif
           endif   !
c
       if (jacobonly .eq. 1) then
c
c...... do we have to re-arrange eigen-solutions in "INCREASING" order
c...... from "lowest" to "highest" eigen-values/eigen-vectors ??
c
       call jacobikjb (xk,xm,z,evalamda,n,tempo2,eps2,nsweeps,np)
c
       write(6,*) 'KJBathe Jacobi evalues ='
       write(6,*) (evalamda(i),i=1,n)
       call errnorm(neig,evalamda,n,xx,z,xm,tempo2,np,xkorig,xitutar)
c......print results
       go to 345
       endif
c......step II: factorize stiffness matrix
       call ludcmp (xk,n,np,indx,d)        ! overwrite xk(-,-)
       write(6,*) 'check point #01: passed factorized'
c      do 61 i=1,n
c      write(6,*) 'factorized K= ',(xk(i,j),j=1,n)
c61    continue
c......step III: choose an arbitrary vector {xx}, any better starting vector ??
       do 9 i=1,n
c      xx(i)=xkorig(i,1)   ! this choice will give xitutar = {0} for certain data
       xx(i)=1
   9   continue
c
       call matvec2 (xm,xx,tempo1,n,np)
c      write(6,*) 'tempo1(-)= ',(tempo1(i),i=1,n)
       call dotprod (xx,tempo1,bb,n)
       bb=dsqrt(bb)
c      write(6,*) 'bb= ',bb
c......get 1-st Lanczos vector
       do 11 i=1,n
       xx(i)=xx(i)/bb
       x(i,1)=xx(i)
  11   continue
c      write(6,*) 'got 1-st Lanczos vector xx(-)= ',(xx(i),i=1,n)
       write(6,*) 'check point #02: got 1-st lanczos vector'
c......step IV: solve for additional Lanczos vectors, with Bi=0.0,
c......        and i = 2,3, ... , r = lanmax
c......part (a): find XBARi vector
c
       lanmax=neig*ntimes
```

```
      if (lanmax .ge. n) lanmax=n  ! can we have lanmax=n, YES !
      do 24 i=1,lanmax
       do 25 j=i,lanmax
       tridiam(i,j)=0.0d0
       tridiam(j,i)=0.0d0
 25    continue
 24    continue
       b(1)=0.0d0
c?     markeig=2*neig
c?     m=2*neig
c?     if (markeig .ge. n) markeig=n  ! can we have markeig=n, YES !
c?     if (m .ge. n) m=n
c?     write(6,*) 'lanmax, markeig,m= ',lanmax,markeig,m
c
C      Lanflag = 0     = 0 for Duc's procedure; = 1 for JQ's way
       IF ( LanFlag .eq. 0 ) THEN
c------ Duc's procedure:
c......step V: obtain eigen solutions in the expanded ( = original ) system
       ELSE
C------ JQ's procedure (after Gramschmidt re_orthogonalization, use "NEW bi":
c      ----------------------------------------
c      Duc's  code        JQ's Code
c      ----------------------------------------
c      4a.               call lubksb
c      Part of 4c.        do loop 3055 :   x = x - b*x(i-1)
c      4b.               call dorprod(..,alf,..)
c      Part of 4c.        do loop 3060:  x = x - alf*x(i-2)
c      4d.               bet2 = .....
c      4c.               call REORTH(...)
c                        Loop 3110 for prepare next iteration
c      ----------------------------------------
       markeig = 2* neig

       if ( markeig .gt. LANMAX ) markeig = LANMAX
       call matvec2(xm,xx,w2,n,np)  ! w2(-) = rhs of Eq.(9.239)
       do 3000 i = 1,n
       w1(i)=0.0d0
3000   w(i) = 0.0D0
       bet = bb             ! computed earlier, scaled factor to normalize 1-st L_vector
       do 3050 JJ = 1, LANMAX   ! ---------- step 4 (for subsequent Lanczos vectors)
c                        (why NOT JJ = 2, LANMAX ??)
       do 1001 I = 1,n
       tempo1(i) = w2(i)       ! rhs of Eq.(9.239)
1001   w4(i) = w2(i)           ! rhs of Eq.(9.239)

c------------------------------------------------------------------------------
       call lubksb(xk,n,np,indx,tempo1)  ! tempo1(-) = xbar_i (see step 4a)

       do 3055 I = 1,n             ! compute parts of step 4c ( = 3-rd term of rhs )
3055   w4(i) = tempo1(i) -  bet*w(i)  ! for first time (loop JJ=1), w(-)=0.

       call dotprod(w2,w4,alf,n)  ! compute a_i-1, Eq.(9.240), see step 4b
```

```
      do 3060 i = 1,n          ! compute step 4c ( = 2-nd term of rhs )
3060  w4(i) = w4(i) - alf*w1(i)  ! array w1(-) has been initialized to ZERO
c                              ! we should have 2-nd term of rhs, even initially ??
c-------------------------------------------------------------------------------------

      call matvec2(xm,w4,w2,n,np)
      call dotprod(w2,w4,bet2,n)
      bet2 = dsqrt(bet2)

      call REORTH(n,np,w2,w4,X,JJ,BET2,eps1)

      call matvec2(xm,w4,w2,n,np)

      call dotprod(w2,w4,bet2,n)
      bet2 = dsqrt(bet2)
      dbet2 = 1.0d0/bet2

      do 3110 i = 1,n
      w(i) = w1(i)
      w1(i) = w4(i) * dbet2
      X(i,JJ+1) = w1(i)
      w2(i) = w2(i) * dbet2
3110  continue
      a(jj) = alf
      b(jj) = bet2
      bet = bet2
      IF ( jj .eq. markeig ) THEN
c     write(6,*) 'JQ tridiagonal matrix dimension= ',JJ
      do 3066 j=1,JJ
        tridiam(j,1) = a(j)
        tridiam(j,2) = b(j)
3066  continue
c
c...... do we have to re-arrange eigen-solutions in "INCREASING" order
c...... from "lowest" to "highest" eigen-values/eigen-vectors ??
c
      write(6,*)'nnnn,jj = ',nnnn,jj
c     call jacobikjb(tridiam,xmi,z,evalamda,JJ,tempo2,eps2,nsweeps,np)
c     call eigsrt(evalamda,z,n,np)  ! sorting eigen-pairs, from large to
Csmall
      do jjj = 1,jj
        do iii = 1,jj
        z(iii,jjj) = 0.0d0
        enddo
        z(jjj,jjj) = 1.0d0
      enddo
      call tq(np,jj,a,b,z,1.0E-15,L)  ! Better version for TRIDIAGONAL
Ceigen-solver
      call eigsrt(a,z,n,np)  ! sorting eigen-pairs, from large to small

      nconv=0

      do 3028 j=1, JJ       ! change into do 28 j=1, m/2 ??
```

```
      zij=z(jj,j)
      errorj=dabs(bet2*zij/a(j))
       tempo1(j) = 0.0d0
       do jjj = 1,jj
         tempo1(jjj) = tempo1(jjj) + abs(z(jjj,j))
       enddo
       tempo1(j) = tempo1(j) * 0.005d0 * errorj
c=================================================================
       if (tempo1(j) .le. eps2) then     ! JQ used this formula for error check
c      if (errorj .le. eps2) then        ! DTN used Eq.(9.248) for error check
c     ! Duc & JQ wants to know if Eq.(9.248) is better or worse ??
c=================================================================
c      write(6,*) 'J , errorj = ',j,errorj,nconv,lossorth(nconv)
       nconv=nconv+1
       lossorth(nconv)=j  ! re-use array lossorth(-) for other purpose
c      write(6,*) 'J , errorj = ',j,errorj,nconv,lossorth(nconv),w(j)
       endif
3028   continue
       nevnyc=neig-nconv
       markeig=markeig+2*nevnyc
       if (markeig .gt. lanmax) markeig=lanmax
c
       if (nconv .ge. neig .OR. JJ .EQ. LANMAX) then
c
       if ( nconv .lt. neig ) then
       write(*,*)'Exceeded Maximum iterations: ',LANMAX
       write(*,*)'Only: ',nconv,' eigenpairs might be accurate.'
       end if
         do 3068 jjj=1,neig
         omega(jjj)=1.00/a(jjj)
         ee = 1.0d0/a(jjj) + ishift
         write(6,*) 'converged eigen # ',jjj, 'omegas = ',ee
3068     continue
         ntsize = jj
         go to 678   ! prepare for successful exit
c
         else
         do iii = 1,jj
           a(iii) = tridiam(iii,1)
           b(iii) = tridiam(iii,2)
         enddo
         endif
         write(6,*)'********** Neq markeig = ', markeig
       END IF
3050   CONTINUE
       END IF
 678   continue
c
       do 32 j=1,neig
        do 33 ii=1,ntsize
        xx(ii)=z(ii,j)
33      continue
        do 3033 II = ntsize+1,n
```

```
          xx(ii) = 0.0d0
3033     continue
      call matvec2 (x,xx,tempo2,n,np)
        do 34 ii=1,n
        phi(ii,j)=tempo2(ii)
 34      continue
 32    continue
c......step VI: compute the "exact, actual" error norm for eigen-solutions
      call errnorm(neig,omega,n,xx,phi,xm,tempo2,np,xkorig,xitutar)
c
345   continue
      stop
      end
C=================================================
      subroutine REORTH(n,np,w2,w4,X,JJ,BET2,eps1)
      REAL*8 HH(1000),W2(*),W4(*),X(np,*)
      real*8 wnorm,eps1
c     wnorm=0.0d0
c     do i=1,n
c     wnorm = wnorm + W2(i)*W2(i)
c     enddo
c     wnorm = dsqrt(wnorm)
c     if ( dabs(wnorm) .gt. 0.999 ) then
c        wnorm = 1.0d0
c     end if
      do 11 K = 1,JJ
      HH(K) = 0.0D0
      do 21 I = 1,n
      HH(K) = HH(K) + W2(I) * X(I,K)
 21     CONTINUE
 11     CONTINUE
c we may want to do selective reortho., by
c checking if hh(k) < eps1 = 10-16
c to decide whether or not skipping loop 41
      icount=0
      DO 41 K = 1,JJ
c     if ( dsqrt(HH(K))/wnorm .gt. eps1 ) then
c     write(*,*)'HH(K),wnorm=',HH(K),wnorm,eps1
      icount = icount + 1
      DO 31 I = 1,N
      W4(I) = W4(i) - HH(K) * X(I,K)
 31     CONTINUE
c     end if
 41     CONTINUE
      write(*,*)'Reorth. done! ',icount,' out of ',JJ,' vectors!'
      RETURN
      END
c%%%%%%%%%%%%%%%%%%%%%%%%%%%%%%%%%%%%%%%%%%%%%
      subroutine dotprod (a,b,answer,n)
      implicit real*8 (a-h,o-z)
      real*8  a(*),b(*)
      answer=0.0D0
      do 1 i=1,n
```

```fortran
      answer=answer+a(i)*b(i)
1     continue
      return
      end
c%%%%%%%%%%%%%%%%%%%%%%%%%%%%%%%%%%%%%%%%%%%%%%%%%%
      subroutine eigsrt(d,v,n,np)
      implicit real*8(a-h,o-z)
c
c.....purpose: sorting eigen-solution vectors (from largest to smallest)
c.....      d(-) = eigen-values
c.....      v(-,-) = eigen-vectors
c
      INTEGER n,np
      REAL*8 d(np),v(np,np)
c     REAL*8 d(100),v(100,100)
      INTEGER i,j,k
      REAL*8 p
      do 13 i=1,n-1
        k=i
        p=d(i)
c...... find the largest eigen-values
        do 11 j=i+1,n
          if(d(j).ge.p)then
            k=j
            p=d(j)
          endif
11      continue
c
        if(k.ne.i)then
          d(k)=d(i)
          d(i)=p
          do 12 j=1,n
            p=v(j,i)
            v(j,i)=v(j,k)
            v(j,k)=p
12        continue
        endif
13    continue
      return
      END
c%%%%%%%%%%%%%%%%%%%%%%%%%%%%%%%%%%%%%%%%%%%%%%%%%%
      subroutine errnorm(neig,omega,n,xx,phi,xm,tempo2,np,xk,xitutar)
      implicit real*8(a-h,o-z)
c.....purpose: error norm check for eigen-solution
      real*8 omega(*),xx(*),phi(np,*),xm(np,*),tempo2(*)
     $,xk(np,*),xitutar(*)
      do 42 j=1,neig
        evalue=omega(j) ! do we have to rearrange eigen-solutions in
CINCREASING order ??
        do 43 ii=1,n
          xx(ii)=phi(ii,j)
43      continue
      call matvec2 (xm,xx,tempo2,n,np)
```

```
       call matvec2 (xk,xx,xitutar,n,np)
         do 44 ii=1,n
         tempo2(ii)=xitutar(ii)-evalue*tempo2(ii)
  44     continue
c......find absolute error norm
         call dotprod (tempo2,tempo2,abserr,n)
         abserr=dsqrt(abserr)
c......find relative error norm
         call dotprod (xitutar, xitutar,value,n)
         value=dsqrt(value)
         relerr=abserr/value
c      ee = 1.0/omega(j)
         write (6,*) 'EIG.#, abserr, relerr = ',j,abserr,relerr
  42     continue
         return
         end
c%%%%%%%%%%%%%%%%%%%%%%%%%%%%%%%%%%%%%%%%%%%%%%%%%%
         subroutine grschmid (n,x,qk,qj,k,xm,rkk,np,tempo2,lossorth)
         implicit real*8(a-h,o-z)
         real*8 x(np,*), qk(*), qj(*), xm(np,*), tempo2(*)
c++++++++++++++++++++++++++++++++++++++++++++++++++++++++++++++++++++++
c      purposes: Grahmscmidt re-orthogonalization process
c                to make sure that current i-th = k-th Lanczos vector
c                will be orthogonal with previous Lanczos vectors
c                Reference: Michael Heath's book "Scientific Computing", pp. 97
c                  However, DTNguyen modified pp. 97, by deleting
c                  "outer-most loop k", since DTNguyen does NOT want
c                  every Lanczos vectors perpendicular to every other
c                  previous Lanczos vectors
c      notes:
c      x(np,*)  = "all" Lanczos vectors upto now
c      qk(*)    = temporary vector, to store "one" current Lanczos vector
c      qj(*)    = previous loss orthogonal vectors (overwritten one-by-one)
c      k        = current i-th Lanczos step
c      tempo2(*)  = temporary vector, to store [MASS] * {current i-th Lanczos vector}
c++++++++++++++++++++++++++++++++++++++++++++++++++++++++++++++++++++++
         integer lossorth(*)
c
         do 1 i=1,n
         qk(i)=x(i,k)
  1      continue
c
         do 2 j=1, k-1
c
         m=lossorth(j)
         write(*,*)'reorthogonalized:  m = ',m
         do 3 i=1,n
c        qj(i)=x(i,j)
         qj(i)=x(i,m)
  3      continue
c
         call matvec2 (xm,qk,tempo2,n,np)
         call dotprod (qj,tempo2,rjk,n)
```

```
c
      do 4 i=1,n
      qk(i)=qk(i)-rjk*qj(i)
4     continue
c
2     continue
c
      call matvec2 (xm,qk,tempo2,n,np)
      call dotprod (qk, tempo2, rkk, n)
      rkk=dsqrt(rkk)
c
      do 6 i=1,n
      qk(i)=qk(i)/rkk
6     continue
      return
      end
c%%%%%%%%%%%%%%%%%%%%%%%%%%%%%%%%%%%%%%%%%%%%%%%
      subroutine jacobikjb (A,B,X,EIGV,N,D,rtol,nsmax,nmax)
c     modified by runesha Nov 27, 1995
c     ...............................................
c     .                                            .
c     . PROGRAM TO SOLVE THE GENERALIZED EIGENPROBLEM .
c     . USING THE GENERALIZED JACOBI ITERATION      .
c     .                                            .
c     . -- INPUT VARIABLES --                       .
c     .                                            .
c     .  A(N,N)  =Stiffness matrix (assumed positive definite) .
c     .  B(N,N)  =Mass matrix (assumed positive definite )   .
c     .  X(N,N)  =Vector storing eigen-vectors on solution exit .
c     .  EIGV(N) =Vector storing eigen-values on solution exit  .
c     .  D(N)    =Working vector                    .
c     .  N       =order of matrices A and B         .
c     .  RTOL    =Convergence tolerance(usually set to 10.**-12 .
c     .  NSMAX   =Maximum number of sweeps allowed   .
c     .          (usually set to 15)                .
c     .                                            .
c     . -- OUTPUT --                                .
c     .                                            .
c     .  A(N,N)  =Diagonalized stiffness matrix     .
c     .  B(N,N)  =Diagonalized mass matrix          .
c     .  X(N,N)  =Eigenvectors stored columnwise    .
c     .  EIGV(N) =Eigen-values                      .
c     .                                            .
c     ...............................................
c
c     Parameter (RTOL=10.**-12,NSMAX=15,IFPR=0,IOUT=6)
c     Parameter (RTOL=0.000000000001,NSMAX=15,IFPR=0,IOUT=6)
      IMPLICIT REAL*8 (A-H,O-Z)
      DIMENSION A(nmax,1),B(nmax,1),X(nmax,1),EIGV(1),D(1)
c     DIMENSION A(100,1),B(100,1),X(100,1),EIGV(1),D(1)
c
*     ABS(X)=DABS(X)
*     SQRT(X)=DSQRT(X)
```

```
c    ..................................................
c    . This program is used in single precision arithmetic on   .
c    . cdc equipment and double precision  arithmetic on ibm    .
c    . or univac machines. activate,deactivate or adjust above  .
c    . card for single or double precision arithmetic           .
c    ..................................................
c
c    INITIALIZE EIGEN-VALUE AND EIGEN-VECTOR MATRICES
c
     DO 10 I=1,N
     IF (A(I,I).GT.0. .AND. B(I,I).GT.0.) GO TO 4
     WRITE (*,2020)
cnguyen (Feb. 2, 2000)     STOP
   4 D(I)=A(I,I)/B(I,I)
  10 EIGV(I)=D(I)
     DO 30 I=1,N
        DO 20 J=1,N
  20    X(I,J)=0.d0
  30 X(I,I)=1.
c.....IF (N.EQ.1) RETURN
     IF (N.EQ.1) then
     write(6,*) 'tridiag. matrix dimension is 1x1 !!??'
     STOP
     endif
c
c    INITIAALIZE SWEEP COUNTER AND BEGIN ITERATION
c
     NSWEEP=0
     NR=N-1
  40 NSWEEP=NSWEEP+1
c    IF (IFPR.EQ.1) WRITE (*,2000)NSWEEP
c
c    CHECK IF PRESENT OFF-DIAGONAL ELEMENT IS LARGE ENOUGH TO
REQUIRE
c    ZEROING
c
     EPS=(.01**NSWEEP)**2
     DO 210 J=1,NR
     JJ=J+1
     DO 210 K=JJ,N
     EPTOLA=(A(J,K)*A(J,K))/(A(J,J)*A(K,K))
     EPTOLB=(B(J,K)*B(J,K))/(B(J,J)*B(K,K))
     IF (( EPTOLA.LT.EPS).AND.(EPTOLB.LT.EPS)) GO TO 210
     AKK=A(K,K)*B(J,K)-B(K,K)*A(J,K)
     AJJ=A(J,J)*B(J,K)-B(J,J)*A(J,K)
     AB=A(J,J)*B(K,K)-A(K,K)*B(J,J)
     CHECK=(AB*AB+4.*AKK*AJJ)/4.
     IF (CHECK)50,60,60
  50 WRITE(*,2020)
     check=-check          !  nguyen
cnguyen    STOP
  60 SQCH=DSQRT(CHECK)
     D1=AB/2.+SQCH
```

```
     D2=AB/2.-SQCH
     DEN=D1
     IF(DABS(D2).GT.DABS(D1)) DEN=D2
     IF(DEN)80,70,80
  70 CA=0.
     CG=-A(J,K)/A(K,K)
     GO TO 90
  80 CA=AKK/DEN
     CG=-AJJ/DEN
c
c    PERFORM THEGENERAL ROTATION TO ZERO THE PRESENT OFF
DIAGGONAL ELEMENT
c
  90 IF(N-2)100,190,100
 100 JP1=J+1
     JM1=J-1
     KP1=K+1
     KM1=K-1
     IF(JM1-1)130,110,110
 110 DO 120 I=1,JM1
     AJ=A(I,J)
     BJ=B(I,J)
     AK=A(I,K)
     BK=B(I,K)
     A(I,J)=AJ+CG*AK
     B(I,J)=BJ+CG*BK
     A(I,K)=AK+CA*AJ
 120 B(I,K)=BK+CA*BJ
 130 IF(KP1-N)140,140,160
 140 DO 150 I=KP1,N
     AJ=A(J,I)
     BJ=B(J,I)
     AK=A(K,I)
     BK=B(K,I)
     A(J,I)=AJ+CG*AK
     B(J,I)=BJ+CG*BK
     A(K,I)=AK+CA*AJ
 150 B(K,I)=BK+CA*BJ
 160 IF (JP1-KM1)170,170,190
 170 DO 180 I=JP1,KM1
     AJ=A(J,I)
     BJ=B(J,I)
     AK=A(I,K)
     BK=B(I,K)
     A(J,I)=AJ+CG*AK
     B(J,I)=BJ+CG*BK
     A(I,K)=AK+CA*AJ
 180 B(I,K)=BK+CA*BJ
 190 AK=A(K,K)
     BK=B(K,K)
     A(K,K)=AK+2.*CA*A(J,K)+CA*CA*A(J,J)
     B(K,K)=BK+2.*CA*B(J,K)+CA*CA*B(J,J)
     A(J,J)=A(J,J)+2.*CG*A(J,K)+CG*CG*AK
```

```
     B(J,J)=B(J,J)+2.*CG*B(J,K)+CG*CG*BK
     A(J,K)=0.d0
     B(J,K)=0.d0
c
c  UPDATE THE EIGEN-VECTOR MATRIX AFTER EACH ROTATION
c
     DO 200 I=1,N
     XJ=X(I,J)
     XK=X(I,K)
     X(I,J)=XJ+CG*XK
 200 X(I,K)=XK+CA*XJ
 210 CONTINUE
c
c  UPDATE THE EIGEN-VALUE AFTER EACH SWEEP
C
     DO 220 I=1,N
     IF (A(I,I).GT.0. .AND.B(I,I).GT.0.)GO TO 220
     WRITE(*,2020)
cnguyen    STOP
 220 EIGV(I)=A(I,I)/B(I,I)
c    IF (IFPR.EQ.0)GO TO 230
c    WRITE(IOUT,2030)
c    WRITE(IOUT,2010) (EIGV(I),I=1,N)
c
c  CHECK FOR CONVERGENCE
c
 230 DO 240 I=1,N
     TOL=RTOL*D(I)
     DIF=DABS(EIGV(I)-D(I))
     IF(DIF.GT.TOL) GO TO 280
 240 CONTINUE

c
c  CHECK ALL OFF-DIAGONAL ELEMENTA TO SEE IF ANOTHER SWEEP IS
c  REQUIRED
c
     EPS=RTOL**2
     DO 250 J=1,NR
     JJ=J+1
     DO 250 K=JJ,N
     EPSA=(A(J,K)*A(J,K))/(A(J,J)+A(K,K))
     EPSB=(B(J,K)*B(J,K))/(B(J,J)+B(K,K))
     IF ((EPSA.LT.EPS) .AND. (EPSB.LT.EPS))GO TO 250
     GO TO 280
 250 CONTINUE
c
c  FILL OUT BOTTOM TRIANGLE OF RESULTANT MATRICES AND SCALE
c  EIGEN-VECTORS
c
 255 DO 260 I=1,N
     DO 260 J=1,N
     A(J,I)=A(I,J)
 260 B(J,I)=B(I,J)
```

```
      DO 270 J=1,N
      if (b(j,j) .lt. 0.0) b(j,j)=-b(j,j)      ! nguyen
      BB=DSQRT(B(J,J))
      DO 270 K=1,N
  270 X(K,J)=X(K,J)/BB



*     write(*,*) 'THE EIGEN-VECTORS ARE :'
*     write(*,*) ((x(i,j),j=1,n),i=1,n)
      return
c     .................................................
c
c     UPDATE D MATRIX AMD START NEW SWEEP,IF ALLOWED
c
  280 DO 290 I=1,N
  290 D(I)=EIGV(I)
      IF(NSWEEP.LT.NSMAX)GO TO 40
      GO TO 255
 2000 FORMAT(27HOSWEEP NUMBER IN *JACOBI* = ,I4)
 2010 FORMAT(1HO,6E20.12)
 2020 FORMAT(55H0***ERROR SOLUTION STOP, MATRICES NOT POSITIVE
      DEFINITE)
 2030 FORMAT(36HOCURRENT EIGEN-VALUES IN *JACOBI* ARE,/)
      END


c%%%%%%%%%%%%%%%%%%%%%%%%%%%%%%%%%%%%%%%%%%%%%%%%%%
      subroutine lubksb(a,n,np,indx,b)
c......Following are FORTRAN subroutines obtained from NUMERICAL RECIPE
      implicit real*8(a-h,o-z)
      INTEGER n,np,indx(n)
      REAL*8 a(np,1),b(*)
c     REAL*8 a(100,100),b(100)
      INTEGER i,ii,j,ll
      REAL*8 sum
      ii=0
      do 12 i=1,n
        ll=indx(i)
        sum=b(ll)
        b(ll)=b(i)
        if (ii.ne.0)then
          do 11 j=ii,i-1
            sum=sum-a(i,j)*b(j)
11        continue
        else if (sum.ne.0.) then
          ii=i
        endif
        b(i)=sum
12    continue
      do 14 i=n,1,-1
        sum=b(i)
        do 13 j=i+1,n
          sum=sum-a(i,j)*b(j)
13      continue
```

```
      b(i)=sum/a(i,i)
14    continue
      return
      END
c%%%%%%%%%%%%%%%%%%%%%%%%%%%%%%%%%%%%%%%%%%%%%%%%
      subroutine ludcmp(a,n,np,indx,d)
c......Following are FORTRAN subroutines obtained from NUMERICAL RECIPE
      implicit real*8(a-h,o-z)
      INTEGER n,np,indx(np),NMAX
      REAL*8 d,a(np,np),TINY
c     REAL*8 d,a(100,100),TINY
c     PARAMETER (NMAX=500,TINY=1.0e-20)
      PARAMETER (NMAX=100,TINY=1.0d-20)
      INTEGER i,imax,j,k
      REAL*8 aamax,dum,sum,vv(np)
      d=1.
      do 12 i=1,n
       aamax=0.0D0
       do 11 j=1,n
        if (dabs(a(i,j)).gt.aamax) aamax=dabs(a(i,j))
11      continue
       if (aamax.eq.0.) pause 'singular matrix in ludcmp'
       vv(i)=1.0D0/aamax
12    continue
      do 19 j=1,n
       do 14 i=1,j-1
        sum=a(i,j)
        do 13 k=1,i-1
         sum=sum-a(i,k)*a(k,j)
13       continue
        a(i,j)=sum
14     continue
       aamax=0.0D0
       do 16 i=j,n
        sum=a(i,j)
        do 15 k=1,j-1
         sum=sum-a(i,k)*a(k,j)
15       continue
        a(i,j)=sum
        dum=vv(i)*dabs(sum)
        if (dum.ge.aamax) then
         imax=i
         aamax=dum
        endif
16     continue
       if (j.ne.imax)then
        do 17 k=1,n
         dum=a(imax,k)
         a(imax,k)=a(j,k)
         a(j,k)=dum
17       continue
        d=-d
        vv(imax)=vv(j)
```

```
      endif
      indx(j)=imax
      if(a(j,j).eq.0.)a(j,j)=TINY
      if(j.ne.n)then
        dum=1.0D0/a(j,j)
        do 18 i=j+1,n
        a(i,j)=a(i,j)*dum
18      continue
      endif
19    continue
      return
      END
c%%%%%%%%%%%%%%%%%%%%%%%%%%%%%%%%%%%%%%%%%%%%%%
      subroutine matvec2(phi,avec,vec,n,nmax)
      implicit real*8(a-h,o-z)
c......purpose: to multiply matrix [phi] * {avec} = {vec}
      real*8 phi(nmax,1),avec(1),vec(1)
c---------------------------
c......initialize
      do 5 i=1,n
5     vec(i)=0.0D0
c------------------------
      do 1 icol=1,n
      do 2 irow=1,n
      vec(irow)=vec(irow)+phi(irow,icol )*avec(icol)
2     continue
1     continue
c-----------------------------------------
c     write(6,*) 'passed matvec.f'
      return
      end
c%%%%%%%%%%%%%%%%%%%%%%%%%%%%%%%%%%%%%%%%%%%%%%
      subroutine tq(N1,N,B,C,Q,eps,L)
      implicit real*8(a-h,o-z)
C*********************************************************************
C This is a subroutine to solve eigenproblem of triangial matrices By
C QR method
C
C B-- diaginal terms of Trianginal matrix T, return the eigenvalue of T
C C-- offdiginalk terms of T
C [T]=
C    b(1) c(1)
C    c(1) b(2) c(2)
C        c(2) b(3) c(3)
C        . . .
C          . . .
C            c(n-2) b(n-1) c(n-1)
C                 c(n-1) b(n)
C
C Q--identity matrix, return eigen-vector
C*********************************************************************
      dimension B(n1),C(n1),Q(n1,n1)
C
```

```
C
      C(N)=0.0
      D=0.0
      F=0.0
      DO 50 J=1,N
       IT=0
       H=EPS*(ABS(B(J))+ABS(C(J)))
       IF(H.GT.D) D=H
       M=J-1
10     M=M+1
       IF(M.LE.N)THEN
        IF(ABS(C(M)).GT.D) GOTO 10
       ENDIF
C
       IF(M.NE.J)THEN
15      IF(IT.EQ.60) THEN
        L=0
        WRITE(6,18)
18       FORMAT(1X,'FAIL')
         RETURN
        ENDIF
        IT=IT+1
        G=B(J)
        P=(B(J+1)-G)/(2.0*C(J))
        R=SQRT(P*P+1.0)
        IF(P.GE.0.0) THEN
         B(J)=C(J)/(P+R)
        ELSE
         B(J)=C(J)/(P-R)
        ENDIF
        H=G-B(J)
        DO 20 I=J+1,N
         B(I)=B(I)-H
20       continue
        F=F+H
        P=B(M)
        E=1.0
        S=0.0
        DO 40 I=M-1,J,-1
         G=E*C(I)
         H=E*P
         IF(ABS(P).GE.ABS(C(I))) THEN
          E=C(I)/P
          R=SQRT(E*E+1.0)
          C(I+1)=S*P*R
          S=E/R
          E=1.0/R
         ELSE
          E=P/C(I)
          R=SQRT(E*E+1.0)
          C(I+1)=S*C(I)*R
          S=1.0/R
          E=E/R
```

```
          ENDIF
          P=E*B(I)-S*G
          B(I+1)=H+S*(E*G+S*B(I))
          DO 30 K=1,N
           H=Q(K,I+1)
           Q(K,I+1)=S*Q(K,I)+E*H
           Q(K,I)=E*Q(K,I)-S*H
30        CONTINUE
40        CONTINUE
          C(J)=S*P
          B(J)=E*P
          IF(ABS(C(J)).GT.D) GOTO 15
          ENDIF
          B(J)=B(J)+F
C
50        CONTINUE
          DO 80 I=1,N
          K=I
          P=B(I)
          IF((I+1).LE.N) THEN
          J=I
60        J=J+1
          IF(J.LE.N) THEN
           IF(B(J).LE.P) THEN
            K=J
            P=B(J)
            GOTO 60
           ENDIF
          ENDIF
          ENDIF
          IF(K.NE.I) THEN
            B(K)=B(I)
            B(I)=P
            DO 70 J=1,N
             P=Q(J,I)
             Q(J,I)=Q(J,K)
             Q(J,K)=P
70          CONTINUE
          ENDIF
80        CONTINUE
          L=1
          RETURN
          END
c%%%%%%%%%%%%%%%%%%%%%%%%%%%%%%%%%%%%%%%%%%%%%%
```

5.7.8 Efficient Software for Lanczos Eigen-Solver

A brief description of the highly efficient sparse eigen-solver is given in the following section.

User Manual for Sparse Lanczos Eigen-Solver

COMPLEX NUMBERS of SPARSEPACK97

(This file is stored under ~/cee/SPARSEPACK97-98
/complex_version_incore_SPARSEPACK97)
(under file name complex_SPARSEPACKusermanual)

[01] To solve the generalized eigen-problem in the form
      [A] * {x} = (Lamda) * [B] * {x}
where

[A] =   complex numbers, symmetrical (positive) "STIFFNESS" matrix

[B] =   complex numbers, symmetrical "MASS" matrix Note: [B] can be either
        "diagonal" (lumped) mass, or have the same non-zero patterns as [A]
        (consistent) mass matrix

        Lamda = eigen-values

        {x}   = corresponding eigen-vectors

[02] How to prepare the input data (assuming matrices [A] and [B] are known) ??

$$
\text{Assuming } [A] =
\begin{matrix}
11.0 & 0.0 & 0.0 & 41.0 & 0.0 & 52.0 \\
 & 44.0 & 0.0 & 0.0 & 63.0 & 0.0 \\
 & & 66.0 & 0.0 & 74.0 & 0.0 \\
 & \text{SYM} & & 88.0 & 85.0 & 0.0 \\
 & & & & 110.0 & 97.0 \\
 & & & & & 112.0
\end{matrix}
$$

Here, N   = size of matrix [A] = 6
      NCOEF = total number of non-zero off-diagonal terms of the upper
              portion of matrix [A] = 6

$$
\text{Assuming } [B] =
\begin{matrix}
22.0 & 0.0 & 0.0 & -41.0 & 0.0 & 12.0 \\
 & 77.0 & 0.0 & 0.0 & -3.0 & 0.0 \\
 & & 66.0 & 0.0 & 74.0 & 0.0 \\
 & \text{SYM} & & 88.0 & 85.0 & 0.0 \\
 & & & & 100.0 & 27.0 \\
 & & & & & 12.0
\end{matrix}
$$

(a)   K.INFO file, should contain two lines with the following
        information:

        Descriptions of Your Problem: format(a60), not exceeding 60
        characters

        nreord, neig, lump, n, n2, ncoef, itime, ishift, iblock, mread
        where:

    nreord = 3 (use MMD reorder algorithm), 0 ("NOT" use re-order
      algorithm)
    neig = number of requested eigen-values/vectors (above threshold
      -shift)
    lump = 1 (lumped, diagonal mass matrix [B], 0 (consistent mass
      matrix [B])
    n = size of matrix [A(n,n)] = same definition as given above
    n2 = n
    ncoef = same definition as given above
    itime = 0 (save memory), 1 (save time)
    ishift = 0 (default, noshift), NONZERO (perform a shift of value
      ISHIFT)
    iblock = -1 (sub-space iteration algorithm), 0 (regular Lanczos
      algorithm)
    mread = negative integer (default, read ascii K* files)
      positive integer (read fort* files)

(b) K.PTRS file, to store # non-zero terms per row (EXCLUDING
    diagonal terms) of the STIFFNESS [A] matrix ("integer" numbers,
    should have N values)

    2, 1, 1, 1, 1, 0

(c) K11.INDXS file, to store column numbers associated with the
    non-zero, off-diagonal terms (in a row-by-row fashion) of the
    STIFFNESS [A] matrix ("integer" numbers, should have NCOEF
    values)

    4, 6, 5, 5, 5, 6

(d) K.DIAG file, to store numerical values of DIAGONAL terms of
    matrix [A] (could be REAL or COMPLEX numbers, should have
    N values)

    (11.0,0.0), (44.0,0.0), (66.0,0.0), (88.0,0.0), (100.0,0.0), (12.0,0.0)

(e) K11.COEFS file, to store numerical values of the non-zero, off-
    diagonal terms (in a row-by-row fashion) of the STIFFNESS [A]
    matrix (could be REAL or COMPLEX numbers, should have
    NCOEF values)

    (41.0,0.0), (52.0,0.0), (63.0,0.0), (74.0,0.0), (85.0,0.0), (27.0,0.0)

For the complete listing of the FORTRAN source codes, instructions on how to
incorporate the highly efficient sparse Lanczos eigen-solver package into any
existing application software (on any computer platforms), and the complete

consulting service in conjunction with this sparse Lanczos eigen-solver, the readers should contact:

Prof. Duc T. Nguyen
Director, Multidisciplinary Parallel-Vector Computation Center
Old Dominion University
Room 1319, ECS Building
Norfolk, VA 23520 (USA)
Tel (757) 683-3761
Fax (757) 683-5354
E-mail dnguyen@odu.edu

5.8 Unsymmetrical Eigen-Solvers

The algorithms for symmetric matrices are highly satisfactory in practice. By contrast, it is impossible to design equally satisfactory algorithms for the nonsymmetric cases, which are needed in Controls-Structures Interaction (CSI), acoustic, and panel flutter applications. There are two reasons for this. First, the eigen-values of a nonsymmetric matrix can be very sensitive to small changes in the matrix elements. Second, the matrix itself can be defective so that there is no complete set of eigen-vectors.

There are several basic building blocks in the QR algorithm, which is generally regarded as the most effective algorithm, for solving all eigen-values of a real, unsymmetric matrix. These basic components of the QR algorithm are reviewed in Sections 5.7.10–5.7.17. Several sub-routines mentioned in Sections 5.7.10–5.7.17 can be downloaded from the cited "Numerical Recipes" Web site and Text Books [5.12].

5.9 Balanced Matrix

The idea of balancing is to use similarity transformations to make corresponding rows and columns of the matrix have comparable norms, thus reducing the overall norm of the matrix while leaving the eigen-values unchanged.

The time taken by the balanced procedure is insignificant as compared to the total time required to find the eigen-values. For this reason, it is strongly recommended that a nonsymmetric matrix be balanced before attempting to solve for eigen-solutions.

The numerical recipe's subroutine balance is partially listed in Table 5.8.

Table 5.8 Numerical Recipe's Subroutine BALANC

```
c      This subroutine is extracted from NUMERICAL RECIPES
c      (FORTRAN version, 1990, ISBN # 0-521-38330-7),
c      Authors = Press, Flannery, Teukolsky and Vetterling
c      pages 366-367
c
c      Given an N by N matrix [A] stored in an array of
```

```
c       physical dimensions NP by NP. This routine replaces
c       it by a balanced matrix with identical eigen-values.
c       A symmetric matrix is already balanced and is unaffected
c       by this procedure. The parameter RADIX should be the
c       machine's floating point radix
c
        SUBROUTINE balanc(a,n,np)
        implicit real*8(a-h,o-z)                 ! nguyen
        INTEGER n,np
c       REAL a(np,np),RADIX,SQRDX
        PARAMETER (RADIX=2.,SQRDX=RADIX**2)
        INTEGER i,j,last
c       REAL c,f,g,r,s                           ! nguyen
```

5.10 Reduction to Hessenberg Form

The strategy for finding the eigen-solution of an unsymmetric matrix is similar to that of the symmetric case. First we reduce the matrix to a simple Hessenberg form, and then we perform an iterative procedure on the Hessenberg matrix. An *upper Hessenberg* matrix has zeros everywhere below the diagonal except for the first sub-diagonal. For example, in the 6×6 case, the non-zero elements are:

$$
\begin{bmatrix}
X & X & X & X & X & X \\
X & X & X & X & X & X \\
0 & X & X & X & X & X \\
0 & 0 & X & X & X & X \\
0 & 0 & 0 & X & X & X \\
0 & 0 & 0 & 0 & X & X
\end{bmatrix}
$$

Thus, a procedure analogous to Gaussian elimination can be used to convert a general unsymmetric matrix to an upper Hessenberg matrix. The detailed coding of the Hessenberg reduction procedure is listed in Numerical Recipe's sub-routine ELMHES, which is partially listed in Table 5.9.

Once the unsymmetric matrix has already been converted into the Hessenberg form, the QR algorithm itself can be applied on the Hessenberg matrix to find all the real and complex eigen-values. For completeness, the QR algorithm will be described in the following sections.

Table 5.9 Numerical Recipe's Subroutine ELMHES

```
c       This subroutine is extracted from NUMERICAL RECIPES
c       (FORTRAN version, 1990, ISBN # 0-521-38330-7),
c       Authors = Press, Flannery, Teukolsky and Vetterling
c       pages 368-369
c
c       Reduction to Hessenberg form by the elimination method.
c       The real, nonsymmetric N by N matrix [A], stored in an
c       array of physical dimensions NP by NP is replaced by an
c       upper Hessenberg matrix with identical eigen-values.
c       Recommended, but not required is that this routine be
```

```
c      preceded by subroutine BALANC. On output, the Hessenberg
c      matrix is in elements A(I,J), with I .le. J+1. Elements
c      with I .gt. J+1 are to be thought of as zero, but are
c      returned with random values
c
       SUBROUTINE elmhes(a,n,np)
       implicit real*8(a-h,o-z)                    ! nguyen
       INTEGER n,np
c      REAL a(np,np)
        dimension a(np,np)                         ! nguyen
       INTEGER i,j,m
c      REAL x,y                                    ! nguyen
```

5.11 QR Factorization

Assuming a matrix [A] with the dimension m×n, where m ≥ n, we are looking for an m×m orthogonal matrix [Q], such that

$$A = Q\begin{bmatrix} R \\ 0 \end{bmatrix} \tag{5.249}$$

where [R] is an upper-triangular matrix, with the dimension n×n.

5.12 Householder QR Transformation

A Householder transformation H is a matrix, which can be given by the following formula:

$$[H] = [I] - \frac{2\vec{v} * \vec{v}^T}{(\vec{v}^T * \vec{v})} \tag{5.250}$$

where $\vec{v}$ is a non-zero vector:
From Eq.(5. 250), one has:

$$[H]^T = [I] - \frac{2(\vec{v} * \vec{v}^T)^T}{(\vec{v}^T * \vec{v})^T} = [H]$$

The product of $[H]*[H]^T = [H]^T[H]$ can be computed as:

$$[H][H]^T = [I - (\frac{2}{v^T v}) * v * v^T][I - (\frac{2}{v^T v}) * v * v^T]$$

$$= I - (\frac{2}{v^T v}) * v * v^T + \frac{4}{(v^T v)^2} * vv^T * vv^T - (\frac{2}{v^T v}) * v * v^T$$

$$[H][H]^T = [I] \Rightarrow [H]^T = [H]^{-1}$$

Thus, the matrix [H] is symmetrical (because $H^T = H$) and orthogonal (because $H^{-1} = H^T$).

Given a vector $\vec{a}$, we would like to select the vector $\vec{v}$ such that

$$[H]\vec{a} = \begin{Bmatrix} \gamma \\ 0 \\ 0 \\ \vdots \\ \vdots \\ 0 \end{Bmatrix} = \gamma \begin{Bmatrix} 1 \\ 0 \\ 0 \\ \vdots \\ \vdots \\ 0 \end{Bmatrix} = \gamma \hat{e}^{(1)} \tag{5.251}$$

Substituting the formula for [H] into the above equation, one obtains:

$$[H]\vec{a} = \gamma \hat{e}^{(1)} = [I - (\frac{2}{v^T v}) * vv^T]\vec{a}$$

or

$$[H]\vec{a} = \vec{a} - (\frac{2*v^T a}{v^T v}) * \vec{v} = \gamma \hat{e}^{(1)} \tag{5.252}$$

From Eq.(5. 252), one can solve for vector $\vec{v}$:

$$\vec{v} = (\vec{a} - \gamma \hat{e}^{(1)}) * (\frac{v^T v}{2v^T a}) \tag{5.253}$$

The "scalar" factor $(\frac{v^T v}{2v^T a})$ is unimportant since it will be cancelled out when matrix [H] is computed. Hence

$$\vec{v} = \vec{a} - \gamma \hat{e}^{(1)} \tag{5.254}$$

Remarks

#[1] To avoid possibilities for numerical overflow or underflow for the case components of $\vec{a}$ are very small or very large, the vector $\vec{a}$ should be normalized (by dividing $\vec{a}$ with its largest magnitude). Such a scale factor will not change the resulting transformation matrix [H].

#[2] To preserve the norm, one should have $\gamma = \pm \|\vec{a}\|_2$, and the (+), or (-) sign should be selected to avoid cancellation.

#[3] The FORTRAN code to implement the Householder transformation is given in Table 5.10.

#[4] There is no need to explicitly form [H] since [H] $\vec{a}$ can be obtained from $\vec{v}$ and $\vec{a}$ directly.

#[5] The Householder transformation can be generalized to produce zero terms below the diagonals of columns of a given rectangular matrix [A] as illustrated by the following symbolic explanation.

$$H_1(A) = \begin{bmatrix} r_{11} & r_{12} & r_{13} \\ 0 & & * \\ 0 & & * \\ 0 & a^{(2)} & * \\ 0 & & * \\ 0 & & * \end{bmatrix}$$

$$H_2(H_1A) = (I - \frac{2v^{(2)}v^{(2)^T}}{v^{(2)^T}v^{(2)}})(H_1A) = \begin{bmatrix} r_{11} & r_{12} & r_{13} \\ 0 & r_{22} & r_{23} \\ 0 & 0 & \\ 0 & 0 & a^{(3)} \\ 0 & 0 & \\ 0 & 0 & \end{bmatrix}$$

$$H_3(H_2H_1A) = (I - \frac{2v^{(3)}v^{(3)^T}}{v^{(3)^T}v^{(3)}})(H_2H_1A) = \begin{bmatrix} r_{11} & r_{12} & r_{13} \\ 0 & r_{22} & r_{23} \\ 0 & 0 & r_{33} \\ 0 & 0 & 0 \\ 0 & 0 & 0 \\ 0 & 0 & 0 \end{bmatrix}$$

where $v^{(2)}$ can be computed from Eq.(5. 254):

$$v^{(2)} = a^{(2)} - \gamma^{(2)}\hat{e}^{(1)}$$

with

$$a^{(2)} = \left\{ \begin{matrix} 0 \\ * \\ * \\ * \\ * \\ * \end{matrix} \right\}$$

Similarly:

$$v^{(3)} = a^{(3)} - \gamma^{(3)}\hat{e}^{(1)}$$

and

$$a^{(3)} = \begin{Bmatrix} 0 \\ 0 \\ \boxed{*} \\ * \\ * \\ * \end{Bmatrix}$$

Table 5.10 Householder Transformation

```
C%%%%%%%%%%%%%%%%%%%%%%%%%%%%%%%%%%%%%%%%%%%%%%%%%%%%%%%%%%%%%%%%%%%

      implicit real*8 (a-h,o-z)
      dimension a(100), v(100), e(100),Ha(100)
      dimension floatk(100,60), rhs(100)
c......stored under cd ~/cee/*odu*clas*/householder*.f
c
c......input data: size of rectangular (or square) float
c......            (stiffness) matrix [floatk] m by n,
c......            where m .ge. n, and vector {rhs} m by 1
c
      mp=100
c
      read(5,*) m,n
      write(6,*) 'm rows, n columns = ',m,n
c
      do 1 j=1,n
      read(5,*) (floatk(i,j),i=1,m)
      write(6,*) 'column # ',j
      write(6,*) 'vector = ',(floatk(i,j),i=1,m)
 1    continue
c
      read(5,*) (rhs(i),i=1,m)
      write(6,*) 'rhs vector = ',(rhs(i),i=1,m)
c
      do 2 j=1,n
      ithzero=j
c
      do 12 i=1,m
 12   a(i)=0.0d0
c
      do 4 i=ithzero,m
      a(i)=floatk(i,j)
 4    continue
c
      if (j .eq. n) then
      write(6,*) '+++++++++++++++++++++++++++'
      write(6,*) 'new float(i,3) = ',(a(i),i=1,m)
      write(6,*) '+++++++++++++++++++++++++++'
      endif
c
      call householder(mp,m,n,a,ithzero,v,e,Ha,floatk,rhs)
 2    continue
c
      stop
```

```
        end
c%%%%%%%%%%%%%%%%%%%%%%%%%%%%%%%%%%%%%%%%%%%%%%%%%%%%%%%%%%%%%%%%

        subroutine householder(mp,m,n,a,ithzero,v,e,Ha,floatk,rhs)
        implicit real*8 (a-h,o-z)
        dimension a(*), v(*), e(*),Ha(*),floatk(mp,*),rhs(*)
c......stored under cd ~/cee/*odu*clas*/householder*.f
c
c......generating the unit vector {e}
c
        do 2 i=1,m
 2      e(i)=0.0d0
        e(ithzero)=1.0d0
c
c......compute the L2 norm of vector {a}
c
        sum=0.0d0
        do 1 i=1,m
        sum=sum+a(i)*a(i)
 1      continue
        sum=dsqrt(sum)
c
c......select the sign, to avoid cancellation during {v} = {a} - gamma
*  {e1}
c
        if ( a(ithzero) .gt. 0.0d0 ) signgamma=-1.0d0
        if ( a(ithzero) .lt. 0.0d0 ) signgamma=+1.0d0
        gamma=signgamma*sum
c
        do 4 i=1,m
        v(i)=a(i)-gamma*e(i)
 4      continue
c
c......Confirm the calculated vector {v} by computing:
c....  [H] * {a} = {a} - (2.0d0 * {v transpose} * {a} / {v transpose} *
{v}) * {v}
c
            write(6,*) '                              '
            write(6,*) '----------------------'
            write(6,*) '                              '
c
c           do 12 j=1,n             !         correct answer but much less
efficient
            do 12 j=ithzero,n       ! same, correct answer but much more
efficient
c
                do 14 i=1,m
                a(i)=floatk(i,j)
 14             continue
c
        call dotprod(v,a,m,scalar1)
        call dotprod(v,v,m,scalar2)
        scalar=2.0d0*scalar1/scalar2
c
        do 6 i=1,m
        Ha(i)=a(i)-scalar*v(i)
 6      continue
        write(6,*) 'Ha(i) = ',(Ha(i),i=1,m)
c
                do 15 i=1,m
                floatk(i,j)=Ha(i)
```

```
 15          continue
c
 12      continue
c......now, also do householder transformation rhs vector
            do 24 i=1,m
            a(i)=rhs(i)
 24         continue
c
        call dotprod(v,a,m,scalar1)
        call dotprod(v,v,m,scalar2)
        scalar=2.0d0*scalar1/scalar2
c
c
        do 26 i=1,m
        Ha(i)=a(i)-scalar*v(i)
 26     continue
c
            do 25 i=1,m
            rhs(i)=Ha(i)
 25         continue
        write(6,*) 'rhs(i) = ',(rhs(i),i=1,m)
c
c......outputs
c
        write(6,*) 'norm (with appropriated sign) = gamma = ',gamma
        write(6,*) 'v(i) = ',(v(i),i=1,m)
        return
        end
c%%%%%%%%%%%%%%%%%%%%%%%%%%%%%%%%%%%%%%%%%%%%%%%%%%%%%%%%%%%%%%%%%
        subroutine dotprod(a,b,n,scalar)
        implicit real*8 (a-h,o-z)
        dimension a(*),b(*)
c
        scalar=0.0d0
        do 1 i=1,n
        scalar=scalar+a(i)*b(i)
 1      continue
c
        return
        end
c%%%%%%%%%%%%%%%%%%%%%%%%%%%%%%%%%%%%%%%%%%%%%%%%%%%%%%%%%%%%%%%%%
```

Example 1

Given the vector $\vec{a} = \begin{Bmatrix} 4 \\ 1 \\ 8 \end{Bmatrix}$,

find a Householder transformation, such that $[H]\vec{a} = \begin{Bmatrix} x \\ 0 \\ 0 \end{Bmatrix}$

Solution

$$\gamma = \pm \|\vec{a}\|_2 = -9$$

$$\vec{v} = \vec{a} - (\gamma)\hat{e}^{(1)} = \begin{Bmatrix} 4 \\ 1 \\ 8 \end{Bmatrix} - (-9)\begin{Bmatrix} 1 \\ 0 \\ 0 \end{Bmatrix} = \begin{Bmatrix} 13 \\ 1 \\ 8 \end{Bmatrix}$$

$$[H]\vec{a} = \vec{a} - (\frac{2 * v^T a}{v^T v}) * \vec{v} = \begin{Bmatrix} -9 \\ 0 \\ 0 \end{Bmatrix} = \gamma \hat{e}^{(1)}$$

Example 2 [5.13]

Using the Householder QR factorization to solve the quadratic polynomial data fitting problem:

$$[A]_{5\times3} * \{\vec{x}\}_{3\times1} = \{\vec{b}\}_{5\times1}$$

where

$$[A] = \begin{bmatrix} 1 & -1.0 & 1.00 \\ 1 & -0.5 & 0.25 \\ 1 & 0.0 & 0.00 \\ 1 & 0.5 & 0.25 \\ 1 & 1.0 & 1.00 \end{bmatrix} \text{ and } \vec{x} = \begin{Bmatrix} 1.0 \\ 0.5 \\ 0.0 \\ 0.5 \\ 2.0 \end{Bmatrix}$$

The computerized results for the above problem are shown below:

Example $2^{[5.13]}$ Result

m rows, n columns = 5   3

column #   1

vector =   1.0 1.0 1.0 1.0 1.0

column #   2

vector =   −1.0 −0.5 0.5 0.0 1.0

column #   3

vector =   1.0 0.25 0.0 0.25 1.0

rhs vector =   1.0 0.5 0.0 0.5 2.0

———————————————————————

Ha(i) =   −2.236074998 0.0 0.0 0.0 0.0

Ha(i) =   2.22045D−16 −0.19098 0.30902 0.80902 1.30902

Ha(i) =   −1.11803 −0.40451 −0.65451 −0.40451 0.34592

rhs(i) =   −1.78885 −0.36180 −0.86180 −0.36180 1.13820

norm (with appropriated sign)   = gamma =   − 2.23607

v(i) =   3.23607 1.0 1.0 1.0 1.0

———————————————————————

Ha(i) =   2.22045D−16 1.58114 5.55112D−17 1.11022D−16 2.22045D−16

Ha(i) =   −1.11803 0.0 −0.72505 −0.58918 4.66924D−02

rhs(i) =   −1.78885 0.63246 −1.03518 −0.81571 0.40377

norm (with appropriated sign)   = gamma =   1.58114

v(i) =   0.0 −1.77212 0.30902 0.80902 1.30902

+++++++++++++++++++++++++++

new float(1, 3) =   0.0 0.0 −0.72505 −0.58912 4.66924D−02

+++++++++++++++++++++++++++

———————————————————————

Ha(i) =   −1.11803 0.0 0.93541 −2.22045D−16 1.38778D−17

rhs(i) =   −1.78885 0.63246 1.33631 2.57609 0.33708

norm (with appropriated sign)   = gamma =   0.93541

v(i) =   0.0 0.0 −1.66046 −0.58918 4.66924D−02

5.13 "Modified" Gram-Schmidt Re-orthogonalization

In real-life numerical calculation, the computed vectors $q^{(k)}$ by the "classical" Gram-Schmidt procedure do not turn out to be orthogonal. The idea presented in Eq.(5.231) is unstable. With a little modification, the so-called "modified" Gram-Schmidt procedure will be more stable and efficient as explained in the following step-by-step algorithms.

Step 1   Compute

$$a^{(k),1} = a^{(k)} - \{q^{(1)^T} a^{(k)}\} q^{(1)} \tag{5.255}$$

Step 2 Project the vector $a^{(k),1}$ (instead of the original vector $a^{(k)}$) onto $q^{(2)}$, and subtract that projection to obtain:

$$a^{(k),2} = a^{(k),1} - \{q^{(2)^T} a^{(k),1}\} q^{(2)} \qquad (5.256)$$

Remarks

[1]   Substituting Eq.(5.255) into Eq.(5.256), one gets:

$$a^{(k),2} = a^{(k)} - q^{(1)^T} a^{(k)} q^{(1)} - q^{(2)^T} \{a^{(k)} - q^{(1)^T} a^{(k)} q^{(1)}\} q^{(2)}$$

or

$$a^{(k),2} = a^{(k)} - q^{(1)^T} a^{(k)} q^{(1)}$$
$$- q^{(2)^T} a^{(k)} q^{(2)}$$
$$+ q^{(2)^T} q^{(1)^T} a^{(k)} q^{(1)} q^{(2)}$$

scalar

scalar $= 0$

so

$$a^{(k),2} = a^{(k)} - \{q^{(1)^T} a^{(k)}\} q^{(1)} - \{q^{(2)^T} a^{(k)}\} q^{(2)} - \cdots$$

[2]   Eqs.(5.255 – 5.256) in the "modified" Gram-Schmidt are, therefore, equivalent to the "classical" Gram-Schmidt (= to project the original vector $a^{(k)}$ onto both $q^{(1)}$ and $q^{(2)}$).

[3]   In the "classical" Gram-Schmidt, the current vector $q^{(k)}$ is orthogonalized with all previous vectors $q^{(1)}, q^{(2)}, ..., q^{(k-1)}$. However, in the "modified" Gram-Schmidt, the current vector, say $q^{(1)}$, is orthogonalized with subsequent vectors $a^{(k),1} = q^{(2)}$, $a^{(k),2} \equiv q^{(3)}$, etc.

[4]   Memory usage in the "modified" classical Gram-Schmidt is more efficient than the "classical" version because the computed vector $a^{(k),1}$, $a^{(k),2}$, etc. will overwrite the original columns 2, 3... of vectors $a^{(2)}$, $a^{(3)}$..., respectively. This is not possible for the "classical" Gram-Schmidt since the original vectors $a^{(k)}$ are also needed in the inner loop!

[5]   In the "modified" Gram-Schmidt, the scalars r_{ij} which are used to form matrix [R] (from [A] = [Q][R]), are obtained row-wise while the "classical" Gram-Schmidt generates the scalars r_{ij} in column-wise fashion.

A more detailed step-by-step procedure for the "modified" Gram-Schmidt is described in Table 5.11. Computer implementation and its outputs are given in Tables 5.12 and 5.13, respectively.

Table 5.11 "Modified" GramSchmidt Algorithms

Step 1: Given the mass matrix [M] with the dimension n by n, and an arbitrary number of vectors $a^{(k)}$.

Step 2: For k = 1, n

(a) $r_{kk} = \left\| a^{(k)^T} M a^{(k)} \right\|$

(b) $q^{(k)} = a^{(k)} \Big/ r_{kk}$

(c)   For j = k+1, n

(d) $r_{kj} = q^{(k)^T} [M] a^{(j)}$

(e) $a^{(j)} = a^{(j)} - r_{kj} q^{(k)}$

(f)   End

(g) End

The unknown vector $\vec{x}$, corresponding to the least square problem $[A]_{5\times3} * \{\vec{x}\}_{3\times1} = \{\vec{b}\}_{5\times1}$, can be solved from:

$$[R]_{3\times3} \vec{x}_{3\times1} = \vec{y}_{3\times1}$$

where

$$[R] = \text{upper-triangular matrix} = \begin{bmatrix} r_{11} & r_{12} & r_{13} \\ 0 & r_{22} & r_{23} \\ 0 & 0 & r_{33} \end{bmatrix}$$

$$[R] = \begin{bmatrix} 2.236 & -5.55*10^{-17} & 1.118 \\ 0 & 1.581 & 0 \\ 0 & 0 & 0.935 \end{bmatrix}$$

$$\vec{y}_{3\times1} = [Q^T]_{3\times5} * \{\vec{b}\}_{5\times1} = \begin{Bmatrix} 1.789 \\ 0.632 \\ 1.336 \end{Bmatrix}$$

Thus

$$\vec{x} = \begin{Bmatrix} 0.086 \\ 0.400 \\ 1.429 \end{Bmatrix}$$

5.14 QR Iteration for Unsymmetrical Eigen-Solutions

The eigen-solutions for a given (unsymmetrical) matrix [A] can be obtained by repeatedly using QR factorization, such as:

$$\text{Initialize } [A_o] = [A] \tag{5.257}$$

Then, where k represents the iteration number, one generates the following sequences of operations:

$$[A_k] = [Q_k] * [R_k] \qquad (5.258)$$

$$[A_{k+1}] = [R_k] * [Q_k] \qquad (5.259)$$

Thus

Step 1 From a known matrix $[A_o] = [A]$, one obtains $[Q_o]$ and $[R_o]$

Step 2 Compute the updated matrix
$$[A_1] = [R_o] [Q_o]$$
Step 3 Obtains $[Q_1]$ and $[R_1]$ from $[A_1]$

Step 4 Compute the updated matrix
$$[A_2] = [R_1] [Q_1]$$

The above steps are repeated until the updated matrix is (or nearly) diagonal. The diagonal terms of $[A_{k+1}]$ will closely approximate the eigen-values of $[A]$, and the product of orthogonal matrices Q_k will closely approximate the eigen-vectors of $[A]$.

Example

Use the QR iteration to find the approximated eigen-solutions of a given matrix

$$[A] = \begin{bmatrix} 2 & -1 \\ -1 & 1 \end{bmatrix} \qquad (5.260)$$

Let

$$[A_o] = [A] = \begin{bmatrix} 2 & -1 \\ -1 & 1 \end{bmatrix}$$

Apply the QR factorization to obtain:
$$[Q_o] = ...$$
$$[R_o] = ...$$
Compute the reverse product
$$[A_1] = [R_o] [Q_o] = ...$$

Table 5.12 FORTRAN Coding for "Modified" GramSchmidt Algorithm

```
C%%%%%%%%%%%%%%%%%%%%%%%%%%%%%%%%%%%%%%%%%%%%%%%%%%%%%%%%%%%%%%%%%%%%%%%%%%%
c        subroutine dummy02
c*************************************************************************
c......Coded by:      Duc T. Nguyen
c......Date:          October 14, 1999
c......Purpose:       Given L vectors [A] in the space R^n, use
c......               the "modified" Gramschmidt (as explained on
c......               page 98, of HEATH's Scientific Computing book)
c......               to generate the set L orthogonal vectors [Q],
c......               where [Q] will overwrite [A] to save computer
c......               memory
c......Notes:         The answers obtained from this code has been
c......               successfully verified with page 99 of Heath's book
```

```
c......Stored at:    cd ~/cee/*odu*clas*/modified*.f
c*****************************************************************
        implicit real*8(a-h,o-z)
        dimension a(100,20)
        maxrow=100
        maxcol=20
        read(5,*) n,l
        write(6,*) 'n,l = ',n,l
        read(5,*) ((a(i,j),i=1,n),j=1,L)   ! read column-wise
        do 1 irow=1,n
        write(6,*) 'given vectors= ', (a(irow,j),j=1,L)    ! write row-
wise
1       continue
        call modgschmidt(a,n,l,maxrow,maxcol)
        do 2 irow=1,n
        write(6,*) 'orthogo vectors= ', (a(irow,j),j=1,L)  ! write row-
wise
2       continue
        stop
        end
c%%%%%%%%%%%%%%%%%%%%%%%%%%%%%%%%%%%%%%%%
c       sample of input data file
c       5    3
c       1.0  1.0  1.0  1.0  1.0
c      -1.0 -0.5  0.0  0.5  1.0
c       1.0  0.25 0.0  0.25 1.0
c%%%%%%%%%%%%%%%%%%%%%%%%%%%%%%%%%%%%%%%%
        subroutine modgschmidt(a,n,L,maxrow,maxcol)
        implicit real*8(a-h,o-z)
        dimension a(maxrow,maxcol)
c
        do 1 kcol=1,L
        rkk=0.0
         do 2 irow=1,n
         rkk=rkk+a(irow,kcol)**2
2        continue
        rkk=dsqrt(rkk)
        write(6,*) 'rkk= ',rkk
c
         do 3 irow=1,n
         a(irow,kcol)=a(irow,kcol)/rkk
3        continue
        write(6,*) 'vector Qk= ',(a(irow,kcol),irow=1,n)
c
         do 4 j=kcol+1,L
          rkj=0.0
          do 5 i=1,n
          rkj=rkj+a(i,kcol)*a(i,j)
5         continue
         write(6,*) 'rkj= ',rkj
c
          do 6 i=1,n
          a(i,j)=a(i,j)-rkj*a(i,kcol)
6         continue
         write(6,*) 'vector Qk= ',(a(irow,j),irow=1,n)
4        continue
c
1       continue
        return
        end
c%%%%%%%%%%%%%%%%%%%%%%%%%%%%%%%%%%%%%%%%%%%%%%%%%%
```

Table 5.13 Outputs for "Modified" GramSchmidt Algorithm

```
n, l =   5   3
given vectors =   1.0 −1.0 1.0
given vectors =   1.0 −0.5 0.25
given vectors =   1.0 0.0 0.0
given vectors =   1.0 0.5 0.25
given vectors =   1.0 1.0 1.0
rkk = 2.23607
vector Qk =   0.44721 0.44721 0.44721 0.44721 0.44721
rkj = −5.55112D −17
vector Qk =   −1.0 −0.5 2.48253D −17 0.5 1.0
rkj = 1.11803
vector Qk =   0.5 −0.25 −0.5 −0.25 0.5
rkk = 1.58114
vector Qk =   −0.63246 −0.31623 1.57009D −17 0.31623 0.63246
rkj = 0.0
vector Qk =   0.5 −0.25 −0.5 −0.25 0.5
rkk = 0.93541
vector Qk =   0.53452 −0.26726 −0.53452 −0.26726 −0.53452
```

$$
\text{orthogo vectors} = \begin{bmatrix} 0.44721 & -0.63246 & 0.53452 \\ 0.44721 & -0.31623 & -0.26726 \\ 0.44721 & 1.57009D-17 & -0.53452 \\ 0.44721 & 0.31623 & -0.26726 \\ 0.44721 & 0.63246 & 0.53452 \end{bmatrix} = [Q]_{5\times3}
$$

5.15 QR Iteration with Shifts for Unsymmetrical Eigen-Solutions

The convergence rate for QR eigen-iterations can be improved by applying a shift to Eqs.(5.258 − 5.259), as follows:

$$[A_k] - \sigma_k[I] = [Q_k][R_k] \tag{5.261}$$

$$[A_{k+1}] = [R_k][Q_k] + \sigma_k[I] \tag{5.262}$$

Two choices for computing the shifted value σ_k are suggested [5.13, page 123]:

(a) σ_k = lower-right corner of $[A_k]$

(b)   A better choice for σ_k can be obtained by computing the eigen-values of the 2×2 submatrix in the lower-right corner of $[A_k]$.

Example: For the same data shown in Eq.(5.260), one computes:

$$A_0 - (\sigma_0 = 1)I = \begin{bmatrix} 2 & -1 \\ -1 & 1 \end{bmatrix} - \begin{bmatrix} 1 & 0 \\ 0 & 1 \end{bmatrix} = \begin{bmatrix} 1 & -1 \\ -1 & 0 \end{bmatrix}$$

Thus

$$[Q_0] = \ldots$$
$$[R_0] = \ldots$$

Then

$$[A_1] = [R_0][Q_0] + (\sigma_0 = 1)[\,I\,]$$

or

$$[A_1] = \ldots$$

Now

$$[A_1] - (\sigma_1 = \ldots\ldots)I = \ldots$$

Thus

$$[Q_1] = \ldots$$
$$[R_1] = \ldots$$

Table 5.14 Numerical Recipe's Subroutine HQR

```
C
c      This subroutine is extracted from NUMERICAL RECIPES
c      (FORTRAN version, 1990, ISBN # 0-521-38330-7),
c      Authors = Press, Flannery, Teukolsky and Vetterling
c      pages 374-376
c
c      Find all eigenvalues of an N by N upper Hessenberg
c      matrix [A], that is stored in an NP by NP array. On
c      input [A] can be exactly as output from routine ELMHES,
c      on output it is destroyed. The real & imaginary parts
c      of the eigenvalues are returned in WR & WI, respectively.
c
       SUBROUTINE hqr(a,n,np,wr,wi)
        implicit real*8(a-h,o-z)             ! nguyen
       INTEGER n,np
c      REAL a(np,np),wi(np),wr(np)           ! nguyen
       dimension a(np,np),wi(np),wr(np)      ! nguyen
       INTEGER i,its,j,k,l,m,nn
c      REAL anorm,p,q,r,s,t,u,v,w,x,y,z      ! nguyen
```

Table 5.15 Main Program for Unsymmetrical Eigen-Solver

```
c%%%%%%%%%%%%%%%%%%%%%%%%%%%%%%%%%%%%%%%%%%
       implicit real*8(a-h,o-z)
       dimension a(500,500), wi(500), wr(500)
C===========================================================
c      This main program/subroutine will compute all
c      all eigen-values (only) of an UNSYMMETRICAL (real) matrix.
c      The output eigenvalues (from Numerical Recipe)
c      are stored in arrays {wr(-)} = REAL part of eigenvalues
c      and                  {wi(-)} = IMAGINARY part of eigenvalues
c
c      These numerical recipe's subroutines (and Dr. Nguyen's main
c      program) are stored at
c
c      cd ~/cee/*odu*clas*/recipe_unsym_evalues.f
c
C===========================================================
c
c      April 17, 2002:  Duc T. Nguyen made minor changes in
c                       three Recipe's routines to have DOUBLE
```

```
c                          PRECISIONS
c
c                          Also, "correct" eigen-values have been obtained
c                          for the following examples:
c
c                                 [ 1.0      -4.0 ]
c                          [A] = [                ]
c                                 [ 5.0       2.0 ]
c
c                          1-st eval = (1.5, -4.444) = 1.5 - 4.444 i
c                          2-nd eval = (1.5, +4.444) = 1.5 + 4.444 i
C+++++++++++++++++++++++++++++++++++++++++++++++++++++++++++++++++++++++++++++
c                                 [ 2.0      -6.0 ]
c                          [A] = [                ]
c                                 [ 8.0       1.0 ]
c
c                          1-st eval = (1.5, +6.91) = 1.5 + 6.91 i
c                          2-nd eval = (1.5, -6.91) = 1.5 - 6.91 i
C===========================================================
c
        np=500
c
        n=2
c
        do 1 i=1,n
        do 2 j=1,n
   2    a(i,j)=0.
   1    continue
c
c       a(1,1)=1.0
c       a(2,2)=2.0
c       a(1,2)=-4.
c       a(2,1)= 5.
        a(1,1)=2.0d0
        a(1,2)=-6.0d0
        a(2,1)=8.0d0
        a(2,2)=1.0d0
c
        call balanc(a,n,np)        ! balancing a given matrix [A]
        call elmhes(a,n,np)        ! transforming to upper Hessenberg
matrix
        call hqr(a,n,np,wr,wi)     ! QR iterations for eigen-pair
solutions
c
        write(6,*) 'real parts of eval= ',(wr(i),i=1,n)
        write(6,*) 'imag parts of eval= ',(wi(i),i=1,n)
c
        stop
        end
```

5.16 Panel Flutter Analysis

Panel flutter is the self-excited or self-sustained oscillation of an external panel of a flight vehicle when exposed to supersonic or hypersonic air flow.

There is a critical value of the non-dimensional dynamic pressure parameter, λ_{cr} (or flow velocity), above which the panel motion becomes unstable and grows exponentially with time. Below this critical dynamic pressure, any disturbance to the panel results in motion that decays exponentially with time.

The aero-dynamic equations of motion can be given as:

$$M\ddot{w} + G\dot{w} + Kw = 0 \tag{5.263}$$

where

> M is the n by n system mass matrix
> G is the n by n system aerodynamic damping matrix
> K contains both the system stiffness and aerodynamic influence matrices (dimension is n by n matrix)
> w, $\dot{w}$, $\ddot{w}$ are the displacement, velocity, and acceleration vector, respectively.

Eq.(5. 263) can also be expressed as:

$$\begin{bmatrix} M & 0 \\ 0 & I \end{bmatrix} \begin{Bmatrix} \ddot{w} \\ \dot{w} \end{Bmatrix} + \begin{bmatrix} G & K \\ -I & 0 \end{bmatrix} \begin{Bmatrix} \dot{w} \\ w \end{Bmatrix} = \{0\} \tag{5.264}$$

If the vector $\begin{Bmatrix} \dot{w} \\ w \end{Bmatrix}$ is sought in the form of

$$\begin{Bmatrix} \dot{w} \\ w \end{Bmatrix} = \tilde{c} \begin{Bmatrix} \Phi_1 \\ \Phi_2 \end{Bmatrix} e^{\Omega t} \tag{5.265}$$

where

$\{\Phi_1\}$ and $\{\Phi_2\}$ are complex eigen-vectors that are arranged as a single column vector.

$\Omega = \alpha + i\omega$ is the complex eigen-value

$\tilde{c}$ is a non-zero constant displacement amplitude

Then, taking the derivative with respect to time of Eq.(5.265), one obtains:

$$\begin{Bmatrix} \ddot{w} \\ \dot{w} \end{Bmatrix} = \tilde{c}\Omega \begin{Bmatrix} \Phi_1 \\ \Phi_2 \end{Bmatrix} e^{\Omega t} \tag{5.266}$$

Substituting Eqs.(5.265 – 5.266) into Eq.(5.264), one gets:

$$\tilde{c}\Omega \begin{bmatrix} M & 0 \\ 0 & I \end{bmatrix} \begin{Bmatrix} \Phi_1 \\ \Phi_2 \end{Bmatrix} e^{\Omega t} + \tilde{c} \begin{bmatrix} G & K \\ -I & 0 \end{bmatrix} \begin{Bmatrix} \Phi_1 \\ \Phi_2 \end{Bmatrix} e^{\Omega t} = \{0\} \tag{5.267}$$

Pre-multiplying both sides of Eq.(5.267) by $\begin{bmatrix} G & K \\ -I & 0 \end{bmatrix}^{-1}$, one obtains:

$$\Omega \begin{bmatrix} G & K \\ -I & 0 \end{bmatrix}^{-1} \begin{bmatrix} M & 0 \\ 0 & I \end{bmatrix} \begin{Bmatrix} \Phi_1 \\ \Phi_2 \end{Bmatrix} + \begin{Bmatrix} \Phi_1 \\ \Phi_2 \end{Bmatrix} = \{0\}$$

or

$$\Omega \begin{bmatrix} G & K \\ -I & 0 \end{bmatrix}^{-1} \begin{bmatrix} -M & 0 \\ 0 & -I \end{bmatrix} \begin{Bmatrix} \Phi_1 \\ \Phi_2 \end{Bmatrix} - \begin{Bmatrix} \Phi_1 \\ \Phi_2 \end{Bmatrix} = \{0\} \qquad (5.268)$$

Let

$$[A] \equiv \begin{bmatrix} G & K \\ -I & 0 \end{bmatrix}^{-1} \begin{bmatrix} -M & 0 \\ 0 & -I \end{bmatrix} \qquad (5.269)$$

and

$$\{x\} \equiv \begin{Bmatrix} \Phi_1 \\ \Phi_2 \end{Bmatrix} \qquad (5.270)$$

Then, Eq.(5.268) becomes:

$$\Omega [A]\{x\} - \{x\} = \{0\} \qquad (5.271)$$

or

$$[A]\{x\} = \frac{1}{\Omega}\{x\}$$

Let

$$\mu \equiv \frac{1}{\Omega} \qquad (5.272)$$

Then, Eq.(5.271) becomes:

$$[A]\{x\} = \mu\{x\} \qquad (5.273)$$

Thus, Eq.(5.273) represents the "standard" eigen-value problem where the coefficient matrix [A] is unsymmetrical.

In nonlinear panel flutter analysis, the above standard eigen-value equations have to be solved repeatedly during the iteration. Fortunately, only a few of the lowest eigen-pairs are required for the solution during each iteration. This feature makes it possible to develop an efficient eigen-solver for either large-amplitude vibration or the nonlinear flutter analysis.

Define

$$[B] \equiv \begin{bmatrix} G & K \\ -I & 0 \end{bmatrix} \qquad (5.274)$$

and

$$[C] \equiv \begin{bmatrix} -M & 0 \\ 0 & -I \end{bmatrix} \qquad (5.275)$$

Then Eq.(5.269) becomes:

$$[A] \equiv [B]^{-1} * [C] \qquad (5.276)$$

and the original solution can be obtained as:

$$\Omega = \frac{1}{\mu} \tag{5.277}$$

$$\{\Phi\} = \begin{Bmatrix} \Phi_1 \\ \Phi_2 \end{Bmatrix} = \{x\} \tag{5.278}$$

Since only a few of the lowest eigen-pairs ($\Omega, \vec{\Phi}$) are required, a reduction scheme that makes the largest ones ($\mu, \vec{x}$) converge first is preferred. This can be done by defining the orthogonal transformation matrix Q_m ($1 < m \le n$) as:

$$AQ_m = Q_m H_m + h_{m+1,m} q_{m+1} \tag{5.279}$$

$$Q_m^T Q_m = I_m = \text{Identity matrix} \tag{5.280}$$

where $Q_m = \{q_1, q_2, ..., q_m\}$, $\|q_i\|_2 = 1$, q_1 is an arbitrary starting vector, and q_{i+1} is determined through

$$h_{i+1,i} q_{i+1} = A q_i - \sum_{k=1}^{i} h_{k,i} q_k \tag{5.281}$$

$$\left. \begin{array}{c} (h_{k,i} = q_i^T A^T q_k) \quad (i = 1,2,...,m) \\ \text{or } h_{k,i} = [Aq^{(i)}]^T q^{(k)} \end{array} \right\} \tag{5.282}$$

Eqs.(5.279 – 5.282) can be executed according to the following sequences:

For i = 1

$\vec{q}^{(1)} = $ arbitrary vector, with $\left\| q^{(1)} \right\|_2 = 1$

From Eq.(5.282), one has:

$$h_{1,1} = q^{(1)^T} A^T q^{(1)} = \text{scalar} \tag{5.283}$$

From Eq.(5.281), one has:

$$h_{2,1} q^{(2)} = A q^{(1)} - h_{1,1} q^{(1)} \tag{5.284}$$

or

$$q^{(2)^T} h_{2,1} q^{(2)} = q^{(2)^T} A q^{(1)} - q^{(2)^T} h_{1,1} q^{(1)} \tag{5.285}$$

Using the properties of Eq.(5.280), Eq.(5.285) becomes:

$$h_{2,1} = q^{(2)^T} A q^{(1)} \tag{5.286}$$

Substituting Eq.(5.286) into Eq.(5.284), one has:

$$(q^{(2)^T} A q^{(1)}) q^{(2)} = A q^{(1)} - h_{1,1} q^{(1)} \tag{5.287}$$

Thus, vector $q^{(2)}$ can be obtained in two steps:

$$q^{(2)} = A q^{(1)} - h_{1,1} q^{(1)}$$

$$q^{(2)} = \frac{q^{(2)}}{(q^{(2)^T} A q^{(1)})}$$

For i = 2

From Eq.(5.282), one obtains:

$$h_{1,2} = q^{(2)^T} A^T q^{(1)}$$

$$h_{2,2} = q^{(2)^T} A^T q^{(2)}$$

From Eq.(5.281):

$$h_{3,2} q^{(3)} = A q^{(2)} - (h_{1,2} q^{(1)} + h_{2,2} q^{(2)}) \tag{5.288}$$

Pre-multiplying the above equation by $q^{(3)^T}$, one gets:

$$h_{3,2} = q^{(3)^T} A \; q^{(2)}$$

Hence, $q^{(3)}$ from Eq.(5.288) can be computed in two steps:

$$q^{(3)} = A q^{(2)} - (h_{1,2} q^{(1)} + h_{2,2} q^{(2)})$$

$$q^{(3)} = \frac{q^{(3)}}{q^{(3)^T} A q^{(2)}}$$

etc.

The reduced, upper Hessenberg matrix can be formed as:

$$H_m = \begin{bmatrix} h_{1,1} & h_{1,2} & h_{1,3} & \cdots & h_{1,m} \\ h_{2,1} & h_{2,2} & h_{2,3} & \cdots & h_{2,m} \\ & h_{3,2} & h_{3,3} & \cdots & h_{3,m} \\ & & h_{4,3} & \cdots & h_{4,m} \\ & O & & \vdots & \vdots \\ & & & h_{m-1,m} & h_{m,m} \end{bmatrix} \tag{5.289}$$

The reduced eigen-equation

$$H_m Z = \theta Z \tag{5.290}$$

can be solved to provide good approximated lowest eigen-pairs for the original, unsymmetrical, standard eigen-value problem as described in Eq.(5.273)

Remarks

1. The reduced eigen-value equations can be solved by the QR method, which is available on most computers.
2. As in the Lanczos process, re-orthogonalization is required.
3. Since m ≤ n, (say n = 1500, m = 50) the computational cost for solving eigen-equation (5.290) is significantly reduced.

4. Vectorized codes [1.9] on a vector/cache computer can be much faster than a scalar code.
5. Key steps involved in the unsymmetrical eigen-solution procedure are summarized in Table 5.16.

Table 5.16 Step-by-Step Procedures for Unsymmetrical Eigen-Equations

Step 0: Initialization

The starting vector $q^{(1)}$ is selected, with $\left\| q^{(1)} \right\|_2 = 1$

Let nmax = maximum number of steps = (2 or 3)* m

Where m = # lowest eigen-pair solutions desired

Step 1: Reduction to Hessenberg form:

(1.1)   For I = 1 to nmax

(1.2) $\tilde{q}^{(I+1)} = Aq^{(I)}$ (5.291)

Thus, unsymmetrical equation solver is needed in Eq.(5.291), also refer to Eq.(5.276)

(1.3)   ⌠For K = 1 to I

(1.4)   ⎨ $H(K,I) = \tilde{q}^{(I+1)^T} * q^{(k)}$ (5.292)

(1.5)   ⌡End

Thus, Eq.(5.292) is the computer implementation of Eq.(5.282)

(1.6)   ⌠For K = 1 to I

(1.7)   ⎨ $\tilde{q}^{(I+1)} = \tilde{q}^{(I+1)} - H(K,I)q^{(k)}$ (5.293)

(1.8)   ⌡End

Thus, Eq.(5.293) is the computer implementation (for GramSchmidt process) of the right-hand-side of Eq.(5.281)

Then:

(1.9) $q^{(I+1)} = \dfrac{\tilde{q}^{(I+1)}}{\left\| \tilde{q}^{(I+1)} \right\|}$ (5.294)

(1.10) $H(I+1,I) = \left\| \tilde{q}^{(I+1)} \right\|_2$ (5.295)

Eq.(5.295) can be derived (or verified) by the following section:

Thus, for I = 1, then from Eq.(5.281), one has

$$H(2,1)q^{(2)} = Aq^{(1)} - H(1,1) * q^{(1)}$$ (5.296)

Pre-multiplying both sides of the above equation by $q^{(2)^T}$, one obtains:

$$H(2,1) = q^{(2)^T} Aq^{(1)}$$ (5.297)

Applying Eq.(5.291), one has:

$$\tilde{q}^{(2)} = Aq^{(1)}$$

Thus, Eq.(5.297) becomes:

$$H(2,1) = q^{(2)^T} \tilde{q}^{(2)}$$

Using Eq.(5.294), then the above equation becomes:

$$H(2,1) = \frac{\tilde{q}^{(2)^T}}{\left\|\tilde{q}^{(2)}\right\|_2} * \tilde{q}^{(2)}$$

or

$$H(2,1) = \frac{\tilde{q}^{(2)^T} * \tilde{q}^{(2)}}{\sqrt{\tilde{q}^{(2)^T} * \tilde{q}^{(2)}}} = \sqrt{\tilde{q}^{(2)^T} * \tilde{q}^{(2)}}$$

or

$$H(2,1) = \left\|\tilde{q}^{(2)}\right\|_2, \text{ which is the same as Eq.(5.295)}$$

(1.11)   Re-orthogonalize (if necessary) of $q^{(l+1)}$ with respect to previous vectors $q^{(1)}, q^{(2)}, \ldots, q^{(l)}$

(1.12)   End

<u>Step 2</u>: Solve the reduced eigen-equation:

$$[H_m]\vec{Z} = \theta\vec{Z} \qquad\qquad (5.290, \text{ repeated})$$

The QR method can be used in this step

<u>Step 3</u>: Compute the eigen-solution of the original eigen-equations:

$$\left.\begin{array}{c} \mu \cong \theta \\ \vec{x} \approx [Q]\vec{Z} \\ \Omega \approx \dfrac{1}{\theta} \\ \{\Phi\} \approx [Q]\vec{Z} \approx \vec{x} \end{array}\right\} \qquad\qquad (5.298)$$

A listing of the parallel (MPI) GramSchmidt code and its numerical performance (on small problems' sizes) are presented in Tables 5.17 – 5.18.

Table 5.17 Parallel (MPI) GramSchmidt QR

```
c
c=======================================================================
c        Table 5.17:   Parallel (MPI) Gramschmidt QR
c
c                      Stored under cd ~/cee/*odu*clas*/mpi*QR.f
c=======================================================================
c
      program main
      include 'mpif.h'
      include 'mpif.h'
c     Author(s) : Yusong HU, and D.T. Nguyen
c
c***NR : number of rows in matrix A, which is also the length of a
```

```
vector
c***NC : number of columns in matrix A, which is also the number of the
vectors

      parameter (nr = 10)
      parameter (nc = 10)

      implicit real*8(a-h,o-z)
      real*8    a(nr,nc), q(nr)
      real      time0, time1, time2,time3

      call MPI_INIT( ierr )
      call MPI_COMM_RANK( MPI_COMM_WORLD, myid, ierr )
      call MPI_COMM_SIZE( MPI_COMM_WORLD, numprocs, ierr )
      print *, "Process ", myid, " of ", numprocs, " is alive"

      open(unit=10,file='record.output',
     $status='old',form='formatted')


c        write(myid,*)'output from proc#',myid

         time0 =  MPI_WTIME()
C****Initialize A
      if (taskid .eq. MASTER) then
         do 30 i=1, nr
           do 40 j=1, nc
              if (j .ne. i) then
                 a(i,j)=i+j
              else
                 a(i,j)=(i+i)*10 000.
              endif
40         continue
30       continue
         endif

         time1 =  MPI_WTIME()
      call
MPI_BCAST(a,nr*nc,MPI_DOUBLE_PRECISION,0,MPI_COMM_WORLD,ierr)
         time2 =  MPI_WTIME()

      jj=0
      do 100 k=1,nc

      if(k .eq. 1) then
         kk=2
         else
         kk=k
      endif
c*********************
      if ( myid .eq. mod(kk-2,numprocs) ) then

      r=0.
      do 110 i=1,nr
       r=r+a(i,k)*a(i,k)
110      continue
      r=sqrt(r)

      do 115 i=1,nr
        q(i)=a(i,k)/r
115      continue
```

```
        endif

     call MPI_BCAST(q(1),nr,MPI_DOUBLE_PRECISION,mod(kk-2,numprocs),
     &MPI_COMM_WORLD,ierr)

           if ( myid .eq. 0 ) then
              do 222 im=1,nr
               a(im,k)=q(im)
222           continue
              write(*,*) 'result(a)', k,'= ',(a(i,k),i=1,nr)
           endif


        if (mod(k-2,numprocs) .eq. myid) then
          jj=jj+numprocs
        endif
        do 120 j=myid+2+jj,nc,numprocs

             rkj=0.
         do 130 i=1,nr
          rkj=rkj+q(i)*a(i,j)
130      continue
         do 140 i=1,nr
          a(i,j)=a(i,j)-rkj*q(i)
140      continue

120     continue
100     continue


        time3 = MPI_WTIME()
        if ( myid .eq. 0 ) then

          do 150 j=1,nc
            write(10,*) 'a =',(a(i,j),i=1,nr)
150       continue

          write(10,*) ' Total Time =', time3 - time1
        endif

        call MPI_FINALIZE(rc)
        stop
        end
```

Table 5.18 Numerical Performance (on ODU/SUN-10,000) of Parallel
Gramschmidt QR

TIME (seconds)

	300×300	500×500	1000×1000	1500×1500
1 procs:	9	42	323	1122
2 procs:	6	24	181	619
3 procs:	5	19	138	457
4 procs:	5	19	134	437

SPEEDUP

	300×300	500×500	1000×1000	1500×1500
2 procs:	1.5	1.75	1.78	1.81
3 procs:	1.8	2.21	2.3	2.46
4 procs:	1.8	2.21	2.41	2.57

From the algorithms presented in Table 5.16, one concludes that major time-consuming operations will be involved with:

(a) Since Eq.(5.291) involves with

$$[A] = [B]^{-1} * [C] \qquad\qquad (5.276, \text{ repeated})$$

Hence

$$[A]q^{(I)} = [B]^{-1} * [C] * q^{(I)}$$

Thus, <u>matrix-vector multiplication</u> for $[C]q^{(I)}$ is required. Factorization of the unsymmetrical matrix [B] is required and the <u>forward/backward</u> solution is also required.

(b) Dot product operations of two vectors is required (see Eq.[5.292]).
(c) Re-orthononalizaton process may be required (see step 1.11).
(d) QR method (see step 2, Eq.[5.290]). Hu-Nguyen's parallel version of QR method can be used here to reduce the QR solution time (see Table 5.17).

<u>Numerical Examples</u>

Numerical recipe's <u>subroutine balanc</u> (see Table 5.8.), <u>Subroutine elmhes</u> (see Table 5.9.), and <u>subroutine hqr</u> (see Table 5.14) are combined by the main program (see Table 5.15) to solve for the unsymmetrical eigen-solutions of the following two examples.

<u>Example 1</u>
Find all eigen-values of the following 2×2 unsymmetrical matrix

$$A = \begin{bmatrix} 2 & -6 \\ 8 & 1 \end{bmatrix}$$

The analytical eigen-value solution for this problem is:

$$\lambda_1 = 1.5 + 6.91\hat{i}$$
$$\lambda_1 = 1.5 - 6.91\hat{i}$$

which also matches with the computer solution (see Table 5.15).

<u>Example 2</u>
Find all eigen-values of the following 2×2 unsymmetrical matrix

$$A = \begin{bmatrix} 1 & -4 \\ 5 & 2 \end{bmatrix}$$

The analytical eigen-value solution for this problem is:

$$\lambda_1 = 1.5 - 4.444\hat{i}$$

$$\lambda_1 = 1.5 + 4.444\hat{i}$$

which also matches with the computer solution (see Table 5.15).

5.17 Block Lanczos Algorithms

The generalized eigen-equation, shown in Eq.(5.1), can be re-written as:

$$K^{-1} K \phi = \lambda K^{-1} M \phi \tag{5.299}$$

or

$$\left(\frac{1}{\lambda}\right) \phi = K^{-1} M \phi \tag{5.300}$$

or

$$\gamma \phi = K^{-1} M \phi \tag{5.301}$$

Then, one needs to find matrices Q, Z, and T, such that

$$T Z = \gamma Z \tag{5.302}$$

$$\phi = Q Z \tag{5.303}$$

$$Q^T M Q = I \tag{5.304}$$

In theory, the tri-diagonal matrix T can be computed as:

$$T = Q^T M K^{-1} M Q \tag{5.305}$$

If Q is **NOT** a square matrix of n×n, say Q is a m×n rectangular matrix (where m << n), then we will only get the approximated eigen-pair solution (γ, ϕ) to the original eigen-problem (see Eq.[5.1]):

$$\delta = K^{-1} M Q_m - Q_m T \tag{5.306}$$

Equation (5.305) can be derived as follows:

Step 1: Substituting Eq.(5.303) into Eq.(5.301), one gets:

$$\gamma (Q Z) = K^{-1} M (Q Z) \tag{5.307A}$$

Step 2: Pre-multiplying both sides of Eq.(5.307A) by $Q^T M$, one obtains:

$$\gamma (Q^T M Q) Z = Q^T M K^{-1} M Q Z \tag{5.307B}$$

Utilizing Eq.(5.304), Eq.(5.307B) becomes:

$$\gamma Z = (Q^T M K^{-1} M Q) Z \tag{5.308}$$

Step 3: Comparing Eq.(5.308) with Eq.(5.302), then Eq.(5.305) is obviously satisfied!

The "error" term δ, shown in Eq.(5.305), can be derived as follows:

Step 1: Pre-multiplying both sides of Eq.(5.306) by Q, one gets:

$$Q\,T = (Q\,Q^T M)\,K^{-1}M\,Q \qquad (5.309)$$

Step 2: Let $R \equiv Q\,Q^T M$ \qquad (5.310)

Post-multiplying both sides of Eq.(5.310) by Q, one has:

$$R\,Q = Q\,(Q^T M\,Q) \qquad (5.311)$$

Step 3: Utilizing Eq.(5.304), Eq.(5.311) will be simplified to:

$$R\,Q = Q \qquad (5.312)$$

Thus, if Q^{-1} exists, then $R \equiv Q\,Q^T M = I = $ Identity Matrix \qquad (5.313)

Hence, Eq.(5.309) becomes:

$$Q\,T \cong K^{-1}M\,Q \qquad (5.314)$$

or

$$K^{-1}M\,Q - Q\,T \equiv \delta \approx 0 \qquad \text{(5.305, repeated)}$$

where

$$Q \equiv \left\{ q^1, q^2, q^3, \cdots, q^m \right\} \qquad (5.315)$$

and

$$q^i \equiv vector\ of\ length\ n \qquad (5.316)$$

One of the major advantages of using the "Block" Lanczos as compared to the "regular" Lanczos algorithms is that the former involves "matrix times matrix" operations. The major differences between "block" and "regular" Lanczos algorithms are listed in the following paragraphs:

(a) There is more than one starting vector. Furthermore, at each Lanczos iteration step, more than one Lanczos vector q^i is generated.

(b) In each Lanczos iteration step, one needs to orthogonalize the vectors in the current "block Lanczos" vectors.

5.17.1 Details of "Block Lanczos" Algorithms

Assuming that there are "b" vectors in each "block Lanczos" vector, the three-term relation can be expressed as:

$$r^j \equiv q^{j+1}\beta_{j+1} = K^{-1}M\,q^j - q^j\alpha_j - q^{j-1}(\beta_j^{\ T}) \qquad (5.317)$$

where r^j, q^{j+1}, q^j, and q^{j-1} are all matrices with dimensions n×b and β_{j+1}, β_j and α_j are all matrices with dimensions b×b.

Derivation of the above equation will be given in the following paragraphs.

From Eq.(5.306), one has:

$$K^{-1}M\,Q_m = Q_m\,T + \delta \tag{5.318}$$

where $m \ll n$, and assuming $n = 8$; $m = 4$.

Eq.(5.318) can be expressed in the "long format" as:

$$[K^{-1}M]_{8\times8} *[q^1 q^2 q^3 q^4]_{8\times4} = [q^1 q^2 q^3 q^4]_{8\times4} * \begin{bmatrix} \alpha_1 & \beta_2 & 0 & 0 \\ \beta_2 & \alpha_2 & \beta_3 & 0 \\ 0 & \beta_3 & \alpha_3 & \beta_4 \\ 0 & 0 & \beta_4 & \alpha_4 \end{bmatrix}_{4\times4} \tag{5.319}$$

$$[K^{-1}M]_{8\times8} *[q^1 q^2 q^3 q^4]_{8\times4} =$$

$$\begin{bmatrix} q_1^1\alpha_1 + q_1^2\beta_2 + 0 + 0 & q_1^1\beta_2 + q_1^2\alpha_2 + q_1^3\beta_3 + 0 & \cdots & 0 + 0 + q_1^3\beta_4 + q_1^4\alpha_4 \\ \vdots & \vdots & \cdots & \vdots \\ q_8^1\alpha_1 + q_8^2\beta_2 + 0 + 0 & q_8^1\beta_2 + q_8^2\alpha_2 + q_8^3\beta_3 + 0 & \cdots & 0 + 0 + q_8^3\beta_4 + q_8^4\alpha_4 \end{bmatrix} \tag{5.320}$$

Thus, if we only want to compute $K^{-1}M\,q^{(j=2)}$, then Eq.(5.320) gives:

$$K^{-1}M\,q^{(j=2)} = \begin{Bmatrix} q_1^1\beta_2 + q_1^2\alpha_2 + q_1^3\beta_3 + 0 \\ \vdots \\ q_8^1\beta_2 + q_8^2\alpha_2 + q_8^3\beta_3 + 0 \end{Bmatrix} \tag{5.321}$$

Hence, Eq.(5.321) can be expressed as:

$$[K^{-1}M]\,q^j = q^{j-1}*\beta_j + q^j*\alpha_j + q^{j+1}*\beta_{j+1} \tag{5.322}$$

or

$$q^{j+1}\beta_{j+1} = K^{-1}M\,q^j - q^j\alpha_j - q^{j-1}\beta_j \equiv r^j \tag{5.316, repeated}$$

In matrix form, Eq.(5.322) can be expressed as:

$$K^{-1}M\,Q_m = Q_m T_m + [0 \quad 0 \quad \cdots \quad r_m] \tag{5.323}$$

where

$$T_m = \begin{bmatrix} \alpha_1 & \beta_2^T & 0 & 0 & 0 & 0 & \cdots & 0 \\ \beta_2 & \alpha_2 & \beta_3^T & 0 & 0 & 0 & \cdots & 0 \\ & \beta_3 & \alpha_3 & \beta_4^T & 0 & 0 & \cdots & 0 \\ & & \ddots & \ddots & \ddots & & & \\ & & & \ddots & \ddots & \ddots & & \\ & & & & \ddots & \ddots & \ddots & \\ & & & & & \ddots & \ddots & \beta_m^T \\ & & & & & & \beta_m & \alpha_m \end{bmatrix} \tag{5.324}$$

Upon solving the following reduced, tri-diagonal "block" eigen-equations:

$$T_m\,Z_m = \gamma\,Z_m \tag{5.325}$$

One gets the original eigen-solutions (approximately, if m << n):

$$\lambda = \frac{1}{\gamma} \qquad (5.326)$$

$$\phi = Q_m \, Z_m \qquad (5.302, \text{ repeated})$$

The step-by-step, "block Lanczos" algorithms can be summarized in Table 5.19.

Table 5.19 Step-by-Step "Block Lanczos" Algorithms

Step 0: Given an arbitrary block vectors $r^0 = n \times b$,
 Where b=block size >= 1, then
 Step 1:
      (a) $q^0 = 0$
      (b)   Find $q^1 = r^0 \, \beta_1^{-1}$; and also determine β_1
             so that:
                 $q^{1^T} M \, q^1 = I \equiv$ Identity Matrix
      (c) $p^1 = M \, q^1$
 Step 2:
      For j=1, 2, 3, ... until "maxiter" (or "converged")
      (a) $\bar{r}^j = K^{-1} p^j$
      (b) $\hat{r}^j = \bar{r}^j - q^{j-1} \, \beta_j^T$
      (c) $\alpha_j = \left\{ q^j \right\}^T M \, \hat{r}^j \equiv \left\{ p^j \right\}^T \hat{r}^j$
      (d) $r^j = \hat{r}^j - q^j \, \alpha_j$
      (e) $\bar{p}^j = M \, r^j$
      (f)   Determine β_{j+1} , such that:
             $q^{j+1} \beta_{j+1} = r^j$; and $\left\{ q^{j+1} \right\}^T M \, q^{j+1} = I$
      (g)   If enough vectors generated, then terminated the do-loop
      (h) $q^{j+1} = r^j \, \beta_{j+1}^{-1}$
             Re-orthogonalized q^{j+1} against $q^1, q^2, \cdots, q^j$ with respect to [M]
             (hence $\left\{ q^i \right\}^T M \, q^j \equiv \delta_{ij}$)
      (i) $p^{j+1} = \bar{p}^j \, \beta_{j+1}^{-1} \equiv (M \, r^j) \, \beta_{j+1}^{-1} \equiv M \, (q^{j+1})$

How to Compute Step # (1b) and/or Step # (2f)

The computational tasks involved in Steps 1b and 2f can be re-casted as the following problem:

Given: the matrices $[B]_{n\times b} \equiv [r^j]$ and $[M]_{n\times n}$

Find: the matrices $[C]_{b\times b} \equiv [\beta_{j+1}]$ and $[A]_{n\times b} \equiv [q^{j+1}]$

such that:

$$[B]_{n\times b} = [A]_{n\times b} * [C]_{b\times b} \tag{5.327}$$

$$[A]^T [M][A] = [I]_{b\times b} \tag{5.328}$$

Solution:

From Eq.(5.327), one has:

$$B^T M B = (C^T \underleftarrow{A^T}) M (\underrightarrow{A} C) \tag{5.329}$$

Utilizing Eq.(5.328), the above equation becomes:

$$D \equiv B^T M B = C^T C \tag{5.330}$$

If one selects the matrix $[C]_{b\times b}$ to be an upper-triangular matrix, and recognizing that:

$$[D]_{b\times b} \equiv [B^T]_{b\times n} [M]_{n\times n} [B]_{n\times b} = \text{known, square matrix} \tag{5.331}$$

Then Eq.(5.330), $D = C^T C$, will represent the "Choleski" factorization of [D] (assuming [D] is a symmetrical, positive definite [or SPD] matrix!).

Remarks

(a) Other formats of [C] are also possible. Thus, an upper-triangular matrix for [C] is not the only choice!

(b) Besides the Choleski algorithm, which can be used to find [C] (if [D] is a SPD matrix), other procedures, such as QR algorithms, etc., can also be employed to find [C].

(c) The following "pseudo FORTRAN" codes can be used to compute the Choleski factorization matrix $[C]_{b\times b}$, from a given, symmetrical matrix $[D]_{b\times b}$

$$
\begin{bmatrix}
\begin{aligned}
&\text{Do } 1 \quad i = 1,b \\
&C(i,i) = \text{dsqrt}\left(D(i,i) - \sum_{m=1}^{i-1} C(m,i)**2 \right) \\
&\quad \begin{cases}
\text{Do } 2 \quad j = i+1,b \\
C(i,j) = \left(D(i,j) - \sum_{m=1}^{i-1} C(m,j)*C(m,i) \right) / C(i,i) \\
\end{cases} \\
&2 \quad \text{Continue} \\
&1 \quad \text{Continue}
\end{aligned}
\end{bmatrix}
$$

(d) Instead of using Choleski algorithms to solve for the unknown matrices $[\beta_{j+1}^T] \equiv [C]_{b \times b}$ and $[q^{j+1}] \equiv [A]_{n \times b}$, which can only handle the case of SPD matrix, the following problem can be solved by a different algorithm.

Given: $[r^{j+1}]_{n \times b}$ and $[M]_{n \times n}$

Find: $[q^{j+1}]_{n \times b}$ and $[\beta_{j+1}]_{b \times b}$

Such that: $[q^{j+1}]_{n \times b} * [\beta_{j+1}]_{b \times b} = [r^{j+1}]_{n \times b}$ (5.332)

and $[q^j]^T M [q^i] = \delta_{ij} = \text{Identity Matrix}$ (5.333)

Solution

Assuming $[\beta] \equiv \begin{bmatrix} \searrow & \rightarrow \\ 0 & \searrow \end{bmatrix} = \text{upper} - \text{triangular matrix}$ (5.334)

Then Eq.(5.332) can be expressed as:

$$
\begin{bmatrix} \downarrow & \downarrow & & \downarrow \\ q^1 & q^2 & \cdots & q^b \\ \downarrow & \downarrow & & \downarrow \end{bmatrix}_{n \times b} * \begin{bmatrix} \searrow & \times & \times \\ & \searrow & \times \\ & & \searrow \end{bmatrix}_{b \times b} = \begin{bmatrix} \downarrow & \downarrow & & \downarrow \\ r^1 & r^2 & \cdots & r^b \\ \downarrow & \downarrow & & \downarrow \end{bmatrix}_{n \times b}
$$ (5.335)

The explicit, long forms of Eq.(5.335) and Eq.(5.333) can be written as:

$$q^1 \, \beta_{11} = r^1$$ (5.336A)

with $$q^{1^T} M q^1 = 1$$ (5.336B)

Hence $$\beta_{11} = q^{1^T} M r^1$$ (5.337)

Similarly, one has:

$$q^1 \, \beta_{12} + q^2 \, \beta_{22} = r^2$$ (5.338A)

with $$q^{2^T} M q^2 = 1$$ (5.338B)

Hence
$$q^{2^T} M q^1 = 0 \qquad (5.338C)$$

$$\beta_{12} = q^{1^T} M r^2 \qquad (5.339A)$$

$$\beta_{22} = q^{2^T} M r^2 \qquad (5.339B)$$

Finally, one has:
$$q^1 \beta_{1b} + q^2 \beta_{2b} + \cdots + q^b \beta_{bb} = r^b \qquad (5.340A)$$

with
$$q^{b^T} M q^b \quad = 1 \qquad (5.340B)$$

$$q^{b^T} M q^{b-1} \quad = 0 \qquad (5.340C)$$

$$q^{b^T} M q^{b-2} \quad = 0 \qquad (5.340D)$$

$$\vdots \qquad \vdots$$

$$q^{b^T} M q^1 \quad = 0 \qquad (5.340E)$$

Hence
$$\beta_{1b} = q^{1T} M r^b \qquad (5.341A)$$

$$\beta_{21} = q^{2T} M r^b \qquad (5.341B)$$

$$\vdots \qquad \vdots$$

$$\beta_{bb} = q^{bT} M r^b \qquad (5.341C)$$

5.17.2 A Numerical Example for "Block Lanczos" Algorithms

The following simple data is given:

$$K = \begin{bmatrix} 1 & 0 & 0 & 0 & 0 & 0 \\ 0 & 2 & 0 & 0 & 0 & 0 \\ 0 & 0 & 3 & 0 & 0 & 0 \\ 0 & 0 & 0 & 5 & 0 & 0 \\ 0 & 0 & 0 & 0 & 6 & 0 \\ 0 & 0 & 0 & 0 & 0 & 8 \end{bmatrix}; M = \begin{bmatrix} 1 & & & & & \\ & 1 & & & & \\ & & 1 & & & \\ & & & 1 & & \\ & & & & 1 & \\ & & & & & 2 \end{bmatrix} \qquad (5.342)$$

Using the block size b = 2, find the eigen-values from the generalized eigen-equation (5.1).

Solution:

Given an arbitrary vector

$$r^0 = \begin{bmatrix} 1 & 0 \\ 1 & 0 \\ 1 & 0 \\ 1 & 0 \\ 1 & 0 \\ 1 & 1 \end{bmatrix} \tag{5.343}$$

Step # 1a:

$$q^0 = 0 \tag{5.344}$$

Step # 1b:

Here $[B] = [r^0]$; $[A] = [q^1]$; and $[C] = [\beta_1]$ (5.345)

Thus, $D = B^T M B = \begin{bmatrix} 7 & 2 \\ 2 & 2 \end{bmatrix}$ (5.346)

The Choleski factorization of $D = C^T C$ gives:

$$C = \begin{bmatrix} \sqrt{7} & \dfrac{2}{\sqrt{7}} \\ 0 & \sqrt{\dfrac{10}{7}} \end{bmatrix} = [\beta_1] \tag{5.347}$$

Hence

$$C^{-1} = [\beta_1]^{-1} = \begin{bmatrix} \dfrac{1}{\sqrt{7}} & \dfrac{-2}{\sqrt{70}} \\ 0 & \sqrt{\dfrac{7}{10}} \end{bmatrix} \tag{5.348}$$

and

$$q^1 = [r^0][\beta_1]^{-1} = \begin{bmatrix} 1/\sqrt{7} & -2/\sqrt{70} \\ 1/\sqrt{7} & -2/\sqrt{70} \\ 1/\sqrt{7} & -2/\sqrt{70} \\ 1/\sqrt{7} & -2/\sqrt{70} \\ 1/\sqrt{7} & -2/\sqrt{70} \\ 1/\sqrt{7} & 5/\sqrt{70} \end{bmatrix} \equiv [q^{11}, q^{12}] \tag{5.349}$$

Step # 1c:

$$[p^1] = [M][q^1] = \begin{bmatrix} \frac{1}{\sqrt{7}} & \frac{-2}{\sqrt{70}} \\ \frac{1}{\sqrt{7}} & \frac{-2}{\sqrt{70}} \\ \frac{1}{\sqrt{7}} & \frac{-2}{\sqrt{70}} \\ \frac{1}{\sqrt{7}} & \frac{-2}{\sqrt{70}} \\ \frac{1}{\sqrt{7}} & \frac{-2}{\sqrt{70}} \\ \frac{2}{\sqrt{7}} & \sqrt{\frac{10}{7}} \end{bmatrix} \equiv [q^{11}, q^{12}] \tag{5.350}$$

For j = 1,2,3, ... until "maxiter" (or "converged")

$$\boxed{j = 1}$$

Step # 2a:

$$\bar{r}^1 = K^{-1} p^1 = \begin{bmatrix} \frac{1}{\sqrt{7}} & \frac{-2}{\sqrt{70}} \\ \frac{0.5}{\sqrt{7}} & \frac{-1}{\sqrt{70}} \\ \frac{1}{(3\sqrt{7})} & \frac{-2}{(3\sqrt{70})} \\ \frac{0.2}{\sqrt{7}} & \frac{-0.4}{\sqrt{70}} \\ \frac{1}{(6\sqrt{7})} & \frac{-1}{(3\sqrt{70})} \\ \frac{1}{(4\sqrt{7})} & \frac{\sqrt{10}}{(8\sqrt{7})} \end{bmatrix} \tag{5.351}$$

Step # 2b:

$$\hat{r}^1 = \bar{r}^1 - q^0 \beta_1^T = \bar{r}^1 \quad \text{(because } q^0 = 0\text{)} \tag{5.352}$$

Step # 2c:

$$\alpha_1 = q^{1T} M \hat{r}^1 = p^{1T} \hat{r}^1 = \begin{bmatrix} 0.385714286 & -0.085833251 \\ -0.085833251 & 0.304285714 \end{bmatrix} \tag{5.353}$$

Notice: $[\alpha_1]$ = symmetrical matrix

Step # 2d:

$$r^1 = \hat{r}^1 - q^1 \alpha_1 = \begin{bmatrix} 0.211660105 & -0.133865604 \\ 0.022677869 & -0.014342743 \\ -0.04031621 & 0.025498211 \\ -0.090711473 & 0.057370974 \\ -0.103310289 & 0.065339164 \\ 0 & 0 \end{bmatrix} \tag{5.354}$$

Notice: $0 \approx 10^{-10}$

Step # 2e:

$$\overline{p^1} = M\, r^1 \tag{5.355}$$

Step # 2f:

Find β_2, such that $q^2 \beta_2 = r^1$ and $[q^2]^T M\, q^2 = I_{2\times 2}$

Compute $[D] = B^T M\, B = [r^1]^T M\, [r^1] = [r^1]^T \overline{p^1}$

Hence $[D] = \begin{bmatrix} 0.065842093 & -0.041641675 \\ \text{sym.} & 0.026336508 \end{bmatrix} \tag{5.356}$

Using Choleski factorization, one gets:

$$D = [C]^T [C]$$

where $[C] = [\beta_2] = \begin{bmatrix} 0.256597142 & -0.162294251 \\ 0 & 0.000574239 \end{bmatrix} \tag{5.357}$

Hence, $[C]^{-1} = [\beta_2]^{-1} = \begin{bmatrix} 3.89715954 & 1101.36725 \\ 0 & 1741.436319 \end{bmatrix} \tag{5.358}$

Step # 2h:

$$q^2 = r^1\, \beta_2^{-1} = \begin{bmatrix} 0.824873197 & -0.002916922 \\ 0.088379274 & -0.00011358 \\ -0.157118702 & 0.000557367 \\ -0.353517082 & 0.001252219 \\ -0.402616678 & 0.00142435 \\ 0 & 0 \end{bmatrix} \equiv [q^{21}_{old}, q^{22}_{old}] \tag{5.359}$$

where $q^{ij} \equiv j^{th}$ vector of the ith block.

Re-orthogonolized tasks:

$$q^{21} = q^{21}_{old} - \left(q^{11^T} M\, q^{21}_{old} \right) q^{11} - \left(q^{12^T} M\, q^{21}_{old} \right) q^{12} \tag{5.360}$$

$$q^{22} = q^{22}_{old} - \left(q^{11^T} M\, q^{22}_{old} \right) q^{11} - \left(q^{12^T} M\, q^{22}_{old} \right) q^{12} - \left(q^{21^T} M\, q^{22}_{old} \right) q^{21} \tag{5.361}$$

The operations involved in Eqs.(5.360 – 5.361) can be symbolically visualized as:

$$[q^{11}, q^{12}] \qquad ; \qquad [q^{21}, q^{22}]$$

Block Lanczos [#]1   Block Lanczos [#]2

(= 2 column vectors)   (= 2 column vectors)

In other words, we want to make sure that vector q^{21} is orthogonal to all previous vectors q^{11} and q^{12}. Similarly, we want to make sure that vector q^{22} is orthogonal to all previous vectors q^{11}, q^{12}, and q^{21}.

$$q^{11^T} M q^{21}_{old} = 3.401680257*10^{-9} \approx 0 \qquad (5.362)$$

$$q^{12^T} M q^{21}_{old} = -2.151411497*10^{-9} \approx 0 \qquad (5.363)$$

$$q^{11^T} M q^{22}_{old} = 0.000076891 \qquad (5.364)$$

$$q^{12^T} M q^{22}_{old} = -0.00004863 \qquad (5.365)$$

$$q^{21^T} M q^{22}_{old} = -0.003519850 \qquad (5.366)$$

Hence

$$[q^2] \equiv [q^{21}, q^{22}] = \begin{bmatrix} 0.824873197 & -0.000054179 \\ 0.088379274 & 0.000156815 \\ -0.157118702 & -0.000036354 \\ -0.353517082 & -0.000032795 \\ -0.402616678 & -0.000033487 \\ 0 & 0 \end{bmatrix} \qquad (5.367)$$

Now, check if:

$$\left\{ \begin{array}{l} q^{21^T} M q^{21} = 1, \quad \text{otherwise } q^{21} = \dfrac{q^{21}}{\left(q^{21^T} M q^{21} \right)^{1/2}} \qquad (5.368) \\[4em] q^{22^T} M q^{22} = 1, \quad \text{otherwise } q^{22} = \dfrac{q^{22}}{\left(q^{22^T} M q^{22} \right)^{1/2}} \qquad (5.369) \end{array} \right.$$

In this example, one gets:

$$[q^2] = \left[\dfrac{q^{21}}{\left(q^{21^T} M q^{21} \right)^{1/2}}, \dfrac{q^{22}}{\left(q^{22^T} M q^{22} \right)^{1/2}} \right] \qquad (5.370)$$

$$\text{or} \qquad [q^2] = \begin{bmatrix} 0.82487836 & -0.30749387 \\ 0.088379827 & 0.890006298 \\ -0.157119685 & -0.206327768 \\ -0.353519295 & -0.186128601 \\ -0.402619198 & -0.190056059 \\ 0 & \approx 0 \end{bmatrix} \qquad (5.371)$$

Step # 2i:

$$p^2 = \overline{p^1}\, \beta_2^{-1} \equiv M\, q^2 = \begin{bmatrix} 1 & & & & & \\ & 1 & & & & \\ & & 1 & & & \\ & & & 1 & & \\ & & & & 1 & \\ & & & & & 2 \end{bmatrix} q^2 = q^2 \qquad (5.372)$$

$\boxed{j = 2}$

Step # 2a:

$$\overline{r^2} = K^{-1}\, p^2 \equiv K^{-1} M\, q^2 \qquad (5.373)$$

$$\text{or} \qquad \overline{r^2} = \begin{bmatrix} 0.82487836 & -0.30749387 \\ 0.044189914 & 0.445003149 \\ -0.052373229 & -0.068775923 \\ -0.070703859 & -0.03722572 \\ -0.0671032 & -0.03167601 \end{bmatrix} \qquad (5.374)$$

Step # 2b:

$$\hat{r}^2 = \overline{r^2} - q^1 \beta_2^T \qquad (5.375)$$

$$= [\overline{r^2}] - \begin{bmatrix} 1/\sqrt{7} & -2/\sqrt{70} \\ 1/\sqrt{7} & -2/\sqrt{70} \\ 1/\sqrt{7} & -2/\sqrt{70} \\ 1/\sqrt{7} & -2/\sqrt{70} \\ 1/\sqrt{7} & -2/\sqrt{70} \\ 1/\sqrt{7} & 5/\sqrt{70} \end{bmatrix} \begin{bmatrix} 0.256597142 & 0 \\ -0.1622844251 & 0.000574239 \end{bmatrix} \qquad (5.376)$$

$$\hat{r}^2 = \begin{bmatrix} 0.689100401 & -0.307356601 \\ -0.091588045 & 0.445140418 \\ -0.188151187 & -0.068638654 \\ -0.206481818 & -0.037088451 \\ -0.202881159 & -0.03153872 \\ -0.000001214 & -0.000343173 \end{bmatrix} \tag{5.377}$$

Step # 2c:

$$\alpha_j = \alpha_2 = p^{2^T} \hat{r}^2 = \begin{bmatrix} 0.602443862 & -0.225597946 \\ \text{sym.} & 0.29969951 \end{bmatrix} \tag{5.378}$$

After two steps, we have:

$$T_m = T_2 = \begin{bmatrix} [\alpha_1] & [\beta_2]^T \\ [\beta_2] & [\alpha_2] \end{bmatrix} \tag{5.379}$$

$$T_m = T_2 \equiv \left[\begin{array}{cc|cc} 0.385714286 & -0.085833251 & 0.256597142 & 0 \\ -0.085833251 & 0.304285714 & -0.162284251 & 0.000574239 \\ \hline & & 0.602443862 & -0.225597946 \\ \text{sym.} & & -0.225597946 & 0.29969951 \end{array} \right]$$

$$\tag{5.380}$$

The above T_2 matrix was used as input for the following eigen-problem:

$$T_2 Z = \gamma Z \tag{5.381}$$

The above "reduced" eigen-equation can be "occasionally" solved to obtain the "approximated" eigen-value solution $\lambda = \frac{1}{\gamma}$. The corresponding "approximated" eigen-vector ϕ can be obtained from Eq.(5.303).

5.18 Summary

The most popular and effective procedure to solve all eigen-values of an unsymmetric matrix involve two major tasks, namely Hessenberg reduction form and QR algorithm on the Hessenberg matrix. In general, the QR algorithm requires between two to three times more computational effort than the Hessenberg reduction algorithm.

For "small-scale," dense, academic unsymmetrical eigen-problems, there is no need to reduce the size of the Hessenberg matrix and Numerical Recipe's sub-routines [5.12] can be used.

For "medium-scale," banded, unsymmetrical eigen-problems, one should reduce the size of the Hessenberg matrix first before solving the reduced eigen-problem by the QR method.

For "large-scale," sparse unsymmetric eigen-problems, one should incorporate the "unsymmetrical" sparse equation solver (see Chapter 3) and other "sparse technologies" [1.13] into Table 5.16.

5.19 Exercises

5.1   Given $[K] = \begin{bmatrix} 2 & -1 \\ -1 & 1 \end{bmatrix}$; and $[M] = \begin{bmatrix} 2 & 0 \\ 0 & 1 \end{bmatrix}$, using the determinant approach (discussed in Section 5.2), find all the eigen-pair solution for the generalized eigen-equation shown in Eq.(5.1).

5.2   Using the data for $[K]_{4 \times 4}$ and $[M]_{4 \times 4}$ as shown in Eqs.(5.28 – 5.29), find all the eigen-pair solutions for the generalized eigen-equation by the "Generalized Jacobi" method.

5.3   Re-do problem 5.2 using the "Sub-space Iteration" method, with the initial guess for the matrix $[X_k]$ as:

$$[X_k]_{4 \times 2} = \begin{bmatrix} 1 & 0 \\ 0 & 1 \\ 0 & 0 \\ 0 & 0 \end{bmatrix} \text{ (see Eq.[5.103])}$$

to obtain the lowest two eigen-pair solutions.

5.4   Re-do problem 5.2 using the "basic Lanczos algorithms" as indicated in Table 5.4 with the initial guess for the vector { x } as:

$$\{x\} = \begin{Bmatrix} 1 \\ 0 \\ 0 \\ 0 \end{Bmatrix}$$

5.5   Re-do problem 5.2 using "Block Lanczos Algorithms" (see Table 5.19) with the initial guess:

$$r^0 = \begin{bmatrix} 1 & 0 \\ 0 & 1 \\ 0 & 0 \\ 0 & 0 \end{bmatrix} \qquad \text{(see Section 5.8.2)}$$

6 Finite Element Domain Decomposition Procedures

6.1 Introduction

The finite element (statics) equilibrium equations can be given as:

$$[K(b)] \cdot \vec{z} = \vec{S}(b) \tag{6.1}$$

where

$\vec{b}$ = design variable vector such as cross-sectional areas of truss members, moment of inertia of beam members, and/or thickness of plate (or shell) members.

$\vec{z}$ = nodal displacement vector

$[K]$ = stiffness matrix

$\vec{S}$ = nodal load vector

Eq.(6.1) can be partitioned as:

$$\begin{bmatrix} K_{BB} & K_{BI} \\ K_{IB} & K_{II} \end{bmatrix} \times \begin{Bmatrix} z_B \\ z_I \end{Bmatrix} = \begin{Bmatrix} s_B \\ s_I \end{Bmatrix} \tag{6.2}$$

where subscripts B and I denote Boundary and Interior terms, respectively.

From the 2nd part of Eq.(6.2), one has:

$$K_{IB}z_B + K_{II}z_I = s_I \tag{6.3}$$

or

$$\vec{z_I} = [K_{II}]^{-1} \cdot \left(\vec{s_I} - K_{IB} \vec{z_B} \right) \tag{6.4}$$

From the 1st part of Eq.(6.2), one has:

$$K_{BB}z_B + K_{BI}z_I = s_B \tag{6.5}$$

Substituting Eq.(6.4) into Eq.(6.5), one obtains:

$$K_{BB}z_B + K_{BI} \cdot [K_{II}]^{-1} \cdot (s_I - K_{IB}z_B) = s_B \tag{6.6}$$

or

$$\left[K_{BB} - K_{BI}K_{II}^{-1}K_{IB} \right] \cdot \vec{z_B} = \left(s_B - K_{BI}K_{II}^{-1}s_I \right) \tag{6.7}$$

Let $\quad \overline{K}_B \equiv K_{BB} - K_{BI}K_{II}^{-1}K_{IB} \equiv$ Effective Boundary Stiffness $\qquad$ (6.8)

and $\quad \overline{s}_B \equiv s_B - K_{BI}K_{II}^{-1}s_I \equiv$ Effective Boundary Load $\qquad$ (6.9)

Eq.(6.8) for $\overline{K}_B$ is also called the Schur complement matrix.

Then, Eq.(6.7) can be expressed as:

$$\left[\overline{K}_B\right] \cdot z_B = \overline{s}_B \qquad (6.10)$$

Having solved the boundary displacement vector z_B from Eq.(6.10), the interior displacement vector z_I can be solved from Eq.(6.4).

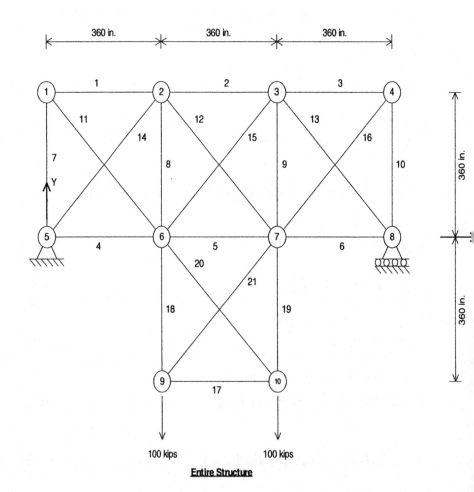

Entire Structure

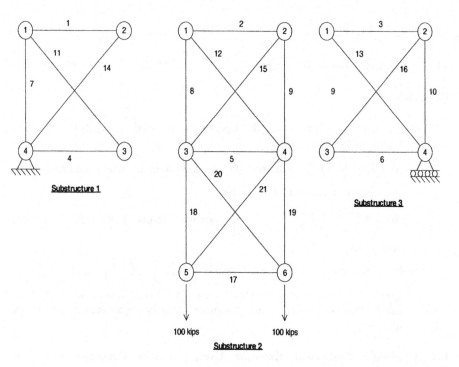

Figure 6.1: Structure Divided Into Three Substructures

For very large-scale problems, the original structure can be partitioned into many smaller sub-structures (or sub-domains); see Figure 6.1. Thus, for a typical r^{th} sub-domain, one obtains:

$$z_I^{(r)} = \left[K_{II}^{(r)} \right]^{-1} \cdot \left(s_I^{(r)} - K_{IB}^{(r)} z_B^{(r)} \right) \tag{6.11}$$

Similarly, one defines

$$\overline{K_B}^{(r)} \equiv K_{BB}^{(r)} - K_{BI}^{(r)} \left[K_{II}^{(r)} \right]^{-1} K_{IB}^{(r)} \tag{6.12}$$

$$\overline{s_B}^{(r)} \equiv s_B^{(r)} - K_{BI}^{(r)} \left[K_{II}^{(r)} \right]^{-1} s_I^{(r)} \tag{6.13}$$

The overall (or system) effective boundary stiffness matrix can be assembled and solved as:

$$\overline{K_B} = \sum_{r=1}^{NSU} \overline{K_B}^{(r)} \tag{6.14}$$

and

$$\overline{s_B} = \sum_{r=1}^{NSU} \overline{s_B}^{(r)} \tag{6.15}$$

$$\overline{K}_B \cdot \overline{z}_B = \overline{s}_B \qquad\qquad (6.16)$$

where NSU $\equiv$ $\underline{N}$umber of $\underline{SU}$bstructures

Remarks

(a) For large-scale problems, the operations involved in Eq.(6.12) are quite expensive because

    (i) $\left[K_{II}^{(r)} \right]^{-1} \cdot K_{IB}^{(r)}$ requires solving a system of linear equations with "many" right-hand-side vectors.

    (ii) $\left[K_{BI}^{(r)} \right] \cdot \left[\left[K_{II}^{(r)} \right]^{-1} \cdot K_{IB}^{(r)} \right]$ requires "sparse matrix*dense matrix" operations.

(b) Even though each individual matrix $\left[K_{BI}^{(r)} \right], \left[K_{II}^{(r)} \right]$, and $\left[K_{IB}^{(r)} \right]$ is sparse, the triple matrix products (shown in the right-hand side of Eq.(6.12)) can be "nearly dense." Thus, computer memory requirements can also be quite large.

6.2 A Simple Numerical Example Using Domain Decomposition (DD) Procedures

A simple 2-D truss structure [6.1], shown in Figures 6.2 – 6.4, will be used to illustrate detailed steps involved in the DD procedures (E = Kips/in.^{2}, A = in.^{2}, for all members).

For each rth substructure, the boundary nodes (or boundary dofs) are numbered first (and start with 1), then the interior nodes (or interior dofs) are subsequently numbered. Element numbers for each rth substructure should also start with 1. In Figure 6.3 while nodes 1 and 2 must be considered as boundary nodes, any interior node of the rth substructure can also be selected as a boundary node. The arrow-directions, shown in Figure 6.3, indicate that a particular truss member is connected from node "i" to node "j." The mapping between the local (sub-structuring) boundary degree-of-freedom (dof) and global (overall) boundary dof can be identified by comparing Figures 6.3 and 6.4. Thus, in sub-structure 1 (r = 1), boundary nodes 1 and 2 will correspond to the system (overall) boundary nodes 2 and 1, respectively.

Two vertical forces are applied at nodes 2 and 3 (as shown in Figure 6.2) while the prescribed displacements (say, due to support settlements) are shown at node 4 (of sub-structure r = 1, see Figure 6.3) and at node 4 (of sub-structure r = 2, see Figure 6.3).

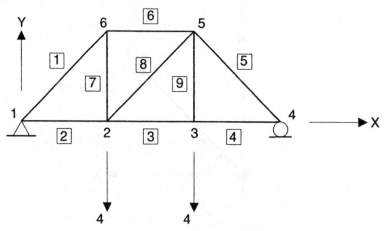

Figure 6.2 Overall Original System

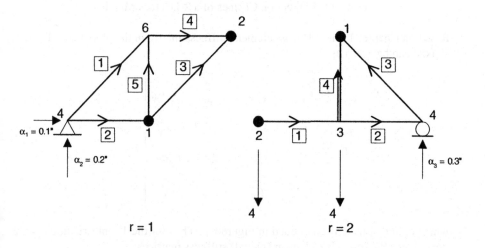

Figure 6.3 Local (Sub-structuring) Numbering System

Figure 6.4 Overall Boundary Nodes Numbering System

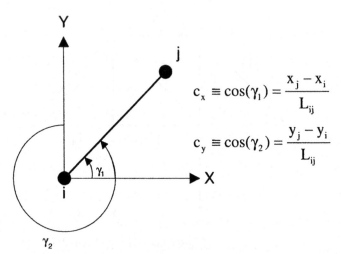

Figure 6.5 Direction Cosines of a 2-D Truss Member

Based on Chapter 1, the 2-D truss element stiffness matrix in the global coordinate X, Y axis, can be given as:

$$\left[k^{(e)}\right] = \left(\frac{EA}{L}\right) \begin{bmatrix} c_x^2 & c_x c_y & -c_x^2 & -c_x c_y \\ c_x c_y & c_y^2 & -c_x c_y & -c_y^2 \\ -c_x^2 & -c_x c_y & c_x^2 & c_x c_y \\ -c_x c_y & -c_y^2 & c_x c_y & c_y^2 \end{bmatrix} \tag{6.17}$$

where c_x and c_y have been defined in Figure 6.5. Thus for the 1st substructure (r = 1), one obtains the following element (global) stiffness matrices:

$$
k_G^1 = \begin{array}{c c} & \begin{array}{cccc} 7 & 8 & 5 & 6 \end{array} \\ \begin{array}{c} 7 \\ 8 \\ 5 \\ 6 \end{array} & \left| \begin{array}{cccc} 42.7209 & 42.7209 & -42.7209 & -42.7209 \\ 42.7209 & 42.7209 & -42.7209 & -42.7209 \\ -42.7209 & -42.7209 & 42.7209 & 42.7209 \\ -42.7209 & -42.7209 & 42.7209 & 42.7209 \end{array} \right| \end{array} \tag{6.18}
$$

$$
k_G^2 = \begin{array}{c c} & \begin{array}{cccc} 7 & 8 & 1 & 2 \end{array} \\ \begin{array}{c} 7 \\ 8 \\ 1 \\ 2 \end{array} & \left| \begin{array}{cccc} 120.8333 & 0 & -120.8333 & 0 \\ 0 & 0 & 0 & 0 \\ -120.8333 & 0 & 120.8333 & 0 \\ 0 & 0 & 0 & 0 \end{array} \right| \end{array} \tag{6.19}
$$

$$
k_G^3 = \begin{array}{c} \\ 1 \\ 2 \\ 3 \\ 4 \end{array} \begin{array}{|cccc|} 1 & 2 & 3 & 4 \\ 42.7209 & 42.7209 & -42.7209 & -42.7209 \\ 42.7209 & 42.7209 & -42.7209 & -42.7209 \\ -42.7209 & -42.7209 & 42.7209 & 42.7209 \\ -42.7209 & -42.7209 & 42.7209 & 42.7209 \end{array} \qquad (6.20)
$$

$$
k_G^4 = \begin{array}{c} \\ 5 \\ 6 \\ 3 \\ 4 \end{array} \begin{array}{|cccc|} 5 & 6 & 3 & 4 \\ 120.8333 & 0 & -120.8333 & 0 \\ 0 & 0 & 0 & 0 \\ -120.8333 & 0 & 120.8333 & 0 \\ 0 & 0 & 0 & 0 \end{array} \qquad (6.21)
$$

$$
k_G^5 = \begin{array}{c} \\ 1 \\ 2 \\ 5 \\ 6 \end{array} \begin{array}{|cccc|} 1 & 2 & 5 & 6 \\ 0 & 0 & 0 & 0 \\ 0 & 120.8333 & 0 & -120.8333 \\ 0 & 0 & 0 & 0 \\ 0 & -120.8333 & 0 & 120.8333 \end{array} \qquad (6.22)
$$

Sub-matrices $\left[K_{BB}^{(r=1)} \right]$, $\left[K_{BI}^{(r=1)} \right]$, $\left[K_{IB}^{(r=1)} \right]$ and $\left[K_{II}^{(r=1)} \right]$ of sub-structure 1 can be assembled as:

$$
K_{BB}^{(r=1)} = \begin{array}{c} \\ 1 \\ 2 \\ 3 \\ 4 \end{array} \begin{array}{|cccc|} 1 & 2 & 3 & 4 \\ 163.5542 & 42.7209 & -42.7209 & -42.7209 \\ 42.7209 & 163.5542 & -42.7209 & -42.7209 \\ -42.7209 & -42.7209 & 163.5542 & 42.7209 \\ -42.7209 & -42.7209 & 42.7209 & 42.7209 \end{array} \qquad (6.23)
$$

$$
K_{BI}^{(r=1)} = \begin{array}{c} \\ 1 \\ 2 \\ 3 \\ 4 \end{array} \begin{array}{|cccc|} 5 & 6 & 7 & 8 \\ 0 & 0 & -120.8333 & 0 \\ 0 & -120.8333 & 0 & 0 \\ -120.8333 & 0 & 0 & 0 \\ 0 & 0 & 0 & 0 \end{array} \qquad (6.24)
$$

$$
K_{II}^{(r=1)} = \begin{array}{c} \\ 5 \\ 6 \\ 7 \\ 8 \end{array} \begin{array}{|cccc|} 5 & 6 & 7 & 8 \\ 163.5542 & 42.7209 & -42.7209 & -42.7209 \\ 42.7209 & 163.5542 & -42.7209 & -42.7209 \\ -42.7209 & -42.7209 & 163.5542 & 42.7209 \\ -42.7209 & -42.7209 & 42.7209 & 42.7209 \end{array} \qquad (6.25)
$$

$$
K_{IB}^{(r=1)} = \begin{array}{c} \\ 5 \\ 6 \\ 7 \\ 8 \end{array} \begin{array}{|cccc|} 1 & 2 & 3 & 4 \\ 0 & 0 & -120.8333 & 0 \\ 0 & -120.8333 & 0 & 0 \\ -120.8333 & 0 & 0 & 0 \\ 0 & 0 & 0 & 0 \end{array} \qquad (6.26)
$$

After imposing the support boundary conditions, the above four equations become:

$$
K_{BB}^{(r=1)}{}_{b.c.} = \begin{array}{c} \\ 1 \\ 2 \\ 3 \\ 4 \end{array}
\begin{array}{cccc}
1 & 2 & 3 & 4 \\
\left| \begin{array}{cccc}
163.5542 & 42.7209 & -42.7209 & -42.7209 \\
42.7209 & 163.5542 & -42.7209 & -42.7209 \\
-42.7209 & -42.7209 & 163.5542 & 42.7209 \\
-42.7209 & -42.7209 & 42.7209 & 42.7209
\end{array} \right|
\end{array}
\tag{6.27}
$$

$$
K_{BI}^{(r=1)}{}_{b.c.} = \begin{array}{c} \\ 1 \\ 2 \\ 3 \\ 4 \end{array}
\begin{array}{cccc}
5 & 6 & 7 & 8 \\
\left| \begin{array}{cccc}
0 & 0 & 0 & 0 \\
0 & -120.8333 & 0 & 0 \\
-120.8333 & 0 & 0 & 0 \\
0 & 0 & 0 & 0
\end{array} \right.
\end{array}
\tag{6.28}
$$

$$
K_{II}^{(r=1)}{}_{b.c.} = \begin{array}{c} \\ 5 \\ 6 \\ 7 \\ 8 \end{array}
\begin{array}{cccc}
5 & 6 & 7 & 8 \\
\left| \begin{array}{cccc}
163.5542 & 42.7209 & 0 & 0 \\
42.7209 & 163.5542 & 0 & 0 \\
0 & 0 & 1 & 0 \\
0 & 0 & 0 & 1
\end{array} \right.
\end{array}
\tag{6.29}
$$

$$
K_{IB}^{(r=1)}{}_{b.c.} = \begin{array}{c} \\ 5 \\ 6 \\ 7 \\ 8 \end{array}
\begin{array}{cccc}
1 & 2 & 3 & 4 \\
\left| \begin{array}{cccc}
0 & 0 & -120.8333 & 0 \\
0 & -120.8333 & 0 & 0 \\
0 & 0 & 0 & 0 \\
0 & 0 & 0 & 0
\end{array} \right.
\end{array}
\tag{6.30}
$$

The boundary and interior load vectors can be obtained as:

$$
F_B^{(r=1)} = \begin{array}{c} 1 \\ 2 \\ 3 \\ 4 \end{array} \left\{ \begin{array}{c} 0 \\ 0 \\ 0 \\ 0 \end{array} \right\}, \quad
F_I^{(r=1)} = \begin{array}{c} 5 \\ 6 \\ 7 \\ 8 \end{array} \left\{ \begin{array}{c} 0 \\ 0 \\ 0 \\ 0 \end{array} \right\}
\tag{6.31}
$$

After imposing the support (prescribed) boundary conditions, Eq.(6.31) becomes:

$$
F_B^{(r=1)}{}_{b.c.} = \begin{array}{c} 1 \\ 2 \\ 3 \\ 4 \end{array} \left\{ \begin{array}{c} 12.08333 \\ 0 \\ 0 \\ 0 \end{array} \right\}, \quad
F_I^{(r=1)}{}_{b.c.} = \begin{array}{c} 5 \\ 6 \\ 7 \\ 8 \end{array} \left\{ \begin{array}{c} 12.81627 \\ 12.81627 \\ 0.1 \\ 0.2 \end{array} \right\}
\tag{6.32}
$$

The effective boundary stiffness matrix and load vector can be computed from Eqs.(6.12 – 6.13) and, and are given as:

$$\overline{K}_B^{(r=1)} = \begin{array}{c} \\ 3 \\ 4 \\ 1 \\ 2 \end{array} \begin{array}{cccc} 3 & 4 & 1 & 2 \\ \left| \begin{array}{cccc} 163.5542 & 42.7209 & -42.7209 & -42.7209 \\ 42.7209 & 67.7463 & -17.6955 & -42.7209 \\ -42.7209 & -17.6955 & 67.7463 & 42.7209 \\ -42.7209 & -42.7209 & 42.7209 & 42.7209 \end{array} \right| \end{array}$$

$$(6.33)$$

$$\overline{F}_B^{(r=1)} = \left\{ \begin{array}{c} 12.08333 \\ 7.5076 \\ 7.5076 \\ 0 \end{array} \right\}$$

$$(6.34)$$

Similarly, for substructure $r = 2$, element stiffness matrices (in global coordinate references) can be computed from Eq.(6.17) as:

$$k_G^1 = \begin{array}{c} \\ 3 \\ 4 \\ 5 \\ 6 \end{array} \begin{array}{cccc} 3 & 4 & 5 & 6 \\ \left| \begin{array}{cccc} 120.8333 & 0 & -120.8333 & 0 \\ 0 & 0 & 0 & 0 \\ -120.8333 & 0 & 120.8333 & 0 \\ 0 & 0 & 0 & 0 \end{array} \right| \end{array}$$

$$(6.35)$$

$$k_G^2 = \begin{array}{c} \\ 5 \\ 6 \\ 7 \\ 8 \end{array} \begin{array}{cccc} 5 & 6 & 7 & 8 \\ \left| \begin{array}{cccc} 120.8333 & 0 & -120.8333 & 0 \\ 0 & 0 & 0 & 0 \\ -120.8333 & 0 & 120.8333 & 0 \\ 0 & 0 & 0 & 0 \end{array} \right| \end{array}$$

$$(6.36)$$

$$k_G^3 = \begin{array}{c} \\ 7 \\ 8 \\ 1 \\ 2 \end{array} \begin{array}{cccc} 7 & 8 & 1 & 2 \\ \left| \begin{array}{cccc} 42.7209 & -42.7209 & -42.7209 & 42.7209 \\ -42.7209 & 42.7209 & 42.7209 & -42.7209 \\ -42.7209 & 42.7209 & 42.7209 & -42.7209 \\ 42.7209 & -42.7209 & -42.7209 & 42.7209 \end{array} \right| \end{array}$$

$$(6.37)$$

$$k_G^4 = \begin{array}{c} \\ 5 \\ 6 \\ 1 \\ 2 \end{array} \begin{array}{cccc} 5 & 6 & 1 & 2 \\ \left| \begin{array}{cccc} 0 & 0 & 0 & 0 \\ 0 & 120.8333 & 0 & -120.8333 \\ 0 & 0 & 0 & 0 \\ 0 & -120.8333 & 0 & 120.8333 \end{array} \right| \end{array}$$

$$(6.38)$$

Sub-matrices $\left[K_{BB}^{(r=2)} \right]$, $\left[K_{BI}^{(r=2)} \right]$, $\left[K_{IB}^{(r=2)} \right]$ and $\left[K_{II}^{(r=2)} \right]$ can be assembled as:

$$K_{BB}^{(r=2)} = \begin{array}{c} \\ 1 \\ 2 \\ 3 \\ 4 \end{array} \begin{array}{cccc} 1 & 2 & 3 & 4 \\ \left| \begin{array}{cccc} 42.7209 & -42.7209 & 0 & 0 \\ -42.7209 & 163.5542 & 0 & 0 \\ 0 & 0 & 120.8333 & 0 \\ 0 & 0 & 0 & 0 \end{array} \right| \end{array}$$

$$(6.39)$$

$$K_{BI}^{(r=2)} = \begin{array}{c} \\ 1 \\ 2 \\ 3 \\ 4 \end{array} \begin{array}{cccc} 5 & 6 & 7 & 8 \\ \hline 0 & 0 & -42.7209 & 42.7209 \\ 0 & -120.8333 & 42.7209 & -42.7209 \\ -120.8333 & 0 & 0 & 0 \\ 0 & 0 & 0 & 0 \end{array} \qquad (6.40)$$

$$K_{IB}^{(r=2)} = \begin{array}{c} \\ 5 \\ 6 \\ 7 \\ 8 \end{array} \begin{array}{cccc} 1 & 2 & 3 & 4 \\ \hline 0 & 0 & -120.8333 & 0 \\ 0 & -120.8333 & 0 & 0 \\ -42.7209 & 42.7209 & 0 & 0 \\ 42.7209 & -42.7209 & 0 & 0 \end{array} \qquad (6.41)$$

$$K_{II}^{(r=1)} = \begin{array}{c} \\ 5 \\ 6 \\ 7 \\ 8 \end{array} \begin{array}{cccc} 5 & 6 & 7 & 8 \\ \hline 241.6666 & 0 & -120.8333 & 0 \\ 0 & 120.8333 & 0 & 0 \\ -120.8333 & 0 & 163.5542 & -42.7209 \\ 0 & 0 & -42.7209 & 42.7209 \end{array} \qquad (6.42)$$

After imposing the support boundary conditions, Eqs.(6.39 – 6.42) become:

$$K_{BB\ b.c.}^{(r=2)} = \begin{array}{c} \\ 1 \\ 2 \\ 3 \\ 4 \end{array} \begin{array}{cccc} 1 & 2 & 3 & 4 \\ \hline 42.7209 & -42.7209 & 0 & 0 \\ -42.7209 & 163.5542 & 0 & 0 \\ 0 & 0 & 120.8333 & 0 \\ 0 & 0 & 0 & 0 \end{array} \qquad (6.43)$$

$$K_{BI\ b.c.}^{(r=2)} = \begin{array}{c} \\ 1 \\ 2 \\ 3 \\ 4 \end{array} \begin{array}{cccc} 5 & 6 & 7 & 8 \\ \hline 0 & 0 & -42.7209 & 0 \\ 0 & -120.8333 & 42.7209 & 0 \\ -120.8333 & 0 & 0 & 0 \\ 0 & 0 & 0 & 0 \end{array} \qquad (6.44)$$

$$K_{IB\ b.c.}^{(r=2)} = \begin{array}{c} \\ 5 \\ 6 \\ 7 \\ 8 \end{array} \begin{array}{cccc} 1 & 2 & 3 & 4 \\ \hline 0 & 0 & -120.8333 & 0 \\ 0 & -120.8333 & 0 & 0 \\ -42.7209 & 42.7209 & 0 & 0 \\ 0 & 0 & 0 & 0 \end{array} \qquad (6.45)$$

$$K_{II\ b.c.}^{(r=1)} = \begin{array}{c} \\ 5 \\ 6 \\ 7 \\ 8 \end{array} \begin{array}{cccc} 5 & 6 & 7 & 8 \\ \hline 241.6666 & 0 & -120.8333 & 0 \\ 0 & 120.8333 & 0 & 0 \\ -120.8333 & 0 & 163.5542 & 0 \\ 0 & 0 & 0 & 1 \end{array} \qquad (6.46)$$

The boundary and interior load vectors of substructure r = 2 can be computed as:

$$F_B^{(r=2)} = \begin{matrix} 1 \\ 2 \\ 3 \\ 4 \end{matrix} \left\{ \begin{matrix} 0 \\ 0 \\ 0 \\ -4 \end{matrix} \right\}, \quad F_I^{(r=2)} = \begin{matrix} 5 \\ 6 \\ 7 \\ 8 \end{matrix} \left\{ \begin{matrix} 0 \\ -4 \\ 0 \\ 0 \end{matrix} \right\} \tag{6.47}$$

After imposing the support boundary conditions, Eq.(6.47) becomes:

$$F_B^{(r=2)}{}_{b.c.} = \begin{matrix} 1 \\ 2 \\ 3 \\ 4 \end{matrix} \left\{ \begin{matrix} 0 \\ -4 \\ -12.81627 \\ -12.81627 \end{matrix} \right\}, \quad F_I^{(r=2)}{}_{b.c.} = \begin{matrix} 5 \\ 6 \\ 7 \\ 8 \end{matrix} \left\{ \begin{matrix} 0 \\ -4 \\ 12.81627 \\ 0.3 \end{matrix} \right\} \tag{6.48}$$

Applying Eqs.(6.12 – 6.13), the effective boundary stiffness matrix and load vector can be computed as:

$$\overline{K}_B^{(r=2)} = \begin{matrix} & 1 & 2 & 3 & 4 \\ 1 & 25.0254 & -25.0254 & -25.0254 & 0 \\ 2 & -25.0254 & 25.0254 & 25.0254 & 0 \\ 3 & -25.0254 & 25.0254 & 25.0254 & 0 \\ 4 & 0 & 0 & 0 & 0 \end{matrix} \tag{6.49}$$

$$\overline{F}_B^{(r=2)} = \left\{ \begin{matrix} 7.5074 \\ -4 \\ -7.5074 \\ 3.5074 \end{matrix} \right\} \tag{6.50}$$

The overall (system) effective boundary load vector and stiffness matrix can be assembled as:

$$\overline{F}_B = \left\{ \begin{matrix} 19.5907 \\ 3.5076 \\ 0.0002 \\ 3.5074 \end{matrix} \right\} \tag{6.51}$$

$$\overline{K}_B^{(r=2)} = \begin{matrix} & 1 & 2 & 3 & 4 \\ 1 & 92.7711 & 17.6955 & -67.7463 & -17.6955 \\ 2 & 17.6955 & 67.7463 & -17.6955 & -42.7209 \\ 3 & -67.7463 & -17.6955 & 188.5796 & 42.7209 \\ 4 & -17.6955 & -42.7209 & 42.7209 & 67.7463 \end{matrix} \tag{6.52}$$

All boundary displacements can be solved from Eq.(6.16) as:

$$
z_B = \begin{matrix} 3 \\ 4 \\ 1 \\ 2 \end{matrix} \begin{Bmatrix} 0.133 \\ 0.054 \\ 0.089 \\ 0.098 \end{Bmatrix}
\tag{6.53}
$$

The interior displacements of each sub-structure can be recovered from Eq.(6.11)

$$
z_B^{(r=1)} = \begin{matrix} 1 \\ 2 \\ 3 \\ 4 \end{matrix} \begin{Bmatrix} 0.089 \\ 0.098 \\ 0.133 \\ 0.054 \end{Bmatrix}
\qquad
z_I^{(r=1)} = \begin{matrix} 5 \\ 6 \\ 7 \\ 8 \end{matrix} \begin{Bmatrix} 0.121 \\ 0.085 \\ 0.1 \\ 0.2 \end{Bmatrix}
\tag{6.54}
$$

$$
z_B^{(r=2)} = \begin{matrix} 1 \\ 2 \\ 3 \\ 4 \end{matrix} \begin{Bmatrix} 0.133 \\ 0.054 \\ 0.089 \\ 0.098 \end{Bmatrix}
\qquad
z_I^{(r=2)} = \begin{matrix} 5 \\ 6 \\ 7 \\ 8 \end{matrix} \begin{Bmatrix} 0.165 \\ 0.065 \\ 0.198 \\ 0.3 \end{Bmatrix}
\tag{6.55}
$$

6.3 Imposing Boundary Conditions on "Rectangular" Matrices $\left[K_{BI}^{(r)} \right]$

Imposing the Dirichlet boundary conditions on the "square" sub-structuring stiffness matrices $\left[K_{BB\ b.c.}^{(r=1)} \right]$ and $\left[K_{II\ b.c.}^{(r=1)} \right]$ (see Eqs.6.27, 6.29) are rather straightforward; this topic has already been explained in great detail in Chapters 1 and 4. Thus, the following paragraphs are intended to explain how to impose the Dirichlet boundary conditions on the rectangular (in general) sub-structuring stiffness matrix $\left[K_{BI\ b.c.}^{(r=1)} \right]$ such as the one shown in Eq.(6.28). To facilitate the discussion, let's consider the entire stiffness matrix of the first sub-structure ($r = 1$) as shown in Figure 6.3.

$$
\left[K^{(r=1)} \right] =
\begin{array}{c}
\\
B \\ B \\ B \\ B \\ I \\ I \\ I \\ I
\end{array}
\begin{array}{cccc|cccc}
B & B & B & B & I & I & I & I \\
\hline
K_{11} & K_{12} & K_{13} & K_{14} & K_{15} & K_{16} & K_{17} & K_{18} \\
K_{21} & K_{22} & K_{23} & K_{24} & K_{25} & K_{26} & K_{27} & K_{28} \\
K_{31} & K_{32} & K_{33} & K_{34} & K_{35} & K_{36} & K_{37} & K_{38} \\
K_{41} & K_{42} & K_{43} & K_{44} & K_{45} & K_{46} & K_{47} & K_{48} \\
\hline
K_{51} & K_{52} & K_{53} & K_{54} & K_{55} & K_{56} & K_{57} & K_{58} \\
K_{61} & K_{62} & K_{63} & K_{64} & K_{65} & K_{66} & K_{67} & K_{68} \\
K_{71} & K_{72} & K_{73} & K_{74} & K_{75} & K_{76} & K_{77} & K_{78} \\
K_{81} & K_{82} & K_{83} & K_{84} & K_{85} & K_{86} & K_{87} & K_{88}
\end{array}
=
\begin{bmatrix}
K_{BB} & K_{BI} \\
K_{IB} & K_{II}
\end{bmatrix}
\quad (6.56)
$$

Since the Dirichlet boundary conditions are imposed on the 7th and 8th degree-of-freedom, Eq.(6.56) becomes:

$$
\left[K_{b.c.}^{(r=1)} \right] =
\begin{array}{c}
\\
B \\ B \\ B \\ B \\ I \\ I \\ I \\ I
\end{array}
\begin{array}{cccc|cccc}
B & B & B & B & I & I & I & I \\
\hline
\times & \times & \times & \times & \times & \times & 0 & 0 \\
\times & \times & \times & \times & \times & \times & 0 & 0 \\
\times & \times & \times & \times & \times & \times & 0 & 0 \\
\times & \times & \times & \times & \times & \times & 0 & 0 \\
\hline
\times & \times & \times & \times & \times & \times & 0 & 0 \\
\times & \times & \times & \times & \times & \times & 0 & 0 \\
0 & 0 & 0 & 0 & 0 & 0 & 1 & 0 \\
0 & 0 & 0 & 0 & 0 & 0 & 0 & 1
\end{array}
=
\begin{bmatrix}
K_{BB} & K_{BI} \\
K_{IB} & K_{II}
\end{bmatrix}
\quad (6.57)
$$

Thus

$$
\left[K_{BI_{b.c.}}^{(r=1)} \right] =
\begin{array}{c}
\\
B \\ B \\ B \\ B
\end{array}
\begin{array}{cccc}
I & I & I & I \\
\times & \times & 0 & 0 \\
\times & \times & 0 & 0 \\
\times & \times & 0 & 0 \\
\times & \times & 0 & 0
\end{array}
\quad (6.58)
$$

It should be noted that Eq.(6.28) has the same form (last two columns) as Eq.(6.58)

6.4 How to Construct a Sparse Assembly of "Rectangular" Matrix $\left[K_{BI}^{(r)} \right]$

The numerical procedures for finite element sparse assembly of the rectangular matrix $\left[K_{BI}^{(r)} \right]$ can be summarized in the following two-step procedure:

Step 1: For each rth sub-structure, identify <u>how many</u> and <u>which</u> finite elements will have contributions to <u>both</u> <u>Boundary</u> and <u>Interior</u> nodes (assuming there are 2 dof per node and 3 nodes per element). Also, assume there are 6 boundary nodes, and 8 interior nodes in the rth substructure.

At the end of step 1, say elements 5, 10, and 11 are recorded. Also, assume the element nodes connectivity are:

El. #	node-I	node-j	node-k
5	6 = B	1 = B	8 = I
10	5 = B	6 = B	9 = I
11	1 = B	13 = I	11 = I

Note: B ≡ Boundary node, if ≤ node 6. I ≡ Interior, if > node 6.

Step 2: Based on the information provided by step 1, one obtains:

		I.Node1 (=7-6)	I.Node2 (=8-6)	3	4	5	6	7 (=13-6)	8 (=14-6)
	B.Node1		1,2 7,8			5,6 11,12		3,4 9,10	
$K_{BI}=$	B.Node2								
	3								
	4								
	5								
	6								

The above table, obtained for K_{BI}, can be generated by the following "pseudo" FORTRAN algorithms:

icount = 0
Do loop
  J = Boundary Node 1
  El. 5 = [6, 1, 8] ➔ B. node has coupling with I. node 2 (= 8-6) ➔ icount = 1
  El. 10 = [5, 6, 9] ➔ B. node 1 has no coupling with element 10.
  El. 11 = [1, 13 11] ➔ B. node 1 has coupling
                 with I. node 7 (= 13-6) ➔ icount = icount+1 = 2
                 with I. node 5 (=11-6). Hence, icount = 3
  so: $IE_{bi}(1) = 1$ (6.59)
        $IE_{bi}(2) = IE_{bi}(1)+(icount = 3)*(ndofpn=2)^2 = 13$ (6.60)

$$JE_{bi} = \{\text{columns } 3, 4, 13, 14, 9, 10, 3, 4, 13, 14, 9, 10\} \qquad (6.61)$$

J = Boundary Node 2
El. 5 = [6, 1, 8] ➜ Boundary node 2 has no coupling with El. 5
El. 10 = [5, 6, 9] ➜ Boundary node 2 has no coupling with El. 10
El. 11 = [1, 13 11] ➜ Boundary node 2 has no coupling with El. 11

Enddo

Remarks

In Eq.(6.61), column numbers 3, 4, 13, 14, 9, and 10 are repeated again, because each node is assumed to have 2 dof.

6.5 Mixed Direct-Iterative Solvers for Domain Decomposition

Using the domain decomposition (D. D.) formulation, one first needs to solve the unknown boundary displacement vector $\vec{z}_B$ from Eq.(6.16). However, the assembly process for obtaining the effective boundary stiffness matrix $\overline{K}_B$ (see Eq.(6.14)) will require the computation of the triple matrix products of $K_{BI}^{(r)} \left[K_{II}^{(r)} \right]^{-1} K_{IB}^{(r)}$ (for each rth substructure), as indicated in Eq.(6.12). For large-scale-applications, Eq.(6.12) is both computational and memory intensive since the related system of linear equations has a lot of right-hand-side vectors. Thus, forward and backward solution phases need to be done repeatedly. Furthermore, although each individual matrix $\left[K_{BI}^{(r)} \right]$, $\left[K_{II}^{(r)} \right]$, and $\left[K_{IB}^{(r)} \right]$ can be sparse, the triple product of $K_{BI}^{(r)} \left[K_{II}^{(r)} \right]^{-1} K_{IB}^{(r)}$ is usually dense, and therefore, a lot of computer memory is required. For the above reasons, mixed direct-iterative solver is suggested for solving Eq.(6.16).

Pre-conditioning Matrix

Consider the linear system $[A]\vec{x} = \vec{b}$, where the matrix $[A]$ is assumed to be symmetric positive definite (SPD). If the solution vector $\vec{x}$ is sought by an iterative solver, then one normally prefers to improve the condition number of the coefficient matrix ($= \dfrac{\lambda_{max}}{\lambda_{min}}$ = ratio of largest over smallest eigen-values of the coefficient matrix) by the "pre-conditioning process" as described in the following paragraphs:

Option 1 (Symmetrical property is preserved)

Let P be a pre-conditioning matrix, which is assumed to be non-singular. Then PAP^T is SPD. Instead of solving $[A]\vec{x} = \vec{b}$, one solves:

$$[P \, A \, \underbrace{P^T}_{\text{Identity Matrix}}] * \{P^{-T} \, \vec{x}\} = [P]\vec{b}$$

or

$$[A^*] * \{\vec{y}\} = \{\vec{b^*}\}$$

where

$$[A^*] \equiv [P \, A \, P^T] = \text{symmetrical matrix}$$

$$\vec{y} \equiv \{P^{-T} \, \vec{x}\}, \, hence \, \vec{x} = [P^T] \, \vec{y}$$

$$\vec{b^*} \equiv [P]\vec{b}$$

The pre-conditioner [P] should be selected so that:

1.   Matrix $[A^*]$ will have a better condition number than its original matrix [A].
2   Matrix [P] should be "cheap" to factorize.

As an example, the Cholesky factorization of [A] can be obtained as:

$$[A] = [U]^T [U]$$

where [U] is an upper triangular, factorized matrix.

Suppose one selects $[P] = U^{-T}$, then:

$$A^* \equiv P[A]P^T = \underbrace{U^{-T} [U^T}_{} \underbrace{U]U^{-1}}_{} = [I]$$

The condition number of $[A^*] = \dfrac{\lambda_{max}}{\lambda_{min}} = \dfrac{1.0}{1.0} = 1$

Therefore, the Conjugate Gradient method, when applied to $[A^* = I]\vec{y} = \vec{b^*}$ will converge in one iteration. However, in this case it is "too expensive" to factorize [P]! The compromised strategies will be:

$$[P] = [U_a]^{-T}$$

where $[U_a] \equiv$ inexpensive approximation of [U], and the amount of fill-in terms occurred in $[U_a]$ can be controlled (or specified) by the user. Various strategies for "incomplete Cholesky factorization" have been suggested to obtain $[U_a]$ for preconditioning purposes. The original CG and its PCG algorithms are summarized in the following section:

CG Without Preconditioner	CG With Preconditioner
Solving $[A]\vec{x} = \vec{b}$	Solving $[A^*]\vec{y} = \vec{b}^*$
Given an initial guess $\vec{x}^{(0)}$	Given $\vec{x}^{(0)}$
	Compute incomplete factor $[P]$
Compute initial residual $\vec{r}^{(0)} = \vec{b} - A x^{(0)}$	Compute $\vec{r}^{(0)} = \{Pb\} - \left[PAP^T\right]\vec{x}^{(0)}$
Set $\vec{d}^{(0)} = 0$; $\rho_{-1} = 1$	Set $\vec{d}^{(0)} = 0$; $\rho_{-1} = 1$
DO i=1, 2, ...	DO i=1, 2, ...
$\rho_{i-1} = \left\{r^{(i-1)}\right\}^T \left\{r^{(i-1)}\right\}$	$\rho_{i-1} = \left\{r^{(i-1)}\right\}^T \left\{r^{(i-1)}\right\}$
$\beta_{i-1} = \rho_{i-1} \Big/ \rho_{i-2}$	$\beta_{i-1} = \rho_{i-1} \Big/ \rho_{i-2}$
$d^{(i)} = r^{(i-1)} + \beta_{i-1} d^{(i-1)}$	$d^{(i)} = r^{(i-1)} + \beta_{i-1} d^{(i-1)}$
$q^{(i)} = [A]d^{(i)}$	$q^{(i)} = [PAP^T]d^{(i)}$
$\alpha_i = \rho_{i-1} \Big/ \left\{d^{(i)}\right\}^T \left\{q^{(i)}\right\}$	$\alpha_i = \rho_{i-1} \Big/ \left\{d^{(i)}\right\}^T \left\{q^{(i)}\right\}$
$x^{(i)} = x^{(i-1)} + \alpha_i d^{(i)}$	$x^{(i)} = x^{(i-1)} + \alpha_i d^{(i)}$
$r^{(i)} = r^{(i-1)} - \alpha_i q^{(i)}$	$r^{(i)} = r^{(i-1)} - \alpha_i q^{(i)}$
Converge ??	Converge ??
END DO	END DO
	If converged, then set $\vec{x} = [P]^T \vec{x}$

Option 2 (Symmetrical property may be destroyed)

Instead of solving $[A]\vec{x} = \vec{b}$, one solves:

$$\underbrace{[P]^{-1}[A]}\vec{x} = \underbrace{[P]^{-1}\vec{b}}$$

or

$$[A^*]\vec{x} = \left\{\vec{b}^*\right\}$$

where

$$[A^*] \equiv [P]^{-1}[A] = \text{may \underline{NOT} be symmetrical}$$

$$\{\vec{b^*}\} \equiv [P]^{-1}\vec{b}$$

This formulation is simpler than the one discussed in Option 1. However, $[A^*]$ may NOT be a symmetrical matrix.

Remarks

1.   If the original matrix [A] is <u>un</u>symmetrical, then Option 2 may be a preferable choice.

2.   If one selects $[P] = [A]$, then $[A^*] = [I]$ and the iterative solver will converge in one iteration.

In Table 6.1, The Preconditioned Conjugate Gradient (PCG) algorithm for solving a system of symmetrical linear equations $[A]\vec{x} = \vec{b}$, with the preconditioned matrix [B], is summarized.

Table 6.1 Preconditioned Conjugate Gradient Algorithm for Solving $[A]\vec{x} = \vec{b}$

Step 1: Initialized $\vec{x_o} = \vec{0}$

Step 2: Residual vector $\vec{r_o} = \vec{b}$ (or $\vec{r_o} = \vec{b} - A\vec{x_0}$, for "any" initial guess $\vec{x_o}$)

Step 3: "Inexpensive" preconditioned $\vec{z_0} = [B]^{-1} \cdot \vec{r_0}$

Step 4: Search direction $\vec{d_0} = \vec{z_0}$

For i = 0, 1, 2, ..., maxiter

Step 5: $\alpha_i = \dfrac{r_i^T z_i}{d_i^T \{A \cdot d_i\}}$

Step 6: $x_{i+1} = x_i + \alpha_i d_i$

Step 7: $r_{i+1} = r_i - \alpha_i [Ad_i]$

Step 8: Convergence check: if $\|r_{i+1}\| < \|r_0\| \cdot \varepsilon$ ➔ stop

Step 9: $z_{i+1} = B^{-1} r_{i+1}$

Step 10: $\beta_i = \dfrac{r_{i+1}^T z_{i+1}}{r_i^T z_i}$

Step 11: $d_{i+1} = z_{i+1} + \beta_i d_i$

End for

6.6 Pre-conditioned Matrix for PCG Algorithm with DD Formulation

The difficulty in constructing an efficient preconditioned matrix [B] in conjunction with PCG algorithm with D. D. formulation is compounded by the fact the

coefficient matrix $\left[\overline{K_B}\right] = \sum_{r=1}^{NSU} \overline{K_B}^{(r)} = \sum_{r=1}^{NSU} \left(K_{BB}^{(r)} - K_{BI}^{(r)} \left[K_{II}^{(r)} \right]^{-1} K_{IB}^{(r)} \right)$ has not

been assembled explicitly. In other words, how can we construct a pre-conditioned matrix [B] when the original coefficient matrix $\left[\overline{K_B}\right]$ has not even been formed? Even with the most simple "diagonal preconditioned" scheme, one still has to introduce "some approximation" about $\left[\overline{K_B}\right]$. The following two options can be considered:

Option 1: let $[B] \approx \left[\overline{K_B}\right] \approx \sum_{r=1}^{NSU} K_{BB,Diag}^{(r)}$

Option 2: For the pre-conditioned purpose only, approximate:

$$\left[K_{II}^{(r)} \right] \approx diagonal\ of \left[K_{II}^{(r)} \right]$$

Hence $\left[K_{II}^{(r)} \right]_{Approx}^{-1}$ is inexpensive, and

$$[B] \approx \left[\overline{K_B}\right] \approx \sum_{r=1}^{NSU} \left(K_{BB,Diag}^{(r)} - K_{BI}^{(r)} \left[K_{II}^{(r)} \right]_{Approx}^{-1} K_{IB}^{(r)} \right)$$

In Table 6.2, the corresponding 11-step procedure for the Pre-Conditioned Conjugate Gradient Algorithm within the context of Domain Decomposition (DD) formulation is summarized.

Table 6.2: Pre-Conditioned Conjugate Gradient D.D. Algorithm for Solving

$$\left[\overline{K_B}\right]\overrightarrow{z_B} = \overrightarrow{f_B}$$

Initialized Phase
Step 1: $\overrightarrow{z_{B_i}} = \vec{0}$

Step 2: Residual vector $\quad \overrightarrow{r_i} = \overrightarrow{f_B} - \overline{K_B}\,\overrightarrow{z_{B_i}} = \sum_{r=1}^{NSU} \overrightarrow{f_B}^{(r)}$ or $$\overrightarrow{r_i} = \sum_{r=1}^{NSU} \left(f_B^{(r)} - K_{BI}^{(r)} \left[K_{II}^{(r)} \right]^{-1} f_I^{(r)} \right)$$ DO 2 r=1,NSU (in parallel computation)
  Step 2.1: Direct sparse solver F/B solution with 1 RHS interior load
         vector

$$\overrightarrow{T_1} = \left[K_{II}^{(r)} \right]^{-1} \cdot f_I^{(r)}$$

Step 2.2: Sparse matrix times vector

$$\overrightarrow{T_2} = K_{BI}^{(r)} \cdot \overrightarrow{T_1}$$

Step 2.3: Compute sub-domain residual

$$\overrightarrow{T_1} = f_B^{(r)} - \overrightarrow{T_2}$$

Step 2.4: Insert sub-domain residual into proper location of $\overrightarrow{r_i}$

The partial resulted vector $\overrightarrow{T_1}$ from each processor will be sent to the Master processor, together with the mapping information about local-global boundary dofs.

Receive & Copy $\overrightarrow{r_i} = \overrightarrow{T_1}^{(r)}$, by the Master Processor.

2 Continue

Step 3: "Inexpensive" preconditioning, by the Master Processor.

$$\overrightarrow{z_i} = [B]^{-1} \cdot \overrightarrow{r_i}$$

Step 4: Search direction, by the Master Processor.

$$\overrightarrow{d_i} = \overrightarrow{z_i}$$

Iteration loop begins (for i = 0, 1, 2, ..., maxiter)

Step 5: Compute scalar $\alpha_i = \dfrac{r_i^T z_i}{d_i^T \left\{ \left[\overline{K_B} \right] \cdot d_i \right\}}$

Step 5.1: $up = r_i^T \cdot z_i$, by the Master Processor

Master Processor broadcasts $\overrightarrow{d_i}$ to all other processors

Step 5.2: compute

$$\overline{K_B} \cdot \overrightarrow{d_i} = \left(\sum_{r=1}^{NSU} \overline{K_B}^{(r)} \right) \cdot \overrightarrow{d_i} = \sum_{r=1}^{NSU} \left(K_{BB}^{(r)} - K_{BI}^{(r)} \left[K_{II}^{(r)} \right]^{-1} K_{IB}^{(r)} \right) \cdot \overrightarrow{d_i}$$

DO 5 r=1,NSU (in parallel computation)

Step 5.2a: Sparse matrix times vector

$$\overrightarrow{T_1} = K_{IB}^{(r)} \cdot \overrightarrow{d_i}$$

Step 5.2b: Direct sparse solver F/B with 1 RHS vector

$$\overrightarrow{T_2} = \left[K_{II}^{(r)} \right]^{-1} \cdot \overrightarrow{T_1}$$

Step 5.2c: Sparse matrix times vector

$$\overrightarrow{T_1} = K_{BI}^{(r)} \cdot \overrightarrow{T_2}$$

Step 5.2d: Sparse matrix times vector

$$\vec{T_2} = K_{BB}^{(r)} \cdot \vec{d_i}$$

Step 5.2e:

$$\vec{T_2} = \vec{T_2} - \vec{T_1}$$

Step 5.2f: Put vector $\vec{T_2}$ into proper location of

$$\left[\overline{K_B} \right] \cdot \vec{d_i} = \overrightarrow{stored}$$

Each processor will send its own $\vec{T_2}$ to the Master Processor.

5 Continue

The following steps will be done by the Master Processor.

Step 5.2g: Received vectors $\vec{T_2}$ (from each processor) and assembled vector $\overrightarrow{stored}$. Then, compute:

$$down = \vec{d_i} \cdot \overrightarrow{stored}$$

$$\alpha_i = \frac{up}{down}$$

Step 6: Compute new, improved solution

$$\overrightarrow{z_{B_{i+1}}} = \overrightarrow{z_{B_i}} + \alpha_i \vec{d_i}$$

Step 7: Compute new residual vector

$$\vec{r_{i+1}} = \vec{r_i} - \alpha_i \cdot \overrightarrow{stored}$$

Step 8: Convergence check:

$$r_{0,norm} = \left\| \vec{r_0} \right\|$$

$$r_{i+1,norm} = \left\| \vec{r_{i+1}} \right\|$$

Iteration steps will stop when $r_{i+1,norm} < \varepsilon \cdot r_{0,norm}$. Where ε is user input parameter

Step 9: "Inexpensive" preconditioning

$$\overrightarrow{z_{i+1}} = [B]^{-1} \cdot \overrightarrow{r_{i+1}}$$

Step 10: Compute $\beta_i = \dfrac{r_{i+1}^T z_{i+1}}{r_i^T z_i}$

$$up = r_{i+1}^T z_{i+1}$$

$$down = r_i^T z_i$$

$$\beta_i = \frac{up}{down}$$

Step 11: New search direction

$$\overrightarrow{d_{i+1}} = \overrightarrow{z_{i+1}} + \beta_i \overrightarrow{d_i}$$

Based upon the "primal" DD formulation, discussed in Sections 6.1 – 6.6, the MPI/Fortran software package DIPSS (Direct Iterative Parallel Sparse Solver) has been developed to solve few large scale, practical engineering problems that are summarized in the following paragraphs:

Example 1 – Three-dimensional acoustic finite element model. In this example, DIPSS is exercised to study the propagation of plane acoustic pressure waves in a 3-D hard wall duct without end reflection and airflow.

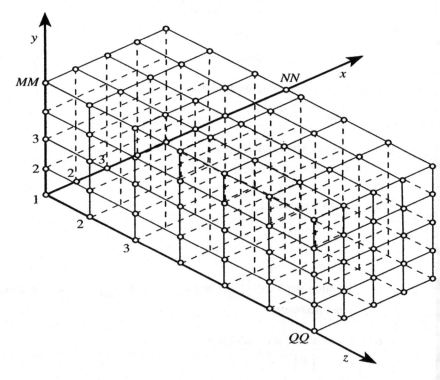

Figure 6.6: Finite Element Model for a Three-Dimensional Hard Wall Duct

The duct is shown in Figure 6.6 and is modeled with brick elements. The source and exit planes are located at the left and right boundary, respectively. The matrix, K, contains complex coefficients and the dimension of K is determined by the product of NN, MM, and QQ (N = MMxNNxQQ). Results are presented for two grids (N = 751,513 and N = 1,004,400) and the finite element analysis procedure for generation of the complex stiffness matrix, K, is presented in [6.2].

DIPSS [6.3] memory and wallclock statistics were also compared to those obtained using the platform specific SGI parallel sparse solver (e.g., ZPSLDLT). These statistics were computed on an SGI ORIGIN 2000 computer platform located at the NASA Langley Research Center. The SGI platform contained 10 gigabytes of memory and 8 ORIGIN 2000 processors were used. It should be noted that the ZPSLDLT is part of the SCSL library (version 1.4 or higher) and is considered to be one of the most efficient commercialized direct sparse solvers capable of performing complex arithmetic. Due to the 3-D nature of the hard wall duct example problem, K encounters lots of fill-in during the factorization phase. Thus, only the small grid (N = 751,513) could fit within the allocated memory on the ORIGIN~2000. ZPSLDLT required 6.5 wallclock hours to obtain the solution on the small grid whereas DIPSS wallclock took only 2.44 hours. DIPSS also required nearly 1 gigabyte less memory than ZPSLDLT, and the DIPSS and ZPSLDLT solution vector were in excellent agreement.

Because DIPSS uses MPI for interprocess communications, it can be ported to other computer platforms. To illustrate this point, the DIPSS software was ported to the SUN 10000 platform at Old Dominion University and used to solve the large grid duct acoustic problem (N = 1,004,400). Wallclock statistics and speedup factors were obtained using as many as 64~SUN~10000 processors. Results are presented in Table 6.3. It should be noted that a super-linear speedup factor of 85.95 has been achieved when 64 SUN 10000 processors are used. This super-linear speedup factor is due to two primary reasons:

1. The large finite element model has been divided into 64 sub-domains. Since each processor is assigned to each smaller sub-domain, the number of operations performed by each processor has been greatly reduced. Note that the number of operations are proportional to $\left(n^{(r)}\right)^3$ for the dense matrix, or $n^{(r)} \cdot BW^2$ for the banded, sparse matrix, where BW represents the half $\underline{B}$and $\underline{W}$idth of the coefficient stiffness matrix.

2. When the entire finite element model is analyzed by a direct, conventional sparse solver, more computer "paging" is required due to a larger problem size.

Table 6.3 Performance of DIPSS Software for 3-D Hard Wall Duct
(N = 1,004,400 complex equations)

# Processor (SUN 10000@ODU)	1	2	4	8	16	32	64
Sparse Assembly Time (seconds)	19.38	10.00	5.08	2.49	1.26	0.70	0.27
Sparse Factorization (seconds)	131,229	58,976	26,174	10,273	3,260	909	56
Total time (entire FEA)	131,846	61,744	27,897	11,751	3,817	1,967	1,534
Total Speed-Up Factor	1.00	2.14	4.73	11.22	34.54	67.03	85.95

Example 2 – Three-dimensional structural bracket finite element model. The DD formulation has also been applied to solve the 3-D structural bracket problem shown in Figure 6.7. The finite element model contains 194,925 degrees of freedom (N = 194,925) and the elements in the matrix, K, are real numbers. Results were computed on a cluster of 1 – 6 personal computers (PCs) running under Windows environments with Intel Pentium. It should be noted that the DIPSS software was not ported to the PC cluster, but the DD formulation was programmed (from scratch, in C^{++}) on the PC cluster processors [6.3].

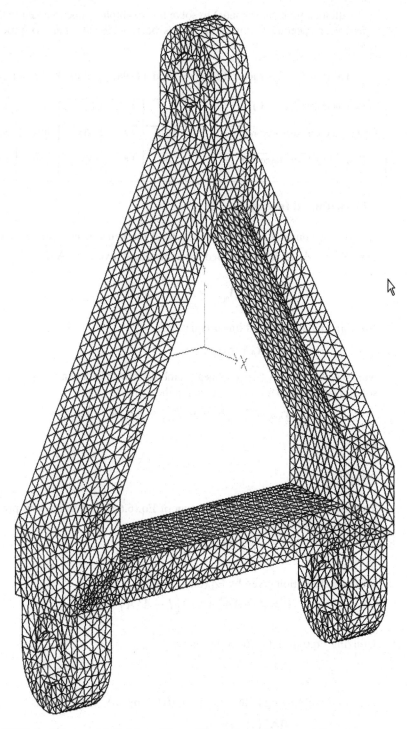

Figure 6.7 Finite Element Model for a Three-Dimensional Structural Bracket

The wallclock time (in seconds) to solve this example is documented in Table 6.4. A super-linear speedup factor of 10.35 has been achieved when six processors were used.

Table 6.4 3-D Structural Bracket Model (194,925 dofs, K = real numbers)

# Processor (Intel PC @ ASU)	1	2	3	4	5	6
Total Walll Clock Time (seconds)	2,670	700	435	405	306	258
Total Speed-Up Factor (seconds)	1.00	3.81	6.14	6.59	8.73	10.35

6.7 Generalized Inverse

First, let us consider some key concepts of the generalized inverse. Given a matrix $A \in R^{m \times n}$, $A^+ \in R^{n \times m}$ is called the generalized inverse of A if

$$A_{m \times n} A^+_{n \times m} A_{m \times n} = A_{m \times n} \qquad (6.62)$$

Now, given the system of linear equations

$$A_{m \times n} \vec{x}_{n \times 1} = \vec{b}_{m \times 1} \qquad (6.63)$$

with $\vec{b} \in$ Range of A ($\vec{b}$ is a linear combinations of independent columns of A), the solution(s) of Eq.(6.63) can be given in the form

$$\vec{x}_{n \times 1} = A^+_{n \times m} \vec{b}_{m \times 1} + \left(I_{n \times n} - A^+_{n \times m} A_{m \times n} \right) \vec{y}_{n \times 1} \qquad (6.64)$$

for $\vec{y} \in R^n$

Proof

To prove that Eq.(6.64) is the solution of Eq.(6.63), one starts with pre-multiplying both sides of Eq.(6.64) with A, thus

$$A\vec{x} = AA^+\vec{b} + A\left(I - A^+ A \right) \vec{y} \qquad (6.65)$$

From the definition given by Eq.(6.62), one has:

$$0 = A - AA^+ A = A\left(I - A^+ A \right) = 0 \qquad (6.66)$$

Utilizing Eq.(6.66), Eq.(6.65) becomes:

$$A\vec{x} = AA^+ b \qquad (6.67)$$

Also, pre-multiplying both sides of Eq.(6.63) by AA^+, one obtains:

$$AA^+ \left(A\vec{x} \right) = AA^+ b \qquad (6.68)$$

or, using the definition of generalized inverse (given by Eq.6.62), one gets:
$$A\vec{x} = AA^+ b \tag{6.69}$$

since
$$A\vec{x} = b \tag{6.70}$$

Comparing the right-hand side of Eq.(6.69) and (6.70), one concludes:
$$AA^+ b = b \tag{6.71}$$

Substituting Eq.(6.71) into Eq.(6.67) will validate the fact that the solution $\vec{x}$, given by Eq.(6.64), does satisfy Eq.(6.63).

Remarks

(a) Let $A = \begin{bmatrix} A_{11} & A_{12} \\ A_{21} & A_{22} \end{bmatrix}$ \hfill (6.72)

Let A = singular matrix, where sub-matrix A_{11} has full rank

Then A^+ = generalized inverse = $\begin{bmatrix} A_{11}^{-1} & 0 \\ 0 & 0 \end{bmatrix}$ \hfill (6.73)

(b) For structural engineering applications, the 2nd part of the solution in Eq.(6.64) represents the "rigid body" displacements.

(c) Let us define
$$(I - A^+ A)\vec{b} \equiv \vec{v}$$

Pre-multiplying both sides of the above equation by matrix A, one obtains:
$$A(I - A^+ A)\vec{b} = A\vec{v}$$
or $(A - AA^+ A)\vec{b} = A\vec{v}$

or $(A - A)\vec{b} = A\vec{v}$

or $\vec{0} = A\vec{v}$, hence $\vec{v} \equiv (I - A^+ A)\vec{b} = \vec{0}$ for an "arbitrary" matrix [A] \hfill (6.74)

(d) Using the definition of the generalized inverse
$$A^+ = \begin{bmatrix} A_{11}^{-1} & 0 \\ 0 & 0 \end{bmatrix}$$
verify that $AA^+ A = A$. \hfill (6.62, repeated)

Proof

$$AA^+A = \begin{bmatrix} A_{11} & A_{12} \\ A_{21} & A_{22} \end{bmatrix} \begin{bmatrix} A_{11}^{-1} & 0 \\ 0 & 0 \end{bmatrix} \begin{bmatrix} A_{11} & A_{12} \\ A_{21} & A_{22} \end{bmatrix}$$

$$AA^+A = \begin{bmatrix} A_{11} & A_{12} \\ A_{21} & A_{21}A_{11}^{-1}A_{12} \end{bmatrix} \tag{6.75}$$

The right-hand side of Eq.(6.75) does not look like matrix A at first glance! However, the following section will prove that $A_{21}A_{11}^{-1}A_{12}$ is indeed the same as A_{22}, and therefore, AA^+A is equal to A.

One starts with:

$$A = \begin{bmatrix} A_{11} & A_{12} \\ A_{21} & A_{22} \end{bmatrix} \tag{6.76}$$

Through the standard Gauss elimination (or factorization) procedures, we can convert Eq.(6.76) into upper-triangular blocks by first pre-multiplying the top portion of Eq.(6.76) with $-A_{21}A_{11}^{-1}$. Then, the results will be added to the bottom portion of Eq.(6.76) to obtain:

$$A_{factorized} = \begin{bmatrix} A_{11} & A_{12} \\ 0 & A_{22} - A_{21}A_{11}^{-1}A_{12} \end{bmatrix} \tag{6.77}$$

Since matrix $[A]$ is singular, and A_{11} has the full rank (or rank of $[A]$ = rank of A_{11}), it implies that A has some dependent row(s). Therefore, during the factorization process, the remaining (dependent) row(s) will have all zero values, hence

$$A_{22} - A_{21}A_{11}^{-1}A_{12} = [0] \tag{6.78}$$

$$\text{or} \quad A_{21}A_{11}^{-1}A_{12} = A_{22} \tag{6.79}$$

Therefore, the right-hand side of Eq.(6.75) is identical to $[A]$.

Example 1

Considering the following system:

$$[A]_{1\times3}\ \vec{x}_{3\times1} = \vec{b}_{1\times1} \tag{6.80}$$

where

$$A \in R^{1 \times 3}, \ x \in R^3$$

Eq.(6.80) represents one equation with three unknowns. Hence, an infinite number of solutions can be expected.

Given $A = \begin{bmatrix} 1, 2, 3 \end{bmatrix}$ (6.81)

$$A^+ = \begin{Bmatrix} 1/6 \\ 1/6 \\ 1/6 \end{Bmatrix} \ \rightarrow \ A^+A = \left(\frac{1}{6}\right)\begin{bmatrix} 1 & 2 & 3 \\ 1 & 2 & 3 \\ 1 & 2 & 3 \end{bmatrix}$$ (6.82)

Hence $I - A^+A = \left(\frac{1}{6}\right)\begin{bmatrix} 5 & -2 & -3 \\ -1 & 4 & -3 \\ -1 & -2 & 3 \end{bmatrix}$ (6.83)

From Eq.(6.64), one obtains:

$$\vec{x} = \left(\frac{b}{6}\right)\begin{Bmatrix} 1 \\ 1 \\ 1 \end{Bmatrix} + \left(\frac{1}{6}\right)\begin{bmatrix} 5 & -2 & -3 \\ -1 & 4 & -3 \\ -1 & -2 & 3 \end{bmatrix}\begin{Bmatrix} y_1 \\ y_2 \\ y_3 \end{Bmatrix}$$ (6.84)

For $\vec{y} = \begin{Bmatrix} y_1 \\ y_2 \\ y_3 \end{Bmatrix} = \begin{Bmatrix} 0 \\ 0 \\ 0 \end{Bmatrix}$ (6.85)

then a solution is:

$$\vec{x} = \left(\frac{b}{6}\right)\begin{Bmatrix} 1 \\ 1 \\ 1 \end{Bmatrix}$$ (6.86)

For another $\vec{y} = \begin{Bmatrix} 1 \\ 0 \\ 0 \end{Bmatrix}$ (6.87)

then $\vec{x} = \left(\frac{b}{6}\right)\begin{Bmatrix} 1 \\ 1 \\ 1 \end{Bmatrix} + \left(\frac{1}{6}\right)\begin{Bmatrix} 5 \\ -1 \\ -1 \end{Bmatrix} = \begin{Bmatrix} (b+5)/6 \\ (b-1)/6 \\ (b-1)/6 \end{Bmatrix}$ (6.88)

Example 2

Considering the system of linear equation $Ax = b$, where

$$A = \begin{bmatrix} 1 & 0 & 1 \\ 0 & 1 & 1 \\ 0 & 0 & 0 \end{bmatrix} \quad \text{and} \quad \vec{b} = \begin{Bmatrix} b_1 \\ b_2 \\ 0 \end{Bmatrix} \quad\quad (6.89)$$

It should be noted that the right-hand-side vector $\vec{b} = \begin{Bmatrix} b_1 \\ b_2 \\ 0 \end{Bmatrix}$ is chosen so that

$\vec{b} \in$ range of $[A] = Ker\left[A^T\right]^{\perp}$. In other words, $\vec{b}$ is a linear combination of the first two linearly independent columns of [A], thereby guaranteeing the existence of a solution $\vec{x}$. Furthermore, since Eq.(6.89) represents a system of linear equations, with two independent equations and three unknowns, its solution will not be unique.

The given matrix A can be partitioned as:

$$A = \begin{bmatrix} [A_{11}] & [A_{12}] \\ [A_{21}] & [A_{22}] \end{bmatrix} = \begin{bmatrix} \begin{bmatrix} 1 & 0 \\ 0 & 1 \end{bmatrix} & \begin{bmatrix} 1 \\ 1 \end{bmatrix} \\ [0 \quad 0] & [0] \end{bmatrix} \quad\quad (6.90)$$

Thus

$$A^+ = \begin{bmatrix} [A_{11}]^{-1} & [0] \\ [0] & [0] \end{bmatrix} = \begin{bmatrix} \begin{bmatrix} 1 & 0 \\ 0 & 1 \end{bmatrix} & \begin{bmatrix} 0 \\ 0 \end{bmatrix} \\ [0 \quad 0] & [0] \end{bmatrix} \quad\quad (6.91)$$

Hence

$$I - A^+A = \begin{bmatrix} 0 & 0 & -1 \\ 0 & 0 & -1 \\ 0 & 0 & +1 \end{bmatrix} \quad\quad (6.92)$$

From Eq.(6.64), one obtains:

$$\vec{x} = \begin{bmatrix} 1 & 0 & 0 \\ 0 & 1 & 0 \\ 0 & 0 & 0 \end{bmatrix} \begin{Bmatrix} b_1 \\ b_2 \\ 0 \end{Bmatrix} + \begin{bmatrix} 0 & 0 & -1 \\ 0 & 0 & -1 \\ 0 & 0 & +1 \end{bmatrix} \begin{Bmatrix} y_1 \\ y_2 \\ y_3 \end{Bmatrix} \quad\quad (6.93)$$

$$\text{or} \quad \begin{Bmatrix} x_1 \\ x_2 \\ x_3 \end{Bmatrix} = \begin{Bmatrix} b_1 \\ b_2 \\ 0 \end{Bmatrix} + \begin{Bmatrix} -y_3 \\ -y_3 \\ +y_3 \end{Bmatrix} = \begin{Bmatrix} b_1 \\ b_2 \\ 0 \end{Bmatrix} + (y_3) \begin{Bmatrix} -1 \\ -1 \\ 1 \end{Bmatrix} \tag{6.94}$$

6.8 FETI Domain Decomposition Formulation [6.4 – 6.7]

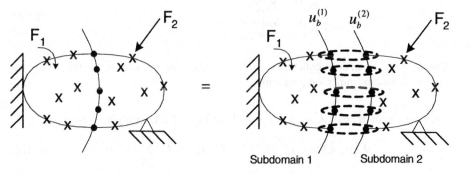

Figure 6.8. Large Finite Element Model Is Decomposed into Two Disconnected
Sub-Domains

Figure 6.8 illustrates a large finite element model that is decomposed into two disconnected sub-domains. The boundary (or interfaced) degree-of-freedom (dof) of the two sub-domains are represented by the vectors $u_b^{(1)}$ and $u_b^{(2)}$, respectively. Similarly, the interior dof and the load vector for the two sub-domains are represented by $u_i^{(1)}$, $f_i^{(1)}$ and $u_i^{(2)}$, $f_i^{(2)}$, respectively. The matrix equilibrium equations for the entire domain can be represented in the partitioned form as:

$$\begin{bmatrix} K_{ii}^{(1)} & K_{ib}^{(1)} & & \\ K_{bi}^{(1)} & K_{bb}^{(1)} & & \\ & & K_{bb}^{(2)} & K_{bi}^{(2)} \\ & & K_{ib}^{(2)} & K_{ii}^{(2)} \end{bmatrix} \begin{Bmatrix} u_i^{(1)} \\ u_b^{(1)} \\ u_b^{(2)} \\ u_i^{(2)} \end{Bmatrix} = \begin{Bmatrix} f_i^{(1)} \\ f_b^{(1)} \\ f_b^{(2)} \\ f_i^{(2)} \end{Bmatrix} \tag{6.95}$$

with the following constrained equations

$$u_b^{(1)} = u_b^{(2)} \tag{6.96}$$

For a more general case, where a large domain is decomposed into "s" sub-domains, Eqs.(6.95 – 6.96) become:

$$K^e u^e = f^e \tag{6.97}$$

and $\quad \sum_{j=1}^{s} B^{(j)} u^{(j)} = 0$ (6.98)

where

$$K^e = \left[K^{(1)}, K^{(2)}, ..., K^{(s)} \right]$$ (6.99)

$$K^{(1)} \equiv \begin{bmatrix} K_{ii}^{(1)} & K_{ib}^{(1)} \\ K_{bi}^{(1)} & K_{bb}^{(1)} \end{bmatrix}$$ (6.100)

The system of matrix equations (6.97 – 6.98) can be casted as the following unconstrained optimization problem:

Find the vector $\vec{u}$ and $\vec{\lambda}$ such that it will minimize the Lagrangian function:

$$L(u,\lambda) \equiv \frac{1}{2} u^{e^T} K^e u^e - f^{e^T} u^e + \lambda^T \sum_{j=1}^{s} B^{(j)} u^{(j)}$$ (6.101)

Applying the necessary conditions on the Lagrangian function, one gets:

$$\frac{\partial L}{\partial \vec{u}} = K^e u^e - f^e + B^{e^T} \lambda = \vec{0}$$ (6.102)

$$\frac{\partial L}{\partial \vec{\lambda}} = B^e u^e = \vec{0}$$ (6.103)

Eqs.(6.102 – 6.103) can be represented as:

$$\begin{bmatrix} K^e & B^{e^T} \\ B^e & 0 \end{bmatrix} \begin{Bmatrix} u^e \\ \lambda \end{Bmatrix} = \begin{Bmatrix} f^e \\ 0 \end{Bmatrix}$$ (6.104)

It should be emphasized that the coefficient matrix of Eq.(6.104) is indefinite and non-singular. Hence, a unique solution for u^e & λ can be expected.

Remarks

It is _not_ recommended to solve the sparse matrix Eq.(6.104) by the "direct sparse" equation solver, due to the following reasons:

(a) The coefficient matrix of Eq.(6.104) has a large dimension, thus the number of fill-in terms introduced by the direct equation solver can be quite large, which may require a lot of computer memory and computational efforts [1.9].

(b) The coefficient matrix is also _indefinite,_ thus special pivoting strategies are normally required.

(c) The vector $\vec{\lambda}$, shown in Eq.(6.104), represents the interacting forces between disconnected sub-domains.

Let N_s and N_f denote the total number of sub-domains, and the number of "floating" sub-domains, respectively. We must deal with N_f $\left(\leq N_s\right)$ of the local Newmann problems (see the top portion of Eq.[6.104]):

$$\mathbf{K}^{(s)}\mathbf{u}^{(s)}=\mathbf{f}^{(s)}-\mathbf{B}^{(s)^T}\lambda \text{, for s = 1, 2, ..., } N_f \tag{6.105}$$

The above matrix equation is ill-posed since the "floating" stiffness matrix of the $s^{\underline{th}}$ sub-domain is <u>singular</u>. To guarantee the solvability of Eq.(6.105), we require that the right-hand-side vector (also refer to Eq.6.63) will satisfy the following condition

$$f^{(s)}-B^{(s)^T}\lambda \in Range\left(K^{(s)}\right)=Kernel\left(K^{(s)^T}\right)^{\perp}=Kernel\left(K^{(s)}\right)^{\perp} \tag{6.106}$$

where the last equality follows from the symmetry of $K^{(s)}$. In other words, we require the right-hand-side vector $\left\{f^{(s)}-B^{(s)^T}\lambda\right\}$ to be a linear combination of independent columns of $K^{(s)}$. The condition Eq.(6.106) guarantees the existence of a solution $u^{(s)}$ in Eq.(6.104).

For "fixed" sub-domain, one can solve for $u^{(s)}$ from Eq.(6.102)

$$u^{(s)}=\left[K^{(s)}\right]^{-1}\cdot\left\{f^{(s)}-B^{(s)^T}\lambda\right\} \tag{6.107}$$

For "float" sub-domain, $u^{(s)}$ can be computed by (refer to Eq.[6.64])

$$u^{(s)}=\left[K^{(s)}\right]^{+}\cdot\left\{f^{(s)}-B^{(s)^T}\lambda\right\}+R^{(s)}\alpha^{(s)} \tag{6.108}$$

where

$$K^{(s)}=\begin{bmatrix} K_{11}^{(s)} & K_{12}^{(s)} \\ K_{21}^{(s)} & K_{22}^{(s)} \end{bmatrix} \tag{6.109}$$

and the generalized inverse $\left[K^{(s)}\right]^{+}$ is defined as (see Eq.[6.73])

$$\left[K^{(s)}\right]^{+}=\begin{bmatrix} K_{11}^{(s)^{-1}} & 0 \\ 0 & 0 \end{bmatrix} \tag{6.110}$$

By referring to Eq.(6.64), one has:

$$\left[R^{(s)}\right]=I-K^{(s)^{+}}K^{(s)} \tag{6.111}$$

$$= I - \begin{bmatrix} K_{11}^{(s)^{-1}} & 0 \\ 0 & 0 \end{bmatrix} \begin{bmatrix} K_{11}^{(s)} & K_{12}^{(s)} \\ K_{21}^{(s)} & K_{22}^{(s)} \end{bmatrix}$$

$$= \begin{bmatrix} I & 0 \\ 0 & I \end{bmatrix} - \begin{bmatrix} I & K_{11}^{(s)^{-1}} \cdot K_{12}^{(s)} \\ 0 & 0 \end{bmatrix}$$

$$\left[R^{(s)} \right] = \begin{bmatrix} 0 & -K_{11}^{(s)^{-1}} \cdot K_{12}^{(s)} \\ 0 & I \end{bmatrix} \tag{6.112}$$

The additional unknown $\alpha^{(s)}$ are determined by using the extra condition in Eq.(6.106) or by referring to Eq.(6.74)

$$\left(I - K^{(s)^+} K^{(s)} \right) \cdot \left\{ f^{(s)} - B^{(s)^T} \lambda \right\} = \vec{0} \; ; s = 1, 2, ..., N_f \tag{6.113}$$

or

$$\left[R^{(s)} \right]^T \cdot \left\{ f^{(s)} - B^{(s)^T} \lambda \right\} = \vec{0} \tag{6.114}$$

Although Eq.(6.113) can be explained from Eq.(6.74), it also has the following physical meaning:

The first term of Eq.(6.113), $\left(I - K^{(s)^+} K^{(s)} \right) = R^{(s)^T}$, represents the "pseudo" rigid body motions while the second term of Eq.(6.113), $\left\{ f^{(s)} - B^{(s)^T} \lambda \right\}$, represents the "pseudo" force. Therefore, the product of these two terms represent the "pseudo" energy (or work) due to rigid body motion, hence the "pseudo" work done is zero.

Substituting Eq.(6.108) into Eq.(6.98), one gets:

$$\sum_{s=1}^{N_s} B^{(s)} \left(K^{(s)^+} \cdot \left\{ f^{(s)} - B^{(s)^T} \lambda \right\} + R^{(s)} \alpha^{(s)} \right) = 0 \qquad \text{or} \tag{6.115}$$

$$\sum_{s=1}^{N_s} B^{(s)} K^{(s)^+} f^{(s)} = \left[\sum_{s=1}^{N_s} B^{(s)} K^{(s)^+} B^{(s)^T} \right] \lambda - \sum_{s=1}^{N_s} B^{(s)} R^{(s)} \alpha^{(s)} \tag{6.116}$$

From Eq.(6.114), one has:

$$-R^{(s)^T} f^{(s)} = -R^{(s)^T} B^{(s)^T} \lambda \text{, where s} = 1, 2, ..., N_f \tag{6.117}$$

Eqs.(6.116, 6.117) can be combined to become:

$$
\begin{bmatrix} F_I & -G_I \\ -G_I^T & 0 \end{bmatrix} \begin{Bmatrix} \lambda \\ \alpha \end{Bmatrix} = \begin{Bmatrix} d \\ -e \end{Bmatrix} \tag{6.118}
$$

where

$$
F_I \equiv \sum_{s=1}^{N_s} B^{(s)} K^{(s)^+} B^{(s)^T} \equiv BK^+ B^T \tag{6.119}
$$

$$
G_I \equiv \left[B^{(1)} R^{(1)}, B^{(2)} R^{(2)}, ..., B^{(N_s)} R^{(N_s)} \right] \equiv BR \tag{6.120}
$$

$$
\alpha \equiv \begin{Bmatrix} \alpha^{(1)} \\ \alpha^{(2)} \\ \cdot \\ \cdot \\ \cdot \\ \alpha^{(N_s)} \end{Bmatrix} \tag{6.121}
$$

$$
d \equiv \sum_{s=1}^{N_s} B^{(s)} K^{(s)^+} f^{(s)} \equiv BK^+ f \tag{6.122}
$$

$$
e \equiv \left[f^{(1)^T} R^{(1)}, f^{(2)^T} R^{(2)}, ..., f^{(N_f)^T} R^{(N_f)} \right]^T \tag{6.123}
$$

and

$$
K^+ = \begin{bmatrix} K^{(1)^+} & & & 0 \\ & \cdot & & \\ & & \cdot & \\ & & & \cdot \\ 0 & & & K^{(N_s)^+} \end{bmatrix} \tag{6.124}
$$

$$R = \begin{bmatrix} R^{(1)} & & & & 0 \\ & \cdot & & & \\ & & \cdot & & \\ & & & \cdot & \\ 0 & & & & R^{(N_s)} \end{bmatrix} \qquad (6.125)$$

Eq.(6.118) is called the <u>dual</u> interface problem[6.4] as λ (Lagrange multiplier vector) is a dual variable.

Table 6.5 Step-by-Step FETI-1 Algorithms

---

0.   Some definitions
ndofs = Number of degree of freedoms (dofs) in a substructure.
nbdof = Number of boundary dofs in a substructure.
nbdofall = Number of total boundary dofs.
ninfeq = Number of total interface equations.
irank = rank of K_{11} matrix of a substructure.

1.   Obtain the stiffness matrix ($K^{(s)}$(ndofs,ndofs)) and force vector ($F^{(s)}$(ndofs)) for each substructure.

Since there is no partitioning in the substructure's stiffness matrix ($K^{(s)}$) (there is no need to distinguish boundary and interior dofs), only $K^{(s)}$ of each substructure will be stored. However, to make use of the old METIS partitioning code, we will put boundary interface dofs at the beginning of the dofs list.

2.   Find the rank of each substructure's stiffness matrix.

We will need a subroutine to calculate the rank of the matrix K for each substructure.

3.   Find the Boolean Transformation integer matrix (B) for each substructure separately (equivalent to find interface equations).

The size of B will be (ninfeq, ndofs). Before we know the information about this matrix, we will have to find the number of substructures attached to each interface dofs. If 2 substructures are attached to a dof, we will need 1 equation to satisfy the interface displacement at that dof. If 3 substructures are attached to a dof, we will need 2 equations to satisfy the interface displacement at that dof. In our current METIS implementation [3.3] for breaking up the original, large structure into "s" smaller sub-domains, we have already obtained this information. By using this information, we can create 3 integer arrays ib, jb, ibvalue to completely describe the sparse matrix $B^{(s)}$ (see Chapter 4).

$$\left[\begin{bmatrix} B^{(1)} \end{bmatrix} \begin{bmatrix} B^{(2)} \end{bmatrix} \begin{bmatrix} B^{(3)} \end{bmatrix} \dots \begin{bmatrix} B^{(s)} \end{bmatrix} \right] \cdot \{x\} = \{0\}$$

4.   Find K_{11}, K_{12}, K_{21} and K_{22} of each substructure.

After obtaining the rank of the stiffness matrix for each substructure, we can now create K_{11}(rank, rank), K_{12}(rank, ndofs-rank), K_{21}(ndofs-rank, rank) and K_{22}(ndofs-rank, ndofs-rank). In this step, we need to identify which rows (and columns) of $K^{(s)}$ are independent (contributed to the full rank matrix). According to our initial experience, selecting different set of independent dofs will make no difference in the final answer.

5.   Find the generalized inverse K^+ of each substructure.

For floating substructures,

$$K^{(s)+} = \begin{bmatrix} K_{11}^{(s)-1} & 0 \\ 0 & 0 \end{bmatrix}$$

For stable substructures, $K^{(s)+}$ is just the regular inverse of the $K^{(s)}$ matrix.

In this step, we will factorize K_{11} of each substructure. ODU's symmetrical direct sparse solver, MA28, MA47 or SuperLU [1.9,4.2,3.2,6.8] can be used for this purpose.

6.   Find the rigid body modes [R] of each substructure.

In this step, there will be no [R] for stable substructures.

From $R^{(s)} = \begin{bmatrix} -K_{11}^{(s)-1} K_{12}^{(s)} \\ I_{22} \end{bmatrix}_{ndofs \times (ndofs-rank)}$

The size of identity matrix block will be the same as K_{22}.
By performing forward and backward solutions on the "factorized $K_{ii}^{(s)}$", we can get [R] matrix. Also, number of required forward and backward (F/B) solutions in this step is (ndofs-rank), which is equal to the number of columns in matrix $K_{12}^{(s)}$.

7.   Perform $B_{ninfeq \times ndofs}^{(s)} \cdot R_{ndofs \times \{ndofs-rank\}}^{(s)}$ for each substructure.

We need sparse matrix times matrix subroutine for this operation. The size of the resulted matrix will be (ninfeq, ndofs-rank)

8.   Find F_1, G_1, d and e

$$F_I = \sum_{s=1}^{Ns} B^{(s)} K^{(s)^+} B^{(s)^T}$$ Hence, F_I will be a square matrix whose size is (ninfeq,

ninfeq). Matrix times matrix subroutine will be required for the above operations.

$$G_I = \left[\left[B^{(1)} R^{(1)} \right] \left[B^{(2)} R^{(2)} \right] .. \left[B^{(Nf)} R^{(Nf)} \right] \right]$$. Thus, G_I will be a matrix whose size is
(ninfeq, summation of (ndofs-rank) for all substructures).

$$d = \sum_{s=1}^{Ns} B^{(s)} K^{(s)^+} f^{(s)}$$. Thus, the size of d will be the number of interface dofs

$$e^{(s)} = \left[f^{(s)^T} R^{(s)} \right]^T \quad ; s = 1, ..., N_f . \text{ Hence, } e = \left[e^{(1)} \quad e^{(2)} \quad ... \quad e^{(Nf)} \right]^T$$

In this step, each processor will calculate its own F_I, G_I ,d and e. Then, there will be
communication between processors to combine the processors' results.
9.   Iterative solver or direct solver for λ and α .

From $\begin{bmatrix} F_I & -G_I \\ -G_I^T & 0 \end{bmatrix} \begin{bmatrix} \lambda \\ \alpha \end{bmatrix} = \begin{bmatrix} d \\ -e \end{bmatrix}$, we will need a solver that can handle indefinite

system [3.2] to solve the problem. The size of the zero sub-matrix should be the
summation of (ndofs-rank) for each substructure.

For the iterative solver, PCPG is recommended. (see Ref. 6.4, or Section 6.9)

10.   Find all displacements of each substructure.

After λ and α are found, we can find the sub-domains' displacements by the
following formula.

$$u^{(s)} = K^{(s)^+} (f^{(s)} - B^{(s)^T} \lambda) + R^{(s)} \alpha^{(s)}$$

The above formula is for a floating substructure. For stable substructures, the last
term will be dropped to calculate the interior displacements.

The operation in this step will require Forward/Backward (F/B) solution and matrix
times vector subroutines.

6.9 Preconditioned Conjugate Projected Gradient (PCPG) of the Dual Interface Problem [6.4]

The dual interface problem, expressed by Eq.(6.104), has some undesirable properties for a "direct" sparse equation solution:

(a) The coefficient matrix is indefinite, hence special pivoting strategies are usually required[1.9].

(b) The dimension of the coefficient matrix is quite large, which may introduce even more non-zero (fill-in) terms during the factorization process. Therefore, excessive computer storage and computational efforts are required.

Because of the above reasons, an "iterative" equation solver, such as the popular family of Conjugate Gradient (CG) optimization algorithms[6.9] is recommended. However, before applying the CG algorithms, we have to transform Eq.(6.104) into a semi-positive definite system and eliminate the (constrained) Eq.(6.118) $G_I^T \lambda = e$ (since CG is an unconstrained optimization algorithm). This is the motivation of the Conjugate Projected Gradient (CPG) and its close relative Pre-Conditioned Conjugate Projected Gradient (PCPG) algorithm [6.4].

In the CPG algorithm, a projector matrix P removes from λ^k the component that does not satisfy the constraint. To do this, we define an orthogonal projection P onto Kernel(G_I^T)

$$P \equiv I - G_I \left(G_I^T G_I \right)^{-1} G_I^T \tag{6.126}$$

Notice that

$$G_I^T P = G_I^T \left\{ I - G_I \left(G_I^T G_I \right)^{-1} G_I^T \right\} \tag{6.127}$$

$$= G_I^T - \left(G_I^T G_I \right) \left(G_I^T G_I \right)^{-1} G_I^T \tag{6.128}$$

$$= G_I^T - G_I^T$$

$$G_I^T P = 0 \tag{6.129}$$

Similarly

$$PG_I = \left\{ I - G_I \left(G_I^T G_I \right)^{-1} G_I^T \right\} G_I \tag{6.130}$$

$$= G_I - G_I \left(G_I^T G_I \right)^{-1} \left(G_I^T G_I \right) \tag{6.131}$$

$$= G_I - G_I \tag{6.132}$$

$$PG_I = 0 \tag{6.133}$$

Now, partition $\lambda \in Kernel\left(G_I^T\right) + Range\left(G_I\right)$ accordingly and write

$$\lambda = \lambda_1 + \lambda_2 \qquad (6.134)$$

where

$$\lambda_1 \in Kernel\left(G_I^T\right) \text{ and } \lambda_2 \in Range\left(G_I\right) \qquad (6.135)$$

Pre-multiplying the top portion of Eq.(6.118) by P, one gets:

$$PF_I\lambda - \left(PG_I\right)\alpha = Pd \qquad (6.136)$$

Utilizing Eq.(6.133), the above equation will reduce to

$$PF_I\lambda = Pd \qquad (6.137)$$

Thus, the dual interface Eq.(6.118) is transformed to

$$PF_I\lambda = Pd$$
$$G_I^T\lambda = e \qquad (6.138)$$

The key idea of the CPG is to select the proper initial value λ^0 so that it will automatically guarantee to satisfy the (constrained) condition (shown in the 2nd half of [Eq.6.138]) in subsequent iterations. A particular solution for λ^0 is of the form:

$$\lambda^0 = G_I\left(G_I^TG_I\right)^{-1}e \qquad (6.139)$$

Remarks

(a) If the constrained condition is satisfied by λ_a and λ_b, respectively, then we will get $G_I^T\lambda_a = e$ and $G_I^T\lambda_b = e$, respectively. However, $\lambda_a + \lambda_b$ will **not** satisfy the constrained condition, since

$$G_I^T\left(\lambda_a + \lambda_b\right) = 2e \neq e$$

(b) In Eq.(6.139), the matrix product $G_I^TG_I$ is symmetrical and positive semi-definite.

If G_I is invertible, then $G_I^TG_I$ is positive definite

(c) The initial selection of λ^0 (according to Eq.[6.139]) is very clever since the constraint will be automatically satisfied!

$$G_I^T\lambda^0 = G_I^T\left\{G_I\left(G_I^TG_I\right)^{-1}e\right\} = \left(G_I^TG_I\right)\left(G_I^TG_I\right)^{-1}e = e$$

Given the initial value λ^0, when the CPG algorithm is applied to $PF_I\lambda = Pd$, then the initial residual with λ^0 is computed as:

$$\omega^0 \equiv Pd - PF_I\lambda^0 = P\left(d - F_I\lambda^0\right) \qquad (6.140)$$

Note that

$$G_I^T \omega^0 = G_I^T \left\{ P\left(d - F_I \lambda^0\right) \right\} = 0 \qquad (6.141)$$

The right-hand side of Eq.(6.141) is zero, due to the consequence of Eq.(6.129), hence $G_I^T P = 0$.

The subsequent λ^k belongs to (or a linear combinations of vectors)

$$\text{span} \left\{ \omega^0, \left(PF_I\right)\omega^0, \left(PF_I\right)^2 \omega^0, ..., \left(PF_I\right)^{k-1} \omega^0 \right\}$$

Therefore, if λ^k is generated by CPG algorithm and applied to $PF_I \lambda = Pd$, then $G_I^T \lambda = 0$. Consequently, a solution for Eq.(6.138) can be obtained iteratively by applying CPG to the following homogeneous problem:

$$\begin{aligned} PF_I \lambda &= Pd \\ G_I^T \lambda &= 0 \end{aligned} \qquad (6.142)$$

Equation(6.142) can be solved iteratively, based on the fact that PF_I is symmetrical on the space of Lagrange multipliers λ, satisfying the constraint $G_I^T \lambda = 0$.

(d) The homogeneous form of the constraint $G_I^T \lambda = 0$ (see Eq.[6.142]) implies that if λ_a, and λ_b satisfies the homogeneous constraint (meaning $G_I^T \lambda_a = 0 = G_I^T \lambda_b$), then so does $\lambda_a + \lambda_b$ (meaning $G_I^T \left\{ \lambda_a + \lambda_b \right\} = 0$, also),

(e) The above remark is important, especially in referring to Eq.(6.134).

(f) Note that $\lambda^0 \in Range\left(G_I\right)$, so that, according to Eq.(6.134), we can write:

$$\lambda = \lambda^0 + \overline{\lambda} \qquad (6.143)$$

for some $\overline{\lambda} \in Kernel\left(G_I\right)$.

Since P is a projection onto Kernel (G_I^T), hence

$$P\overline{\lambda} = \overline{\lambda} \qquad (6.144)$$

Inserting Eq.(6.143) into the top portion of Eq.(6.118), one gets:

$$F_I \left(\lambda = \lambda^0 + \overline{\lambda} \right) - G_I \alpha = d \qquad (6.145)$$

or

$$F_I \overline{\lambda} = \left(d - F_I \lambda^0\right) + G_I \alpha \tag{6.146}$$

Pre-multiplying both sides of the above equation by P^T, one obtains:

$$P^T F_I \overline{\lambda} = P^T \left(d - F_I \lambda^0\right) + P^T G_I \alpha \tag{6.147}$$

or

$$P^T F_I \overline{\lambda} = P^T \left(d - F_I \lambda^0\right) + \left(G_I^T P\right)^T \alpha \tag{6.148}$$

Utilizing Eq.(6.129), then the above equation will simplify to:

$$\left[P^T F_I\right] \overline{\lambda} = P^T \left(d - F_I \lambda^0\right) \tag{6.149}$$

Equation(6.149) represents an alternative interface problem

$$\left[P^T F_I\right] \overline{\lambda} = P^T \left(d - F_I \lambda^0\right)$$
$$G_I^T \overline{\lambda} = 0 \tag{6.150}$$

(g) Since the subsequent λ^k is a linear combination of elements from the Krylov space

$$\left\{\omega^0, \left(PF_I\right)\omega^0, \left(PF_I\right)^2 \omega^0, ..., \left(PF_I\right)^{k-1} \omega^0\right\}$$

and since $G_I^T \omega^0 = 0$ (see Eq.[6.141]), $G_I^T P = 0$ (see Eq.[6.129]), therefore $G_I^T \lambda^k = 0$.

In summary:

If our initial value for λ^0 is given by Eq.(6.139), then it will automatically satisfy the constraint condition

$$G_I^T \lambda = e$$

In the subsequent kth iteration, $G_I^T \lambda^k$ (see the bottom part of Eq.[6.118]) = 0, thus the initially satisfied constraint condition $G_I^T \lambda = e$ will remain to be satisfied since the unwanted component λ^k is zero ($G_I^T \lambda^k = 0$).

In this section, the conjugate projected gradient (CPG) method is described in Table 6.6

Table 6.6 The CPG Iterative Solver [6.4, 6.11]

1. Initialize
$$\lambda^0 = G_I \left(G_I^T G_I \right)^{-1} e$$
$$r^0 = d - F_I \lambda^0 \quad \dots \dots \dots \text{(residual)}$$
2. Iterate k = 1, 2, … until convergence
$$w^{k-1} = P r^{k-1} \quad \dots \dots \dots \dots \text{(projected residual)}$$
$$\varsigma^k = w^{k-1^T} w^{k-1}/w^{k-2^T} w^{k-2} \quad \left(\varsigma^1 = 0 \right)$$
$$p^k = w^{k-1} + \varsigma^k p^{k-1} \quad \left(p^1 = w^0 \right)$$
$$v^k = w^{k-1^T} w^{k-1}/p^{k^T} F_I p^k$$
$$\lambda^k = \lambda^{k-1} + v^k p^k$$
$$r^k = r^{k-1} - v^k F_I p^k$$
Since solving the projected problem is equivalent to the original interface problem,
$$F_I \lambda^k - G_I \alpha^k = d \qquad \Rightarrow G_I \alpha^k = -r^k$$
$$\Rightarrow G_I^T G_I \alpha^k = -G_I^T r^k$$
$$\Rightarrow \alpha^k = -\left(G_I^T G_I \right)^{-1} G_I^T r^k$$
Using the obtained (λ^k, α^k), an approximated solution to $u^{(s)}$ can be found.

6.10 Automated Procedures for Computing Generalized Inverse and Rigid Body Motions

The stiffness matrix of the "floating" sub-domain is singular, due to the fact that there are not enough (support) constraints to prevent its rigid body motion. To facilitate the discussion, let's assume the "floating" sub-domain's stiffness matrix is given as:

$$
\left[K_{float}\right] =
\begin{bmatrix}
1 & 2 & -3 & 2 & -2 & -3 & -2 \\
2 & 4 & -6 & 4 & -4 & -6 & -4 \\
-3 & -6 & 9 & -6 & 6 & 9 & 6 \\
2 & 4 & -6 & 5 & -1 & -5 & -7 \\
-2 & -4 & 6 & -1 & 13 & 9 & -5 \\
-3 & -6 & 9 & -5 & 9 & 13 & 9 \\
-2 & -4 & 6 & -7 & -5 & 9 & 27
\end{bmatrix}
\begin{matrix}
\\ = 2 \cdot row1 \\ = -3 \cdot row1 \\ \\ = -8 \cdot row1 + 3 \cdot row4 \\ \\ \\
\end{matrix}
\tag{6.151}
$$

In Eq.(6.151), it can be observed that rows number 2, 3, and 5 are dependent rows (and columns). Thus, Eq.(6.151) can be re-arranged as:

$$
\left[K_{float}\right] \equiv
\begin{bmatrix}
K_{11} & K_{12} \\
K_{21} & K_{22}
\end{bmatrix}
\tag{6.152}
$$

where

$$
\left[K_{11}\right] =
\begin{bmatrix}
1 & 2 & -3 & -2 \\
2 & 5 & -5 & -7 \\
-3 & -5 & 13 & 9 \\
-2 & -7 & 9 & 27
\end{bmatrix}
\tag{6.153}
$$

$$
\left[K_{12}\right] = \left[K_{21}\right]^{T} =
\begin{bmatrix}
2 & -3 & -2 \\
4 & -6 & -1 \\
-6 & 9 & 9 \\
-4 & 6 & -5
\end{bmatrix}
\tag{6.154}
$$

$$
\left[K_{22}\right] =
\begin{bmatrix}
4 & -6 & -4 \\
-6 & 9 & 6 \\
-4 & 6 & 13
\end{bmatrix}
\tag{6.155}
$$

Sub-matrix $\left[K_{11}\right]$ has a full rank (rank = 4).

Once sub-matrices $\begin{bmatrix} K_{11} \end{bmatrix}$ and $\begin{bmatrix} K_{12} \end{bmatrix}$ can be identified, the generalized inverse matrix, and the corresponding rigid body matrix can be computed from Eq.(6.110), and Eq.(6.112), respectively. Both Eqs.(6.110, 6.112) require the computation of the "factorized $\begin{bmatrix} K_{11} \end{bmatrix}$." For efficient computation of the "factorized $\begin{bmatrix} K_{11} \end{bmatrix}$," and recovering the original sub-matrix $\begin{bmatrix} K_{12} \end{bmatrix}$, the following step-by-step procedure is recommended [6.11].

Step 1: The "floating" sub-domain's stiffness matrix $\begin{bmatrix} K_{float} \end{bmatrix}$ (see Eq.6.151) is given in the sparse formats.

Step 2: The symmetrical, floating stiffness matrix (shown in Eq.6.151) can be factorized by the familiar sparse LDL^T algorithms (see [1.9]), with the following <u>minor</u> modifications:

    (a) Whenever a <u>dependent</u> i^{th} <u>row</u> is encountered (such as the factorized $u_{ii} = 0$), then the following things need to be done:

        a.1 Record the dependent row number(s). For the data given by Eq.(6.151), the dependent rows are rows number 2, 3, and 5.

        a.2 Set all the non-zero terms of the factorized i^{th} row (of L^T) to zero.

        a.3 Set $\dfrac{1}{u_{ii}} \equiv D_{ii} = 0$ (6.156)

    (b) Whenever an independent i^{th} row is encountered, the factorized i^{th} row will have contributions from all appropriate previously factorized rows. However, contributions from the previously factorized (dependent) rows will be ignored.

    Thus, when Step 2 is completed, the (LDL^T) factorized matrix (for the data shown in Eq.[6.151]) can be computed as:

$$[U] \equiv \begin{bmatrix} 1 & 2 & -3 & 2 & -2 & -3 & -2 \\ . & 0 & 0 & 0 & 0 & 0 & 0 \\ . & . & 0 & 0 & 0 & 0 & 0 \\ . & . & . & 1 & 3 & 1 & -3 \\ . & . & . & . & 0 & 0 & 0 \\ . & . & . & . & . & 0.333 & 2 \\ . & . & . & . & . & . & 0.5 \end{bmatrix} \qquad (6.157)$$

Step 3: Extract the "factorized $\begin{bmatrix} K_{11} \end{bmatrix}$" from Eq.(6.157), and obtain sub-matrices $\begin{bmatrix} K_{12} \end{bmatrix}$ and $\begin{bmatrix} K_{22} \end{bmatrix}$.

$$IU = \{1, 4, 6, 7, 7, 11, 16, 22\}^T \qquad (6.158)$$

$$JU = \left\{ (2,3,4),(3,4),(4),(1,2,3,4),(1,2,3,4,5),(1,2,3,4,5,6) \right\}^{T} \text{ (6.159)}$$

$\underleftrightarrow{}$ Column numbers of factorized matrix [K$_{11}$] $\underleftrightarrow{}$ Row numbers of matrices [K$_{12}$] and [K$_{22}$]

Equations(6.158 – 6.159) describe the non-zero locations for the factorized matrix [K$_{11}$] and for the matrices [K$_{12}$] and [K$_{22}$] (see Eqs.6.154 – 6.155). Eqs.(6.158 – 6.159) can be further understood by referring to the subsequent Eqs.(6.160 – 6.162).

$$[U] = \begin{bmatrix} \frac{1}{1} & 2 & -3 & -2 & 2 & -3 & -2 \\ . & \frac{1}{1} & 1 & -3 & 4 & -6 & -1 \\ . & . & \frac{1}{0.333} & 2 & -6 & 9 & 9 \\ . & . & . & \frac{1}{0.5} & -4 & 6 & -5 \\ \hline & & & & 4 & -6 & -4 \\ & & & & . & 9 & 6 \\ & & & & . & . & 13 \end{bmatrix} \text{ (6.160)}$$

or, Eq.(6.160) can be symbolically expressed as:

$$[U] = \begin{bmatrix} \left[K_{11} factorized \right] & \left[K_{12} \right] \\ & \left[K_{22} \right] \end{bmatrix} \text{ (6.161)}$$

It should be noted that the factorized $[K_{11}]$ is stored in <u>row-by-row sparse formats</u>. However, the combined matrices $[K_{12}]$ and $[K_{22}]$ are stored in <u>column-by-column formats</u>. Thus, Eq.(6.160) is stored in a 1-Dimensional array as:

$$IU = \begin{bmatrix} \times & 1^{st} & 2^{nd} & 3^{rd} & 7^{th} & 11^{th} & 16^{th} \\ & \times & 4^{th} & 5^{th} & 8^{th} & 12^{th} & 17^{th} \\ & & \times & 6^{th} & 9^{th} & 13^{th} & 18^{th} \\ & & & \times & 10^{th} & 14^{th} & 19^{th} \\ & & & & \times & 15^{th} & 20^{th} \\ & & & & & \times & 21^{th} \\ & & & & & & \times \end{bmatrix} \quad (6.162)$$

The reader should refer to Eq.(6.162) in order to better understand Eqs.(6.158) and (6.159)

Remarks For a better understanding of the details of storage schemes, see the following paragraphs:

1) The computerized, calculated LDL^T for $[K_{11}]$ can be verified by substituting all numerical values into the following relationship:

$$[K_{11}] = [L][D][L]^T$$

$$\begin{bmatrix} 1 & 2 & -3 & -2 \\ 2 & 5 & -5 & -7 \\ -3 & -5 & 13 & 9 \\ -2 & -7 & 9 & 27 \end{bmatrix} = \begin{bmatrix} 1 & 0 & 0 & 0 \\ 2 & 1 & 0 & 0 \\ -3 & 1 & 1 & 0 \\ -2 & -3 & 2 & 1 \end{bmatrix} \begin{bmatrix} \frac{1}{1} & 0 & 0 & 0 \\ 0 & \frac{1}{1} & 0 & 0 \\ 0 & 0 & \frac{1}{0.333} & 0 \\ 0 & 0 & 0 & \frac{1}{0.5} \end{bmatrix} \begin{bmatrix} 1 & 2 & -3 & -2 \\ 0 & 1 & 1 & 3 \\ 0 & 0 & 1 & 2 \\ 0 & 0 & 0 & 1 \end{bmatrix} \quad (6.163)$$

2) Using MATLAB software, the eigen-values of matrix Eq.(6.151) can be computed as:

$$\vec{\lambda} = \{0.0, 0.0, 0.0, 0.2372, 4.9375, 24.9641, 41.8612\}^T$$

Since there are 3 zero eigen-values, it implies there are 3 rigid body modes (or 3 dependent rows/columns) in Eq.(6.151).
If the row-by-row Choleski factorization scheme is applied to Eq.(6.151), we will encounter that the factorized

$u_{22} = 0 = u_{33} = u_{55}$, which indicated that row numbers 2, 3, and 5 are dependent rows. Thus, if we set all factorized terms of rows number 2, 3, and 5 at zero, and "ignoring" these three rows in the factorization of subsequent rows, one obtains the following Choleski factorized matrix $[U]$:

$$[U] = \begin{bmatrix} 1 & 2 & -3 & 2 & -2 & -3 & -2 \\ & 0 & 0 & 0 & 0 & 0 & 0 \\ & & 0 & 0 & 0 & 0 & 0 \\ & & & 1 & 3 & 1 & -3 \\ & & & & 0 & 0 & 0 \\ & & & & & 1.7321 & 3.464 \\ & & & & & & 1.4145 \end{bmatrix}$$

If we delete rows (and columns) number 2, 3, and 5, then the Choleski factorized sub-matrix $[U_{11}]$ can be identified as:

$$[U_{11}] = \begin{bmatrix} 1 & 2 & -3 & -2 \\ 0 & 1 & 1 & -3 \\ 0 & 0 & 1.7321 & 3.464 \\ 0 & 0 & 0 & 1.415 \end{bmatrix}$$

It can also be verified by MATLAB that the following relationship holds:

$$[K_{11}] = [U_{11}]^T \cdot [U_{11}]$$

$$\begin{bmatrix} 1 & 2 & -3 & -2 \\ 2 & 5 & -5 & -7 \\ -3 & -5 & 13 & 9 \\ -2 & -7 & 9 & 27 \end{bmatrix} = \begin{bmatrix} 1 & 0 & 0 & 0 \\ 2 & 1 & 0 & 0 \\ -3 & 1 & 1.7321 & 0 \\ -2 & -3 & 3.464 & 1.415 \end{bmatrix} \cdot \begin{bmatrix} 1 & 2 & -3 & -2 \\ 0 & 1 & 1 & -3 \\ 0 & 0 & 1.7321 & 3.464 \\ 0 & 0 & 0 & 1.415 \end{bmatrix}$$

3) During the LDL^T (or Choleski) factorized process to identify the "dependent rows" (hence the rank of sub-matrix $[K_{11}]$ can be identified), the factorized sub-matrix $[U_{11}]$ of the "floating" sub-domain stiffness matrix $K^{(s)} = \begin{bmatrix} K_{11} & K_{12} \\ K_{21} & K_{22} \end{bmatrix}$ can also be simultaneously identified.

4) The complete FORTRAN source code for "Automated Procedures for Obtaining Generalized Inverse for FETI Formulation" has been developed and described in [6.10].

This FORTRAN source code and its outputs are listed in Table 6.7. Further information about the generalized inverse in conjunction with the Singular Value Decomposition (SVD) is conveniently summarized in Appendix A of this book.

Table 6.7 Generalized Inverse by LDL^T Factorization

```
c
         Implicit real*8 (a-h, o-z)
c
c=================================================================
c
c        Remarks :
c        (a)       Identifying which are dependent rows of a "floating" substructure
c        (b)       Factorizing (by LDL_transpose) of a floating substructure stiffness
c                  Whenever a dependent row is encountered during LDL factored
c                  process, then we just :
c                  [1] set all factorized values of the dependent row to be ZEROES
c                  [2] ignore the dependent row(s) in all future faztorized rows
c        (c)       [K "float"] = [K11]    [K12]
c                               [K21]    [K22]
c                  where [K11] = full rank ( =non-singular )
c        (d)       The LDL_transpose of [K11] can be obtained by taking the results
c                  of part (b) and deleting the dependent rows/columns
c        Author(s) : Prof. Duc T. Nguyen
c        Version :   04-30-2004 (EDUCATIONAL purpose, LDL/FULL matrix is assumed)
c        Stored at : cd ~/cee/*odu*clas*/generalized_inverse_by_ldl.f
c
c=================================================================
c
         dimension u(99,99), idepenrows(99), tempo1(99)
c
         iexample=1                    ! can be 1, or 2, or 3
c
         if (iexample . eq. 1)   n=3
c
         if (iexample . eq. 2)   n=12
c
         if (iexample . eq. 3)   n=7
c
         do 1 i=1,n
         do 2 j=1,n
         u(i,j)=0
2        continue
1        continue
c
         if (iexample . eq. 1)   then
c
         u(1,1)= 2.         ! non-singular case
c        u(1,1)= 1.         !     singular case
         u(1,2)= -1.
         u(2,2)= 2.
         u(2,3)= -1.
         u(3,3)= 1.
```

```
c
      elseif (iexample . eq. 2)    then
c
      u(1,1)= 1.88*10**5
      u(1,2)= -4.91*10**4
      u(1,3)= -1.389*10**5
      u(1,7)= -4.91*10**4
      u(1,8)= 4.91*10**4
c
      u(2,2)= 1.88*10**5
      u(2,6)= -1.389*10**5
      u(2,7)= 4.91*10**4
      u(2,8)= -4.91*10**4
c
      u(3,3)= 1.88*10**5
      u(3,4)= 4.91*10**4
      u(3,5)= -4.91*10**4
      u(3,6)= -4.91*10**4
c
      u(4,4)= 1.88*10**5
      u(4,5)= -4.91*10**4
      u(4,6)= -4.91*10**4
      u(4,8)= -1.389*10**5
c
      u(5,5)= 2.371*10**5
      u(5,7)= -1.389*10**5
      u(5,11)= -4.91*10**4
      u(5,12)= 4.91*10**4
c
      u(6,6)= 3.76*10**5
      u(6,10)= -1.389*10**5
      u(6,11)= 4.91*10**4
      u(6,12)= -4.91*10**4
c
      u(7,7)= 2.371*10**5
      u(7,9)= -4.91*10**4
      u(7,10)= -4.91*10**4
c
      u(8,8)= 3.76*10**5
      u(8,9)= -4.91*10**4
      u(8,10)= -4.91*10**4
      u(8,12)= -1.389*10**5
c
      u(9,9)= 1.88*10**5
      u(9,10)= 4.91*10**4
      u(9,11)= -1.389*10**5
c
      u(10,10)= 1.88*10**5
c
      u(11,11)= 1.88*10**5
      u(11,12)= -4.91*10**4
c
      u(12,12)= 1.88*10**5
```

```
c
        elseif (iexample . eq. 3)    then
c
        u(1,1)= 1.
        u(1,2)= 2.
        u(1,3)= -3.
        u(1,4)= 2.
        u(1,5)= -2.
        u(1,6)= -3.
        u(1,7)= -2.
c
        u(2,2)= 4.
        u(2,3)= -6.
        u(2,4)= 4.
        u(2,5)= -4.
        u(2,6)= -6.
        u(2,7)= -4.
c
        u(3,3)= 9.
        u(3,4)= -6.
        u(3,5)= 6.
        u(3,6)= 9.
        u(3,7)= 6.
c
        u(4,4)= 5.
        u(4,5)= -1.
        u(4,6)= -5.
        u(4,7)= -7.
c
        u(5,5)= 13.
        u(5,6)= 9.
        u(5,7)= -5.
c
        u(6,6)= 13.
        u(6,7)= 9.
c
        u(7,7)= 27.
        Endif
c
        do 4 i=1,n
        do 5 j=1,n
        u(j,i)=u(i,j)
5       continue
4       continue
c
        call generalized_inverse_ldl (n, u, idependrows, ndependrows)
c
        write(6,*) '# dependent rows = ' ,ndependrows
        if (ndependrows .ge. 1)    then
        write(6,*) ' dependent rows = ' ,(idependrows(i) ,i=1, ndependrows)
        endif
c       write(6,*) 'LDL factorized u(- , -) =', ((u(i,j) ,j=i,n) ,i=1,n)
c       extracting & writing the LDL factorized of full rank of [K11]
```

```
c         by deleting the dependent row(s) /column(s) of [u]
          do 52 i=1,n
          iskiprow=0
            do 53 j=1,ndependrows
            if (idependrows(j) .eq. i)    iskiprow=1
53          continue
          if (iskiprow .eq. 1)    go to 52
            icount=0
            do 54 j=i,n
            iskipcol=0
              do 55 k=1,ndependrows
              if (idepenrows(k) .eq. 0)    iskipcol=1
55            continue
            if (iskipcol .eq. 0)    then
            icount=icount+1
            tempo1(icount)=u(i,j)
            endif
54          continue
            write(6,*) 'LDL of [K11] = ' ,(tempo1(k) ,k=1,icount)
52        continue
c
          stop
          end
c
c%%%%%%%%%%%%%%%%%%%%%%
c
          subroutine generalized_inverse_ldl (n, u, idependrows, ndependrows)
          Implicit real*8 (a-h, o-z)
          dimension u(99,*), idepenrows(*)
c
c=====================================================================
c
c         Remarks :
c         (a)      Identifying which are dependent rows of a "floating" substructure
c         (b)      Factorizing (by LDL_transpose) of a floating substructure stiffness
c                  Whenever a dependent row is encountered during LDL factored
c                  process, then we just :
c                  [1] set all factorized values of the dependent row to be ZEROES
c                  [2] ignore the dependent row(s) in all future factorized rows
c         (c)      [K "float"] = [K11]    [K12]
c                                [K21]    [K22]
c                  where [K11] = full rank ( =non-singular )
c         (d)      The LDL_transpose of [K11] can be obtained by taking the results
c                  of part (b) and deleting the dependent rows/columns
c         Author(s) : Prof. Duc T. Nguyen
c         Version :   04-30-2004
c         Stored at : cd ~/cee/*odu*clas*/generalized_inverse_by_ldl.f
c
c=====================================================================
c
          eps=0.0000000001
          do 11 i=2,n
          do 22 k=1,i-1
```

```
              if (dabs( u(k,k) ) .lt. eps)     go to 22   ! check for "previous"
c                                                          ! dependent row(s)
              xmult=u(k,i) /u(k,k)
                 do 33 j=i,n
                 u(i,j)=u(i,j) –xmult*u(k,j)
33               continue
              u(k,i)=xmult
22            continue
c
c=============================================
c
c       to zero out entire dependent row
              if (dabs( u(i,i) ) .lt. eps)     then
              write(6,*) 'dependent row # i, u(i,i) = ' ,i ,u(i,i)
              ndependrows= ndependrows+1
              idependrows(ndependrows)= i
                 do 42 j=i,n
42               u(i,j)=0.
                 do 44 k=1,i-1
44               u(k,i)=0.
              endif
c
c=============================================
c
11            continue
c
              return
              end
c
c%%%%%%%%%%%%%%%%%%%%%%%%
```

Table 6.7   L D L_t factorized of the "full rank" sub-matrix [K11] of Example 3

```
dependent row # i, u(i,i) =     2     0.0E+0
dependent row # i, u(i,i) =     3     0.0E+0
dependent row # i, u(i,i) =     5     0.0E+0
# dependent rows  =    3
dependent rows =     2     3     5

LDL of [K11]   =   1.0  2.0  -3.0  -2.0
LDL of [K11]   =   1.0  1.0  -3.0
LDL of [K11]   =   3.0  2.0
LDL of [K11]   =   2.0
```

```
++++++++++++++++++++++++++++++++++++++++++++++++
```

Table 6.7   L D L_t factorized of the "full rank" sub-matrix [K11] of Example 2

```
dependent row # i, u(i,i) =     10    -8.731149137020111E-11
dependent row # i, u(i,i) =     11    -5.820766091346741E-11
dependent row # i, u(i,i) =     12    -2.9103830456733703E-11
# dependent rows  =    3
dependent rows =     10    11    12
```

LDL of [K11] = 188000.0 -0.26117002127659574 -0.7388297872340426 0.0E+0
0.0E+0 0.0E+0 -0.26117002127659574 0.26117002127659574 0.0E+0
LDL of [K11] = 175176.54255319148 -0.2070856178827499 0.0E+0 0.0E+0
-0.7929143821172502 0.2070856178827499 -0.2070856178827499 0.0E+0
LDL of [K11] = 77864.19232391396 0.6305851063829787 -0.6305851063829787 -1.0
-0.36941489361702123 0.36941489361702123 0.0E+0
LDL of [K11] = 157038.27127659574 -0.11550223476828982 0.0E+0
0.11550223476828982 -1.0 0.0E+0
LDL of [K11] = 204043.26040931543 -0.24063524519998494 -0.7593647548000151
0.0E+0 0.0E+0
LDL of [K11] = 176184.80946068066 -0.211623292466662 -2.0648651936724454E-17
0.0E+0
LDL of [K11] = 78494.47532361932 0.6255217300016221 -0.6255217300016221
LDL of [K11] = 157286.88305692037 -0.11690029517761087
LDL of [K11] = 155137.45100017014

++

Table 6.7   L D L_t factorized of the "full rank" sub-matrix [K11] of Example 1
                                                               (singular case)

dependent row # i, u(i,i) =   3   0.0E+0
dependent rows =   1
dependent rows =   3

LDL of [K11] = 1.0 -1.0
LDL of [K11] = 1.0

++

Table 6.7   L D L_t factorized of the "full rank" sub-matrix [K11] of Example 1
                                                               (non-singular case)

dependent rows =   0

LDL of [K11] = 2.0 -0.5 0.0E+0
LDL of [K11] = 1.5 -0.6666666666666666
LDL of [K11] = 0.33333333333333337

++

6.11 Numerical Examples of a 2-D Truss by FETI Formulation

To illustrate the numerical details of FETI formulation, a 2-D truss structure with its known geometry, element connectivity, support boundary conditions, applied joint loads, etc., is shown in Figure 6.1. All truss members have the same cross-sectional area (A = 5) and Young modulus (E = 10,000,000). Element connectivity information for all 21-truss members can be summarized by:

$$node\_i = \{1,2,3,5,6,7,1,2,3,4,1,2,3,2,3,4,9,6,7,6,7\}^T \quad (6.164)$$

$$node\_j = \{2,3,4,6,7,8,5,6,7,8,6,7,8,5,6,7,10,9,10,10,9\}^T \quad (6.165)$$

The orientation for each of the truss member, specified by the angels γ_1 (shown in Figure 6.5), are given as (in degrees):

$$\gamma_1 = (0\ 0\ 0\ 0\ 0\ 0\ -90\ -90\ -90\ -90\ -45\ -45\ -45\ -135\ -135\ -135\ 0\ -90\ -90\ -45\ -135)^T \quad (6.166)$$

Element stiffness matrices (in global coordinate references) for each of the 21 truss members are computed as (see [Eq.6.17]):

$$K^{e_1} = \begin{bmatrix} 138888.888889 & 0 \\ 0 & 0 \end{bmatrix}$$

$$K^{e_2} = \begin{bmatrix} 138888.888889 & 0 \\ 0 & 0 \end{bmatrix}$$

$$K^{e_3} = \begin{bmatrix} 138888.888889 & 0 \\ 0 & 0 \end{bmatrix}(6.167)$$

$$K^{e_4} = \begin{bmatrix} 138888.888889 & 0 \\ 0 & 0 \end{bmatrix}$$

$$K^{e_5} = \begin{bmatrix} 138888.888889 & 0 \\ 0 & 0 \end{bmatrix}$$

$$K^{e_6} = \begin{bmatrix} 138888.888889 & 0 \\ 0 & 0 \end{bmatrix}(6.168)$$

$$K^{e_7} = \begin{bmatrix} 0 & 0 \\ 0 & 138888.888889 \end{bmatrix}$$

$$K^{e_8} = \begin{bmatrix} 0 & 0 \\ 0 & 138888.888889 \end{bmatrix}$$

$$K^{e_9} = \begin{bmatrix} 0 & 0 \\ 0 & 138888.888889 \end{bmatrix}(6.169)$$

$$K^{e_{10}} = \begin{bmatrix} 0 & 0 \\ 0 & 138888.888889 \end{bmatrix}$$

$$K^{e_{11}} = \begin{bmatrix} 4.91 \times 10^4 & -4.91 \times 10^4 \\ -4.91 \times 10^4 & 4.91 \times 10^4 \end{bmatrix}$$

$$K^{e_{12}} = \begin{bmatrix} 4.91 \times 10^4 & -4.91 \times 10^4 \\ -4.91 \times 10^4 & 4.91 \times 10^4 \end{bmatrix} \quad (6.170)$$

$$K^{e_{13}} = \begin{bmatrix} 4.91 \times 10^4 & -4.91 \times 10^4 \\ -4.91 \times 10^4 & 4.91 \times 10^4 \end{bmatrix}$$

$$K^{e_{14}} = \begin{bmatrix} 4.91 \times 10^4 & 4.91 \times 10^4 \\ 4.91 \times 10^4 & 4.91 \times 10^4 \end{bmatrix}$$

$$K^{e_{15}} = \begin{bmatrix} 4.91 \times 10^4 & 4.91 \times 10^4 \\ 4.91 \times 10^4 & 4.91 \times 10^4 \end{bmatrix} \quad (6.171)$$

$$K^{e_{16}} = \begin{bmatrix} 4.91 \times 10^4 & 4.91 \times 10^4 \\ 4.91 \times 10^4 & 4.91 \times 10^4 \end{bmatrix}$$

$$K^{e_{17}} = \begin{bmatrix} 1.389 \times 10^5 & 0 \\ 0 & 0 \end{bmatrix}$$

$$K^{e_{18}} = \begin{bmatrix} 0 & 0 \\ 0 & 1.389 \times 10^5 \end{bmatrix} \quad (6.172)$$

$$K^{e_{19}} = \begin{bmatrix} 0 & 0 \\ 0 & 1.389 \times 10^5 \end{bmatrix}$$

$$K^{e_{20}} = \begin{bmatrix} 4.91 \times 10^4 & -4.91 \times 10^4 \\ -4.91 \times 10^4 & 4.91 \times 10^4 \end{bmatrix}$$

$$K^{e_{21}} = \begin{bmatrix} 4.91 \times 10^4 & 4.91 \times 10^4 \\ 4.91 \times 10^4 & 4.91 \times 10^4 \end{bmatrix} \quad (6.173)$$

It should be noted here that Eqs.(6.167 – 6.173) only represent a 2x2 sub-matrix, shown in the upper-left corner of Eq.(6.17). However, it is obvious to see that the entire 4x4 element stiffness matrices can be easily generated.

The applied nodal loads for the 2-D truss are given as:

$$F = (0\ 0\ 0\ 0\ 0\ 0\ 0\ 0\ 0\ 0\ 0\ 0\ 0\ 0\ 0\ 0\ 0\ 0\ -100000\ 0\ -100000)^T \quad (6.174)$$

The total global stiffness matrix, after imposing the boundary (Dirichlet) conditions at degree-of-freedoms 9, 10, and 16 can be assembled and given as:

$$[K]_{20 \times 20} \to \text{Matrix K is shown below}: \quad (6.175)$$

K =	1	2	3	4	5
1	$1.88*10^5$	$-4.91*10^4$	$-1.389*10^5$	0	**0**
2	$-4.91*10^4$	$1.88*10^5$	0	0	**0**
3	$-1.389*10^5$	0	$3.76*10^5$	0	**$-1.389*10^5$**
4	0	0	0	$2.371*10^5$	**0**
5	0	0	$-1.389*10^5$	0	**$3.76*10^5$**
6	0	0	0	0	**0**
7	0	0	0	0	**$-1.389*10^5$**
8	0	0	0	0	**0**
9	0	0	0	0	**0**
10	0	0	0	0	**0**
11	$-4.91*10^4$	$4.91*10^4$	0	0	**$-4.91*10^4$**
12	$4.91*10^4$	$-4.91*10^4$	0	$-1.389*10^5$	**$-4.91*10^4$**
13	0	0	$-4.91*10^4$	$4.91*10^4$	**0**
14	0	0	$4.91*10^4$	$-4.91*10^4$	**0**
15	0	0	0	0	**$-4.91*10^4$**
16	0	0	0	0	**0**
17	0	0	0	0	**0**
18	0	0	0	0	**0**
19	0	0	0	0	**0**
20	**0**	**0**	**0**	**0**	**0**

K =	6	7	8	9	10
1	0	0	0	0	**0**
2	0	0	0	0	**0**
3	0	0	0	0	**0**
4	0	0	0	0	**0**
5	0	$-1.389*10^5$	0	0	**0**
6	$2.371*10^5$	0	0	0	**0**
7	0	$1.88*10^5$	$4.91*10^4$	0	**0**
8	0	$4.91*10^4$	$1.88*10^5$	0	**0**
9	0	0	0	1	**0**
10	0	0	0	0	**1**
11	$-4.91*10^4$	0	0	0	**0**
12	$-4.91*10^4$	0	0	0	**0**
13	0	$-4.91*10^4$	$-4.91*10^4$	0	**0**
14	$-1.389*10^5$	$-4.91*10^4$	$-4.91*10^4$	0	**0**
15	$4.91*10^4$	0	0	0	**0**
16	0	0	0	0	**0**
17	0	0	0	0	**0**
18	0	0	0	0	**0**
19	0	0	0	0	**0**
20	**0**	**0**	**0**	**0**	**0**

K =	11	12	13	14	15
1	$4.91*10^4$	$-4.91*10^4$	0	0	**0**
2	$-4.91*10^4$	$4.91*10^4$	0	0	**0**
3	0	0	$-4.91*10^4$	$4.91*10^4$	**0**
4	0	$-1.389*10^5$	$4.91*10^4$	$-4.91*10^4$	**0**
5	$-4.91*10^4$	$-4.91*10^4$	0	0	$\mathbf{-4.91*10^4}$
6	$-4.91*10^4$	$-4.91*10^4$	0	$-1.389*10^5$	$\mathbf{4.91*10^4}$
7	0	0	$-4.91*10^4$	$-4.91*10^4$	**0**
8	0	0	$-4.91*10^4$	$-4.91*10^4$	**0**
9	0	0	0	0	**0**
10	0	0	0	0	**0**
11	$4.251*10^5$	$-4.91*10^4$	$-1.389*10^5$	0	**0**
12	$-4.91*10^4$	$4.251*10^5$	0	0	**0**
13	$-1.389*10^5$	0	$4.251*10^5$	$4.91*10^4$	$\mathbf{-1.389*10^5}$
14	0	0	$4.91*10^4$	$4.251*10^5$	**0**
15	0	0	$-1.389*10^5$	0	$\mathbf{1.88*10^5}$
16	0	0	0	0	**0**
17	0	0	$-4.91*10^4$	$-4.91*10^4$	**0**
18	0	$-1.389*10^5$	$-4.91*10^4$	$-4.91*10^4$	**0**
19	$-4.91*10^4$	$4.91*10^4$	0	0	**0**
20	$\mathbf{4.91*10^4}$	$\mathbf{-4.91*10^4}$	**0**	$-1.389*10^5$	**0**

K =	16	17	18	19	20
1	0	0	0	0	**0**
2	0	0	0	0	**0**
3	0	0	0	0	**0**
4	0	0	0	0	**0**
5	0	0	0	0	**0**
6	0	0	0	0	**0**
7	0	0	0	0	**0**
8	0	0	0	0	**0**
9	0	0	0	0	**0**
10	0	0	0	0	**0**
11	0	0	0	$-4.91*10^4$	$\mathbf{4.91*10^4}$
12	0	0	$-1.389*10^5$	$4.91*10^4$	$\mathbf{-4.91*10^4}$
13	0	$-4.91*10^4$	$-4.91*10^4$	0	**0**
14	0	$-4.91*10^4$	$-4.91*10^4$	0	$\mathbf{-1.389*10^5}$
15	0	0	0	0	**0**
16	1	0	0	0	**0**
17	0	$1.88*10^5$	$4.91*10^4$	$-1.389*10^5$	**0**
18	0	$4.91*10^4$	$1.88*10^5$	0	**0**
19	0	$-1.389*10^5$	0	$1.88*10^5$	$\mathbf{-4.91*10^4}$
20	**0**	**0**	**0**	$-4.91*10^4$	$\mathbf{1.88*10^5}$

The unknown displacement vector {disp} can be solved from $[K]\{disp\} = \{F\}$ where $[K]$ and $\{F\}$ have been given by Eq.(6.175) and Eq.(6.174), respectively:

$$\{disp\} = \begin{cases} 1.331 & -0.355 & 0.976 & -2.01 & 0.203 & -2.01 & -0.151 & -0.355 \\ 0 & 0 & 0.365 & -2.323 & 0.814 & -2.323 & 1.18 & 0 \\ 0.699 & -2.824 & 0.481 & -2.824 \end{cases}^T \quad (6.176)$$

The solution given by Eq.(6.176) will be compared with FETI domain decomposition solution, which is discussed in the following paragraphs. The global stiffness matrix and nodal load vector of sub-domain 1 (see Figure 6.1) can be assembled from Eqs.(6.164 – 6.174) and given as:

$$K1 = \begin{bmatrix} 1.88{\times}10^5 & -4.91{\times}10^4 & -1.389{\times}10^5 & 0 & -4.91{\times}10^4 & 4.91{\times}10^4 & 0 & 0 \\ -4.91{\times}10^4 & 1.88{\times}10^5 & 0 & 0 & 4.91{\times}10^4 & -4.91{\times}10^4 & 0 & 0 \\ -1.389{\times}10^5 & 0 & 1.88{\times}10^5 & 4.91{\times}10^4 & 0 & 0 & 0 & 0 \\ 0 & 0 & 4.91{\times}10^4 & 4.91{\times}10^4 & 0 & 0 & 0 & 0 \\ -4.91{\times}10^4 & 4.91{\times}10^4 & 0 & 0 & 1.88{\times}10^5 & -4.91{\times}10^4 & 0 & 0 \\ 4.91{\times}10^4 & -4.91{\times}10^4 & 0 & 0 & -4.91{\times}10^4 & 4.91{\times}10^4 & 0 & 0 \\ 0 & 0 & 0 & 0 & 0 & 0 & 1 & 0 \\ 0 & 0 & 0 & 0 & 0 & 0 & 0 & 1 \end{bmatrix} \quad (6.177)$$

The rank of K1 is 7.

$$F1 = \begin{pmatrix} 0 & 0 & 0 & 0 & 0 & 0 & 0 & 0 \end{pmatrix}^T \quad (6.178)$$

Similarly, for sub-domain 2 (see Figure 6.1), the global stiffness matrix and nodal load vector can be assembled as:

K2 =	1	2	3	4	5	6
1	$1.88*10^5$	$-4.91*10^4$	$-1.389*10^5$	0	0	**0**
2	$-4.91*10^4$	$1.88*10^5$	0	0	0	**$-1.389*10^5$**
3	$-1.389*10^5$	0	$1.88*10^5$	$4.91*10^4$	$-4.91*10^4$	**$-4.91*10^4$**
4	0	0	$4.91*10^4$	$1.88*10^5$	$-4.91*10^4$	**$-4.91*10^4$**
5	0	0	$-4.91*10^4$	$-4.91*10^4$	$2.371*10^5$	**0**
6	0	$-1.389*10^5$	$-4.91*10^4$	$-4.91*10^4$	0	**$3.76*10^5$**
7	$-4.91*10^4$	$4.91*10^4$	0	0	$-1.389*10^5$	**0**
8	$4.91*10^4$	$-4.91*10^4$	0	$-1.389*10^5$	0	**0**
9	0	0	0	0	0	**0**
10	0	0	0	0	0	**$-1.389*10^5$**
11	0	0	0	0	$-4.91*10^4$	**$4.91*10^4$**
12	**0**	**0**	**0**	**0**	**$4.91*10^4$**	**$-4.91*10^4$**

K2 =	7	8	9	10	11	12
1	$-4.91*10^4$	$4.91*10^4$	0	0	0	**0**
2	$4.91*10^4$	$-4.91*10^4$	0	0	0	**0**
3	0	0	0	0	0	**0**
4	0	$-1.389*10^5$	0	0	0	**0**
5	$-1.389*10^5$	0	0	0	$-4.91*10^4$	**$4.91*10^4$**
6	0	0	0	$-1.389*10^5$	$4.91*10^4$	**$-4.91*10^4$**
7	$2.371*10^5$	0	$-4.91*10^4$	$-4.91*10^4$	0	**0**
8	0	$3.76*10^5$	$-4.91*10^4$	$-4.91*10^4$	0	**$-1.389*10^5$**
9	$-4.91*10^4$	$-4.91*10^4$	$1.88*10^5$	$4.91*10^4$	$-1.389*10^5$	**0**
10	$-4.91*10^4$	$-4.91*10^4$	$4.91*10^4$	$1.88*10^5$	0	**0**
11	0	0	$-1.389*10^5$	0	$1.88*10^5$	**$-4.91*10^4$**
12	0	**$-1.389*10^5$**	0	0	**$-4.91*10^4$**	**$1.88*10^5$**

$$(6.179)$$

The rank of K2 is 9.

$$F2 = \begin{pmatrix} 0 & 0 & 0 & 0 & 0 & 0 & 0 & 0 & 0 & -100000 & 0 & -100000 \end{pmatrix}^T \quad (6.180)$$

Finally, for sub-domain 3 (see Figure 6.1), one obtains:

$$K3 = \begin{bmatrix} 1.88\times10^5 & -4.91\times10^4 & -1.389\times10^5 & 0 & 0 & 0 & -4.91\times10^4 & 0 \\ -4.91\times10^4 & 4.91\times10^4 & 0 & 0 & 0 & 0 & 4.91\times10^4 & 0 \\ -1.389\times10^5 & 0 & 1.88\times10^5 & 4.91\times10^4 & -4.91\times10^4 & -4.91\times10^4 & 0 & 0 \\ 0 & 0 & 4.91\times10^4 & 1.88\times10^5 & -4.91\times10^4 & -4.91\times10^4 & 0 & 0 \\ 0 & 0 & -4.91\times10^4 & -4.91\times10^4 & 1.88\times10^5 & 4.91\times10^4 & -1.3889\times10^5 & 0 \\ 0 & 0 & 4.91\times10^4 & -4.91\times10^4 & 4.91\times10^4 & 4.91\times10^4 & 0 & 0 \\ -4.91\times10^4 & 4.91\times10^4 & 0 & 0 & -1.3889\times10^5 & 0 & 1.88\times10^5 & 0 \\ 0 & 0 & 0 & 0 & 0 & 0 & 0 & 1 \end{bmatrix} \quad (6.181)$$

The rank of K3 is 6.

$$F3 = \begin{pmatrix} 0 & 0 & 0 & 0 & 0 & 0 & 0 & 0 \end{pmatrix}^T \quad (6.182)$$

Also, the Boolean transformation matrix of all sub-structures is given as:

$$B1 = \begin{bmatrix} 0 & 0 & 1 & 0 & 0 & 0 & 0 & 0 \\ 0 & 0 & 0 & 1 & 0 & 0 & 0 & 0 \\ 0 & 0 & 0 & 0 & 1 & 0 & 0 & 0 \\ 0 & 0 & 0 & 0 & 0 & 1 & 0 & 0 \\ 0 & 0 & 0 & 0 & 0 & 0 & 0 & 0 \\ 0 & 0 & 0 & 0 & 0 & 0 & 0 & 0 \\ 0 & 0 & 0 & 0 & 0 & 0 & 0 & 0 \\ 0 & 0 & 0 & 0 & 0 & 0 & 0 & 0 \end{bmatrix} \quad (6.183)$$

$$B2 = \begin{bmatrix} -1 & 0 & 0 & 0 & 0 & 0 & 0 & 0 & 0 & 0 & 0 & 0 \\ 0 & -1 & 0 & 0 & 0 & 0 & 0 & 0 & 0 & 0 & 0 & 0 \\ 0 & 0 & 0 & 0 & -1 & 0 & 0 & 0 & 0 & 0 & 0 & 0 \\ 0 & 0 & 0 & 0 & 0 & -1 & 0 & 0 & 0 & 0 & 0 & 0 \\ 0 & 0 & 1 & 0 & 0 & 0 & 0 & 0 & 0 & 0 & 0 & 0 \\ 0 & 0 & 0 & 1 & 0 & 0 & 0 & 0 & 0 & 0 & 0 & 0 \\ 0 & 0 & 0 & 0 & 0 & 0 & 1 & 0 & 0 & 0 & 0 & 0 \\ 0 & 0 & 0 & 0 & 0 & 0 & 0 & 1 & 0 & 0 & 0 & 0 \end{bmatrix} \quad (6.184)$$

$$B3 = \begin{bmatrix} 0 & 0 & 0 & 0 & 0 & 0 & 0 & 0 \\ 0 & 0 & 0 & 0 & 0 & 0 & 0 & 0 \\ 0 & 0 & 0 & 0 & 0 & 0 & 0 & 0 \\ 0 & 0 & 0 & 0 & 0 & 0 & 0 & 0 \\ -1 & 0 & 0 & 0 & 0 & 0 & 0 & 0 \\ 0 & -1 & 0 & 0 & 0 & 0 & 0 & 0 \\ 0 & 0 & 0 & 0 & -1 & 0 & 0 & 0 \\ 0 & 0 & 0 & 0 & 0 & -1 & 0 & 0 \end{bmatrix} \qquad (6.185)$$

From Eq.(6.177), the partitioned sub-matrices [K1_11] and [K1_12] of sub-domain 1 can be identified as (with row 1 of Eq.[6.177] considered as the dependent row):

$$K1\_11 = \begin{bmatrix} 1.88 \times 10^5 & 0 & 0 & 4.91 \times 10^4 & -4.91 \times 10^4 & 0 & 0 \\ 0 & 1.88 \times 10^5 & 4.91 \times 10^4 & 0 & 0 & 0 & 0 \\ 0 & 4.91 \times 10^4 & 4.91 \times 10^4 & 0 & 0 & 0 & 0 \\ 4.91 \times 10^4 & 0 & 0 & 1.88 \times 10^5 & -4.91 \times 10^4 & 0 & 0 \\ -4.91 \times 10^4 & 0 & 0 & -4.91 \times 10^4 & 4.91 \times 10^4 & 0 & 0 \\ 0 & 0 & 0 & 0 & 0 & 1 & 0 \\ 0 & 0 & 0 & 0 & 0 & 0 & 1 \end{bmatrix} \quad (6.186)$$

$$K1\_12 = \begin{bmatrix} -4.91 \times 10^4 \\ -1.389 \times 10^5 \\ 0 \\ -4.91 \times 10^4 \\ 4.91 \times 10^4 \\ 0 \\ 0 \end{bmatrix} \qquad (6.187)$$

Similarly, the partitioned sub-matrices [K2_11] and [K2_12] of sub-domain 2 can be identified from Eq.(6.179) as:

K2_11 =	1	2	3	4	5
1	$1.88*10^5$	$-4.91*10^4$	$-4.91*10^4$		**0**
2	$-4.91*10^4$	$2.371*10^5$	0	$-1.389*10^5$	**0**
3	$-4.91*10^4$	0	$3.76*10^5$	0	**0**
4	0	$-1.389*10^5$	0	$2.371*10^5$	**0**
5	0	0	0	0	**$3.76*10^5$**

6	0	0	0	$-4.91*10^4$	$\mathbf{-4.91*10^4}$
7	0	0	$-1.389*10^5$	$-4.91*10^4$	$\mathbf{-4.91*10^4}$
8	0	$4.91*10^4$	$-4.91*10^4$	0	$\mathbf{-1.389*10^5}$
9	**0**	$\mathbf{-4.91*10^4}$	$\mathbf{4.91*10^4}$	**0**	**0**

K2_11 =	6	7	8	9	
1	0	0	0	**0**	
2	0	0	$-4.91*10^4$	$\mathbf{4.91*10^4}$	
3	0	$-1.389*10^5$	$4.91*10^4$	$\mathbf{-4.91*10^4}$	
4	$-4.91*10^4$	$-4.91*10^4$	0	**0**	(6.188)
5	$-4.91*10^4$	$-4.91*10^4$	0	$\mathbf{-1.389*10^5}$	
6	$1.88*10^5$	$4.91*10^4$	$-1.389*10^5$	**0**	
7	$4.91*10^4$	$1.88*10^5$	0	**0**	
8	$-1.389*10^5$	0	$1.88*10^5$	$\mathbf{-4.91*10^4}$	
9	**0**	**0**	$\mathbf{-4.91*10^4}$	$\mathbf{1.88*10^5}$	

$$K2\_12 = \begin{bmatrix} -1.389 \times 10^5 & 0 & 4.91 \times 10^4 \\ 0 & 0 & -4.91 \times 10^4 \\ 0 & -1.389 \times 10^5 & -4.91 \times 10^4 \\ -4.91 \times 10^4 & 4.91 \times 10^4 & 0 \\ 4.91 \times 10^4 & 4.91 \times 10^4 & -1.389 \times 10^5 \\ 0 & 0 & 0 \\ 0 & 0 & 0 \\ 0 & 0 & 0 \\ 0 & 0 & 0 \end{bmatrix} \qquad (6.189)$$

Finally, for sub-domain 3, one obtains the following partitioned sub-matrices (from Eq.[6.181]):

$$K3\_11 = \begin{bmatrix} 1.88 \times 10^5 & -4.91 \times 10^4 & 0 & 0 & 0 & 0 \\ -4.91 \times 10^4 & 4.91 \times 10^4 & 0 & 0 & 0 & 0 \\ 0 & 0 & 1.88 \times 10^5 & -4.91 \times 10^4 & -4.91 \times 10^4 & 0 \\ 0 & 0 & -4.91 \times 10^4 & 1.88 \times 10^5 & 4.91 \times 10^4 & 0 \\ 0 & 0 & -4.91 \times 10^4 & 4.91 \times 10^4 & 4.91 \times 10^4 & 0 \\ 0 & 0 & 0 & 0 & 0 & 1 \end{bmatrix} \quad (6.190)$$

$$K3\_12 = \begin{bmatrix} -1.389 \times 10^5 & -4.91 \times 10^4 \\ 0 & 4.91 \times 10^4 \\ 4.91 \times 10^4 & 0 \\ -4.91 \times 10^4 & -1.389 \times 10^5 \\ -4.91 \times 10^4 & 0 \\ 0 & 0 \end{bmatrix} \quad (6.191)$$

Applying Eq.(6.110), and referring to Eqs.(6.186, 6.188, 6.190), the generalized inversed matrices for sub-domains 1 – 3 can be computed as:

$$
K1P = \begin{bmatrix}
7.2\times10^{-6} & 0 & 0 & 0 & 7.2\times10^{-6} & 0 & 0 & 0 \\
0 & 7.2\times10^{-6} & -7.2\times10^{-6} & 0 & 0 & 0 & 0 & 0 \\
0 & -7.2\times10^{-6} & 2.756\times10^{-5} & 0 & 0 & 0 & 0 & 0 \\
0 & 0 & 0 & 7.2\times10^{-6} & 7.2\times10^{-6} & 0 & 0 & 0 \\
7.2\times10^{-6} & 0 & 0 & 7.2\times10^{-6} & 3.476\times10^{-5} & 0 & 0 & 0 \\
0 & 0 & 0 & 0 & 0 & 1 & 0 & 0 \\
0 & 0 & 0 & 0 & 0 & 0 & 1 & 0 \\
0 & 0 & 0 & 0 & 0 & 0 & 0 & 0
\end{bmatrix} \quad (6.192)
$$

K2P =	1	2	3	4	5	6
1	$6.446*10^{-6}$	$3.561*10^{-6}$	$7.537*10^{-7}$	$2.885*10^{-6}$	$7.537*10^{-7}$	$\mathbf{3.184*10^{-6}}$
2	$3.561*10^{-6}$	$1.719*10^{-5}$	$-3.561*10^{-6}$	$1.393*10^{-5}$	$3.639*10^{-6}$	$\mathbf{2.257*10^{-5}}$
3	$7.537*10^{-7}$	$-3.561*10^{-6}$	$6.446*10^{-6}$	$-2.885*10^{-6}$	$-7.537*10^{-7}$	$\mathbf{-1.038*10^{-5}}$
4	$2.885*10^{-6}$	$1.393*10^{-5}$	$-2.885*10^{-6}$	$1.652*10^{-5}$	$4.315*10^{-6}$	$\mathbf{2.257*10^{-5}}$
5	$7.537*10^{-7}$	$3.639*10^{-6}$	$-7.537*10^{-7}$	$4.315*10^{-6}$	$6.446*10^{-6}$	$\mathbf{1.122*10^{-5}}$
6	$3.184*10^{-6}$	$2.257*10^{-5}$	$-1.038*10^{-5}$	$2.257*10^{-5}$	$1.122*10^{-5}$	$\mathbf{5.955*10^{-5}}$
7	$6.756*10^{-7}$	$-3.938*10^{-6}$	$6.524*10^{-6}$	$-2.587*10^{-6}$	$-6.756*10^{-7}$	$\mathbf{-1.44*10^{-5}}$
8	$3.262*10^{-6}$	$2.295*10^{-5}$	$-1.046*10^{-5}$	$2.228*10^{-5}$	$1.114*10^{-5}$	$\mathbf{5.636*10^{-5}}$
9	$6.756*10^{-7}$	$3.262*10^{-6}$	$-6.756*10^{-7}$	$4.613*10^{-6}$	$6.524*10^{-6}$	$\mathbf{1.44*10^{-5}}$
10	0	0	0	0	0	**0**
11	0	0	0	0	0	**0**
12	**0**	**0**	**0**	**0**	**0**	**0**

K2P =	7	8	9	10	11	12
1	$6.756*10^{-7}$	$3.262*10^{-6}$	$6.756*10^{-7}$	0	0	**0**
2	$-3.938*10^{-6}$	$2.295*10^{-5}$	$3.262*10^{-6}$	0	0	**0**
3	$6.524*10^{-6}$	$-1.046*10^{-5}$	$-6.756*10^{-7}$	0	0	**0**
4	$-2.587*10^{-6}$	$2.228*10^{-5}$	$4.613*10^{-6}$	0	0	**0**
5	$-6.756*10^{-7}$	$1.114*10^{-5}$	$6.524*10^{-6}$	0	0	**0**
6	$-1.44*10^{-5}$	$5.636*10^{-5}$	$1.44*10^{-5}$	0	0	**0**
7	$1.305*10^{-5}$	$-1.372*10^{-5}$	$-1.351*10^{-6}$	0	0	**0**
8	$-1.372*10^{-5}$	$5.963*10^{-5}$	$1.508*10^{-5}$	0	0	**0**
9	$-1.351*10^{-6}$	$1.508*10^{-5}$	$1.305*10^{-5}$	0	0	**0**
10	0	0	0	0	0	**0**
11	0	0	0	0	0	**0**
12	**0**	**0**	**0**	**0**	**0**	**0**

(6.193)

$$
K3P = \begin{bmatrix}
7.2\times10^{-6} & 7.2\times10^{-6} & 0 & 0 & 0 & 0\ 0\ 0 \\
7.2\times10^{-6} & 2.756\times10^{-5} & 0 & 0 & 0 & 0\ 0\ 0 \\
0 & 0 & 7.2\times10^{-6} & 0 & 7.2\times10^{-6} & 0\ 0\ 0 \\
0 & 0 & 0 & 7.2\times10^{-6} & -7.2\times10^{-6} & 0\ 0\ 0 \\
0 & 0 & 7.2\times10^{-6} & -7.2\times10^{-6} & 3.476\times10^{-5} & 0\ 0\ 0 \\
0 & 0 & 0 & 0 & 0 & 1\ 0\ 0 \\
0 & 0 & 0 & 0 & 0 & 0\ 0\ 0 \\
0 & 0 & 0 & 0 & 0 & 0\ 0\ 0
\end{bmatrix} \quad (6.194)
$$

Applying Eq.(6.112), and referring to Eqs.(6.186 – 6.191), the rigid body matrices for sub-domains 1 – 3 can be computed as:

$$
R1 = \begin{bmatrix}
0 \\
1 \\
-1 \\
0 \\
-1 \\
0 \\
0 \\
1
\end{bmatrix} \quad (6.195)
$$

$$
R2 = \begin{bmatrix}
1 & 0 & 0 \\
1 & -1 & 1 \\
0 & 1 & 0 \\
1 & -1 & 1 \\
0 & 0 & 1 \\
1 & -2 & 2 \\
0 & 1 & 0 \\
1 & -2 & 2 \\
0 & 0 & 1 \\
1 & 0 & 0 \\
0 & 1 & 0 \\
0 & 0 & 1
\end{bmatrix} \quad (6.196)
$$

$$
R3 = \begin{bmatrix} 1 & 0 \\ 1 & -1 \\ 0 & 0 \\ 0 & 1 \\ 1 & -1 \\ 0 & 0 \\ 1 & 0 \\ 0 & 1 \end{bmatrix}
\tag{6.197}
$$

Utilizing Eq.(6.120), one obtains:

$$
BR1 = \begin{bmatrix} 1 \\ -1 \\ 0 \\ -1 \\ 0 \\ 0 \\ 0 \\ 0 \end{bmatrix}
\qquad
BR2 = \begin{bmatrix} -1 & 0 & 0 \\ 0 & -1 & 0 \\ -1 & 1 & -1 \\ 0 & -1 & 0 \\ 1 & 0 & 0 \\ 0 & 0 & 1 \\ 1 & -1 & 1 \\ 0 & 0 & 1 \end{bmatrix}
\qquad
R3 = \begin{bmatrix} 0 & 0 \\ 0 & 0 \\ 0 & 0 \\ 0 & 0 \\ -1 & 0 \\ -1 & 1 \\ 0 & -1 \\ -1 & 1 \end{bmatrix}
\tag{6.198}
$$

Utilizing Eq.(6.119), and referring to Eqs.(6.183 – 6.185, 6.192 – 6.194), one obtains:

$$
F_1 = \begin{bmatrix}
7.2 \cdot 10^{-6} & -7.2 \cdot 10^{-6} & 0 & 0 & 0 & 0 & 0 & 0 \\
-7.2 \cdot 10^{-6} & 2.756 \cdot 10^{-5} & 0 & 0 & 0 & 0 & 0 & 0 \\
0 & 0 & 2.439 \cdot 10^{-5} & 3.639 \cdot 10^{-6} & -3.561 \cdot 10^{-6} & 0 & -1.393 \cdot 10^{-5} & -3.639 \cdot 10^{-6} \\
0 & 0 & 3.639 \cdot 10^{-6} & 4.121 \cdot 10^{-5} & -7.537 \cdot 10^{-7} & 0 & 2.885 \cdot 10^{-6} & 7.537 \cdot 10^{-7} \\
0 & 0 & -3.561 \cdot 10^{-6} & -7.537 \cdot 10^{-7} & 1.365 \cdot 10^{-5} & 7.2 \cdot 10^{-6} & 2.885 \cdot 10^{-6} & 7.537 \cdot 10^{-7} \\
0 & 0 & 0 & 0 & 7.2 \cdot 10^{-6} & 2.756 \cdot 10^{-5} & 0 & 0 \\
0 & 0 & -1.393 \cdot 10^{-5} & 2.885 \cdot 10^{-6} & 2.885 \cdot 10^{-6} & 0 & 2.372 \cdot 10^{-5} & -2.885 \cdot 10^{-6} \\
0 & 0 & -3.639 \cdot 10^{-6} & 7.537 \cdot 10^{-7} & 7.537 \cdot 10^{-7} & 0 & -2.885 \cdot 10^{-6} & 4.121 \cdot 10^{-5}
\end{bmatrix}
\tag{6.199}
$$

Using Eq.(6.120), and referring to Eq.(6.198), sub-matrix $[G_I]$ can be assembled as:

$$G_I = \begin{bmatrix} 1 & -1 & 0 & 0 & 0 & 0 \\ -1 & 0 & -1 & 0 & 0 & 0 \\ 0 & -1 & 1 & -1 & 0 & 0 \\ -1 & 0 & -1 & 0 & 0 & 0 \\ 0 & 1 & 0 & 0 & -1 & 0 \\ 0 & 0 & 0 & 1 & -1 & 1 \\ 0 & 1 & -1 & 1 & 0 & -1 \\ 0 & 0 & 0 & 1 & -1 & 1 \end{bmatrix} \tag{6.200}$$

Utilizing Eqs.(6.122 – 6.123), and referring to Eqs.(6.183 – 6.185, 6.192 – 6.194, 6.178, 6.180, 6.182, 6.195 – 6.197), one obtains:

$$e = \begin{bmatrix} 0 \\ 0 \\ -1 \times 10^5 \\ -1 \times 10^5 \\ 0 \\ 0 \end{bmatrix} \qquad d = \begin{bmatrix} 0 \\ 0 \\ -0.068 \\ 0.585 \\ -0.135 \\ 0 \\ -0.203 \\ -0.585 \end{bmatrix} \tag{6.201}$$

Eq.(6.118) can be solved iteratively for the unknown vector $\begin{Bmatrix} \lambda \\ \alpha \end{Bmatrix}$ by the Projected Conjugate Gradient Method discussed in Section 6.9 and can be summarized as:

$$P = \begin{bmatrix} 0 & 0 & 0 & 0 & 0 & 0 & 0 & 0 \\ 0 & 0.5 & 0 & -0.5 & 0 & 0 & 0 & 0 \\ 0 & 0 & 0 & 0 & 0 & 0 & 0 & 0 \\ 0 & -0.5 & 0 & 0.5 & 0 & 0 & 0 & 0 \\ 0 & 0 & 0 & 0 & 0 & 0 & 0 & 0 \\ 0 & 0 & 0 & 0 & 0 & 0.5 & 0 & -0.5 \\ 0 & 0 & 0 & 0 & 0 & 0 & 0 & 0 \\ 0 & 0 & 0 & 0 & 0 & -0.5 & 0 & 0.5 \end{bmatrix} \tag{6.202}$$

$$\lambda_0 = G_I \cdot \left(G_I^T \cdot G_I \right)^{-1} \cdot e, \qquad r_0 = d - F_I \cdot \lambda_0 \qquad (6.203)$$

$$\lambda_0 = \begin{bmatrix} 1 \times 10^5 \\ 5 \times 10^4 \\ -1 \times 10^5 \\ 5 \times 10^4 \\ 1 \times 10^5 \\ -5 \times 10^4 \\ -1 \times 10^5 \\ -5 \times 10^4 \end{bmatrix}, \qquad r_0 = \begin{bmatrix} -0.36 \\ -0.658 \\ 0.971 \\ -0.71 \\ -1.132 \\ 0.658 \\ 0.199 \\ 0.71 \end{bmatrix} \qquad (6.204)$$

Iteration Phase

Iteration k = 1

$$w_0 = P \cdot r_0, \zeta_k = 0, p_k = w_0, v_k = \frac{w_0 \cdot w_0}{p_k \cdot \left(F_I \cdot p_k \right)}, \lambda_k = \lambda_0 + v_k \cdot p_k, r_k = r_0 - v_k \left(F_I \cdot p_k \right) \quad (6.205)$$

$$w_0 = \begin{bmatrix} 0 \\ 0.026 \\ 0 \\ -0.026 \\ 0 \\ -0.026 \\ 0 \\ 0.026 \end{bmatrix} \qquad v_k = 2.94 \times 10^4 \qquad (6.206)$$

$$\lambda_k = \begin{bmatrix} 1\times10^5 \\ 5.076377\times10^4 \\ -1\times10^5 \\ 4.923623\times10^4 \\ 1\times10^5 \\ -5.076377\times10^4 \\ -1\times10^5 \\ -4.923623\times10^4 \end{bmatrix} \tag{6.207}$$

$$r_k = \begin{bmatrix} -0.355 \\ -0.679 \\ 0.976 \\ -0.679 \\ -1.128 \\ 0.679 \\ 0.203 \\ 0.679 \end{bmatrix} \tag{6.208}$$

Iteration k = 2

$$w_{k-1} = P \cdot r_{k-1}, \; \zeta_k = \frac{w_{k-1} \cdot w_{k-1}}{w_0 \cdot w_0} \tag{6.209}$$

$$p_k = w_{k-1} + \zeta_k \cdot p_{k-1}, v_k = \frac{w_{k-1} \cdot w_{k-1}}{p_k \cdot (F_I \cdot p_k)} \tag{6.210}$$

$$\lambda_k = \lambda_{k-1} + v_k \cdot p_k, \; r_k = r_{k-1} - v_k (F_I \cdot p_k) \tag{6.211}$$

$$w_{k-1} = \begin{bmatrix} 0 \\ 0 \\ 0 \\ 0 \\ 0 \\ 0 \\ 0 \\ 0 \end{bmatrix} \quad v_k = 6.409 \times 10^4 \tag{6.212}$$

$$\lambda_k = \begin{bmatrix} 1 \times 10^5 \\ 5.076377 \times 10^4 \\ -1 \times 10^5 \\ 4.923623 \times 10^4 \\ 1 \times 10^5 \\ -5.076377 \times 10^4 \\ -1 \times 10^5 \\ -4.923623 \times 10^4 \end{bmatrix}, \quad r_k = \begin{bmatrix} -0.355 \\ -0.679 \\ 0.976 \\ -0.679 \\ -1.128 \\ 0.679 \\ 0.203 \\ 0.679 \end{bmatrix}, \quad \lambda_k - \lambda_{k-1} = \begin{bmatrix} 0 \\ 0 \\ 0 \\ 0 \\ 0 \\ 0 \\ 0 \\ 0 \end{bmatrix} \tag{6.213}$$

Iteration k = 3

$$w_{k-1} = P \cdot r_{k-1}, \quad \zeta_k = \frac{w_{k-1} \cdot w_{k-1}}{w_{k-2} \cdot w_{k-2}} \tag{6.214}$$

$$p_k = w_{k-1} + \zeta_k \cdot p_{k-1}, v_k = \frac{w_{k-1} \cdot w_{k-1}}{p_k \cdot (F_1 \cdot p_k)} \tag{6.215}$$

$$\lambda_k = \lambda_{k-1} + v_k \cdot p_k, \quad r_k = r_{k-1} - v_k (F_1 \cdot p_k) \tag{6.216}$$

$$w_{k-1} = \begin{bmatrix} 0 \\ 0 \\ 0 \\ 0 \\ 0 \\ 0 \\ 0 \\ 0 \end{bmatrix} \qquad v_k = 1.645 \times 10^4 \tag{6.217}$$

$$\lambda_k = \begin{bmatrix} 1 \times 10^5 \\ 5.076377 \times 10^4 \\ -1 \times 10^5 \\ 4.923623 \times 10^4 \\ 1 \times 10^5 \\ -5.076377 \times 10^4 \\ -1 \times 10^5 \\ -4.923623 \times 10^4 \end{bmatrix}, \, r_k = \begin{bmatrix} -0.355 \\ -0.679 \\ 0.976 \\ -0.679 \\ -1.128 \\ 0.679 \\ 0.203 \\ 0.679 \end{bmatrix}, \, \lambda_k - \lambda_{k-1} = \begin{bmatrix} 0 \\ 0 \\ 0 \\ 0 \\ 0 \\ 0 \\ 0 \\ 0 \end{bmatrix} \tag{6.218}$$

Iteration k = 4

$$w_{k-1} = P \cdot r_{k-1}, \, \zeta_k = \frac{w_{k-1} \cdot w_{k-1}}{w_{k-2} \cdot w_{k-2}} \tag{6.219}$$

$$p_k = w_{k-1} + \zeta_k \cdot p_{k-1}, v_k = \frac{w_{k-1} \cdot w_{k-1}}{p_k \cdot (F_l \cdot p_k)} \tag{6.220}$$

$$\lambda_k = \lambda_{k-1} + v_k \cdot p_k, r_k = r_{k-1} - v_k (F_l \cdot p_k) \tag{6.221}$$

$$w_{k-1} = \begin{bmatrix} 0 \\ 0 \\ 0 \\ 0 \\ 0 \\ 0 \\ 0 \\ 0 \end{bmatrix} \qquad v_k = 7.414 \times 10^3 \tag{6.222}$$

$$\lambda_k = \begin{bmatrix} 1 \times 10^5 \\ 5.076377 \times 10^4 \\ -1 \times 10^5 \\ 4.923623 \times 10^4 \\ 1 \times 10^5 \\ -5.076377 \times 10^4 \\ -1 \times 10^5 \\ -4.923623 \times 10^4 \end{bmatrix}, \ r_k = \begin{bmatrix} -0.355 \\ -0.679 \\ 0.976 \\ -0.679 \\ -1.128 \\ 0.679 \\ 0.203 \\ 0.679 \end{bmatrix}, \ \lambda_k - \lambda_{k-1} = \begin{bmatrix} 0 \\ 0 \\ 0 \\ 0 \\ 0 \\ 0 \\ 0 \\ 0 \end{bmatrix} \tag{6.223}$$

The value of λ is already converged at k = 4.

$$\alpha = -\left(G_I^T \cdot G_I\right)^{-1} \cdot G_I^T \cdot r_k \qquad \alpha = \begin{bmatrix} 1.331 \\ 0.976 \\ -2.01 \\ -2.01 \\ -0.151 \\ 1.18 \end{bmatrix} \tag{6.224}$$

After solving for α and λ, the displacements for each sub-structure can be computed as:

$$u1 = \begin{bmatrix} -0.355 \\ 0.976 \\ -2.01 \\ 0.365 \\ -2.323 \\ 0 \\ 0 \\ 1.331 \end{bmatrix} \quad u2 = \begin{bmatrix} 0.203 \\ 0.365 \\ -2.323 \\ 0.814 \\ -2.323 \\ 0.699 \\ -2.824 \\ 0.481 \\ -2.824 \\ 0.976 \\ -2.01 \\ -2.01 \end{bmatrix} \quad u3 = \begin{bmatrix} 0.203 \\ -2.01 \\ -0.355 \\ 0.814 \\ -2.323 \\ 0 \\ -0.151 \\ 1.18 \end{bmatrix} \quad (6.225)$$

Using FETI-1 domain decomposition formulation, the mapping between the sub-domains' and the original domain's dof numbering system (see Figure 6.1) can be given as:

$$order1^T = \begin{pmatrix} 2 & 3 & 4 & 5 & 6 & 7 & 8 & 1 \end{pmatrix} \quad (6.226)$$

$$order2^T = \begin{pmatrix} 3 & 5 & 6 & 7 & 8 & 9 & 10 & 11 & 12 & 1 & 2 & 4 \end{pmatrix} \quad (6.227)$$

$$order3^T = \begin{pmatrix} 1 & 2 & 4 & 5 & 6 & 8 & 3 & 7 \end{pmatrix} \quad (6.228)$$

Utilizing Eqs.(6.226 – 6.228), Eq.(6.225) can be re-arranged as:

$$u1n^T = \begin{pmatrix} 1.331 & -0.355 & 0.976 & -2.01 & 0.365 & -2.323 & 0 & 0 \end{pmatrix} \quad (6.229)$$

$$u2n^T = \begin{pmatrix} 0.976 & -2.01 & 0.203 & -2.01 & 0.365 & -2.323 & 0.814 & -2.323 & 0.699 & -2.824 & 0.481 & -2.824 \end{pmatrix} (6.230)$$

$$u3n^T = \begin{pmatrix} 0.203 & -2.01 & -0.151 & -0.355 & 0.814 & -2.323 & 1.18 & 0 \end{pmatrix} \quad (6.231)$$

Eqs.(6.229 – 6.231) can be combined to give;

$$
u_1 n^T =
\begin{array}{cccccccccc}
1 & 2 & 3 & 4 & 5 & 6 & 7 & 8 & 9 & 10 \\
(\ 1.331 & -0.355 & 0.976 & -2.01 & 0.203 & -2.01 & -0.151 & -0.355 & 0 & 0 \ ) \\
11 & 12 & 13 & 14 & 15 & 16 & 17 & 18 & 19 & 20 \\
(\ 0.365 & -2.323 & 0.814 & -2.323 & 1.18 & 0 & 0.699 & -2.824 & 0.481 & -2.824 \ )
\end{array}
\qquad (6.232)
$$

It can be easily observed that the FETI-1 solution, given by Eq.(6.232), is identical to the more conventional (<u>without</u> using FETI-1 domain decomposition) solution as shown earlier in Eq.(6.176).

Another Trivial Example for FETI-1

Figure 6.9 shows an axially loaded (rod) member, with the "unit" axial load applied at node 2.

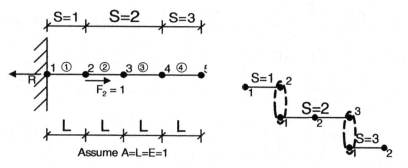

Figure 6.9 Axially Loaded (Rod) Member   **Figure 6.10** Three Sub-Domains

The 2x2 element stiffness matrix in its local coordinate axis also coincides with its global coordinate reference and can be computed from Eq.(6.17) as:

$$
\left[K^{(e)} \right]_{local} = \left[K^{(e)} \right]_{global} = \left[K^{(e)} \right] =
\begin{bmatrix} 1 & -1 \\ -1 & 1 \end{bmatrix}
\qquad (6.233)
$$

Case A: Conventional Finite Element Method

Following the familiar finite element assemble procedures, and after imposing the boundary conditions, one obtains the following global stiffness equilibrium equations:

$$\begin{bmatrix} 1 & -1 & 0 & 0 & 0 \\ -1 & 2 & -1 & 0 & 0 \\ 0 & -1 & 2 & -1 & 0 \\ 0 & 0 & -1 & 2 & -1 \\ 0 & 0 & 0 & -1 & 1 \end{bmatrix} \begin{Bmatrix} u_1 = 0 \\ u_2 \\ u_3 \\ u_4 \\ u_5 \end{Bmatrix} = \begin{Bmatrix} R \\ 1 \\ 0 \\ 0 \\ 0 \end{Bmatrix} \qquad (6.234)$$

The solution for Eqs.(6.234) can be easily obtained as:

$$\{u_2, u_3, u_4, u_5\} = \{1,1,1,1\} \qquad (6.235)$$

Case B: FETI-1 Method

The sub-domains' global stiffness matrices can be assembled and given as (also refer to Figure 6.10):

$$K^{(1)} = \begin{bmatrix} 1 & 0 \\ 0 & 1 \end{bmatrix} \qquad (6.236)$$

$$K^{(2)} = \begin{bmatrix} 1 & -1 & 0 \\ -1 & 2 & -1 \\ 0 & -1 & -1 \end{bmatrix} \qquad (6.237)$$

$$K^{(3)} = \begin{bmatrix} 1 & -1 \\ -1 & 1 \end{bmatrix} \qquad (6.238)$$

The displacement compatibility requirements for the interfaced (or boundary) nodes between the three sub-domains (see Figure 6.10) can be expressed as:

$$u_2^{(s=1)} = u_1^{(s=2)} \qquad (6.239)$$

$$u_3^{(s=2)} = u_1^{(s=3)} \qquad (6.240)$$

Eqs.(6.239 – 6.240) can be expressed in the matrix notation as:

$$[B]\vec{u} = \vec{0} \qquad (6.241)$$

or

$$\left[\begin{bmatrix} 0 & 1 \\ 0 & 0 \end{bmatrix} \begin{bmatrix} -1 & 0 & 0 \\ 0 & 0 & 1 \end{bmatrix} \begin{bmatrix} 0 & 0 \\ -1 & 0 \end{bmatrix}\right] \left\{ \begin{array}{c} \left\{ \begin{array}{c} u_1^{(1)} \\ u_2^{(1)} \end{array} \right\} \\ \left\{ \begin{array}{c} u_1^{(2)} \\ u_2^{(2)} \\ u_3^{(2)} \end{array} \right\} \\ \left\{ \begin{array}{c} u_1^{(3)} \\ u_2^{(3)} \end{array} \right\} \end{array} \right\} = \left\{ \begin{array}{c} 0 \\ 0 \end{array} \right\} \qquad (6.242)$$

It should be noted here that in Eq.(6.241), or in Eq.(6.242), the total number of rows and columns of $[B]$ is equal to the total number of interfaced equations (see Eqs.6.239 – 6.240) and the total degree-of-freedom (dof) of all three sub-domains, respectively.

The Boolean transformation matrices for each sth sub-domain, therefore, can be identified as:

$$B^{(s=1)} = \begin{bmatrix} 0 & 1 \\ 0 & 0 \end{bmatrix} \qquad (6.243)$$

$$B^{(s=2)} = \begin{bmatrix} -1 & 0 & 0 \\ 0 & 0 & 1 \end{bmatrix} \qquad (6.244)$$

$$B^{(s=3)} = \begin{bmatrix} 0 & 0 \\ -1 & 0 \end{bmatrix} \qquad (6.245)$$

The applied nodal load vectors for each sub-domain can be obtained as:

$$f^{(1)} = \left\{ \begin{array}{c} 0 \\ 1 \end{array} \right\}; \ f^{(2)} = \left\{ \begin{array}{c} 0 \\ 0 \\ 0 \end{array} \right\}; \ f^{(3)} = \left\{ \begin{array}{c} 0 \\ 0 \end{array} \right\} \qquad (6.246)$$

From Eqs.(6.236 – 6.238), it can be observed that the ranks of the three sub-domains' stiffness matrices are 2, 2, and 1, respectively. Thus, the following sub-domains' partitioned matrices can be identified:

$$K_{11}^{(s=1)} = K^{(s=1)} \quad \text{and} \quad K_{12}^{(s=1)} = \text{null(or empty)} \qquad (6.247)$$

$$K_{11}^{(2)} = \begin{bmatrix} 1 & -1 \\ -1 & 2 \end{bmatrix} \quad \text{and} \quad K_{12}^{(2)} = \left\{ \begin{array}{c} 0 \\ -1 \end{array} \right\} \qquad (6.248)$$

$$K_{11}^{(3)} = [1] \qquad \text{and} \qquad K_{12}^{(3)} = [-1] \qquad (6.249)$$

Since only sub-domains 2 and 3 are considered as "floating" sub-domains, the "generalized inverse" matrices for these sub-domains can be computed as:

$$\left[K^{(2))} \right]^{+} = \begin{bmatrix} \begin{bmatrix} 2 & 1 \\ 1 & 1 \end{bmatrix} & \begin{bmatrix} 0 \\ 0 \end{bmatrix} \\ [0 \quad 0] & [0] \end{bmatrix} \qquad (6.250)$$

$$\left[K^{(3)} \right]^{+} = \begin{bmatrix} [1] & [0] \\ [0] & [0] \end{bmatrix} \qquad (6.251)$$

The rigid body matrices for sub-domains 2 and 3 can be computed as:

$$R^{(2)} = \begin{bmatrix} -\left[K_{11}^{(2)} \right]^{-1} \left[K_{12}^{(2)} \right] \\ [I] \end{bmatrix} = \left\{ \begin{matrix} \{1\} \\ \{1\} \\ \{1\} \end{matrix} \right\} \qquad (6.252)$$

$$R^{(3)} = \left\{ \begin{matrix} 1 \\ 1 \end{matrix} \right\} \qquad (6.253)$$

Matrices $[F_I]$, $[G_I]$, and the right-hand-side vectors $\{d\}$ and $\{e\}$, can be computed as indicated in Eqs.(6.119, 6.120, 6.122 and 6.123):

$$[F_I] = \sum_{s=1}^{N_s} B^{(s)} K^{(s)^+} B^{(s)^T} = \begin{bmatrix} 3 & 0 \\ 0 & 1 \end{bmatrix} \qquad (6.254)$$

$$[G_I] = \left[B^{(1)} R^{(1)}, B^{(2)} R^{(2)}, B^{(3)} R^{(3)} \right] = \left\{ \begin{matrix} -1 \\ 0 \end{matrix} \right\} \qquad (6.255)$$

$$\{d\} = \sum_{s=1}^{N_s} B^{(s)} K^{(s)^+} f^{(s)} = \left\{ \begin{matrix} 1 \\ 0 \end{matrix} \right\} \qquad (6.256)$$

$$\{e\} = \left[f^{(1)^T} R^{(1)}, f^{(2)^T} R^{(2)}, f^{(3)^T} R^{(3)} \right]^T = [0] \qquad (6.257)$$

Thus, Eq.(6.118) can be assembled as:

$$\begin{bmatrix} \begin{bmatrix} 3 & 0 \\ 0 & 1 \end{bmatrix} & \begin{bmatrix} 1 \\ 0 \end{bmatrix} \\ [1 \quad 0] & [0] \end{bmatrix} \left\{ \begin{matrix} \{ \lambda_1 \\ \lambda_2 \} \\ \alpha_1 \end{matrix} \right\} = \left\{ \begin{matrix} \{ 1 \\ 0 \} \\ \{0\} \end{matrix} \right\} \qquad (6.258)$$

The coefficient matrix on the left-hand side of Eq.(6.258) is an indefinite matrix. Its three sub-determinant values are +3, +3, and -1, respectively. For large-scaled applications, Eq.(6.258) should be solved by the iterative solver such as the Pre-conditioned Conjugate Projected Gradient (PCPG) algorithm [6.4]. However, in this example, both direct and PCPG iterative solvers are used to obtain the solutions.

B.1 Direct Solver: The unknowns for Eq.(6.258) can be found as:

$$
\begin{Bmatrix} \lambda_1 \\ \lambda_2 \\ \alpha_1 \end{Bmatrix} = \begin{Bmatrix} 0 \\ 0 \\ 1 \end{Bmatrix}
$$

Finally, the unknown displacement vector for the sth sub-domain can be computed according to Eq.(6.108) as:

$$
u^{(s=1)} = \begin{Bmatrix} 0 \\ 1 \end{Bmatrix} \equiv \begin{Bmatrix} u_1^{(1)} \\ u_2^{(1)} \end{Bmatrix} \tag{6.259}
$$

$$
u^{(s=2)} = \begin{Bmatrix} 1 \\ 1 \\ 1 \end{Bmatrix} \equiv \begin{Bmatrix} u_1^{(2)} \\ u_2^{(2)} \\ u_3^{(2)} \end{Bmatrix} \tag{6.260}
$$

$$
u^{(s=3)} = \begin{Bmatrix} 1 \\ 1 \end{Bmatrix} \equiv \begin{Bmatrix} u_1^{(3)} \\ u_2^{(3)} \end{Bmatrix} \tag{6.261}
$$

B.2 PCPG Iterative Solver: Based upon the step-by-step PCPG procedure given in Table 6.6, one obtains:

(a) Initialized Phase

From Eqs.(6.126 and 6.255), one has:

$$
P = I - G_I \left(G_I^T G_I \right)^{-1} G_I^T = \begin{bmatrix} 0 & 0 \\ 0 & 0 \end{bmatrix} \tag{6.262}
$$

From Eqs.(6.139 and 6.257), one obtains:

$$
\lambda^0 = G_I \left(G_I^T G_I \right)^{-1} e = \begin{Bmatrix} 0 \\ 0 \end{Bmatrix} \tag{6.263}
$$

From Eq.(6.140), one gets:

$$
r^0 = \vec{d} - F_I \lambda^0 = \begin{Bmatrix} 1 \\ 0 \end{Bmatrix} \tag{6.264}
$$

(b) Iteration Phase

Iteration k = 1

$$w^{k-1} = Pr^{k-1} = Pr^0 \tag{6.265}$$

$$\text{or } w^0 = \begin{Bmatrix} 0 \\ 0 \end{Bmatrix} \tag{6.266}$$

$$\varsigma^k = w^{k-1^T} w^{k-1} / w^{k-2^T} w^{k-2} \tag{6.267}$$

$$\varsigma^1 = 0, \text{ because } w^{k-2} \text{ has not been defined yet} \tag{6.268}$$

$$p^k = w^{k-1} + \varsigma^k p^{k-1} \tag{6.269}$$

$$\text{or } p^1 = w^0 = \begin{Bmatrix} 0 \\ 0 \end{Bmatrix}, \text{ because } p^{k-1} \text{ has not been defined yet} \tag{6.270}$$

$$v^k = w^{k-1^T} w^{k-1} / p^{k^T} F_I p^k = 0 \tag{6.271}$$

$$\lambda^k = \lambda^{k-1} + v^k p^k = \begin{Bmatrix} 0 \\ 0 \end{Bmatrix} \tag{6.272A}$$

$$r^k = r^{k-1} - v^k F_I p^k = \begin{Bmatrix} 1 \\ 0 \end{Bmatrix} \tag{6.272B}$$

Iteration k = 2

$$w^1 = Pr^1 = \begin{Bmatrix} 0 \\ 0 \end{Bmatrix} \tag{6.273}$$

$$\varsigma^2 = w^{1^T} w^1 / w^{0^T} w^0 = 0 \tag{6.274}$$

$$p^2 = w^1 + \varsigma^2 p^1 = \begin{Bmatrix} 0 \\ 0 \end{Bmatrix} \tag{6.275}$$

$$v^2 = w^{1^T} w^1 / p^{2^T} F_I p^2 = 0 \tag{6.276}$$

$$\lambda^2 = \lambda^1 + v^2 p^2 = \begin{Bmatrix} 0 \\ 0 \end{Bmatrix} \tag{6.277}$$

$$r^2 = r^1 - v^2 F_I p^2 = \begin{Bmatrix} 1 \\ 0 \end{Bmatrix} \tag{6.278}$$

The value of λ has already been converged.
Thus, from the top portion of Eq.(6.118), one gets:

$$F_I \lambda^k - G_1 \alpha^k = d \tag{6.279}$$

Hence

$$G_1 \alpha^k = F_I \lambda^k - d \tag{6.280}$$

Using the notation of $\vec{r}$ defined in the initialized phase of Table 6.6, Eq.(6.280) can be re-written as:

$$G_1 \alpha^k \equiv -r^k \tag{6.281}$$

Pre-multiplying both sides of Eq.(6.218) by G_I^T, one obtains:

$$\left(G_I^T G_I\right) \alpha^k = -G_I^T r^k \tag{6.282}$$

Hence

$$\alpha^k = -\left(G_I^T G_I\right)^{-1} G_I^T r^k = \begin{Bmatrix} 1 \\ 1 \end{Bmatrix} \tag{6.283}$$

Once the solutions for λ^k and α^k have been obtained, the unknown sub-domains' displacement vector can be solved from Eq.(6.108) as:

$$u^{(s)} = \left[K^{(s)} \right]^+ \cdot \left\{ f^{(s)} - B^{(s)^T} \lambda \right\} + R^{(s)} \alpha^{(s)} \tag{6.108, repeated}$$

Hence

$$u^{(1)} = \begin{Bmatrix} 0 \\ 1 \end{Bmatrix} \tag{6.284}$$

$$u^{(2)} = \begin{Bmatrix} 1 \\ 1 \\ 1 \end{Bmatrix} \tag{6.285}$$

$$u^{(3)} = \begin{Bmatrix} 1 \\ 1 \end{Bmatrix} \tag{6.286}$$

6.12 A Preconditioning Technique for Indefinite Linear System [6.12]

Following [6.12], the "displacement" constraint equation is described as:

$$u_d = Q u_0 \tag{6.287}$$

The goal is to solve the following large indefinite system of linear equations, which may arise in a finite element analysis

$$K_{gg} u_g = f_g \tag{6.288}$$

where the set $g = d \cup 0$ contains all degrees of freedom. Let

$$K_{gg} = \begin{bmatrix} K_{00} & K_{0d} \\ K_{d0} & K_{dd} \end{bmatrix}; \ u_g = \begin{Bmatrix} u_0 \\ u_d \end{Bmatrix} \text{ and } f_g = \begin{Bmatrix} f_0 \\ f_d \end{Bmatrix} \tag{6.289}$$

Eq.(6.287) can be expressed as:

$$Q\overrightarrow{u_0} - \overrightarrow{u_d} = \vec{0} \tag{6.290}$$

or

$$[Q, -I] \begin{Bmatrix} \overrightarrow{u_o} \\ \overrightarrow{u_d} \end{Bmatrix} = \vec{0} \tag{6.291}$$

or

$$[Q, -I]\vec{u_g} = \vec{0} \tag{6.292}$$

Incorporating all the constraints, the Lagrange multiplier method is to solve

$$\begin{bmatrix} \begin{bmatrix} K_{00} & K_{0d} \\ K_{d0} & K_{dd} \end{bmatrix} & \begin{bmatrix} Q^T \\ -I \end{bmatrix} \\ \begin{bmatrix} Q & -I \end{bmatrix} & [0] \end{bmatrix} \begin{bmatrix} \begin{bmatrix} u_0 \\ u_d \end{bmatrix} \\ [\lambda] \end{bmatrix} = \begin{bmatrix} \begin{bmatrix} f_0 \\ f_d \end{bmatrix} \\ [0] \end{bmatrix} \tag{6.293}$$

where

$$\vec{u} \equiv \begin{Bmatrix} u_0 \\ u_d \\ \lambda \end{Bmatrix} \qquad \text{and} \qquad \vec{f} = \begin{Bmatrix} f_0 \\ f_d \\ 0 \end{Bmatrix} \tag{6.294}$$

Eq.(6.293) can be derived by minimizing the Lagrangian:

$$Min. L \equiv \frac{1}{2} u_g^T K_{gg} u_g - u_g^T f_g + \lambda^T [Q, -I] u_g \tag{6.295}$$

or

$$Min. L \equiv \frac{1}{2} u_g^T K_{gg} u_g - u_g^T f_g + u_g^T [Q, -I]^T \lambda \tag{6.296}$$

or

$$Min. L \equiv \frac{1}{2} u_g^T K_{gg} u_g - u_g^T f_g + u_g^T \begin{bmatrix} Q^T \\ -I \end{bmatrix} \lambda \tag{6.297}$$

By setting $\dfrac{\partial L}{\partial \vec{u_g}} = \vec{0}$ and $\dfrac{\partial L}{\partial \vec{\lambda}} = \vec{0}$, one obtains:

$$K_{gg} \vec{u_g} - \vec{f_g} + \begin{bmatrix} Q^T \\ -I \end{bmatrix} \vec{\lambda} = \vec{0} \tag{6.298}$$

$$[Q, -I]\vec{u_g} = \vec{0} \tag{6.299}$$

Using the notations defined in Eq.(6.289), Eqs.(6.298 and 6.299) are identical to Eq.(6.293) where λ is a fictitious variable of the Lagrange multipliers. The matrix of coefficients will be denoted by

$$K_{11} = \begin{bmatrix} K_{gg} & R \\ R^T & 0 \end{bmatrix} \equiv \begin{bmatrix} K & R \\ R^T & 0 \end{bmatrix} \tag{6.300}$$

where

$$R \equiv \begin{bmatrix} Q^T \\ -I \end{bmatrix} \tag{6.301}$$

so that Eq.(6.293) becomes:

$$K_{11} u = f \tag{6.302}$$

We select a pre-conditioner B for K_{11}, which is largely ill-conditioned for many applications, in the form

$$B = \begin{bmatrix} \overline{K} & R \\ R^T & 0 \end{bmatrix} \tag{6.303}$$

where $\overline{K}$ is defined to be a diagonal matrix so that the pre-conditioned matrix can be "inverted" easily, and therefore can be efficiently integrated into the Pre-Conditioned Conjugate Gradient iterative solver for the solution of Eq.(6.302) in [6.12], the pre-conditioner B is chosen by defining the diagonal elements to be

$$\overline{K}_{ii} = \left(a_i^T a_i \right)^{\frac{1}{2}}, \qquad i = 1, 2, \dots, m \tag{6.304}$$

where a_i is the ith column of K_{11}. Now, we solve, instead of Eq.(6.302)

$$B^{-1} K_{11} u = B^{-1} f \tag{6.305}$$

If we want to approximate the solution u in Eq.(6.305) by the conjugate gradient method, we must follow the CG algorithm for the coefficient matrix $B^{-1} K_{11}$.

Step a: First solve for r_0 in $Br_0 = f$. Partition r_0 and f into

$$r_0 = \begin{bmatrix} r_0^1 \\ r_o^2 \end{bmatrix} \tag{6.306}$$

and $\qquad f = \begin{bmatrix} f^1 \\ f^2 \end{bmatrix} \tag{6.307}$

Then $Br_0 = f$ can be described as:

$$\begin{bmatrix} \overline{K} & R \\ R^T & 0 \end{bmatrix} \begin{bmatrix} r_0^1 \\ r_o^2 \end{bmatrix} = \begin{bmatrix} f^1 \\ f^2 \end{bmatrix}$$

$$\tag{6.308}$$

The top portion of Eq.(6.308) can be expanded as:

$$\overline{K} \vec{r_0^1} + R \vec{r_0^2} = \vec{f^1} \tag{6.309}$$

or $\qquad \vec{r_0^1} = \left[\overline{K} \right]^{-1} \cdot \left\{ \vec{f^1} - R \vec{r_0^2} \right\} \tag{6.310}$

substituting Eq.(6.310) into the bottom portion of Eq.(6.308), one obtains:

$$R^T \cdot \left[\overline{K} \right]^{-1} \cdot \left\{ \vec{f^1} - R \vec{r_0^2} \right\} = \vec{f^2} \tag{6.311}$$

or $\qquad -R^T \cdot \left[\overline{K} \right]^{-1} R \vec{r_0^2} = \vec{f^2} - R^T \cdot \left[\overline{K} \right]^{-1} \vec{f^1} \tag{6.312}$

Eqs.(6.309 and 6.312) can be combined to become:

$$
\begin{bmatrix} \overline{K} & R \\ 0 & -R^T \cdot \left[\overline{K} \right]^{-1} R \end{bmatrix} \begin{Bmatrix} r_0^1 \\ r_o^2 \end{Bmatrix} = \begin{bmatrix} f^1 \\ -R^T \cdot \left[\overline{K} \right]^{-1} \overrightarrow{f^1} + \overrightarrow{f^2} \end{bmatrix} \quad (6.313)
$$

Eq.(6.313) can also be interpreted as a consequence of applying the Gauss elimination procedures on Eq.(6.308). Thus, the unknown vector $\begin{Bmatrix} r_0^1 \\ r_0^2 \end{Bmatrix}$ can be solved by the familiar backward solution. Defining the Schur Complement S as:

$$
S \equiv R^T \cdot \left[\overline{K} \right]^{-1} R \quad (6.314)
$$

Then, the solution for Eq.(6.313) can be given as:

$$
S r_0^2 = R^T \overline{K}^{-1} f^1 - f^2
$$
$$
r_0^1 = \overline{K}^{-1} \left(f^1 - R r_0^2 \right) \quad (6.315)
$$

Since $\overline{K}$ is diagonal, the Schur complement can be computed exactly without expensive computation cost.

Step b: Now, the familiar PCG algorithm (see Table 6.1) can be applied where the pre-conditioned equation (see Step 9 of Table 6.1) can be efficiently solved by using the procedure discussed in Step a.

Remarks

It may be worthy to investigate the possibility for applying Section 6.12 in conjunction with FETI formulation discussed in the Section 6.8 since Eq.(6.293, or 6.300) have the same form as the previous Eq.(6.104).

6.13 FETI-DP Domain Decomposition Formulation [6.6, 6.7]

In the primal domain decomposition (DD) formulation (discussed in Section 6.1), different sub-domains are not completely separated (even in the initial step). Furthermore, nodes (or dof) are labeled as interior and boundary nodes for each rth sub-domain.

Hence

$$\left[K^{(r)} \right] = \begin{bmatrix} K_{BB}^{(r)} & K_{BI}^{(r)} \\ K_{IB}^{(r)} & K_{II}^{(r)} \end{bmatrix} \tag{6.316}$$

The contributions of different sub-domains are integrated through these boundary (or interfaced) dofs. For the most general 3-D structural mechanics applications, each sub-domain will have at most 6 rigid body motions (related to the 3 translational and 3 rotational dof). Thus, each rth sub-domain only needs to have 2 or 3 boundary nodes in order to remove all these rigid body motions. For practical, large-scale engineering applications, each sub-domain will have a lot more than just 2 or 3 boundary nodes, hence the sub-matrix $\left[K_{II}^{(r)} \right]$ is non-singular. For the dual FETI-1 D.D. [6.4] formulation (discussed in Section 6.8), different sub-domains are assumed to be completely separated initially. The coupling effects among sub-domains will be incorporated later by introducing constraint equations (to enforce the compatibility requirement for the interfaced displacements) with the dual unknown Lagrange multipliers $\vec{\lambda}$. Thus, in FETI-1 formulation, there is no need to distinguish between boundary and interior dof for the rth sub-domain. Once the dual unknowns $\vec{\lambda}$ (and $\vec{\alpha}$) have been solved from Eq.(6.118), the primal unknown vector $\overrightarrow{u^{(r)}}$ (no need to distinguish between $\overrightarrow{u_B^{(r)}}$ and $\overrightarrow{u_I^{(r)}}$) can be solved from Eq.(6.108).

While FETI-1 "dual formulation" has certain computational and (parallel) scalable advantages as compared to the "primal formulation," it also introduces some undesirable effects, such as:

(a) The needs to compute the "generalized inverse" for those "floating" sub-domains.

(b) The popular Pre-Conditioned Conjugate Gradient (PCG) iterative solver has to be modified to become the Pre-Conditioned Conjugate Projected Gradient (PCPG) iterative solver.

The FETI-DP formulation, to be discussed in this section, will offer the following advantages, as compared to the primal, and FETI-1 dual formulations:

(a) There is no need to deal with the "generalized inverse" since there will not be any "floating" sub-domains.

(b) The well-documented, popular PCG iterative solver can be directly utilized.

(c) The primal and dual D. D. formulation can be considered as a special case of FETI-DP.

(d) Unlike the primal DD formulation that may lead to a large system of simultaneous equations (especially when many processors are used) to be solved for the unknown boundary dof vector, the system of linear equations associated with the unknown "corner" dof vector (to be defined soon) has a much smaller size.

The FETI-DP can be formulated by referring to Figure 6.11, where three sub-domains are illustrated.

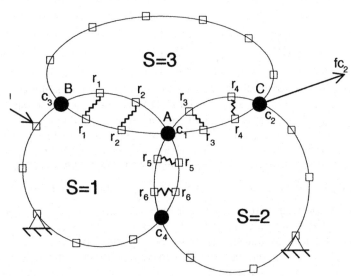

Figure 6.11 FETI-DP Definitions of DOF

The definitions of "corner" dof, shown in Figure 6.11, are given as following [6.17]:

(C1)   The dof belonged to more than c^s sub-domains, (see corner dof c_1 in Figure 6.11 for $c^s > 2$ sub-domains)

(C2)   The dof located at the beginning and the end of each edge of sub-domains (see corner of c_2 and c_3 in Figure 6.11)

Remarks

(1)   The "true" corner dof is c_1 (in Figure 6.11) since it belongs to three sub-domains. Thus, it fits the definition (C_1) for corner dof.

(2)   The constrained "remainder dof" at nodes B and C (see Figure 6.11) can also be selected as "corner dof," based on the definition (C_2). This will assure that the sub-domain's stiffness matrix will be non-singular. This statement will soon be more obvious during the derivation of the FETI-DP formulation.

(3)   The definition (C_2) holds only for finite element meshes composed of 2-D elements.

(4)   If $c^s = 1$, then all interfaced dof will be the corner dof. Thus, the FETI-DP will collapse into the "primal" D. D. formulation.

(5)   If $c^s > n_s$, then the number of corner dof goes to zero. Hence, FETI-DP will collapse into FETI-1 "dual" D. D. formulation.

(6) If $c^s = 2$, then the definition of corner dof will be valid for finite element meshes with both 2-D and 3-D elements.

(7) $c^s = 3$ is employed for uniform meshes with 3-D finite elements to reduce the (possible) large number of corner dof. Since each rth sub-domain only need a few corner nodes (say 2, or 3 corner nodes, or up to 6 corner dofs) to remove all its rigid body motion, the total "corner" dof required for the entire domain will be minimized if we have corner dof that belong to several sub-domains.

(8) For non-uniform finite element meshes with 3-D elements, $c^s = 3$ is selected in conjunction with appropriate procedures to assure the existence of at least three non-collinear "corner" nodes to avoid singular sub-domain matrices [6.6, 6.17].

From Figure 6.11, the sub-domains' stiffness matrices and nodal displacement, vectors can be partitioned as:

$$\left[K^{(r=1)} \right] = \begin{bmatrix} K_{rr}^{(1)} & K_{rc}^{(1)} \\ K_{cr}^{(1)} & K_{cc}^{(1)} \end{bmatrix} \quad \text{and} \quad \left\{ u^{(1)} \right\} = \begin{Bmatrix} u_r^{(1)} \\ u_c^{(1)} \end{Bmatrix} \tag{6.317}$$

$$\left[K^{(r=2)} \right] = \begin{bmatrix} K_{rr}^{(2)} & K_{rc}^{(2)} \\ K_{cr}^{(2)} & K_{cc}^{(2)} \end{bmatrix} \quad \text{and} \quad \left\{ u^{(2)} \right\} = \begin{Bmatrix} u_r^{(2)} \\ u_c^{(2)} \end{Bmatrix} \tag{6.318}$$

$$\left[K^{(r=3)} \right] = \begin{bmatrix} K_{rr}^{(3)} & K_{rc}^{(3)} \\ K_{cr}^{(3)} & K_{cc}^{(3)} \end{bmatrix} \quad \text{and} \quad \left\{ u^{(3)} \right\} = \begin{Bmatrix} u_r^{(3)} \\ u_c^{(3)} \end{Bmatrix} \tag{6.319}$$

Remarks

(1) The prescribed Dirichlet dof, due to the physical supports in sub-domains r = 1 and 2, are assumed to be already incorporated into $\left[K^{(1)} \right]$, $\left\{ u^{(1)} \right\}$, $\left[K^{(2)} \right]$, and $\left\{ u^{(2)} \right\}$ in Eqs. (6.317 and 6.318).

(2) Compatibility requirements can be imposed on the constrained "remainder dof" as follows (see Figure 6.11):

$$r_1^{(s=1)} = r_1^{(s=3)} \tag{6.320}$$

$$r_2^{(s=1)} = r_2^{(s=3)} \tag{6.321}$$

$$r_3^{(s=2)} = r_3^{(s=3)} \tag{6.322}$$

$$r_4^{(s=2)} = r_4^{(s=3)} \tag{6.323}$$

$$r_5^{(s=1)} = r_5^{(s=2)} \tag{6.324}$$

$$r_6^{(s=1)} = r_6^{(s=2)} \tag{6.325}$$

(3) Corner point A is usually selected at a point where more than two sub-domains meet. While these selection criteria will help to <u>minimize</u> the size of the total (= system) "corner dof" (which plays a

similar role as the "boundary dof" in the "primal" domain decomposition formulation), any point on the boundary of a sub-structure can be selected as a "corner point." However, in order to have a more meaningful (and easier) physical interpretation, corner points (or corner dof) are suggested at the beginning and ending of each edge of a sub-domain.

(4)   One of the major purposes of introducing a minimum number of corner dof for each sub-domain is to assure that the sub-domain will have no rigid body motions, and therefore, special procedures for obtaining the generalized inverse and rigid body modes can be avoided. For 2-D and 3-D structural applications, usually 3 and 6 "corner dof" are required, respectively, to remove all rigid body motions.

The Boolean transformation matrix can be established in order to relate the sth sub-domain's local dof to the global (= system) dof of the entire domain. Thus, one can write the following set of equations, which correspond to sub-domains s = 1, 2, and 3, respectively.

$$\left\{u^{(s=1)}\right\} = \left\{\begin{array}{c} u_r^{(s=1)} \\ u_c^{(s=1)} \end{array}\right\} = \begin{bmatrix} I & 0 & 0 & 0 \\ 0 & 0 & 0 & B_c^{(s=1)} \end{bmatrix} \cdot \left\{\begin{array}{c} u_r^{(1)} \\ u_r^{(2)} \\ u_r^{(3)} \\ u_c \end{array}\right\}_{global} \tag{6.326}$$

or, in a more compact form:

$$\left\{u^{(s=1)}\right\} = \left[\overline{I}^{(s=1)}\right] \cdot \{u\} \tag{6.327}$$

In Eq.(6.326), the vector $\{u_c\}$ (appeared on the right-hand side) represents the total (or system) "corner dof" for the entire domain. Matrices [I] and $\left[B_c^{(s=1)}\right]$ represent the identity and Boolean transformation matrix, respectively.

The definitions of $\left[\overline{I}^{(s=1)}\right]$ and $\{u\}$ (see [Eq.6.327]) are obvious, as one compares them with the right-hand side of Eq.(6.326).

Similarly, one obtains:

$$\left\{u^{(s=2)}\right\} = \left\{\begin{array}{c} u_r^{(s=2)} \\ u_c^{(s=2)} \end{array}\right\} = \begin{bmatrix} 0 & I & 0 & 0 \\ 0 & 0 & 0 & B_c^{(s=2)} \end{bmatrix} \cdot \left\{\begin{array}{c} u_r^{(1)} \\ u_r^{(2)} \\ u_r^{(3)} \\ u_c \end{array}\right\} \tag{6.328}$$

or

$$\left\{u^{(s=2)}\right\} = \left[\overline{I}^{(s=2)}\right] \cdot \{u\} \tag{6.329}$$

$$\left\{u^{(s=3)}\right\} = \begin{Bmatrix} u_r^{(s=3)} \\ u_c^{(s=3)} \end{Bmatrix} = \begin{bmatrix} 0 & 0 & I & 0 \\ 0 & 0 & 0 & B_c^{(s=3)} \end{bmatrix} \cdot \begin{Bmatrix} u_r^{(1)} \\ u_r^{(2)} \\ u_r^{(3)} \\ u_c \end{Bmatrix} \tag{6.330}$$

or $\quad \left\{u^{(s=3)}\right\} = \left[\overline{I}^{(s=3)}\right] \cdot \{u\} \tag{6.331}$

Compatibility requirements can be imposed on the constrained "remainder dof" (interface constraints between sub-domains) as follows (similar to Eqs.[6.320 – 6.325]):

$$\sum_s B_r^{(s)} u_r^{(s)} = 0 \tag{6.332}$$

or, for the case of three sub-domains:

$$\begin{bmatrix} \left[B_r^{(1)}\right] & \left[B_r^{(2)}\right] & \left[B_r^{(3)}\right] & [0] \end{bmatrix} \begin{Bmatrix} u_r^{(1)} \\ u_r^{(2)} \\ u_r^{(3)} \\ u_c \end{Bmatrix} = 0 \tag{6.333}$$

or, symbolically

$$\left[B_r\right] \cdot \{u\} = 0 \tag{6.334}$$

The number of rows of $\left[B_r\right]$ will be equal to the total number of constraints.

The total strain energy can be computed as:

$$S.E. = \frac{1}{2} \sum_{s=1}^{3} \left\{u^{(s)}\right\}^T \left[K^{(s)}\right] \left\{u^{(s)}\right\} \tag{6.335}$$

Substituting Eq.(6.321) into Eq.(6.335), one obtains:

$$S.E. = \frac{1}{2} \sum_{s=1}^{3} \left\{ \left(\overline{I}^{(s)} u\right)^T \left[K^{(s)}\right] \left(\overline{I}^{(s)} u\right) \right\} \tag{6.336}$$

$$S.E. = \frac{1}{2} \{u\}^T \sum_{s=1}^{3} \left\{ \left[\overline{I}^{(s)}\right]^T \left[K^{(s)}\right] \left[\overline{I}^{(s)}\right] \right\} \{u\} \tag{6.337}$$

or, symbolically:

$$S.E. = \frac{1}{2} \{u\}^T \left[K_g\right] \{u\} \tag{6.338}$$

Assuming s = 2nd sub-domain, then

$$\left[\bar{I}^{(s)}\right]^T \left[K^{(s)}\right]\left[\bar{I}^{(s)}\right] = \begin{bmatrix} 0 & 0 \\ I & 0 \\ 0 & 0 \\ 0 & B_c^{(s)^T} \end{bmatrix} \cdot \begin{bmatrix} K_{rr}^{(s)} & K_{rc}^{(s)} \\ K_{cr}^{(s)} & K_{cc}^{(s)} \end{bmatrix} \cdot \begin{bmatrix} 0 & I & 0 & 0 \\ 0 & 0 & 0 & B_c^{(s)} \end{bmatrix} \tag{6.339}$$

$$= \begin{bmatrix} 0 & 0 & 0 & 0 \\ 0 & K_{rr}^{(s)} & 0 & K_{rc}^{(s)} B_c^{(s)} \\ 0 & 0 & 0 & 0 \\ 0 & B_c^{(s)^T} K_{cr}^{(s)^T} & 0 & B_c^{(s)^T} K_{cc}^{(s)} B_c^{(s)} \end{bmatrix} \tag{6.340}$$

The system (global) stiffness matrix can be assembled as:

$$\left[K_g\right] = \sum_{s=1}^{3}\left\{\left[\bar{I}^{(s)}\right]^T \left[K^{(s)}\right]\left[\bar{I}^{(s)}\right]\right\} \tag{6.341}$$

$$= \begin{bmatrix} K_{rr}^{(1)} & 0 & 0 & K_{rc}^{(1)} B_c^{(1)} \\ 0 & K_{rr}^{(2)} & 0 & K_{rc}^{(2)} B_c^{(2)} \\ 0 & 0 & K_{rr}^{(3)} & K_{rc}^{(3)} B_c^{(3)} \\ B_c^{(1)^T} K_{cr}^{(1)^T} & B_c^{(2)^T} K_{cr}^{(2)^T} & B_c^{(3)^T} K_{cr}^{(3)^T} & \sum_{s=1}^{3} B_c^{(s)^T} K_{cc}^{(s)} B_c^{(s)} \end{bmatrix} \tag{6.342}$$

The system (global) load vector can also be assembled in a similar way. The external load vector applied on the sth sub-domain can be given as:

$$\left\{f^{(s)}\right\} = \begin{Bmatrix} f_r^{(s)} \\ f_c^{(s)} \end{Bmatrix} \tag{6.343}$$

The scalar external work done by the above given load vector can be computed as:

$$W = \sum_{s=1}^{3}\left\{u^{(s)}\right\}^T \cdot \left\{f^{(s)}\right\} \tag{6.344}$$

$$W = \sum_{s=1}^{3}\left\{\bar{I}^{(s)} u\right\}^T \cdot \left\{f^{(s)}\right\} \tag{6.345}$$

$$W = \left\{u\right\}^T \left(\sum_{s=1}^{3}\bar{I}^{(s)^T} \cdot f^{(s)}\right) \tag{6.346}$$

$$W = \left\{u\right\}^T \cdot \left\{f_g\right\} \tag{6.347}$$

where $\quad \left\{ f_g \right\} \equiv \sum_{s=1}^{3} \overline{I}^{(s)^T} \cdot f^{(s)}$ $\qquad(6.348)$

$$\overline{I}^{(s=2)^T} \cdot f^{(s)} = \begin{bmatrix} 0 & 0 \\ I & 0 \\ 0 & 0 \\ 0 & B_c^{(s)^T} \end{bmatrix} \cdot \begin{Bmatrix} f_r^{(s)} \\ f_c^{(s)} \end{Bmatrix} \qquad(6.349)$$

$$\overline{I}^{(s=2)^T} \cdot f^{(s)} = \begin{Bmatrix} 0 \\ f_r^{(s)} \\ 0 \\ B_c^{(s)^T} f_c^{(s)} \end{Bmatrix} \qquad(6.350)$$

Thus $\quad \left\{ f_g \right\} = \sum_{s=1}^{3} \overline{I}^{(s)^T} \cdot f^{(s)} = \begin{Bmatrix} f_r^{(1)} \\ f_r^{(2)} \\ f_r^{(3)} \\ \displaystyle\sum_{s=1}^{3} B_c^{(s)^T} f_c^{(s)} \end{Bmatrix}$ $\qquad(6.351)$

The total potential energy of the entire domain, subjected to the constraints of Eq.(6.334), can be expressed as:

$$\Pi = \frac{1}{2}\{u\}^T \left[K_g \right]\{u\} - \{u\}^T \{f_g\} + \{\lambda\}^T \left[B_r \right]\{u\}$$

$$\Pi = \frac{1}{2}\{u\}^T \left[K_g \right]\{u\} - \{u\}^T \{f_g\} + \{u\}^T \left[B_r \right]^T \{\lambda\} \qquad(6.352)$$

The equilibrium equations can be obtained as:

$$\frac{\partial \Pi}{\partial u} = \left[K_g \right]\{u\} - \{f_g\} + \left[B_r \right]^T \{\lambda\} = 0 \qquad(6.353)$$

or

$$\left[K_g \right]\{u\} = \{f_g\} - \left[B_r \right]^T \{\lambda\} \qquad(6.354)$$

The dimension for the above equation is $\left(\sum_s n_r^{(s)} \right) + n_c$. The last term in Eq.(6.354)

can be expanded as:

$$\left[B_r \right]^T \{\lambda\} = \left(\left[B_r^{(1)} \right]^T \left[B_r^{(2)} \right]^T \left[B_r^{(3)} \right]^T \left[0 \right] \right) \cdot \{\lambda\} \qquad(6.355)$$

Eq.(6.353) can be symbolically represented as:

$$\begin{bmatrix} K_{rr} & K_{rc} \\ K_{cr} & K_{cc} \end{bmatrix} \begin{Bmatrix} u_r \\ u_c \end{Bmatrix} = \begin{Bmatrix} f_r - B_r^T \lambda \\ f_c \end{Bmatrix} \tag{6.356}$$

where the sub-matrices K_{rr}, K_{rc}, K_{cr}, and K_{cc} of $\begin{bmatrix} K_g \end{bmatrix}$ can be identified by referring to Eq.(6.342). It should be noted that the constraint equations are only associated with the <u>interface</u> "remainder dof," and <u>not</u> directly related to the "<u>corner dof</u>." This observation reflects on the right-hand side of Eq.(6.356).

Eq.(6.354) has to be solved in conjunction with the following n_λ additional equations:

$$\frac{\partial \Pi}{\partial u} = [B_r]\{u\} = \{0\} \tag{6.357}$$

From the top and bottom portions of Eq.(6.356), one obtains:

$$K_{rr}u_r + K_{rc}u_c = f_r - [B_r]^T \lambda \tag{6.358}$$

$$K_{cr}u_r + K_{cc}u_c = f_c \tag{6.359}$$

From Eq.(6.358), one has:

$$u_r = [K_{rr}]^{-1}\left(f_r - [B_r]^T \lambda - [K_{rc}]u_c \right) \tag{6.360}$$

Pre-multiplying both sides of Eq.(6.360) by $[B_r]$, and utilizing Eq.(6.332 or 6.334), one obtains:

$$[B_r]u_r = 0 = [B_r][K_{rr}]^{-1}\left(f_r - [B_r]^T \lambda - [K_{rc}]u_c \right) \tag{6.361}$$

Defining the following parameters:

$$d_r \equiv [B_r][K_{rr}]^{-1} f_r \tag{6.362}$$

$$F_{rr} \equiv [B_r][K_{rr}]^{-1}[B_r]^T \tag{6.363}$$

$$F_{rc} \equiv [B_r][K_{rr}]^{-1}[K_{rc}] \tag{6.364}$$

Then, Eq.(6.361) becomes:

$$0 = d_r - [F_{rr}]\lambda - [F_{rc}]u_c \tag{6.365}$$

Substituting Eq.(6.360) into Eq.(6.359), one has:

$$[K_{cr}][K_{rr}]^{-1}\left(f_r - [B_r]^T \lambda - [K_{rc}]u_c \right) + [K_{cc}]u_c = f_c \tag{6.366}$$

or

$$[K_{cr}][K_{rr}]^{-1} f_r - f_c = [K_{cr}][K_{rr}]^{-1}[B_r]^T \lambda + \left([K_{cr}][K_{rr}]^{-1}[K_{rc}] - [K_{cc}] \right)u_c \tag{6.367}$$

The following parameters are defined:

$$-f_c + [K_{cr}][K_{rr}]^{-1} f_r \equiv -f_c^* \tag{6.368}$$

$$\left[K_{cr}\right]\left[K_{rr}\right]^{-1}\left[K_{rc}\right] - \left[K_{cc}\right] \equiv -\left[K_{cc}^{*}\right] \tag{6.369}$$

$$\left[K_{cr}\right]\left[K_{rr}\right]^{-1}\left[B_{r}\right]^{T} \equiv \left[F_{cr}\right] \tag{6.370}$$

Using the new notations, Eqs.(6.365, 6.367) can be expressed as (in terms of the unknowns λ and u_{c}):

$$\begin{bmatrix} F_{rr} & F_{rc} \\ F_{cr} & -K_{cc}^{*} \end{bmatrix} \begin{Bmatrix} \lambda \\ u_{c} \end{Bmatrix} = \begin{Bmatrix} d_{r} \\ -f_{c}^{*} \end{Bmatrix} \tag{6.371}$$

The unknown vector $\{u_{c}\}$ can be eliminated from the 2nd half of Eq.(6.371) as follows:

$$\{u_{c}\} = \left[K_{cc}^{*}\right]^{-1}\left(\left[F_{cr}\right]\lambda + f_{c}^{*}\right) \tag{6.372}$$

Substituting Eq.(6.372) into the 1st half of Eq.(6.371):

$$\left[F_{rr}\right]\lambda + \left[F_{rc}\right]\left\{\left[K_{cc}^{*}\right]^{-1}\left(\left[F_{cr}\right]\lambda + f_{c}^{*}\right)\right\} = d_{r} \tag{6.373}$$

or

$$\left(F_{rr} + F_{rc}\left[K_{cc}^{*}\right]^{-1}F_{cr}\right)\lambda = d_{r} - \left(F_{rc}\left[K_{cc}^{*}\right]^{-1}f_{c}^{*}\right) \tag{6.374}$$

or, symbolically

$$\left[K_{\lambda\lambda}\right]\lambda = d_{\lambda} \tag{6.375}$$

Remarks

(5)   Matrix $\left[K_{cc}^{*}\right]$ is sparse, symmetric positive definite, and the size of this matrix is relatively small.

(6)   Matrix $\left[K_{rr}\right]$ is non-singular since there will be enough selected "corner dof" to guarantee that $\left[K_{rr}\right]^{-1}$ exists.

The entire FETI-DP step-by-step procedures can be summarized in Table 6.8.

Table 6.8 FETI-DP Step-by-Step Procedures

Step 0	Assuming all data for the "conventional" finite element (without using Primal, FETI-1, or FETI-DP domain decomposition formulation) is already available.
Step 1	Based on the user's specified number of processors, and/or the number of sub-domains, and utilizing the METIS [3..3], the original finite element domain will be partitioned into $NP \equiv NS$ (= Number of Processors = Number of sub-domains)
Step 2	Post processing METIS outputs to define "corner" nodes (or dofs), "remainder" nodes, and other required sub-domains' information.
Step 3	Construct the sth sub-domain's matrices $\left[K_{rr}^{(s)} \right]$, $\left[K_{rc}^{(s)} \right]$, $\left[K_{cr}^{(s)} \right]$ and $\left[K_{cc}^{(s)} \right]$ and load vectors $\left\{ f_r^{(s)} \right\}$ and $\left\{ f_c^{(s)} \right\}$.
Step 4	Construct the following sub-domain's sparse Boolean matrices: $\left[B_c^{(s)} \right]$, for "corner" dofs $\left[B_r^{(s)} \right]$, for "remainder" dofs
Step 5	Construct the following matrices and/or vectors:

$$[K_{rr}] \equiv \begin{bmatrix} K_{rr}^{(s=1)} & & \\ & K_{rr}^{(s=2)} & \\ & & K_{rr}^{(s=3)} \end{bmatrix}, \text{ see Eq.(6.356)} \qquad (6.376)$$

$$[K_{rc}] \equiv \begin{bmatrix} \left[K_{rc}^{(s=1)} B_c^{(s=1)} \right] \\ \left[K_{rc}^{(s=2)} B_c^{(s=2)} \right] \\ \left[K_{rc}^{(s=3)} B_c^{(s=3)} \right] \end{bmatrix}, \text{ see Eq.(6.356)} \qquad (6.377)$$

$$[K_{cr}] \equiv \left(\left[B_c^{(s=1)^T} K_{cr}^{(s=1)} \right] \left[B_c^{(s=2)^T} K_{cr}^{(s=2)} \right] \left[B_c^{(s=3)^T} K_{cr}^{(s=3)} \right] \right) \quad (6.378)$$

$$[K_{cc}] = \sum_{s=1}^{3} B_c^{(s)^T} K_{cc}^{(s)} B_c^{(s)} \qquad (6.379)$$

$$[B_r] \equiv \left(\left[B_r^{(s=1)} \right] \left[B_r^{(s=2)} \right] \left[B_r^{(s=3)} \right] [0] \right) \qquad (6.380)$$

Step 6	Construct

$$d_r \equiv [B_r][K_{rr}]^{-1} f_r \qquad \text{(6.362, repeated)}$$

$$F_{rr} \equiv [B_r][K_{rr}]^{-1} [B_r]^T \qquad \text{(6.363, repeated)}$$

$$F_{rc} \equiv [B_r][K_{rr}]^{-1} [K_{rc}] \qquad \text{(6.364, repeated)}$$

Step 7   Construct

$$-f_c^* \equiv [K_{cr}][K_{rr}]^{-1} f_r - f_c \qquad \text{(6.368, repeated)}$$

$$-[K_{cc}^*] \equiv [K_{cr}][K_{rr}]^{-1}[K_{rc}] - [K_{cc}] \qquad \text{(6.369, repeated)}$$

$$[F_{cr}] \equiv [K_{cr}][K_{rr}]^{-1}[B_r]^T \qquad \text{(6.370, repeated)}$$

Step 8   Construct

$$[K_{\lambda\lambda}] \equiv \left(F_{rr} + F_{rc}\left[K_{cc}^* \right]^{-1} F_{cr} \right) \qquad \text{(6.374, repeated)}$$

$$d_\lambda \equiv d_r - \left(F_{rc}\left[K_{cc}^* \right]^{-1} f_c^* \right) \qquad \text{(6.374, repeated)}$$

Step 9   Solve for λ, from Eq.(6.375):

$$\lambda = [K_{\lambda\lambda}]^{-1} d_\lambda \qquad \text{(6.375, repeated)}$$

Step 10 Solve for u_c, from Eq.(6.372):

$$\{u_c\} = \left[K_{cc}^* \right]^{-1}\left([F_{cr}]\lambda + f_c^* \right) \qquad \text{(6.372, repeated)}$$

Step 11 Solve for u_r, from Eq.(6.360):

$$u_r = [K_{rr}]^{-1}\left(f_r - [B_r]^T \lambda - [K_{rc}]u_c \right) \qquad \text{(6.360, repeated)}$$

Remarks

(7)   Solving for u_r (shown in Eq.[6.360]) is not a major concern, because

    (a) $[K_{rr}]$ is sparse, symmetrical (or unsymmetrical) and it can be factorized by sub-domain fashion.

    (b) Only one right-hand-side vector is required.

(8)   Solving for u_c (shown in Eq.[6.372]) can be moderately computational expensive, because

    (a) $\left[K_{cc}^* \right]$ has relatively small dimensions, hence it is not too expensive to factorize.

    (b) However, during the process to construct $\left[K_{cc}^* \right]$, shown in Eq.(6.369), the number of right-hand-side vectors (associated with $[K_{rr}]^{-1}$) is related to the system "corner dof," which may not be a large number; however, it may not be a small number either!

    (c) Only one right-hand-side (RHS) vector is required for Eq.(6.372).

(9)   Factorizing the matrix $\left[K_{\lambda\lambda}\right]$ (see Eq.[6.374]) for the solution of λ (see Eq.[6.375]) can be a major concern because the construction of $\left[K_{\lambda\lambda}\right]$ will require the computation of $\left[K_{cc}^{*}\right]^{-1}F_{cr}$ or $\left[K_{cc}^{*}\right]^{-1}\cdot\left(\left[K_{cr}\right]\left[K_{rr}\right]^{-1}\left[B_{r}\right]^{T}\right)$. Thus, the number of RHS vectors associated with $\left[K_{cc}^{*}\right]^{-1}$ (hence, associated with $\left[K_{\lambda\lambda}\right]^{-1}$) can be large! Standard Pre-Conditioned Conjugate Gradient (PCG) iterative solver is recommended in [6.6] for solving the linear Eq.(6.375) since the coefficient $\left[K_{\lambda\lambda}\right]$ is a symmetric positive definite matrix. However, a recursive (or multi-level) sub-domain approach [6.18, 6.19, 6.20] with "only direct sparse solvers" (without using "iterative" solvers) have also been recently proposed.

29-Bar Truss Example

In this section, the FETI-DP procedures will be applied on a 2-D truss example as shown in Figures 6.12 - 6.13.

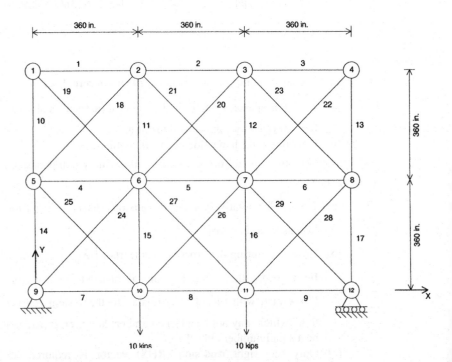

Figure 6.12 Entire Structure of 29-Bar Truss Example

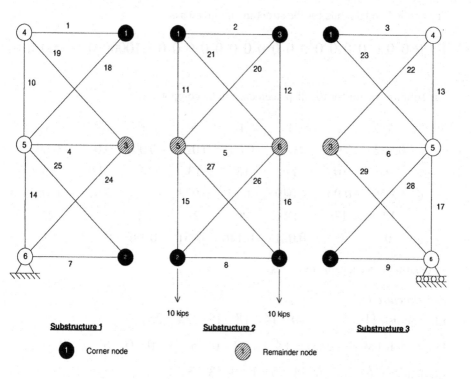

Substructure 1   Substructure 2   Substructure 3

● Corner node   ▨ Remainder node

Figure 6.13 29-Bar Truss Example with Three Sub-Structures

In this example, all truss members have the same cross-sectional area of 5 sq. in. and the Young's modulus of 10,000,000 psi is used.

Element stiffness matrix for members 1 - 9 can be computed as:

$$K = \begin{bmatrix} 138888.888889 & 0 \\ 0 & 0 \end{bmatrix}$$

Element stiffness matrix for members 10 - 17 can be computed as:

$$K = \begin{bmatrix} 0 & 0 \\ 0 & 138888.888889 \end{bmatrix}$$

Element stiffness matrix for members 18, 20, 22, 24, 26, and 28 can be computed as:

$$K = \begin{bmatrix} 4.91 \times 10^4 & 4.91 \times 10^4 \\ 4.91 \times 10^4 & 4.91 \times 10^4 \end{bmatrix}$$

Element stiffness matrix for members 19, 21, 23, 25, 27, and 29 can be computed as:

$$K = \begin{bmatrix} 4.91 \times 10^4 & -4.91 \times 10^4 \\ -4.91 \times 10^4 & 4.91 \times 10^4 \end{bmatrix}$$

The applied nodal loads for the structure are given as:

$$F=\begin{bmatrix}0 & 0 & 0 & 0 & 0 & 0 & 0 & 0 & 0 & 0 & 0 & 0 & 0 & 0 & 0 & 0 & 0 & 0 & 0 & -10000 & 0 & -1000 & 0 & 0\end{bmatrix}$$

The solution vector for the displacement can be computed as:

	1	2	3	4	5	6	7	8
(	0.062	−0.063	0.049	−0.091	0.023	−0.091	0.011	−0.063)
	9	10	11	12	13	14	15	16
$x=$ (	0.055	−0.051	0.049	−0.11	0.023	−0.11	0.018	−0.051)
	17	18	19	20	21	22	23	24
(	0	0	0.021	−0.146	0.051	−0.146	0.073	0)

- Construct K for all substructures

Sub-structure 1

Element list: $\begin{pmatrix}1 & 4 & 7 & 10 & 14 & 18 & 19 & 24 & 25\end{pmatrix}$

Local-Global Mapping array: $\begin{pmatrix}4 & 1 & 0 & 0 & 5 & 3 & 0 & 0 & 6 & 2 & 0 & 0\end{pmatrix}$

Local node i: $\begin{pmatrix}4 & 5 & 6 & 4 & 5 & 1 & 4 & 3 & 5\end{pmatrix}$

Local node j: $\begin{pmatrix}1 & 3 & 2 & 5 & 6 & 5 & 3 & 6 & 2\end{pmatrix}$

Sub-structure K after imposing boundary conditions:

	1	2	3	4	5	6	7	8	9	10	11	12
1	1.88·10⁵	4.91·10⁴	0	0	0	0	-1.389·10⁵	0	-4.91·10⁴	-4.91·10⁴	0	0
2	4.91·10⁴	4.91·10⁴	0	0	0	0	0	0	-4.91·10⁴	-4.91·10⁴	0	0
3	0	0	1.88·10⁵	-4.91·10⁴	0	0	0	0	-4.91·10⁴	4.91·10⁴	0	0
4	0	0	-4.91·10⁴	4.91·10⁴	0	0	0	0	4.91·10⁴	-4.91·10⁴	0	0
5	0	0	0	0	2.371·10⁵	0	-4.91·10⁴	4.91·10⁴	-1.389·10⁵	0	0	0
6	0	0	0	0	0	9.821·10⁴	4.91·10⁴	-4.91·10⁴	0	0	0	0
7	-1.389·10⁵	0	0	0	-4.91·10⁴	4.91·10⁴	1.88·10⁵	-4.91·10⁴	0	0	0	0
8	0	0	0	0	4.91·10⁴	-4.91·10⁴	-4.91·10⁴	1.88·10⁵	0	-1.389·10⁵	0	0
9	-4.91·10⁴	-4.91·10⁴	-4.91·10⁴	4.91·10⁴	-1.389·10⁵	0	0	0	2.371·10⁵	0	0	0
10	-4.91·10⁴	-4.91·10⁴	4.91·10⁴	-4.91·10⁴	0	0	0	-1.389·10⁵	0	3.76·10⁵	0	0
11	0	0	0	0	0	0	0	0	0	0	1	0
12	0	0	0	0	0	0	0	0	0	0	0	1

$K1 =$

Sub-structure Force vector: $\begin{pmatrix}0 & 0 & 0 & 0 & 0 & 0 & 0 & 0 & 0 & 0 & 0 & 0\end{pmatrix}$

Sub-structure 2

Element list: $\begin{pmatrix}2 & 5 & 8 & 11 & 12 & 15 & 16 & 20 & 21 & 26 & 27\end{pmatrix}$

Local-Global Mapping array: $\begin{pmatrix}0 & 1 & 3 & 0 & 0 & 5 & 6 & 0 & 0 & 2 & 4 & 0\end{pmatrix}$

Local node i: $\begin{pmatrix}1 & 5 & 2 & 1 & 3 & 5 & 6 & 3 & 1 & 6 & 5\end{pmatrix}$

Local node j: $\begin{pmatrix}3 & 6 & 4 & 5 & 6 & 2 & 4 & 5 & 6 & 2 & 4\end{pmatrix}$

Sub-structure K after imposing boundary conditions:

$K2 =$

	1	2	3	4	5	6	7	8	9	10	11	12
1	$1.88 \cdot 10^5$	$-4.91 \cdot 10^4$	0	0	$-1.389 \cdot 10^5$	0	0	0	0	0	$-4.91 \cdot 10^4$	$4.91 \cdot 10^4$
2	$-4.91 \cdot 10^4$	$1.88 \cdot 10^5$	0	0	0	0	0	0	0	$-1.389 \cdot 10^5$	$4.91 \cdot 10^4$	$-4.91 \cdot 10^4$
3	0	0	$1.88 \cdot 10^5$	$4.91 \cdot 10^4$	0	0	$-1.389 \cdot 10^5$	0	0	0	$-4.91 \cdot 10^4$	$-4.91 \cdot 10^4$
4	0	0	$4.91 \cdot 10^4$	$1.88 \cdot 10^5$	0	0	0	0	0	$-1.389 \cdot 10^5$	$-4.91 \cdot 10^4$	$-4.91 \cdot 10^4$
5	$-1.389 \cdot 10^5$	0	0	0	$1.88 \cdot 10^5$	$4.91 \cdot 10^4$	0	0	$-4.91 \cdot 10^4$	$-4.91 \cdot 10^4$	0	0
6	0	0	0	0	$4.91 \cdot 10^4$	$1.88 \cdot 10^5$	0	0	$-4.91 \cdot 10^4$	$-4.91 \cdot 10^4$	0	$-1.389 \cdot 10^5$
7	0	0	$-1.389 \cdot 10^5$	0	0	0	$1.88 \cdot 10^5$	$-4.91 \cdot 10^4$	$-4.91 \cdot 10^4$	$4.91 \cdot 10^4$	0	0
8	0	0	0	0	0	0	$-4.91 \cdot 10^4$	$1.88 \cdot 10^5$	$4.91 \cdot 10^4$	$-4.91 \cdot 10^4$	0	$-1.389 \cdot 10^5$
9	0	0	0	0	$-4.91 \cdot 10^4$	$-4.91 \cdot 10^4$	$-4.91 \cdot 10^4$	$4.91 \cdot 10^4$	$2.371 \cdot 10^5$	0	$-1.389 \cdot 10^5$	0
10	0	$-1.389 \cdot 10^5$	0	$-1.389 \cdot 10^5$	$-4.91 \cdot 10^4$	$-4.91 \cdot 10^4$	$4.91 \cdot 10^4$	$-4.91 \cdot 10^4$	0	$3.76 \cdot 10^5$	0	0
11	$-4.91 \cdot 10^4$	$4.91 \cdot 10^4$	$-4.91 \cdot 10^4$	$-4.91 \cdot 10^4$	0	0	0	0	$-1.389 \cdot 10^5$	0	$2.371 \cdot 10^5$	0
12	$4.91 \cdot 10^4$	$-4.91 \cdot 10^4$	$-4.91 \cdot 10^4$	$-4.91 \cdot 10^4$	0	$-1.389 \cdot 10^5$	0	$-1.389 \cdot 10^5$	0	0	0	$3.76 \cdot 10^5$

Sub-structure force vector:

$$(0 \quad 0 \quad 0 \quad -10000 \quad 0 \quad 0 \quad 0 \quad -10000 \quad 0 \quad 0 \quad 0 \quad 0)$$

Sub-structure 3

Element list: $(3 \quad 6 \quad 9 \quad 13 \quad 17 \quad 22 \quad 23 \quad 28 \quad 29)$

Local-Global Mapping array: $(0 \quad 0 \quad 1 \quad 4 \quad 0 \quad 0 \quad 3 \quad 5 \quad 0 \quad 0 \quad 2 \quad 6)$

Local node i: $(1 \quad 3 \quad 2 \quad 4 \quad 5 \quad 4 \quad 1 \quad 5 \quad 3)$

Local node j: $(4 \quad 5 \quad 6 \quad 5 \quad 6 \quad 3 \quad 5 \quad 2 \quad 6)$

Sub-structure K after imposing boundary conditions:

$K3 =$

	1	2	3	4	5	6	7	8	9	10	11	12
1	$1.8799 \cdot 10^5$	$-4.9105 \cdot 10^4$	0	0	0	0	$-1.389 \cdot 10^5$	0	$-4.9105 \cdot 10^4$	$4.9105 \cdot 10^4$	0	0
2	$-4.9105 \cdot 10^4$	$4.9105 \cdot 10^4$	0	0	0	0	0	0	$4.9105 \cdot 10^4$	$-4.9105 \cdot 10^4$	0	0
3	0	0	$1.8799 \cdot 10^5$	$4.9105 \cdot 10^4$	0	0	0	0	$-4.9105 \cdot 10^4$	$-4.9105 \cdot 10^4$	$-1.389 \cdot 10^5$	0
4	0	0	$4.9105 \cdot 10^4$	$4.9105 \cdot 10^4$	0	0	0	0	$-4.9105 \cdot 10^4$	$-4.9105 \cdot 10^4$	0	0
5	0	0	0	0	$2.371 \cdot 10^5$	0	$-4.9105 \cdot 10^4$	$-4.9105 \cdot 10^4$	$-1.389 \cdot 10^5$	0	$-4.9105 \cdot 10^4$	0
6	0	0	0	0	0	$9.8200 \cdot 10^4$	$-4.9105 \cdot 10^4$	$-4.9105 \cdot 10^4$	0	0	$4.9105 \cdot 10^4$	0
7	$-1.389 \cdot 10^5$	0	0	0	$-4.9105 \cdot 10^4$	$-4.9105 \cdot 10^4$	$1.8799 \cdot 10^5$	$4.9105 \cdot 10^4$	0	0	0	0
8	0	0	0	0	$-4.9105 \cdot 10^4$	$-4.9105 \cdot 10^4$	$4.9105 \cdot 10^4$	$1.8799 \cdot 10^5$	0	$-1.389 \cdot 10^5$	0	0
9	$-4.9105 \cdot 10^4$	$4.9105 \cdot 10^4$	$-4.9105 \cdot 10^4$	$-4.9105 \cdot 10^4$	$-1.389 \cdot 10^5$	0	0	0	$2.371 \cdot 10^5$	0	0	0
10	$4.9105 \cdot 10^4$	$-4.9105 \cdot 10^4$	$-4.9105 \cdot 10^4$	$-4.9105 \cdot 10^4$	0	0	0	$-1.389 \cdot 10^5$	0	$3.7500 \cdot 10^5$	0	0
11	0	0	$-1.389 \cdot 10^5$	0	$-4.9105 \cdot 10^4$	$4.9105 \cdot 10^4$	0	0	0	0	$1.8799 \cdot 10^5$	0
12	0	0	0	0	0	0	0	0	0	0	0	1

Sub-structure force vector: $\begin{pmatrix} 0 & 0 & 0 & 0 & 0 & 0 & 0 & 0 & 0 & 0 & 0 & 0 \end{pmatrix}$

- Construct Krr, Krc, Kcr, Kcc, fr and fc for all substructures

Sub-structure 1

$$Kcc1 = \begin{bmatrix} 1.88\times10^5 & 4.91\times10^4 & 0 & 0 \\ 4.91\times10^4 & 4.91\times10^4 & 0 & 0 \\ 0 & 0 & 1.88\times10^5 & -4.91\times10^4 \\ 0 & 0 & -4.91\times10^4 & 4.91\times10^4 \end{bmatrix}$$

$$Kcr1 = \begin{bmatrix} 0 & 0 & -1.389\times10^5 & 0 & -4.91\times10^4 & -4.91\times10^4 & 0 & 0 \\ 0 & 0 & 0 & 0 & -4.91\times10^4 & -4.91\times10^4 & 0 & 0 \\ 0 & 0 & 0 & 0 & -4.91\times10^4 & 4.91\times10^4 & 0 & 0 \\ 0 & 0 & 0 & 0 & 4.91\times10^4 & -4.91\times10^4 & 0 & 0 \end{bmatrix}$$

$$Krc1 = \begin{pmatrix} 0 & 0 & 0 & 0 \\ 0 & 0 & 0 & 0 \\ -1.389\times10^5 & 0 & 0 & 0 \\ 0 & 0 & 0 & 0 \\ -4.91\times10^4 & -4.91\times10^4 & -4.91\times10^4 & 4.91\times10^4 \\ -4.91\times10^4 & -4.91\times10^4 & 4.91\times10^4 & -4.91\times10^4 \\ 0 & 0 & 0 & 0 \\ 0 & 0 & 0 & 0 \end{pmatrix} \qquad Fc1 = \begin{pmatrix} 0 \\ 0 \\ 0 \\ 0 \end{pmatrix}$$

$$Krr1 = \begin{bmatrix} 2.371\times10^5 & 0 & -4.91\times10^4 & 4.91\times10^4 & -1.389\times10^5 & 0 & 0 & 0 \\ 0 & 9.821\times10^4 & 4.91\times10^4 & -4.91\times10^4 & 0 & 0 & 0 & 0 \\ -4.91\times10^4 & 4.91\times10^4 & 1.88\times10^5 & -4.91\times10^4 & 0 & 0 & 0 & 0 \\ 4.91\times10^4 & -4.91\times10^4 & -4.91\times10^4 & 1.88\times10^5 & 0 & -1.389\times10^5 & 0 & 0 \\ -1.389\times10^5 & 0 & 0 & 0 & 2.371\times10^5 & 0 & 0 & 0 \\ 0 & 0 & 0 & -1.389\times10^5 & 0 & 3.76\times10^5 & 0 & 0 \\ 0 & 0 & 0 & 0 & 0 & 0 & 1 & 0 \\ 0 & 0 & 0 & 0 & 0 & 0 & 0 & 1 \end{bmatrix}$$

$$Fr1 = \begin{Bmatrix} 0 \\ 0 \\ 0 \\ 0 \\ 0 \\ 0 \\ 0 \\ 0 \end{Bmatrix}$$

Sub-structure 2

$$Kcc2 = \begin{bmatrix} 1.88\times10^5 & -4.91\times10^4 & 0 & 0 & -1.389\times10^5 & 0 & 0 & 0 \\ -4.91\times10^4 & 1.88\times10^5 & 0 & 0 & 0 & 0 & 0 & 0 \\ 0 & 0 & 1.88\times10^5 & 4.91\times10^4 & 0 & 0 & -1.389\times10^5 & 0 \\ 0 & 0 & 4.91\times10^4 & 1.88\times10^5 & 0 & 0 & 0 & 0 \\ -1.389\times10^5 & 0 & 0 & 0 & 1.88\times10^5 & 4.91\times10^4 & 0 & 0 \\ 0 & 0 & 0 & 0 & 4.91\times10^4 & 1.88\times10^5 & 0 & 0 \\ 0 & 0 & -1.389\times10^5 & 0 & 0 & 0 & 1.88\times10^5 & -4.91\times10^4 \\ 0 & 0 & 0 & 0 & 0 & 0 & -4.91\times10^4 & 1.88\times10^5 \end{bmatrix}$$

$$Fc2 = \begin{Bmatrix} 0 \\ 0 \\ 0 \\ -1\times10^4 \\ 0 \\ 0 \\ 0 \\ -1\times10^4 \end{Bmatrix}$$

$$Kcr2 = \begin{pmatrix} 0 & 0 & -4.91\times10^4 & 4.91\times10^4 \\ 0 & -1.389\times10^5 & 4.91\times10^4 & -4.91\times10^4 \\ 0 & 0 & -4.91\times10^4 & -4.91\times10^4 \\ 0 & -1.389\times10^5 & -4.91\times10^4 & -4.91\times10^4 \\ -4.91\times10^4 & -4.91\times10^4 & 0 & 0 \\ -4.91\times10^4 & -4.91\times10^4 & 0 & -1.389\times10^5 \\ -4.91\times10^4 & 4.91\times10^4 & 0 & 0 \\ 4.91\times10^4 & -4.91\times10^4 & 0 & -1.389\times10^5 \end{pmatrix}$$

$$Fr2 = \begin{pmatrix} 0 \\ 0 \\ 0 \\ 0 \end{pmatrix}$$

$$Krc2 = \begin{bmatrix} 0 & 0 & 0 & 0 & -4.91\times10^4 & -4.91\times10^4 & -4.91\times10^4 & 4.91\times10^4 \\ 0 & -1.389\times10^5 & 0 & -1.389\times10^5 & -4.91\times10^4 & -4.91\times10^4 & 4.91\times10^4 & -4.91\times10^4 \\ -4.91\times10^4 & 4.91\times10^4 & -4.91\times10^4 & -4.91\times10^4 & 0 & 0 & 0 & 0 \\ 4.91\times10^4 & -4.91\times10^4 & -4.91\times10^4 & -4.91\times10^4 & 0 & -1.389\times10^5 & 0 & -1.389\times10^5 \end{bmatrix}$$

$$Krr2 = \begin{pmatrix} 2.371 \times 10^5 & 0 & -1.389 \times 10^5 & 0 \\ 0 & 3.76 \times 10^5 & 0 & 0 \\ -1.389 \times 10^5 & 0 & 2.371 \times 10^5 & 0 \\ 0 & 0 & 0 & 3.76 \times 10^5 \end{pmatrix}$$

Sub-structure 3

$$Kcc3 = \begin{bmatrix} 1.88 \times 10^5 & -4.91 \times 10^4 & 0 & 0 \\ -4.91 \times 10^4 & 4.91 \times 10^4 & 0 & 0 \\ 0 & 0 & 1.88 \times 10^5 & 4.91 \times 10^4 \\ 0 & 0 & 4.91 \times 10^4 & 4.91 \times 10^4 \end{bmatrix}$$

$$Kcr3 = \begin{bmatrix} 0 & 0 & -1.389 \times 10^5 & 0 & -4.91 \times 10^4 & 4.91 \times 10^4 & 0 & 0 \\ 0 & 0 & 0 & 0 & 4.91 \times 10^4 & -4.91 \times 10^4 & 0 & 0 \\ 0 & 0 & 0 & 0 & -4.91 \times 10^4 & 4.91 \times 10^4 & -1.389 \times 10^5 & 0 \\ 0 & 0 & 0 & 0 & 4.91 \times 10^4 & -4.91 \times 10^4 & 0 & 0 \end{bmatrix}$$

$$Krr3 = \begin{bmatrix} 2.371 \times 10^5 & 0 & -4.91 \times 10^4 & -4.91 \times 10^4 & -1.389 \times 10^5 & 0 & -4.91 \times 10^4 \\ 0 & 9.821 \times 10^4 & -4.91 \times 10^4 & -4.91 \times 10^4 & 0 & 0 & 4.91 \times 10^4 \\ -4.91 \times 10^4 & -4.91 \times 10^4 & 1.88 \times 10^5 & 4.91 \times 10^4 & 0 & 0 & 0 \\ -4.91 \times 10^4 & -4.91 \times 10^4 & 4.91 \times 10^4 & 1.88 \times 10^5 & 0 & -1.389 \times 10^5 & 0 \\ -1.389 \times 10^5 & 0 & 0 & 0 & 2.371 \times 10^5 & 0 & 0 \\ 0 & 0 & 0 & -1.389 \times 10^5 & 0 & 3.76 \times 10^5 & 0 \\ -4.91 \times 10^4 & 4.91 \times 10^4 & 0 & 0 & 0 & 0 & 1.88 \times 10^5 \\ 0 & 0 & 0 & 0 & 0 & 0 & 0 \end{bmatrix}$$

$$Fc3 = \begin{Bmatrix} 0 \\ 0 \\ 0 \\ 0 \end{Bmatrix}$$

$$Krc3 = \begin{bmatrix} 0 & 0 & 0 & 0 \\ 0 & 0 & 0 & 0 \\ -1.389 \times 10^5 & 0 & 0 & 0 \\ 0 & 0 & 0 & 0 \\ -4.91 \times 10^4 & 4.91 \times 10^4 & -4.91 \times 10^4 & -4.91 \times 10^4 \\ 4.91 \times 10^4 & -4.91 \times 10^4 & -4.91 \times 10^4 & -4.91 \times 10^4 \\ 0 & 0 & -1.389 \times 10^5 & 0 \\ 0 & 0 & 0 & 0 \end{bmatrix} \qquad Fr3 = \begin{Bmatrix} 0 \\ 0 \\ 0 \\ 0 \\ 0 \\ 0 \\ 0 \\ 0 \end{Bmatrix}$$

- Construct Boolean Transformation Matrix
Sub-structure 1

$$
Br1 = \begin{pmatrix} 1 & 0 & 0 & 0 & 0 & 0 & 0 & 0 \\ 0 & 1 & 0 & 0 & 0 & 0 & 0 & 0 \\ 0 & 0 & 0 & 0 & 0 & 0 & 0 & 0 \\ 0 & 0 & 0 & 0 & 0 & 0 & 0 & 0 \end{pmatrix}
\qquad
Bc1 = \begin{pmatrix} 1 & 0 & 0 & 0 & 0 & 0 & 0 & 0 \\ 0 & 1 & 0 & 0 & 0 & 0 & 0 & 0 \\ 0 & 0 & 1 & 0 & 0 & 0 & 0 & 0 \\ 0 & 0 & 0 & 1 & 0 & 0 & 0 & 0 \end{pmatrix}
$$

Sub-structure 2

$$
Br2 = \begin{pmatrix} -1 & 0 & 0 & 0 \\ 0 & -1 & 0 & 0 \\ 0 & 0 & 1 & 0 \\ 0 & 0 & 0 & 1 \end{pmatrix}
\qquad
Bc2 = \begin{pmatrix} 1 & 0 & 0 & 0 & 0 & 0 & 0 & 0 \\ 0 & 1 & 0 & 0 & 0 & 0 & 0 & 0 \\ 0 & 0 & 1 & 0 & 0 & 0 & 0 & 0 \\ 0 & 0 & 0 & 1 & 0 & 0 & 0 & 0 \\ 0 & 0 & 0 & 0 & 1 & 0 & 0 & 0 \\ 0 & 0 & 0 & 0 & 0 & 1 & 0 & 0 \\ 0 & 0 & 0 & 0 & 0 & 0 & 1 & 0 \\ 0 & 0 & 0 & 0 & 0 & 0 & 0 & 1 \end{pmatrix}
$$

Sub-structure 3

$$
Br3 = \begin{pmatrix} 0 & 0 & 0 & 0 & 0 & 0 & 0 & 0 \\ 0 & 0 & 0 & 0 & 0 & 0 & 0 & 0 \\ -1 & 0 & 0 & 0 & 0 & 0 & 0 & 0 \\ 0 & -1 & 0 & 0 & 0 & 0 & 0 & 0 \end{pmatrix}
\qquad
Bc3 = \begin{pmatrix} 0 & 0 & 0 & 0 & 1 & 0 & 0 & 0 \\ 0 & 0 & 0 & 0 & 0 & 1 & 0 & 0 \\ 0 & 0 & 0 & 0 & 0 & 0 & 1 & 0 \\ 0 & 0 & 0 & 0 & 0 & 0 & 0 & 1 \end{pmatrix}
$$

Construct Krc, Kcr, Kcc and Br

$$
Krc\_1 = \begin{pmatrix} 0 & 0 & 0 & 0 & 0 & 0 & 0 & 0 \\ 0 & 0 & 0 & 0 & 0 & 0 & 0 & 0 \\ -1.389 \times 10^5 & 0 & 0 & 0 & 0 & 0 & 0 & 0 \\ 0 & 0 & 0 & 0 & 0 & 0 & 0 & 0 \\ -4.91 \times 10^4 & -4.91 \times 10^4 & -4.91 \times 10^4 & 4.91 \times 10^4 & 0 & 0 & 0 & 0 \\ -4.91 \times 10^4 & -4.91 \times 10^4 & 4.91 \times 10^4 & -4.91 \times 10^4 & 0 & 0 & 0 & 0 \\ 0 & 0 & 0 & 0 & 0 & 0 & 0 & 0 \\ 0 & 0 & 0 & 0 & 0 & 0 & 0 & 0 \end{pmatrix}
$$

$$
Krc\_2 = \begin{bmatrix}
0 & 0 & 0 & 0 & -4.91\times10^4 & -4.91\times10^4 & -4.91\times10^4 & 4.91\times10^4 \\
0 & -1.389\times10^5 & 0 & -1.389\times10^5 & -4.91\times10^4 & -4.91\times10^4 & 4.91\times10^4 & -4.91\times10^4 \\
-4.91\times10^4 & 4.91\times10^4 & -4.91\times10^4 & -4.91\times10^4 & 0 & 0 & 0 & 0 \\
4.91\times10^4 & -4.91\times10^4 & -4.91\times10^4 & -4.91\times10^4 & 0 & -1.389\times10^5 & 0 & -1.389\times10^5
\end{bmatrix}
$$

$$
Krc\_3 = \begin{pmatrix}
0 & 0 & 0 & 0 & 0 & 0 & 0 & 0 \\
0 & 0 & 0 & 0 & 0 & 0 & 0 & 0 \\
0 & 0 & 0 & 0 & -1.389\times10^5 & 0 & 0 & 0 \\
0 & 0 & 0 & 0 & 0 & 0 & 0 & 0 \\
0 & 0 & 0 & 0 & -4.91\times10^4 & 4.91\times10^4 & -4.91\times10^4 & -4.91\times10^4 \\
0 & 0 & 0 & 0 & 4.91\times10^4 & -4.91\times10^4 & -4.91\times10^4 & -4.91\times10^4 \\
0 & 0 & 0 & 0 & 0 & 0 & -1.389\times10^5 & 0 \\
0 & 0 & 0 & 0 & 0 & 0 & 0 & 0
\end{pmatrix}
$$

$$
Kcr\_1 = \begin{pmatrix}
0 & 0 & -1.389\times10^5 & 0 & -4.91\times10^4 & -4.91\times10^4 & 0 & 0 \\
0 & 0 & 0 & 0 & -4.91\times10^4 & -4.91\times10^4 & 0 & 0 \\
0 & 0 & 0 & 0 & -4.91\times10^4 & 4.91\times10^4 & 0 & 0 \\
0 & 0 & 0 & 0 & 4.91\times10^4 & -4.91\times10^4 & 0 & 0 \\
0 & 0 & 0 & 0 & 0 & 0 & 0 & 0 \\
0 & 0 & 0 & 0 & 0 & 0 & 0 & 0 \\
0 & 0 & 0 & 0 & 0 & 0 & 0 & 0 \\
0 & 0 & 0 & 0 & 0 & 0 & 0 & 0
\end{pmatrix}
$$

$$
Kcr\_2 = \begin{pmatrix}
0 & 0 & -4.91\times10^4 & 4.91\times10^4 \\
0 & -1.389\times10^5 & 4.91\times10^4 & -4.91\times10^4 \\
0 & 0 & -4.91\times10^4 & -4.91\times10^4 \\
0 & -1.389\times10^5 & -4.91\times10^4 & -4.91\times10^4 \\
-4.91\times10^4 & -4.91\times10^4 & 0 & 0 \\
-4.91\times10^4 & -4.91\times10^4 & 0 & -1.389\times10^5 \\
-4.91\times10^4 & 4.91\times10^4 & 0 & 0 \\
4.91\times10^4 & -4.91\times10^4 & 0 & -1.389\times10^5
\end{pmatrix}
$$

$$Kcr\_3 = \begin{pmatrix} 0 & 0 & 0 & 0 & 0 & 0 & 0 & 0 \\ 0 & 0 & 0 & 0 & 0 & 0 & 0 & 0 \\ 0 & 0 & 0 & 0 & 0 & 0 & 0 & 0 \\ 0 & 0 & 0 & 0 & 0 & 0 & 0 & 0 \\ 0 & 0 & -1.389 \times 10^5 & 0 & -4.91 \times 10^4 & 4.91 \times 10^4 & 0 & 0 \\ 0 & 0 & 0 & 0 & 4.91 \times 10^4 & -4.91 \times 10^4 & 0 & 0 \\ 0 & 0 & 0 & 0 & -4.91 \times 10^4 & -4.91 \times 10^4 & -1.389 \times 10^5 & 0 \\ 0 & 0 & 0 & 0 & -4.91 \times 10^4 & -4.91 \times 10^4 & 0 & 0 \end{pmatrix}$$

$$Kcc\_1 = \begin{pmatrix} 1.88 \times 10^5 & 4.91 \times 10^4 & 0 & 0 & 0 & 0 & 0 & 0 \\ 4.91 \times 10^4 & 4.91 \times 10^4 & 0 & 0 & 0 & 0 & 0 & 0 \\ 0 & 0 & 1.88 \times 10^5 & -4.91 \times 10^4 & 0 & 0 & 0 & 0 \\ 0 & 0 & -4.91 \times 10^4 & 4.91 \times 10^4 & 0 & 0 & 0 & 0 \\ 0 & 0 & 0 & 0 & 0 & 0 & 0 & 0 \\ 0 & 0 & 0 & 0 & 0 & 0 & 0 & 0 \\ 0 & 0 & 0 & 0 & 0 & 0 & 0 & 0 \\ 0 & 0 & 0 & 0 & 0 & 0 & 0 & 0 \end{pmatrix}$$

$$Kcc\_2 = \begin{bmatrix} 1.88 \times 10^5 & -4.91 \times 10^4 & 0 & 0 & -1.389 \times 10^5 & 0 & 0 & 0 \\ -4.91 \times 10^4 & 1.88 \times 10^5 & 0 & 0 & 0 & 0 & 0 & 0 \\ 0 & 0 & 1.88 \times 10^5 & 4.91 \times 10^4 & 0 & 0 & -1.389 \times 10^5 & 0 \\ 0 & 0 & 4.91 \times 10^4 & 1.88 \times 10^5 & 0 & 0 & 0 & 0 \\ -1.389 \times 10^5 & 0 & 0 & 0 & 1.88 \times 10^5 & 4.91 \times 10^4 & 0 & 0 \\ 0 & 0 & 0 & 0 & 4.91 \times 10^4 & 1.88 \times 10^5 & 0 & 0 \\ 0 & 0 & -1.389 \times 10^5 & 0 & 0 & 0 & 1.88 \times 10^5 & -4.91 \times 10^4 \\ 0 & 0 & 0 & 0 & 0 & 0 & -4.91 \times 10^4 & 1.88 \times 10^5 \end{bmatrix}$$

$$Kcc\_3 = \begin{pmatrix} 0 & 0 & 0 & 0 & 0 & 0 & 0 & 0 \\ 0 & 0 & 0 & 0 & 0 & 0 & 0 & 0 \\ 0 & 0 & 0 & 0 & 0 & 0 & 0 & 0 \\ 0 & 0 & 0 & 0 & 0 & 0 & 0 & 0 \\ 0 & 0 & 0 & 0 & 1.88 \times 10^5 & -4.91 \times 10^4 & 0 & 0 \\ 0 & 0 & 0 & 0 & -4.91 \times 10^4 & 4.91 \times 10^4 & 0 & 0 \\ 0 & 0 & 0 & 0 & 0 & 0 & 1.88 \times 10^5 & 4.91 \times 10^4 \\ 0 & 0 & 0 & 0 & 0 & 0 & 4.91 \times 10^4 & 4.91 \times 10^4 \end{pmatrix}$$

- Construct dr, Frr, Frc

$$dr\_1 = \begin{pmatrix} 0 \\ 0 \\ 0 \\ 0 \end{pmatrix} \qquad dr\_2 = \begin{pmatrix} 0 \\ 0 \\ 0 \\ 0 \end{pmatrix} \qquad dr\_3 = \begin{pmatrix} 0 \\ 0 \\ 0 \\ 0 \end{pmatrix} \qquad dr\_ = \begin{pmatrix} 0 \\ 0 \\ 0 \\ 0 \end{pmatrix}$$

$$Frr\_1 = \begin{pmatrix} 8.004 \times 10^{-6} & -2.511 \times 10^{-6} & 0 & 0 \\ -2.511 \times 10^{-6} & 1.416 \times 10^{-5} & 0 & 0 \\ 0 & 0 & 0 & 0 \\ 0 & 0 & 0 & 0 \end{pmatrix}$$

$$Frr\_2 = \begin{pmatrix} 6.421 \times 10^{-6} & 0 & -3.761 \times 10^{-6} & 0 \\ 0 & 2.66 \times 10^{-6} & 0 & 0 \\ -3.761 \times 10^{-6} & 0 & 6.421 \times 10^{-6} & 0 \\ 0 & 0 & 0 & 2.66 \times 10^{-6} \end{pmatrix}$$

$$Frr\_3 = \begin{pmatrix} 0 & 0 & 0 & 0 \\ 0 & 0 & 0 & 0 \\ 0 & 0 & 8.501 \times 10^{-6} & 1.458 \times 10^{-6} \\ 0 & 0 & 1.458 \times 10^{-6} & 1.64 \times 10^{-5} \end{pmatrix}$$

$$Frc\_1 = \begin{pmatrix} -0.444 & -0.174 & -0.286 & 0.286 & 0 & 0 & 0 & 0 \\ 0.411 & -0.016 & 0.161 & -0.161 & 0 & 0 & 0 & 0 \\ 0 & 0 & 0 & 0 & 0 & 0 & 0 & 0 \\ 0 & 0 & 0 & 0 & 0 & 0 & 0 & 0 \end{pmatrix}$$

$$Frc\_2 = \begin{pmatrix} 0.185 & -0.185 & 0.185 & 0.185 & 0.315 & 0.315 & 0.315 & -0.315 \\ 0 & 0.369 & 0 & 0.369 & 0.131 & 0.131 & -0.131 & 0.131 \\ -0.315 & 0.315 & -0.315 & -0.315 & -0.185 & -0.185 & -0.185 & 0.185 \\ 0.131 & -0.131 & -0.131 & -0.131 & 0 & -0.369 & 0 & -0.369 \end{pmatrix}$$

$$Frc\_3 = \begin{pmatrix} 0 & 0 & 0 & 0 & 0 & 0 & 0 & 0 \\ 0 & 0 & 0 & 0 & 0 & 0 & 0 & 0 \\ 0 & 0 & 0 & 0 & 0.447 & -0.192 & 0.553 & 0.297 \\ 0 & 0 & 0 & 0 & 0.405 & 0.053 & -0.405 & 0.137 \end{pmatrix}$$

- Construct $-fc^*$, $-Kcc^*$ and Fcr

$$fcs1 = \begin{pmatrix} 0 \\ 0 \\ 0 \\ 0 \\ 0 \\ 0 \\ 0 \\ 0 \end{pmatrix} \qquad fcs2 = \begin{pmatrix} 0 \\ 0 \\ 0 \\ 0 \\ 0 \\ 0 \\ 0 \\ 0 \end{pmatrix} \qquad fcs3 = \begin{pmatrix} 0 \\ 0 \\ 0 \\ 0 \\ 0 \\ 0 \\ 0 \\ 0 \end{pmatrix}$$

$$Kccs1 = \begin{pmatrix} 1.733 \times 10^5 & 3.278 \times 10^4 & 1.311 \times 10^4 & -1.311 \times 10^4 & 0 & 0 & 0 & 0 \\ 3.278 \times 10^4 & 2.342 \times 10^4 & 6.956 \times 10^3 & -6.956 \times 10^3 & 0 & 0 & 0 & 0 \\ 1.311 \times 10^4 & 6.956 \times 10^3 & 2.984 \times 10^4 & -2.984 \times 10^4 & 0 & 0 & 0 & 0 \\ -1.311 \times 10^4 & -6.956 \times 10^3 & -2.984 \times 10^4 & 2.984 \times 10^4 & 0 & 0 & 0 & 0 \\ 0 & 0 & 0 & 0 & 0 & 0 & 0 & 0 \\ 0 & 0 & 0 & 0 & 0 & 0 & 0 & 0 \\ 0 & 0 & 0 & 0 & 0 & 0 & 0 & 0 \\ 0 & 0 & 0 & 0 & 0 & 0 & 0 & 0 \end{pmatrix}$$

$$Kccs2 = \begin{bmatrix} 2.19 \times 10^4 & -2.19 \times 10^4 & 9.07 \times 10^3 & 9.07 \times 10^3 & 9.07 \times 10^3 & -9.07 \times 10^3 & 9.07 \times 10^3 & -2.721 \times 10^4 \\ -2.19 \times 10^4 & 7.32 \times 10^4 & -9.07 \times 10^3 & 4.224 \times 10^3 & 9.07 \times 10^3 & 2.721 \times 10^4 & -2.721 \times 10^4 & 4.535 \times 10^4 \\ 9.07 \times 10^3 & -9.07 \times 10^3 & 2.19 \times 10^4 & 2.19 \times 10^4 & 9.07 \times 10^3 & 2.721 \times 10^4 & 9.07 \times 10^3 & 9.07 \times 10^3 \\ 9.07 \times 10^3 & 4.224 \times 10^3 & 2.19 \times 10^4 & 7.32 \times 10^4 & 2.721 \times 10^4 & 4.535 \times 10^4 & -9.07 \times 10^3 & 2.721 \times 10^4 \\ 9.07 \times 10^3 & 9.07 \times 10^3 & 9.07 \times 10^3 & 2.721 \times 10^4 & 2.19 \times 10^4 & 2.19 \times 10^4 & 9.07 \times 10^3 & -9.07 \times 10^3 \\ -9.07 \times 10^3 & 2.721 \times 10^4 & 2.721 \times 10^4 & 4.535 \times 10^4 & 2.19 \times 10^4 & 7.32 \times 10^4 & 9.07 \times 10^3 & 4.224 \times 10^3 \\ 9.07 \times 10^3 & -2.721 \times 10^4 & 9.07 \times 10^3 & -9.07 \times 10^3 & 9.07 \times 10^3 & 9.07 \times 10^3 & 2.19 \times 10^4 & -2.19 \times 10^4 \\ -2.721 \times 10^4 & 4.535 \times 10^4 & 9.07 \times 10^3 & 2.721 \times 10^4 & -9.07 \times 10^3 & 4.224 \times 10^3 & -2.19 \times 10^4 & 7.32 \times 10^4 \end{bmatrix}$$

$$Kccs3 = \begin{pmatrix} 0 & 0 & 0 & 0 & 0 & 0 & 0 & 0 \\ 0 & 0 & 0 & 0 & 0 & 0 & 0 & 0 \\ 0 & 0 & 0 & 0 & 0 & 0 & 0 & 0 \\ 0 & 0 & 0 & 0 & 0 & 0 & 0 & 0 \\ 0 & 0 & 0 & 0 & 1.733 \times 10^5 & -3.289 \times 10^4 & 1.47 \times 10^4 & 1.318 \times 10^4 \\ 0 & 0 & 0 & 0 & -3.289 \times 10^4 & 2.401 \times 10^4 & -1.622 \times 10^4 & -7.349 \times 10^3 \\ 0 & 0 & 0 & 0 & 1.47 \times 10^4 & -1.622 \times 10^4 & 1.733 \times 10^5 & 3.593 \times 10^4 \\ 0 & 0 & 0 & 0 & 1.318 \times 10^4 & -7.349 \times 10^3 & 3.593 \times 10^4 & 3.01 \times 10^4 \end{pmatrix}$$

$$
Fcr\_1 = \begin{pmatrix}
-0.444 & 0.411 & 0 & 0 \\
-0.174 & -0.016 & 0 & 0 \\
-0.286 & 0.161 & 0 & 0 \\
0.286 & -0.161 & 0 & 0 \\
0 & 0 & 0 & 0 \\
0 & 0 & 0 & 0 \\
0 & 0 & 0 & 0 \\
0 & 0 & 0 & 0
\end{pmatrix} \;.
$$

$$
Fcr\_2 = \begin{pmatrix}
0.185 & 0 & -0.315 & 0.131 \\
-0.185 & 0.369 & 0.315 & -0.131 \\
0.185 & 0 & -0.315 & -0.131 \\
0.185 & 0.369 & -0.315 & -0.131 \\
0.315 & 0.131 & -0.185 & 0 \\
0.315 & 0.131 & -0.185 & -0.369 \\
0.315 & -0.131 & -0.185 & 0 \\
-0.315 & 0.131 & 0.185 & -0.369
\end{pmatrix}
$$

$$
Fcr\_3 = \begin{pmatrix}
0 & 0 & 0 & 0 \\
0 & 0 & 0 & 0 \\
0 & 0 & 0 & 0 \\
0 & 0 & 0 & 0 \\
0 & 0 & 0.447 & 0.405 \\
0 & 0 & -0.192 & 0.053 \\
0 & 0 & 0.553 & -0.405 \\
0 & 0 & 0.297 & 0.137
\end{pmatrix}
$$

- Construct $K_{\lambda\lambda}$ and d_λ

$$
K_{\lambda\lambda=} \begin{pmatrix}
2.025 \times 10^{-5} & 0 & -5.966 \times 10^{-6} & 0 \\
0 & 3.075 \times 10^{-5} & 0 & 4.159 \times 10^{-7} \\
-5.966 \times 10^{-6} & 0 & 2.025 \times 10^{-5} & 0 \\
0 & 4.159 \times 10^{-7} & 0 & 3.075 \times 10^{-5}
\end{pmatrix}
$$

$$
d_\lambda = \begin{pmatrix}
0.029 \\
0.142 \\
0.029 \\
-0.142
\end{pmatrix}
$$

Solve for λ

$$d_\lambda = \begin{pmatrix} 2.062 \times 10^3 \\ 4.681 \times 10^3 \\ 2.062 \times 10^3 \\ -4.681 \times 10^3 \end{pmatrix}$$

By using λ, solve for corner displacements

$$u_c = \begin{pmatrix} 0.049377 \\ -0.091078 \\ 0.021429 \\ -0.146337 \\ 0.023325 \\ -0.091078 \\ 0.051273 \\ -0.146337 \end{pmatrix}$$

Then, solve for remainder displacements

$$u_{r1}^T = \begin{pmatrix} 0.049299 & -0.109908 & 0.06165 & -0.062844 & 0.054988 & -0.050571 & 0 & 0 \end{pmatrix}$$

$$u_{r2}^T = \begin{pmatrix} 0.049299 & -0.109908 & 0.023403 & -0.109908 \end{pmatrix}$$

$$u_{r3}^T = \begin{pmatrix} 0.023403 & -0.109908 & 0.011052 & -0.062844 & 0.017714 & -0.050571 & 0.072702 & 0 \end{pmatrix}$$

Remarks

Changing the location of corner dofs will make no difference in the solution vectors as long as the adjacent substructures are connected by enough corner dofs.

6.14 Multi-Level Sub-Domains and Multi-Frontal Solver [6.13-6.20]

Multi-level sub-structures (or sub-domains) have been utilized and documented in the literature [6.18-6.20]. While the concepts of multi-level sub-structures do create additional complication for programming coding (more complex bookkeeping for intermediate results are required, etc.), it also offers substantial computational and especially communication advantages in a massively parallel computer environment. A simple 2-D, square domain (shown in Figure 6.14) is partitioned into 4 sub-domains, and is assigned to 4 different processors for parallel computation purposes. Assume the "primal" domain decomposition (DD) formulation of Section 6.5 is utilized here.

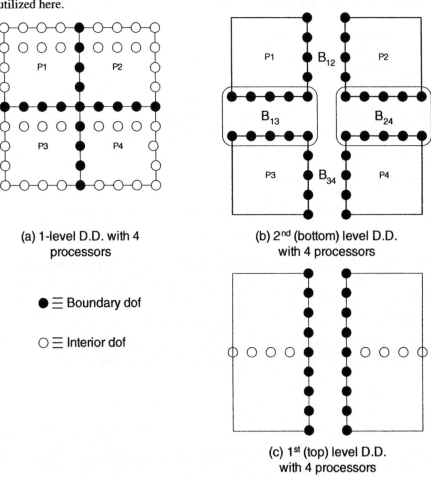

(a) 1-level D.D. with 4 processors

(b) 2nd (bottom) level D.D. with 4 processors

● ≡ Boundary dof

○ ≡ Interior dof

(c) 1st (top) level D.D. with 4 processors

Figure 6.14 Single and Multi-level D.D. Formulation

As can be seen from Figure 6.14a, the total (or system) number of boundary dof (assuming each node has only 1 dof) for a single level D. D. formulation is 17 dof. However, if multi-level D. D. formulation is used, then the maximum number of boundary dof involved in the 2nd (or bottom) level is 13 (see Figure 6.14b), and the maximum number of boundary dof involved in the 1st (or top) level is 9. With respect to the top level, the 5 boundary nodes indicated in Figure 6.14b will be considered as interior node in Figure 6.14c. The total (system) number of boundary dof is directly related to "communication" time among sub-domains' processors. Using multi-level sub-domain concepts will help reduce the total number of boundary dof from 17 (see Figure 6.14a) to only 9 (see Figure 6.14 b and Figure 6.14c).

While the serial version of the sparse solver developed and discussed in Chapter 3 of this book can be used within each sub-domain, shown in Figure 6.14, a multi-frontal sparse solver can also be integrated into the general DD formulations.

The concept of the multi-frontal method was first introduced by Duff and Reid [6.21] as a generalization of the frontal method of Irons [6.22]. The main idea of the frontal solution is to eliminate the variable while assembling the equations. As soon as the coefficients of an equation are completely assembled from the contributions of all relevant elements, the corresponding variables can be eliminated. Therefore, the complete structural stiffness matrix is never formed. The elimination process of the frontal method is illustrated in Figure 6.15.

The difference between the elimination process of the single-frontal method and that of the multi-frontal method can be understood by comparing Figures 6.15 - 6.16. By the single-frontal method, one front is used and spreads out all over the whole domain as fully assembled degrees of freedom (DOFs) are eliminated from the front. By the multi-frontal method, a given domain is divided into two sub-domains recursively, and the internal DOFs of each sub-domain are eliminated first by the frontal method. The remaining interface DOFs of each sub-domain, after eliminating internal DOFs, become new fronts, and they are merged with each other recursively in the reverse order. At each merging stage, fully assembled DOFs are eliminated immediately.

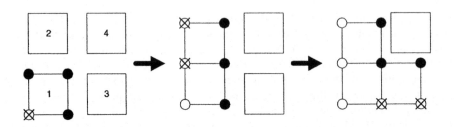

● : active DOF　　　⊠ : DOF eleminated at　　　○ : DOF eleminated at
　　　　　　　　　　　　current step　　　　　　　　previous step

Figure 6.15 Elimination Process of the Frontal Method.

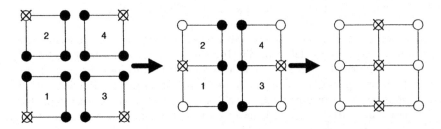

● : active DOF　　　⊠ : DOF eleminated at　　　○ : DOF eleminated at
　　　　　　　　　　　　current step　　　　　　　　previous step

Figure 6.16 Elimination Process of the Multi-Frontal Method

6.15 Iterative Solution with Successive Right-Hand Sides [6.23 – 6.24]

To keep the discussion more general, our objective here is to solve the system of "unsymmetrical," linear equations, which can be expressed in the matrix notations as:

$$[A]\,\vec{x} = \vec{b} \tag{6.381}$$

where [A] is a given N×N unsymmetrical matrix, and $\vec{b}$ is a given, single right-hand-side vector. A system of "symmetrical" linear equations can be treated as a special case of "unsymmetrical" equations. However, more computational efficiency can be realized if specific algorithms (such as the Conjugate Gradient algorithms) that exploit the symmetrical property are used. Successive right-hand-side vectors will be discussed near the end of this section.

Solving Eq.(6.381) for the unknown vector $\vec{x}$ is equivalent to either of the following optimization problems:

$$\text{Minimize} \quad \Psi_1 \equiv \frac{1}{2} x^T A\, x - x^T b \tag{6.382}$$
$$\quad\quad\quad\quad\; x \in R^N$$

or

$$\text{Minimize} \quad \underset{x \in R^N}{\psi_2} \equiv (A \vec{x} - \vec{b})^T * (A \vec{x} - \vec{b}) \tag{6.383}$$

Equation(6.382) is preferred if the case matrix [A] is "symmetrical" while Eq.(6.383) is suggested if matrix [A] is "unsymmetrical."

Remarks

[1]   The necessary condition to minimize a function is to require its gradient (or derivative) to be equal to zero. From Eq.(6.382), one obtains:

$$\nabla \psi_1^T \equiv \left(\frac{\partial \psi_1}{\partial \vec{x}} \right)^T = [A]\vec{x} - b = \vec{0} \tag{6.384}$$

Thus Eq.(6.384) is equivalent to Eq.(6.381).

[2]   From Eq.(6.383), one obtains:

$$\text{Min. } \psi_2^T = x^T A^T A x - x^T A^T b - b^T A x + b^T b \tag{6.385}$$

Since $b^T A x$ is a scalar quantity, hence

$$b^T A x = x^T A^T b \tag{6.386}$$

Therefore, Eq.(6.385) becomes

$$\text{Min. } \psi_2^T = x^T A^T A x - 2 x^T A^T b + b^T b \tag{6.387}$$

The gradient of Eq.(6.387) can be obtained as:

$$\nabla \psi_2^T \equiv 2 A^T A x - 2 A^T b = \vec{0} \tag{6.388}$$

or

$$\nabla \psi_2^T \equiv A^T (A x - b) = \vec{0} \tag{6.389}$$

Eq.(6.389) is also equivalent to Eq.(6.381).

The minimization problem, described by Eq.(6.382) or Eq.(6.383), can be iteratively solved by the following step-by-step procedures as indicated in Table 6.9.

Table 6.9 Step-by-Step Iterative Optimization Procedures

Step 0	Initial guessed vector: $\vec{x} = \overrightarrow{x^{(0)}}$; and set iterarion count i = 0 $\qquad$ (6.390)
Step 1	Set i=i+1 at the current design point, find the "search direction" to travel, $s^{(i-1)}$?
Step 2	Find the step-size, α (how far should we travel along a given direction

---

$s^{(i-1)}$)?

Step 3   Find the updated, improved design, $x^{(i)}$?

$$x^{(i)} = x^{(i-1)} + \alpha s^{(i-1)} \tag{6.391}$$

Step 4   Convergence test ??

Convergence is achieved if :

$$\left\| \nabla \psi_1^T \right\| \leq \text{Tolerance} \tag{6.392}$$

or

$$\left\| \nabla \psi_2^T \right\| \leq \text{Tolerance} \tag{6.393}$$

and / or

$$\left\| x^{(i)} - x^{(i-1)} \right\| \leq \text{Tolerance} \tag{6.394}$$

If convergence is achieved (or, iteration [#] i=max. [#] of iterations allowed) then stop the process.

Else

Return to step1

End if

In Table 6.9, the two most important steps are to find "the search direction" $s^{(i-1)}$ to travel from the current design point (see Step 1) and to find "the step size" α (see Step 2).

[A] How to Find the "Step Size," α, along a Given Direction $\vec{s}$??

Assuming the search direction $\vec{s}^{(i)}$ has already been found, then the new, improved design point can be computed as (see Step 3 of Table 6.9):

$$x^{(i+1)} = x^{(i)} + \alpha s^{(i)} \tag{6.395}$$

Thus, Eqs.(6.382 – 6.383) become minimization problems with only one variable (= α), as follows:

$$\text{Min.} \quad \psi_1 = \frac{1}{2} \left(x^i + \alpha_1 s^i \right)^T A \left(x^i + \alpha_1 s^i \right) - \left(x^i + \alpha_1 s^i \right)^T b \tag{6.396}$$

and

$$\text{Min.} \quad \psi_2 = \left[A \left(x^i + \alpha_2 s^i \right) - b \right]^T * \left[A \left(x^i + \alpha_2 s^i \right) - b \right] \tag{6.397}$$

In order to minimize the function values ψ_1 (or ψ_2), one needs to satisfy the following requirements:

$$\frac{d\psi_1}{d\alpha_1} = 0 \tag{6.398}$$

and

$$\frac{d\psi_2}{d\alpha_2} = 0 \tag{6.399}$$

From Eq.(6.398), and utilizing Eq.(6.396), one obtains:

$$\frac{d\psi_1}{d\alpha_1} = 0 = \left(\frac{1}{2}\right)\left\{\left(s^i\right)^T * A\left(x^i + \alpha_1 s^i\right) + \left(x^i + \alpha_1 s^i\right)^T A \, s^i\right\} - \left(s^i\right)^T b$$

$$\tag{6.400}$$

In this case, since [A] is a symmetrical matrix, hence $[A] = [A]^T$. Furthermore, since $\left(x^i + \alpha_1 s^i\right)^T * A \, s^i$ is a scalar quantity, hence

$$\left(x^i + \alpha_1 s^i\right)^T * A \, s^i = \left(s^i\right)^T *(A^T = A)*\left(x^i + \alpha_1 s^i\right) \tag{6.401}$$

Utilizing Eq.(6.401), Eq.(6.400) becomes:

$$0 = \left(\frac{1}{2}\right)\left\{2\left(s^i\right)^T * A\left(x^i + \alpha_1 s^i\right)\right\} - \left(s^i\right)^T b \tag{6.402}$$

$$0 = \left(s^i\right)^T * A \, x^i + \alpha_1 \left(s^i\right)^T A \, s^i - \left(s^i\right)^T b$$

$$0 = \left(s^i\right)^T *\left(A \, x^i - b\right) + \alpha_1 \left(s^i\right)^T A \, s^i \tag{6.403}$$

or

$$0 = \left(s^i\right)^T *\left(r^i\right) + \alpha_1 \left(s^i\right)^T A \, s^i \tag{6.404}$$

Hence

$$\alpha_1 = \frac{-\left(s^i\right)^T *\left(r^i\right)}{\left(s^i\right)^T A \, s^i} \tag{6.405}$$

Comparing Eq.(6.404) with Eq.(6.403), one clearly sees that

$$r^i = (A \, x^i - b) \tag{6.406}$$

Thus, Eq.(6.406) represents the "residual" (or "error") of Eq.(6.381).

Similarly, from Eq.(6.399) and utilizing Eq.(6.397), one obtains ($A \neq A^T$, since in this case A is an "unsymmetrical matrix"):

$$0 = \frac{d\psi_2}{d\alpha_2} = \left(s^i\right)^T * A^T * \left[A\left(x^i + \alpha_2 s^i\right) - b\right] + \left[A\left(x^i + \alpha_2 s^i\right) - b\right]^T * A\, s^i$$

(6.407)

since the last (product) term of Eq.(6.407) is a scalar quantity, hence "transposing" it will yield the same result. Thus, Eq.(6.407) can be re-arranged as:

$$0 = \left(s^i\right)^T A^T * \left[A\left(x^i + \alpha_2 s^i\right) - b\right] + \left(s^i\right)^T A^T * \left[A\left(x^i + \alpha_2 s^i\right) - b\right]$$

(6.408)

or

$$0 = 2\left(s^i\right)^T A^T * \left[A\left(x^i + \alpha_2 s^i\right) - b\right]$$

(6.409)

or

$$0 = \left(s^i\right)^T A^T * \left[\left(A\, x^i - b\right) + \alpha_2 A\, s^i\right]$$

(6.410)

Utilizing Eq.(6.406), Eq.(6.410) becomes:

$$0 = \left(s^i\right)^T A^T * \left[\left(r^i\right) + \alpha_2 A\, s^i\right]$$

(6.411)

Hence

$$\alpha_2 = \frac{-\left(s^i\right)^T A^T \left(r^i\right)}{\left(s^i\right)^T A^T A\, s^i}$$

(6.412)

[B] How to Find the "Search Direction," s^i ??

The initial direction, s^0, is usually selected as the initial residual:

$$s^0 \equiv -r^0 = -(A\, x^0 - b)$$

(6.413)

The reason for the above selection of the initial search direction was because Eq.(6.413) represents the gradient $\nabla\psi_1(x^0)$, or "steepest descent direction," of the objective ψ_1, defined in Eq.(6.382).

The step size is selected such that

$$\frac{d\psi_1(x^{i+1})}{d\alpha_1} = 0 = \frac{d\psi_1(x^{i+1} + \alpha_1 s^i)}{d\alpha_1} = \frac{d\psi_1}{dx} * \frac{dx}{d\alpha_1}$$

(6.414)

$$0 = \nabla\psi_1(x^{i+1}) * s^i$$

(6.415)

and

$$\frac{d\psi_2(x^{i+1})}{d\alpha_2} = 0 = \frac{d\psi_2(x^{i+1} + \alpha_2 s^i)}{d\alpha_2} = \frac{d\psi_2}{dx} * \left(\frac{dx}{d\alpha_2} = s^i \right) \quad (6.416)$$

Utilizing Eq.(6.409), the above equation becomes:

$$\left(s^i \right)^T A^T * \left[A\left(x^i + \alpha_2 s^i \right) - b \right] = 0 \tag{6.417}$$

or

$$\left(s^i \right)^T A^T * \left[A\left(x^{i+1} \right) - b \right] = 0 \tag{6.418}$$

$$\left(s^i \right)^T A^T * \left[r^{i+1} \right] = 0 \tag{6.419}$$

Since Eq.(6.419) represents the "scalar quantity," hence it can also be presented in its "transposed" form, as follows:

$$\left[r^{i+1} \right]^T * A s^i = 0 \tag{6.420}$$

Comparing Eq.(6.420) with Eq.(6.416), one concludes:

$$\nabla \psi_2 \equiv \frac{\partial \psi_2}{\partial x} \equiv \left[r^{i+1} \right]^T * A \tag{6.421}$$

Thus, Eq.(6.420) can also be presented as:

$$\nabla \psi_2 (x^{i+1}) * s^i = 0 \tag{6.422}$$

One can build a set of "A_conjugate" vectors $(= s^0, s^1, \cdots, s^i, s^{i+1})$ by applying the

Gram-Schmidt procedures to the (new) residual vector:

$$\hat{r}^{i+1} = A x^{i+1} - b = 0 \tag{6.423}$$

for obtaining the search direction:

$$\hat{s}^{i+1} = \hat{r}^{i+1} + \sum_{k=0}^{i} \beta_k \hat{s}^k \tag{6.424}$$

where

$$\beta_k = \frac{-\hat{r}^{i+1} A \hat{s}^k}{\left(\hat{s}^k \right)^T A \hat{s}^k} \tag{6.425}$$

and the following property of "A_conjugate" vectors will be satisfy:

$$\left(\hat{s}^{i+1} \right)^T A \hat{s}^k = 0 \quad ; k = 0, 1, 2, \cdots, i \tag{6.426}$$

Remarks

#[1]   The "hat" notations ($^\wedge$) in Eqs.(6.423 - 6.426) indicate that the involved vectors (such as $\hat{r}$ and $\hat{s}$) have <u>not</u> yet been normalized.

#[2]   Eqs.(6.424 - 6.425) can be derived in a simplified way as follows:

The initial search direction is equated to:

$$\hat{s}^0 = \hat{r}^0 \tag{6.427}$$

The subsequent search directions can be defined as:

$$\hat{s}^1 = \hat{r}^1 + \beta_{i,0} \, \hat{s}^0 \tag{6.428}$$

The unknown constant $\beta_{i,0}$ can be solved by invoking the property of "A_conjugate" vectors:

$$\left(\hat{s}^1\right)^T A \, \hat{s}^0 = 0 \tag{6.429}$$

Substituting Eq.(6.428) into Eq.(6.429), one obtains:

$$\left(\hat{r}^1 + \beta_{1,0} \, \hat{s}^0\right)^T A \, \hat{s}^0 = 0 \tag{6.430}$$

Hence

$$\beta_{i,0} = \frac{-\left(\hat{r}^1\right)^T A \, \hat{s}^0}{\left(\hat{s}^0\right)^T A \, \hat{s}^0} \tag{6.431}$$

Similarly, one has (see Eq.[6.424])

$$\hat{s}^2 = \hat{r}^2 + \beta_{2,0} \, \hat{s}^0 + \beta_{2,1} \, \hat{s}^1 \tag{6.432}$$

Invoking the property of "A_conjugate" vectors, one obtains:

$$\left(\hat{s}^2\right)^T A \, \hat{s}^0 = 0 \tag{6.433}$$

$$\left(\hat{s}^2\right)^T A \, \hat{s}^1 = 0 \tag{6.434}$$

Substituting Eq.(6.432) into Eqs.(6.433 - 6.434), one gets:

$$\beta_{2,0} = \frac{-\hat{r}^2 A \, \hat{s}^0}{\left(\hat{s}^0\right)^T A \, \hat{s}^0} \tag{6.435}$$

and

$$\beta_{2,1} = \frac{-\hat{r}^2 A \, \hat{s}^1}{\left(\hat{s}^1\right)^T A \, \hat{s}^1} \tag{6.436}$$

One can easily recognize that Eq.(6.424) and Eq.(6.425) are the general versions of Eqs.(6.432) and (6.435 - 6.436), respectively.

#[3]   If [A] is a symmetrical and positive definite matrix, then the "summation terms" in Eq.(6.424) can be expressed by a "single term" as:

$$\hat{s}^{i+1} = \hat{r}^{i+1} - \left(\frac{\left(\hat{r}^{i+1}\right)^T \hat{r}^{i+1}}{\left(\hat{r}^i\right)^T \hat{r}^i} \right) \hat{s}^i \tag{6.437}$$

The advantage offered by Eq.(6.437) not only is that the "summation terms" calculation can be avoided, but it also avoids the "explicit dependent on the matrix [A]."

Eq.(6.437) can also be derived by expressing the search direction formula (see Eq.[6.424]) in the different form:

$$\left(\hat{s}^{i+1}\right)^T = \left(\hat{r}^{i+1}\right)^T + \beta_i \left(\hat{s}^i\right)^T \tag{6.438}$$

It should be noted that the "conjugate direction" vector, shown in Eq.(6.438), would become the "steepest direction" had the last term in Eq.(6.438) been dropped.

Post-multiplying Eq.(6.438) by $A\,\hat{s}^i$ and utilizing the conjugate vectors' property, one obtains:

$$\left(\hat{s}^{i+1}\right)^T \left(A\,\hat{s}^i\right) = \left(\hat{r}^{i+1}\right)^T \left(A\,\hat{s}^i\right) + \beta_i \left(\hat{s}^i\right)^T \left(A\,\hat{s}^i\right) \tag{6.439}$$

$$0 = \left(\hat{r}^{i+1}\right)^T \left(A\,\hat{s}^i\right) + \beta_i \left(\hat{s}^i\right)^T \left(A\,\hat{s}^i\right) \tag{6.440}$$

From the new design point, Eq.(6.395), one obtains:

$$\hat{s}^i = \frac{(x^{i+1} - x^i)}{\alpha_1} \tag{6.441}$$

or

$$A\,\hat{s}^i = \frac{(A\,x^{i+1} - A\,x^i)}{\alpha_1} \tag{6.442}$$

or

$$A\,\hat{s}^i = \frac{\left[A\,x^{i+1} + b - (A\,x^i + b)\right]}{\alpha_1} \tag{6.443}$$

or

$$A\,\hat{s}^i = \frac{(\hat{r}^{i+1} - \hat{r}^i)}{\alpha_1} \tag{6.444}$$

Applying Eq.(6.444) into the first term of Eq.(6.440), one gets:

$$0 = \left(\hat{r}^{i+1}\right)^T * \frac{(\hat{r}^{i+1} - \hat{r}^i)}{\alpha_1} + \beta_i \left(\hat{s}^i\right)^T \left(A \hat{s}^i\right) \tag{6.445}$$

or

$$\beta_i = \frac{-\left(\hat{r}^{i+1}\right)^T * (\hat{r}^{i+1} - \hat{r}^i)}{\alpha_1 \left(\hat{s}^i\right)^T \left(A \hat{s}^i\right)} \tag{6.446}$$

From Eq.(6.415), one gets:

$$0 = \left(\hat{r}^{i+1}\right)^T * \hat{s}^i \tag{6.447}$$

or

$$0 = \left(\hat{s}^i\right)^T * \hat{r}^{i+1} \tag{6.448}$$

Now, post-multiplying both sides of Eq.(6.438) by $\hat{r}^{i+1}$, one gets:

$$\left(\hat{s}^{i+1}\right)^T \left(\hat{r}^{i+1}\right) = \left(\hat{r}^{i+1}\right)^T \left(\hat{r}^{i+1}\right) + \beta_i \left(\hat{s}^i\right)^T \left(\hat{r}^{i+1}\right) \tag{6.449}$$

Utilizing Eq.(6.448), the last term of Eq.(6.449) will disappear, hence

$$\left(\hat{s}^{i+1}\right)^T \left(\hat{r}^{i+1}\right) = \left(\hat{r}^{i+1}\right)^T \left(\hat{r}^{i+1}\right) \tag{6.450}$$

or

$$\left(\hat{s}^i\right)^T \left(\hat{r}^i\right) = \left(\hat{r}^i\right)^T \left(\hat{r}^i\right) \tag{6.451}$$

Using Eq.(6.451), Eq.(6.405) can be expressed as:

$$\alpha_1 = \frac{-\left(\hat{r}^i\right)^T (\hat{r}^i)}{\left(\hat{s}^i\right)^T A \hat{s}^i} \tag{6.452}$$

Substituting Eq.(6.452) into Eq.(6.446), one obtains:

$$\beta_i = \frac{-\left(\hat{r}^{i+1}\right)^T * (\hat{r}^{i+1} - \hat{r}^i)}{-\left(\hat{r}^i\right)^T * (\hat{r}^i)} \tag{6.453}$$

or

$$\beta_i = \frac{\left(\hat{r}^{i+1}\right)^T * (\hat{r}^{i+1} - \hat{r}^i)}{\left(\hat{r}^i\right)^T * (\hat{r}^i)} = \text{Polak-Rebiere Algorithm} \tag{6.454}$$

From Eq.(6.438), one gets:

$$\hat{r}^{i} = \hat{s}^{i} - \beta_{i-1}\, \hat{s}^{i-1} \tag{6.455}$$

Thus:

$$\left(\hat{r}^{i+1}\right)^{T} * (\hat{r}^{i}) = \left(\hat{r}^{i+1}\right)^{T} \left(\hat{s}^{i} - \beta_{i-1}\, \hat{s}^{i-1}\right) \tag{6.456}$$

zero, see Eq.(6.447)

or

$$\left(\hat{r}^{i+1}\right)^{T} * (\hat{r}^{i}) = -\beta_{i-1} \left(\hat{r}^{i+1}\right)^{T} \left(\hat{s}^{i-1}\right) \tag{6.457}$$

From Eq.(6.444), one gets:

$$\hat{r}^{i+1} = \alpha_{1} A\, \hat{s}^{i} + \hat{r}^{i} \tag{6.458}$$

or

$$\left(\hat{r}^{i+1}\right)^{T} = \alpha_{1} \left(\hat{s}^{i}\right)^{T} (A^{T} = A) + \left(\hat{r}^{i}\right)^{T} \tag{6.459}$$

Substituting Eq.(6.459) into Eq.(6.457), one gets:

$$\left(\hat{r}^{i+1}\right)^{T} * (\hat{r}^{i}) = -\beta_{i-1}\left[\alpha_{1}\left(\hat{s}^{i}\right)^{T} A + \left(\hat{r}^{i}\right)^{T}\right]\left(\hat{s}^{i-1}\right) \tag{6.460}$$

zero, see Eq.(6.448)

zero, see Eq.(6.426)

Hence:

$$\left(\hat{r}^{i+1}\right)^{T} * (\hat{r}^{i}) = 0 \tag{6.461}$$

Utilizing Eq.(6.461), Polak-Rebiere Algorithm for β_i in Eq.(6.454) becomes:

$$\beta_{i} = \frac{\left(\hat{r}^{i+1}\right)^{T} * (\hat{r}^{i+1})}{\left(\hat{r}^{i}\right)^{T} * (\hat{r}^{i})} = \text{Fletcher-Reeves Algorithm} \tag{6.462}$$

#[4]   If [A] is an "unsymmetrical" matrix, then [6.23] suggests to use " $[A^{T} A]$_conjugate " vectors as:

$$\hat{s}^{i+1} = \hat{r}^{i+1} + \sum_{k=0}^{i} \beta_{k}\, \hat{s}^{k} \tag{6.463}$$

where

$$\beta_{k} = \frac{\left(-A\, \hat{r}^{i+1}\right)^{T} \left(A\, \hat{s}^{k}\right)}{\left(\hat{s}^{k}\right)^{T} A^{T} A \left(\hat{s}^{k}\right)} \tag{6.464}$$

with the following "Conjugate_like" property:

$$\left(A\, \hat{s}^{\,i+1} \right)^T \left(A\, \hat{s}^{\,k} \right) = 0 \tag{6.465}$$

Eqs.(6.464 and 6.465) can be derived as follows:

First, pre-multiplying both sides of Eq.(6.463) by $\left(\hat{s}^{\,k} \right)^T A^T A$, one obtains:

$$\left(\hat{s}^{\,k} \right)^T A^T A\, \hat{s}^{\,i+1} = \left(\hat{s}^{\,k} \right)^T A^T A\, \hat{r}^{\,i+1} + \sum_{k=0}^{i} \beta_k \left(\hat{s}^{\,k} \right)^T A^T A\, \hat{s}^{\,k} \tag{6.466}$$

Since each (product) term of the above equation is "scalar quantity," transposing the 1st term of the right-hand side of Eq.(6.466) will give:

$$\left(A\hat{s}^{\,k} \right)^T \left(A\hat{s}^{\,i+1} \right) = \left(\hat{r}^{\,i+1} \right)^T A^T A \left(\hat{s}^{\,k} \right) + \sum_{k=0}^{i} \beta_k \left(A\hat{s}^{\,k} \right)^T \left(A\hat{s}^{\,k} \right) \tag{6.467}$$

or

$$\beta_k = \frac{\left(A\hat{s}^{\,k} \right)^T \left(A\hat{s}^{\,i+1} \right) - \left(\hat{r}^{\,i+1} \right)^T A^T A \left(\hat{s}^{\,k} \right)}{\left(\hat{s}^{\,k} \right)^T A^T \left(A\hat{s}^{\,k} \right)} \tag{6.468}$$

Since $[A]\vec{x} = \vec{b}$ (where [A] is an unsymmetrical matrix)

Hence $A^T (A\,\vec{x}) = A^T\, \vec{b}$ \tag{6.469}

or $[A^*]\vec{x} = \vec{b^*}$ \tag{6.470}

where

$$[A^*] \equiv [A]^T [A] = \text{symmetrical matrix} \tag{6.471}$$

$$\vec{b^*} \equiv [A]^T\, \vec{b} \tag{6.472}$$

Since $[A^*]$ is a symmetrical matrix, hence the property given by Eq.(6.426) can be applied to get:

$$\left(\hat{s}^{\,i+1} \right)^T [A^*]\hat{s}^{\,k} = 0 \quad ; k = 0,1,2,\cdots,i \tag{6.473}$$

Using Eq.(6.471), Eq.(6.473) becomes:

$$\left(\hat{s}^{\,i+1} \right)^T [A^T A]\hat{s}^{\,k} = 0 \tag{6.474}$$

or $\left(A\hat{s}^{\,i+1} \right)^T \left(A\hat{s}^{\,k} \right) = 0 = \left(A\hat{s}^{\,k} \right)^T \left(A\hat{s}^{\,i+1} \right)$ \tag{6.475}

Utilizing Eq.(6.475), Eq.(6.468) will be simplified to Eq.(6.464).

#[5]   If [A] and $[A]^T [A]$ are positive definite, the A_conjugate and

A^T A_conjugate vectors $\hat{s}^{i+1}$, generated by Eqs.(6.425) and (6.464), respectively, are linearly independent. This can be recognized by proving that the coefficients of the following homogeneous equation are zero:

$$\sum_i a_i \hat{s}^i = 0 \tag{6.476}$$

Pre-multiplying the above equation by $\left(A\hat{s}^j \right)^T$, one has:

$$\sum_i a_i \left(\left(\hat{s}^j \right)^T A^T \right) \hat{s}^i = 0 \tag{6.477}$$

For the case [A] is a symmetrical, positive definite matrix, then:

$$\sum_i a_i \left(\hat{s}^j \right)^T A\left(\hat{s}^i \right) = 0 \tag{6.478}$$

Using the property of A_conjugate vectors, Eq.(6.478) becomes:

$$a_j \left(\hat{s}^j \right)^T A\left(\hat{s}^j \right) = 0 \tag{6.479}$$

Since $\left(\hat{s}^j \right)^T A\left(\hat{s}^j \right) > 0$, hence $a_j = 0$ (for $j = 0, 1, 2, \cdots, n-1$)

Note that [6.23] has relaxed the positive definite condition of A^T A to the positive definite condition of the symmetrical part of A:

$$A = \underbrace{\frac{1}{2}\left(A + A^T \right)}_{\text{symmetrical matrix}} + \frac{1}{2}\left(A - A^T \right) \tag{6.480}$$

#[6]   Because of the conjugate property of $\hat{s}^i$, Eqs.(6.447) and (6.420) can be extended to the following general cases, respectively:

$$(r^i)^T s^j = 0; \text{ where } j < i \tag{6.481}$$

and

$$(r^i)^T (As^j) = 0; \text{ where } j < i \tag{6.482}$$

Equation (6.481) can be proven as follows:
Since $j < i$, one has:

$$x^i = x^{j+1} + \sum_{k=j+1}^{i} \alpha_k s^k \tag{6.483}$$

One can also easily recognized the following relationship:

$$\left(r^i \right)^T \left(s^j \right) = \nabla \psi_1 \left(x^i \right) * \left(s^j \right) \tag{6.484}$$

Utilizing Eq.(6.483), Eq.(6.484) becomes:

$$\left(r^i\right)^T \left(s^j\right) = \nabla \psi_1 \left(x^{j+1} + \sum_{k=j+1}^{i} \alpha_k \, s^k \right) * s^j \qquad (6.485)$$

Since $\nabla \psi_1 = (A\,\vec{x} - \vec{b})^T$, hence the above equation becomes:

$$\left(r^i\right)^T \left(s^j\right) = \left(A\,x^{j+1} + \sum_{k=j+1}^{i} \alpha_k \, A\,s^k - \vec{b} \right)^T * s^j \qquad (6.486)$$

$$= \left\{ \nabla \psi_1 \left(x^{j+1} \right) + \sum_{k=j+1}^{i} \alpha_k \, A\,s^k \right\}^T * s^j \qquad (6.487)$$

$$\left(r^i\right)^T \left(s^j\right) = \underbrace{\nabla \psi_1 \left(x^{j+1} \right) * s^j}_{\text{zero, see Eq.(6.447)}} + \underbrace{\sum_{k=j+1}^{i} \alpha_k \left(s^k\right)^T \left(A^T = A\right) s^j}_{\text{zero, property of conjugate vectors}} \qquad (6.488)$$

Hence $(r^i)^T \, s^j = 0; \; where \; j < i$ \qquad (6.489)

The major consequence of Eq.(6.489) is that the solution can be found at the nth iteration for a system of "n" symmetrical, positive definite equations.

From Eq.(6.489), one has :

$$\left(r^n\right)^T s^j = 0 \quad ; \text{for } j = 0, 1, 2, \cdots, n-1 \qquad (6.490)$$

For the complete linearly independent vectors $\left\{ s^0, s^1, s^2, \cdots, s^{n-1} \right\}$,

hence $\left(r^n\right)^T = 0$

#[7]   If [A] is an unsymmetrical matrix, then from Eq.(6.421), one obtains:

$$\nabla \psi_2 (x^n) = (r^n)^T A \qquad (6.491)$$

Thus, Eq.(6.482) can be expressed as:

$$\nabla \psi_2 (x^n) * s^j = 0 \qquad (6.492)$$

For the complete linearly independent vectors

$$\left\{ s^0, s^1, s^2, \cdots, s^{n-1} \right\}, \quad \text{hence} \quad \nabla \psi_2 (x^n) = 0 \qquad (6.493)$$

[C] Based upon the discussion in previous sections, the Generalized Conjugate Residual (GCR) algorithms can be described in the following step-by-step procedures as indicated in Table 6.10.

Table 6.10 GCR Step-by-Step Algorithms

<u>Step 1:</u> Choose an initial guess for the solution, x^0

Compute $r^0 = b - A x^0$ (see Eq.6.406)

<u>Step 2:</u> Start optimization iteration, for j=1,2,...

Choose an initial search direction, $\hat{s}^j$

where $\hat{s}^j = r^{j-1}$ (see Eq.6.413)

Compute $\hat{v}^j = A\hat{s}^j$; (portion of Eq.6.412)   (6.494)

<u>Step 3:</u> Generate a set of conjugate vectors $\hat{s}^j$ (The Gram-Schmidt Process)

for $\quad k = 1, 2, \cdots j-1$

$\beta = \left(\hat{v}^j\right)^T \left(v^k\right)$; see Eq.(6.464)   (6.495)

$\hat{v}^j = \hat{v}^j - \beta v^k$ (6.496)

$\hat{s}^j = \hat{s}^j - \beta s^k$; see Eq.(6.463)   (6.497)

End for

<u>Step 4:</u> Normalize the vectors

$\gamma = \left(\hat{v}^j\right)^T \left(\hat{v}^j\right)$; complete the denominator of Eq.(6.412)   (6.498)

$v^j = \dfrac{\hat{v}^j}{\gamma}$; "almost" completing Eq.(6.412)   (6.499)

$s^j = \dfrac{\hat{s}^j}{\gamma}$ (6.500)

<u>Step 5:</u> Compute the step size

$\alpha = \left(r^{j-1}\right)^T \left(v^j\right)$; completely done with Eq.(6.437)   (6.501)

<u>Step 6:</u> Compute the updated solution

$x^j = x^{j-1} + \alpha s^j$ (see Eq.6.395)

<u>Step 7:</u> Update the residual vector

$$r^j = r^{j-1} - \alpha\, v^j \tag{6.502}$$

<u>Step 8:</u> Convergence check??

$$\text{If } \left[\left\| r^j \right\| \Big/ \left\| b \right\| \right] \le \varepsilon_{\text{Tol}} \text{ ; Then} \tag{6.503}$$

Stop
Else
$$j = j + 1$$
Go To Step 2

Remarks about Table 6.10

(1) Equations (6.494, 6.496, 6.498, 6.499) can be explained with the help of Figure 6.17.

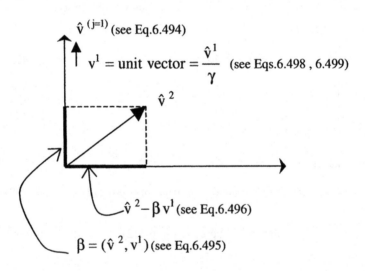

Figure 6.17 Graphical Interpretation of Gram-Schmidt Process

(2) Eq.(6.502) can be derived as the follows:
Pre-multiplying both sides of the equation (shown in Step 6 of Table 6.10) by -[A], one obtains:

$$-A\,x^j = -A\,x^{j-1} - \alpha A\, s^j \tag{6.504}$$

Then, adding the term "+ b" to both sides of the above equation, one gets:

$$\underbrace{b - A\,x^j} = \underbrace{b - A\,x^{j-1}} - \alpha A\, s^j \tag{.6.505}$$

or

$$r^j = r^{j-1} - \alpha A\, s^j \tag{6.506}$$

Using Eq.(6.500), Eq.(6.506) becomes:

$$r^j = r^{j-1} - \alpha\, A \left(s^j = \frac{\hat{s}^{\,j}}{\gamma} \right) \qquad\qquad (6.507)$$

Using Eq.(6.494), Eq.(6.507) becomes:

$$r^j = r^{j-1} - \frac{\alpha}{\gamma} \left(A\hat{s}^{\,j} = \hat{v}^{\,j} \right) \qquad\qquad (6.508)$$

Using Eq.(6.499), Eq.(6.508) becomes:

$$r^j = r^{j-1} - \alpha \left(\frac{\hat{v}^{\,j}}{\gamma} \equiv v^j \right) \qquad\qquad \text{(see Eq.6.502)}$$

(3) Explanation of Eq.(6.495) in Step 3 of Table 6.10:
From Eq.(6.495), one has:

$$\beta = \left(\hat{v}^{\,j} \right)^T \left(v^k \right) \qquad\qquad (6.495)$$

However, utilizing Eq.(6.494), the above equation becomes:

$$\beta = \left(A\hat{s}^{\,j} \right)\left(v^k \right) \qquad\qquad (6.509)$$

Utilizing Eq.(6.413), in Step 2 of Table 6.10, Eq.(6.509) becomes:

$$\beta = \left(A\, r^{\,j-1} \right) v^k \qquad\qquad (6.510)$$

Finally, utilizing Eq.(6.499), Eq.(6.510) can be expressed as:

$$\beta = \left(A\, r^{\,j-1} \right) * \left(\frac{\hat{v}^k}{\gamma} \equiv \frac{A\hat{s}^k}{\left(\hat{s}^k \right)^T A^T A\hat{s}^k} \right) \qquad\qquad (6.511)$$

Thus, Eq.(6.511) represents the implementation of Eq.(6.464)!

[D] How to Efficiently Handle Successive Right-Hand-Side Vectors [6.23-6.24]

Assuming we have to solve for the following problems:

$$[A]\vec{x}_i = \vec{b}_i \qquad\qquad (6.512)$$

In Eq.(6.512), the right-hand-side (RHS) vectors may <u>NOT</u> be all available at the same time, but that $\vec{b}_i$ depends on $\vec{x}_{i-1}$. There are two objectives in this section that will be discussed in subsequent paragraphs:

#(1) Assume the 1st solution $\vec{x}_1$, which corresponds to the 1st RHS vector, $\vec{b}_1$ has already been found in "n_1" iterations. Here, we would like to utilize the first "n_1" generated (and orthogonal) vectors $s_{i=1}^{j=1,2,\cdots,n_1}$ to find a "better initial guess" for the 2nd RHS solution vector $\vec{x}_{i=2}$. The new, improved algorithms will select the initial solution $\vec{x}^0_{i=2}$, to minimize the errors defined by ψ_1 (see Eq.6.382) or ψ_2 (see Eq.[6.383]) in the vector space spanned by the already existing (and expanding) "n_1" conjugate vectors. In other words, one has:

$$\vec{x}^0_{i=2} \equiv [s_1^1, s_1^2, \cdots, s_1^{n_1}]_{n \times n_1} * \{P\}_{n_1 \times 1} = [S_1] * \{P\} \qquad (6.513)$$

Eq.(6.513) expresses that the initial guess vector $\vec{x}^0_2$ is a linear combination of columns $s_1^1, s_1^2, \cdots, s_1^{n_1}$.

By minimizing (with respect to $\{P\}$) ψ_2 (defined in Eq.[6.383]), one obtains:

$$\nabla_p(\psi_2) \equiv \frac{\partial \psi_2}{\partial \vec{P}} = \vec{0} \qquad (6.514)$$

which will lead to:

$$\{P\}_{n_1 \times 1} = \frac{[s_1^T]_{n_1 \times 1} * [A^T]_{n \times n} * \{\vec{b}_2\}_{n \times 1}}{[s_1^T][A^T][A][s_1]} \qquad (6.515)$$

For the "symmetrical" matrix case, one gets:

$$\nabla_p(\psi_1) \equiv \frac{\partial \psi_1}{\partial \vec{P}} = \vec{0} \qquad (6.516)$$

which will lead to:

$$\{P\} = \frac{[s_1^T] * \{\vec{b}_2\}}{[s_1^T][A][s_1]} \qquad (6.517)$$

Remarks

(1.1)   Equation (6.515) can be derived as follows:

$$\text{Min.}\psi_2 = (A x^0 - b_2)^T (A x^0 - b_2) \qquad (6.408, \text{repeated})$$

Substituting Eq.(6.513) into the above equation, one obtains:

$$\text{Min.}\psi_2 = ([A][s_1]P - b_2)^T * ([A][s_1]P - b_2) \qquad (6.518)$$

or

$$\text{Min.}\psi_2 = (P^T[s_1]^T[A]^T - b_2^T) * (A s_1 P - b_2) \qquad (6.519)$$

$$\text{Min.}\psi_2 = P^T s_1^T A^T A s_1 P \boxed{-P^T s_1^T A^T b_2 - b_2^T A s_1 P} + b_2^T b_2 \qquad (6.520)$$

$$\text{Min.}\psi_2 = P^T s_1^T A^T A s_1 P \boxed{-2 P^T s_1^T A^T b_2} + b_2^T b_2 \qquad (6.521)$$

Set

$$\left(\frac{\partial \psi_2}{\partial \vec{P}}\right)^T = \vec{0} = 2(s_1^T A^T A s_1)\vec{P} - 2s_1^T A^T b_2 \qquad (6.522)$$

or $\qquad \vec{P} = \frac{s_1^T A^T b_2}{s_1^T A^T A s_1} \qquad$ (see Eq.6.515)

(1.2) The initial guess $x_{i=2}^0$, suggested by Eqs.(6.513 and 6.515), will be very good if $\vec{b}_2$ is "close" to $\vec{b}_1$. Otherwise, its computational effectiveness will be reduced.

(1.3) If $\vec{b}_2$ is "close" to $\vec{b}_1$, then an initial guess $x_{i=2}^0$ given by Eqs.(6.513 and 6.515) should be even better than selecting $x_{i=2}^0$ = previous converged solution $\vec{x}_{i=1}$. The reason is because the former selection (based on Eqs.[6.12 and 6.515]) will satisfy Eq.(6.516)!

Since $[s_1]_{n \times n_1}$ is a set of conjugate vectors, hence $s_1^T A^T A s_1$ (or $s_1^T A^T s_1$, for the case where [A] is a symmetrical matrix) is a "DIAGONAL" matrix. Therefore, the components of $\vec{P}$ (see Eq.6.515) can be efficiently computed.

The proof that $[s_1^T][A^T][A][s_1]$ is a "diagonal" matrix can be easily recognized by referring to Eq.(6.465), which states if $i+1 \neq k$, then the matrix (4 term) product (= off-diagonal) is zero. However, if $i+1 = k$, then the matrix product (= diagonal) is not zero.

(1.4) The step-by-step algorithms to generate a "good" initial guess $\vec{x}_2^0$ for the 2nd RHS vector can be given as shown in Table 6.11.

Table 6.11 Step-by-Step Algorithms to Generate a "Good" Initial Guess for RHS Vectors

$\vec{x}_2^0 = \vec{0}$	(6.523)
$\vec{r}_2^0 = \vec{b}_2$	(6.524)
for $\quad k = 1, 2, \cdots, n_1$	(6.525)
$\qquad P = \left(r_2^0\right)^T \left(v_1^k\right)$	(6.526)
$\qquad x_2^0 = x_2^0 + P s_1^k$	(6.527 also see Eq.6.513)

c   Recalled : $v_1^k = A s_1^k$ (see Eq.6.494, where s_1^k has already been normalized

c   according to Eq.6.500)

c   Furthermore, kept in mind that v_1^k and s_1^k vectors had already been generated

c   and stored when the 1st-RHS had been processed.

c   Thus, Eq.(6.526) represents the implementation of Eq.(6.515), but corresponding

c   to "ONLY 1 column" of matrix [S_1]. That's why we need to have a "do loop

c   index k" (see Eq.6.525) to completely execute Eq.(6.515) & Eq.(6.513).

$$r_2^0 = r_2^0 - P v_1^k \tag{6.528}$$

c   Eq.(6.528) can be derived as follows:

c   First, pre-multiplying both sides of Eq.(6.527) by (-A), one obtains

c   (note :P=scalar, in Eq.6.526):

c $(-A) x_2^0 = (-A) x_2^0 + P(-A) s_1^k$

c   Then, adding (b_2) to both sides of the above equation, one gets:

c $-A x_2^0 + b_2 = -A x_2^0 + b_2 - P\left(A s_1^k\right)$

c   or

c $r_2^0 = r_2^0 - P\left(A s_1^k\right)$

c   or, referring to Eq.(6.493), then the above Eq. becomes:

c $r_2^0 = r_2^0 - P\left(v_1^k\right)$, which is the same as indicated in Eq.(6.528).

End for

#(2) For successive RHS vectors, the "search vectors" s^j (see Eq.[6.497], in Table 6.10) need to be modified, so that these vectors s^j will not only be orthogonal among themselves (corresponding to the current 2nd RHS vector), but they also will be orthogonal with the existing n_1 vectors [corresponding to ALL "previous" RHS vector(s)]. Thus, the total number of (cumulative) conjugate vectors will be increased; hence "faster convergence" in the GCR algorithm can be expected.

Obviously, there will be a "trade-off" between "faster convergence rate" versus the "undesired" increase in computer memory requirement. In practical computer implementation, the user will specify how many "cumulative" vectors s^j and v^j can be stored in RAM. Then, if these vectors s^j and v^j have already filled up the user's specified incore memory available and convergence is still NOT YET achieved, one may have to "restart" the process (by starting with the new initial guess, etc.). However, a better (more reliable, more stable) approach will be out-of-core strategies, where there is NO LIMIT on how many cumulative vectors s^j and v^j can have. As soon as these vectors fill up the user's specified incore memory available, they are written on the disk space, therefore using incore RAM to store additional cumulative vectors.

In summary, for multiple (successive) RHS vectors, the basic GCR algorithms (shown in Table 6.10) need to be modified as indicated in Table 6.12 for more computational efficiency and practical applications.

Table 6.12 Extended GCR Algorithm for Computing the *ith* Solution of
Equation(6.537)

Compute an initial guess for x_i^0 as follows :
$n_{i-1} = 0$
Do 1111 i=1,nrhs
$x_i^0 = 0;$
$r_i^0 = b_i$
⎡ for j=1,$\cdots$,n_{i-1}

$P = (r_i^0, v^j);$

$x_i^0 = x_i^0 + P s^j;$

$r_i^0 = r_i^0 - P v^j;$

⎣ End for

Apply the modified GCR algorithm :
$m = n_{i-1} + 1$
⎡ for j=1,2,$\cdots$,maxiter

   $\hat{s}^m = r_i^{j-1};$

   $\hat{v}^m = A \hat{s}^m;$

   ⎡ for k=1,$\cdots$,m-1

      $\beta = (\hat{v}^m, v^k) = (\hat{v}^m)^T (v^k);$

      $\hat{v}^m = \hat{v}^m - \beta v^k;$

      $\hat{s}^m = \hat{s}^m - \beta s^k;$

   ⎣ end for

   $\gamma = \left\| \hat{v}^m \right\|;$

   $v^m = \hat{v}^m \big/ \gamma;$

   $s^m = \hat{s}^m \big/ \gamma;$

   $\alpha = (r_i^{j-1}, v^m) = (r_i^{j-1})^T (v^m);$

   $x_i^j = x_i^{j-1} + \alpha s^m;$

   $r_i^j = r_i^{j-1} - \alpha v^m;$

   ⎡ if $\left\| r_i^j \right\| / \left\| b_i \right\| < \varepsilon$ then

      $n_i = m;$

      $n_{i-1} = n_i;$

      Go to 1111

      end if

   ⎣ m = m + 1;

⎣ end for

1111   continue

The amount of work and the amount of memory required are proportional to the total number of search vectors generated. To make the extended GCR algorithm competitive with direct solution algorithms, it is essential to keep the total number of generated search vectors as small as possible. This can be achieved by pre-conditioning the system of equations (See Eq.[6.512]). Another way to reduce the amount of work and the memory is to use the extended GCR algorithm in combination with the Domain Decomposition (DD) approach.

6.16 Summary

In this chapter, various domain decomposition (DD) formulations have been presented. Sparse and dense [6.25] matrix technologies, mixed direct-iterative [6.23–6.24] equation solvers, pre-conditioned algorithms, indefinite sparse solver, generalized sparse inverse (or factorization), parallel computation, etc., have all been integrated into a DD formulation. Several numerical examples with explicit, detailed calculations have been used to verify the step-by-step algorithms.

6.17 Exercises

6.1 Re-do the example discussed in Section 6.2, with the following changes:

$$E = 30,000 \text{ K/in} \quad \text{(for all members)}$$

$$A = 4 \text{ in}^2 \quad \text{(for all members)}$$

Sub-structure 1 consists of members 1, 7, 2, 6, 8, 3, and nodes 1, 2, 3, 5, 6 (see Figure 6.2). Sub-structure 2 consists of members 4, 5, 9, and nodes 3, 5, 4.

6.2 Given the following matrix data:

$$[A] = \begin{bmatrix} 2 & -1 & 0 & 0 \\ -1 & 2 & -1 & 0 \\ 0 & -1 & 2 & -1 \\ 0 & 0 & -1 & 1 \end{bmatrix}$$

$$\vec{b} = \begin{Bmatrix} 1 \\ 0 \\ 0 \\ 0 \end{Bmatrix} \quad ; \quad \vec{x_0} = \begin{Bmatrix} 0 \\ 1 \\ 0 \\ 2 \end{Bmatrix} \quad ; \quad [B] = \begin{bmatrix} 2 & & & \\ & 2 & & \\ & & 2 & \\ & & & 1 \end{bmatrix}$$

Using the PCG algorithm (see Table 6.1) with the given initial guessed vector $\vec{x_0}$ and the pre-conditioned matrix [B], find the solution of $[A]\vec{x} = \vec{b}$.

6.3 Given the following stiffness matrix

$$K = \begin{bmatrix} 1 & 2 & -3 & 2 & -2 \\ 2 & 4 & -6 & 4 & -4 \\ -3 & -6 & 9 & -6 & 6 \\ 2 & 4 & -6 & 5 & -7 \\ -2 & -4 & 6 & -7 & 27 \end{bmatrix}$$

(a)   Identify (if existed) the dependent row(s).
(b)   Find the generalized inverse (if appropriated).

6.4 Using the FETI-DP Algorithms, find all nodal displacements of the following 29-Bar Truss Example, with the "corner" and "remainder" nodes defined in the following figures:

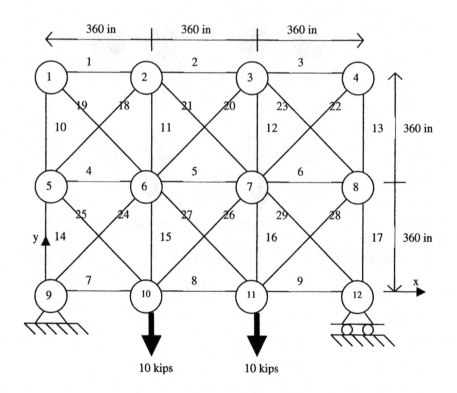

Entire Structure

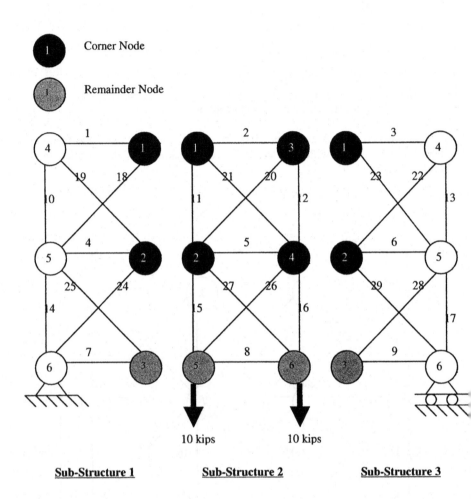

Sub-Structure 1   **Sub-Structure 2**   **Sub-Structure 3**

6.5 Given the following "unsymmetrical" system of equations $[A]\vec{x}=\vec{b}$, where:

$$[A] = \begin{bmatrix} 2 & 1 & 0 & 1 & 2 & 4 & 7 \\ 1 & 3 & 2 & 5 & 7 & 7 & 1 \\ 1 & 0 & 1 & 8 & 1 & 4 & 5 \\ 1 & 2 & 4 & 7 & 2 & 1 & 6 \\ 6 & 4 & 2 & 7 & 9 & 7 & 2 \\ 3 & 2 & 3 & 6 & 9 & 6 & 3 \\ 5 & 8 & 2 & 6 & 9 & 1 & 8 \end{bmatrix}$$

$$\vec{b}^{(1)} = \begin{Bmatrix} 91 \\ 117 \\ 100 \\ 103 \\ 149 \\ 142 \\ 158 \end{Bmatrix} \quad ; \quad \vec{b}^{(2)} = \begin{Bmatrix} 91 \\ 117 \\ 99 \\ 103 \\ 149 \\ 142 \\ 158 \end{Bmatrix}$$

and using the algorithms presented in Table 6.11, with initial guess $\overrightarrow{x^{(1)}} = \vec{0} = \overrightarrow{x_{initial}^{(1)}}$

(a)   Find the unknown vector $\overrightarrow{x^{(1)}}$, associated with the right-hand-side (RHS) vector $\overrightarrow{b^{(1)}}$. [Hint: solution should be $\{1,2,3,4,5,6,7\}^T$]

(b)   Find $\overrightarrow{x^{(2)}}$, associated with the RHS vector $\overrightarrow{b^{(2)}}$.

Appendix A: Singular Value Decomposition (SVD)

Singular Value Decomposition (SVD) has many important, practical applications: search engines [2], computational information retrieval [3], least square problems [1], image compressing [4], etc.

A general (square, or rectangular) matrix A can be decomposed as:

$$A = U \Sigma V^H \tag{A.1}$$

where

$\left[\Sigma \right]$ = diagonal matrix (does <u>not</u> have to be a square matrix)

$$= \begin{cases} \Sigma_{ij} = 0, & \text{for } i \neq j \\ \Sigma_{ij} \geq 0, & \text{for } i = j \end{cases}$$

$[U]$ and $[V]$ = unitary matrices $\begin{cases} U^H = U^T \text{ (for real matrix)} \\ \qquad = U^{-1} \end{cases}$

The SVD procedures to obtain matrices [U], [Σ], and [V] from a given matrix [A] can be illustrated by the following examples:

Example 1:   Given $[A] = \begin{bmatrix} 1 & 2 \\ 3 & 4 \end{bmatrix}$ (A.2)

Step 1: Compute $A A^H = A A^T = \begin{bmatrix} 1 & 2 \\ 3 & 4 \end{bmatrix}\begin{bmatrix} 1 & 3 \\ 2 & 4 \end{bmatrix} = \begin{bmatrix} 5 & 11 \\ 11 & 25 \end{bmatrix}$ (A.3)

Also: $A A^H = (U \Sigma \underbrace{V^H)(V}_{\substack{\text{Identity} \\ \text{Matrix}}} \Sigma^H U^H) = U \Sigma^2 U^H$ (A.4)

$$A^H A = V \Sigma^2 V^H \tag{A.5}$$

[1]   B. Noble, et al., (2nd Edition), <u>Applied Linear Algebra</u>, Prentice Hall.
[2]   M. W. Berry and M. Bworne, "Understanding Search Engines," SIAM.
[3]   M. W. Berry, "Computational Information Retrieval," SIAM.
[4]   J. W. Demmel, "Applied Numerical Linear Algebra," SIAM.

Step 2: Compute the standard eigen-solution of $A A^T$

$$\left[A A^T - \lambda I \right] \vec{u} = \vec{0} \tag{A.6}$$

Hence

$$\det \begin{bmatrix} 5-\lambda & 11 \\ 11 & 25-\lambda \end{bmatrix} = \vec{0} \tag{A.7}$$

or

$$\lambda^2 - 30\lambda + 4 = 0 \tag{A.8}$$

Thus

$$\lambda = 15 \pm \sqrt{221} = \begin{pmatrix} 0.1339 & 29.87 \end{pmatrix} \tag{A.9}$$

Now

$$\sigma = \sqrt{\lambda} = \begin{pmatrix} 0.3660 & 5.465 \end{pmatrix} \tag{A.10}$$

Hence

$$[\Sigma] = \begin{bmatrix} 0.3660 & 0 \\ 0 & 5.465 \end{bmatrix} \tag{A.11}$$

- For $\lambda = 15 - \sqrt{221}$, then Eq.(A.6) becomes:

$$11\, u_1^{(1)} + (25-\lambda)\, u_2^{(1)} = 0 \tag{A.12}$$

or

$$u_1^{(1)} = \frac{(\lambda - 25)\, u_2^{(1)}}{11} \tag{A.13}$$

Let $u_2^{(1)} = 1$, then : $\qquad$ (A.14)

$$u_1^{(1)} = \frac{(15 - \sqrt{221} - 25) * 1}{11} = -2.261 \tag{A.15}$$

So

$$u_1^{(1)} = \begin{Bmatrix} -2.261 \\ 1 \end{Bmatrix} \tag{A.16}$$

Eq.(A.16) can be normalized, so that $\left\| u^{(1)} \right\| = 1$, to obtain:

$$u_{Normalized}^{(1)} = \frac{1}{\sqrt{(-2.261)^2 + (1)^2}} \begin{Bmatrix} -2.261 \\ 1 \end{Bmatrix} = \begin{Bmatrix} -0.9145 \\ 0.4046 \end{Bmatrix} \tag{A.17}$$

- Similarly, for $\lambda = 15 - \sqrt{221}$, and let $u_2^{(2)} = 1$, then

$$u^{(2)} = \begin{Bmatrix} 0.4424 \\ 1 \end{Bmatrix} \tag{A.18}$$

Hence

$$u_{Normalized}^{(2)} = \frac{1}{\sqrt{(0.4424)^2 + (1)^2}} \begin{Bmatrix} 0.4424 \\ 1 \end{Bmatrix} = \begin{Bmatrix} 0.4046 \\ 0.9145 \end{Bmatrix} \tag{A.19}$$

Thus

$$[U] = \begin{bmatrix} u_N^{(1)} & u_N^{(2)} \end{bmatrix} = \begin{bmatrix} -0.9145 & 0.4046 \\ 0.4046 & 0.9145 \end{bmatrix} \qquad (A.20)$$

Step 3: Compute $A^H A = A^T A = \begin{bmatrix} 1 & 3 \\ 2 & 4 \end{bmatrix} \begin{bmatrix} 1 & 2 \\ 3 & 4 \end{bmatrix} = \begin{bmatrix} 10 & 14 \\ 14 & 20 \end{bmatrix}$ $\qquad (A.21)$

The two eigen-values associated with $(A^T A)$ can be computed as:

$$\lambda = 15 \pm \sqrt{221} = (0.1339 \quad 29.87) \qquad (A.22)$$

Hence

$$\sigma = \sqrt{\lambda} = \sqrt{(0.1339 \quad 29.87)} = (0.3660 \quad 5.465) \qquad (A.23)$$

• For $\quad \lambda = 15 - \sqrt{221}$, and let $v_2^{(1)} = 1$, then

$$v^{(1)} = \begin{Bmatrix} -1.419 \\ 1 \end{Bmatrix} \quad ; \text{hence} \quad v_N^{(1)} = \begin{Bmatrix} -0.8174 \\ 0.5760 \end{Bmatrix} \qquad (A.24)$$

• For $\quad \lambda = 15 + \sqrt{221}$, and let $v_2^{(2)} = 1$, then

$$v^{(2)} = \begin{Bmatrix} 0.7047 \\ 1 \end{Bmatrix} \quad ; \text{hence} \quad v_N^{(2)} = \begin{Bmatrix} 0.5760 \\ 0.8174 \end{Bmatrix} \qquad (A.25)$$

Thus

$$[V] = \begin{bmatrix} v_N^{(1)} & v_N^{(2)} \end{bmatrix} = \begin{bmatrix} -0.8174 & 0.5760 \\ 0.5760 & 0.8174 \end{bmatrix} \qquad (A.26)$$

Therefore, the SVD of [A] can be obtained from Eq.(A.1) as:

$$A = U\Sigma V = \begin{bmatrix} -0.9145 & 0.4046 \\ 0.4046 & 0.9145 \end{bmatrix} \begin{bmatrix} 0.3660 & 0 \\ 0 & 0.3660 \end{bmatrix} \begin{bmatrix} -0.8174 & 0.5760 \\ 0.5760 & 0.8174 \end{bmatrix}$$

Example 2: $\qquad$ Given $[A] = \begin{bmatrix} 1 & 1 \\ 2 & 2 \\ 2 & 2 \end{bmatrix}$ $\qquad (A.27)$

• Compute $\quad \underset{3 \times 2 \quad 2 \times 3}{A * A^H} = \begin{bmatrix} 2 & 4 & 4 \\ 4 & 8 & 8 \\ 4 & 8 & 8 \end{bmatrix}$ $\qquad (A.28)$

The corresponding eigen-values and eigen-vectors of Eq.(A.28) can be given as:

$$(\lambda_1 \ \lambda_2 \ \lambda_3) = (18 \ 0 \ 0) \implies \sigma = \sqrt{(\lambda_1 \ \lambda_2 \ \lambda_3)} = (3\sqrt{2} \ 0 \ 0) \qquad (A.29)$$

$$\left(u^{(1)} \quad u^{(2)} \quad u^{(3)} \right) = \begin{bmatrix} \frac{1}{3} & \frac{-2}{\sqrt{5}} & \frac{2\sqrt{5}}{15} \\ \frac{2}{3} & \frac{1}{\sqrt{5}} & \frac{4\sqrt{5}}{15} \\ \frac{2}{3} & 0 & \frac{-5\sqrt{5}}{15} \end{bmatrix} \tag{A.30}$$

- Compute $\quad \underset{2\times3}{A^H} * \underset{3\times2}{A} = \begin{bmatrix} 9 & 9 \\ 9 & 9 \end{bmatrix}$ (A.31)

The corresponding eigen-values and eigen-vectors of Eq.(A.31) can be given as:

$$\left(\lambda_1 \quad \lambda_2 \right) = \left(18 \quad 0 \right) \tag{A.32}$$

$$\sigma = \sigma_1 = \sqrt{\lambda} = \sqrt{18} = 3\sqrt{2} \tag{A.33}$$

$$\left(v^{(1)} \quad v^{(2)} \right) = \begin{bmatrix} \frac{1}{\sqrt{2}} & \frac{1}{\sqrt{2}} \\ \frac{1}{\sqrt{2}} & \frac{-1}{\sqrt{2}} \end{bmatrix} \tag{A.34}$$

Hence $\quad \underset{3\times2}{A} = \underset{3\times3}{U} \quad \underset{3\times2}{\Sigma} \quad \underset{2\times2}{V}$ (A.35)

$$A = \begin{bmatrix} \frac{1}{3} & \frac{-2}{\sqrt{5}} & \frac{2\sqrt{5}}{15} \\ \frac{2}{3} & \frac{1}{\sqrt{5}} & \frac{4\sqrt{5}}{15} \\ \frac{2}{3} & 0 & \frac{-5\sqrt{5}}{15} \end{bmatrix} \begin{bmatrix} 3\sqrt{2} & 0 \\ 0 & 0 \\ 0 & 0 \end{bmatrix} \begin{bmatrix} \frac{1}{\sqrt{2}} & \frac{1}{\sqrt{2}} \\ \frac{1}{\sqrt{2}} & \frac{-1}{\sqrt{2}} \end{bmatrix} \tag{A.36}$$

Example 3:   Given $[A] = \begin{bmatrix} 1 & 2 & 3 \\ 4 & 5 & 6 \\ 7 & 8 & 9 \\ 10 & 11 & 12 \end{bmatrix}$ (A.37)

The SVD of A is:

$$A = U\Sigma V^H = U\Sigma V^T = \begin{bmatrix} 0.141 & 0.825 & -0.420 & -0.351 \\ 0.344 & 0.426 & 0.298 & 0.782 \\ 0.547 & 0.028 & 0.664 & -0.509 \\ 0.750 & -0.371 & -0.542 & 0.079 \end{bmatrix} * \begin{bmatrix} 25.5 & 0 & 0 \\ 0 & 1.29 & 0 \\ 0 & 0 & 0 \\ 0 & 0 & 0 \end{bmatrix}$$

$$* \begin{bmatrix} 0.504 & 0.574 & 0.644 \\ -0.761 & -0.057 & 0.646 \\ 0.408 & -0.816 & 0.408 \end{bmatrix} \tag{A.38}$$

Example 4: (Relationship between SVD and generalized inverse)

"Let the $m \times n$ matrix A of rank k have the SVD

$A = U \Sigma V^H$; with $\sigma_1 \geq \sigma_2 \geq \cdots \geq \sigma_k \rangle 0$.

Then the generalized inverse A^+ of A is the $n \times m$ matrix.

$A^+ = V \Sigma^+ U^H$; where $\Sigma^+ = \begin{bmatrix} [E] & [0] \\ [0] & [0] \end{bmatrix}$ and E is the $k \times k$ diagonal matrix, with

$E_{ii} = \sigma_i^{-1}$ for $1 \leq i \leq k$"

Given SVD of $A = \begin{bmatrix} \dfrac{1}{3} & \dfrac{-2\sqrt{5}}{5} & \dfrac{2\sqrt{5}}{15} \\ \dfrac{2}{3} & \dfrac{\sqrt{5}}{5} & \dfrac{4\sqrt{5}}{15} \\ \dfrac{2}{3} & 0 & \dfrac{-\sqrt{5}}{15} \end{bmatrix} \begin{bmatrix} 3\sqrt{2} & 0 \\ 0 & 0 \\ 0 & 0 \end{bmatrix} \begin{bmatrix} \dfrac{\sqrt{2}}{2} & \dfrac{\sqrt{2}}{2} \\ \dfrac{\sqrt{2}}{2} & \dfrac{-\sqrt{2}}{2} \end{bmatrix}$

Hence : $A^+ = \begin{bmatrix} \dfrac{\sqrt{2}}{2} & \dfrac{\sqrt{2}}{2} \\ \dfrac{\sqrt{2}}{2} & \dfrac{-\sqrt{2}}{2} \end{bmatrix} \begin{bmatrix} \dfrac{1}{3\sqrt{2}} & 0 & 0 \\ 0 & 0 & 0 \end{bmatrix} \begin{bmatrix} \dfrac{1}{3} & \dfrac{2}{3} & \dfrac{2}{3} \\ \dfrac{-2\sqrt{5}}{5} & \dfrac{\sqrt{5}}{5} & 0 \\ \dfrac{2\sqrt{5}}{15} & \dfrac{4\sqrt{5}}{15} & \dfrac{-\sqrt{5}}{15} \end{bmatrix}$

$= \begin{bmatrix} \dfrac{1}{18} & \dfrac{1}{9} & \dfrac{1}{9} \\ \dfrac{1}{18} & \dfrac{1}{9} & \dfrac{1}{9} \end{bmatrix}$

References

[1.1]   J. N. Reddy, *An Introduction to the Finite Element Method,*
         2nd edition, McGraw-Hill (1993)

[1.2]   K. J. Bathe, *Finite Element Procedures,* Prentice Hall (1996)

[1.3]   K. H. Huebner, *The Finite Element Method for Engineers,*
         John Wiley & Sons (1975)

[1.4]   T. R. Chandrupatla and A.D. Belegundu, *Introduction to Finite
         Elements in Engineering,* Prentice-Hall (1991)

[1.5]   D. S. Burnett, *Finite Element Analysis: From Concepts to Applications,*
         Addison-Wesley Publishing Company (1987)

[1.6]   M. A. Crisfield, *Nonlinear Finite Element Analysis of Solids and
         Structures,* volume 2, John Wiley & Sons (2001)

[1.7]   O. C. Zienkiewicz, *The Finite Element Method,* 3rd edition,
         McGraw-Hill (1977)

[1.8]   D. R. Owen and E. Hinton, *Finite Elements in Plasticity: Theory
         and Practice,* Pineridge Press Limited, Swansea, UK (1980)

[1.9]   D. T. Nguyen, *Parallel-Vector Equation Solvers for Finite Element
         Engineering Applications,* Kluwer/Plenum Publishers (2002)

[1.10]   J. Jin, *The Finite Element Method in Electromagnetics,* John Wiley
         & Sons (1993)

[1.11]   P. P. Sivester and R. L. Ferrari, *Finite Elements for Electrical
         Engineers,* 3rd edition, Cambridge University Press (1996)

[1.12]   R. D. Cook, *Concepts and Applications of Finite Element Analysis,*
         2nd edition, John Wiley & Sons (1981)

[1.13]   S. Pissanetzky, *Sparse Matrix Technology,* Academic Press, Inc.
         (1984)

[1.14]   J. A. Adam, "The effect of surface curvature on wound healing in bone:
         II. The critical size defect." Mathematical and Computer Modeling, 35
         (2002), p 1085 - 1094.

[2.1]   W. Gropp, "Tutorial on MPI: The Message Passing Interface,"
         Mathematics and Computer Science Division, Argonne National
         Laboratory, Argonne, IL 60439

[3.1]   SGI sparse solver library sub-routine, Scientific Computing Software Library (SCSL) User's Guide, document number 007-4325-001, published Dec. 30, 2003

[3.2]   I. S. Duff and J. K. Reid, "MA47, a FORTRAN Code for Direct Solution of Indefinite Sparse Symmetric Linear Systems," RAL (Report) #95-001, Rutherford Appleton Laboratory, Oxon, OX11 OQX (Jan 1995)

[3.3]   G. Karypis and V. Kumar, "ParMETiS: Parallel Graph Partitioning and Sparse Matrix Ordering Library," University of Minnesota, CS Dept., Version 2.0 (1998)

[3.4]   J. W. H. Liu, "Reordering Sparse Matrices For Parallel Elimination," Technical Report #87-01, Computer Science, York University, North York, Ontario, Canada (1987)

[3.5]   D. T. Nguyen, G. Hou, B. Han, and H. Runesha, "Alternative Approach for Solving Indefinite Symmetrical System of Equation," *Advances in Engineering Software*, Vol. 31 (2000), pp. 581 – 584, Elsevier Science Ltd.

[3.6]   I. S. Duff, and G. W. Stewart (editors), Sparse Matrix Proceedings 1979, SIAM (1979)

[3.7]   I. S. Duff, R. G. Grimes, and J. G. Lewis, "Sparse Matrix Test Problems," ACM Trans. Math Software, 15, pp. 1 – 14 (1989)

[3.8]   G. H. Golub and C. F. VanLoan, "Matrix Computations," Johns Hopkins University Press, Baltimore, MD, 2nd edition (1989)

[3.9]   A. George and J. W. Liu, *Computer Solution of Large Sparse Positive Definite Systems,* Prentice-Hall (1981)

[3.10]   E. Ng and B. W. Peyton, "Block Sparse Choleski Algorithm on Advanced Uniprocessor Computer," *SIAM J. of Sci. Comput.,* volume 14, pp. 1034 - 1056 (1993).

[3.11]   H. B. Runesha and D. T. Nguyen, "Vectorized Sparse Unsymmetrical Equation Solver for Computational Mechanics," *Advances in Engr. Software,* volume 31, nos. 8 - 9, pp. 563 - 570 (Aug. - Sept. 2000), Elsevier

[4.1]   J. A. George, "Nested Disection of a Regular Finite Element Mesh," *SIAM J. Numer. Anal.,* volume 15, pp. 1053 - 1069 (1978)

[4.2]   I. S. Duff and J. K. Reid, "The Design of MA48: A Code for the Direct Solution of Sparse Unsymmetric Linear Systems of Equations," *ACM Trans. Math. Software.,* 22 (2): 187 - 226 (June 1996)

[4.3] I. S. Duff and J. Reid, "MA27: A Set of FORTRAN Subroutines for Solving Sparse Symmetric Sets of Linear Equations," AERE Technical Report, R-10533, Harwell, England (1982)

[5.1] Nguyen, D. T., Bunting, C., Moeller, K. J., Runesha H. B., and Qin, J., "Subspace and Lanczos Sparse Eigen-Solvers for Finite Element Structural and Electromagnetic Applications," *Advances in Engineering Software*, volume 31, nos. 8 - 9, pages 599 - 606 (August - Sept. 2000)

[5.2] Nguyen, D. T. and Arora, J. S., "An Algorithm for Solution of Large Eigenvalue Problems," *Computers & Structures*, vol. 24, no. 4, pp. 645 - 650, August 1986.

[5.3] Arora, J. S. and Nguyen, D. T., "Eigen-solution for Large Structural Systems with Substructures," *International Journal for Numerical Methods in Engineering*, vol. 15, 1980, pp. 333 - 341.

[5.4] Qin, J. and Nguyen, D. T., "A Vector Out-of-Core Lanczons Eigensolver for Structural Vibration Problems," presented at the 35th Structures, Structural Dynamics, and Material Conference, Hilton Head, SC, (April 18 - 20, 1994).

[5.5] K. J. Bathe, *Finite Element Procedures,* Prentice Hall (1996)

[5.6] G. Golub, R. Underwood, and J. H. Wilkinson, "The Lanczos Algorithm for Symmetric Ax=Lamda*Bx Problem," Tech. Rep. STAN-CS-72-720, Computer Science Dept., Stanford University (1972)

[5.7] B. Nour-Omid, B. N. Parlett, and R. L. Taylor, "Lanczos versus Subspace Iteration for Solution of Eigenvalue Problems," *IJNM in Engr.,* volume 19, pp. 859 - 871 (1983)

[5.8] B. N. Parlett and D. Scott, "The Lanczos Algorithm with Selective Orthogonalization," *Mathematics of Computation,* volume 33, no. 145, pp. 217 - 238 (1979)

[5.9] H.D. Simon, "The Lanczos Algorithm with Partial Reorthogonalization", Mathematics of Computation, 42, no. 165, pp. 115-142 (1984)

[5.10] J. J. Dongarra, C. B. Moler, J. R. Bunch, and G. W. Stewart, LINPACK Users' Guide, SIAM, Philadelphia (1979)

[5.11] S. Rahmatalla and C. C. Swan, "Continuum Topology Optimization of Buckling-Sensitive Structures," *AIAA Journal,* volume 41, no. 6, pp. 1180 - 1189 (June 2003)

[5.12]   W. H. Press, B. P. Flannery, S. A. Teukolsky, and W. T. Vetterling, *Numerical Recipes (FORTRAN Version)*, Cambridge University Press (1989)

[5.13]   M. T. Heath, *Scientific Computing: An Introductory Survey*, McGraw-Hill (1997)

[6.1]   Tuna Baklan, "CEE711/811: Topics in Finite Element Analysis," Homework #5, Old Dominion University, Civil & Env. Engr. Dept., Norfolk, VA (private communication)

[6.2]   W. R. Watson, "Three-Dimensional Rectangular Duct Code with Application to Impedance Eduction," *AIAA Journal*, 40, pp. 217-226 (2002)

[6.3]   D. T. Nguyen, S. Tungkahotara, W. R. Watson, and S. D. Rajan. "Parallel Finite Element Domain Decomposition for Structural/ Acoustic Analysis," *Journal of Computational and Applied Mechanics*, volume 4, no. 2, pp. 189 - 201 (2003)

[6.4]   C. Farhat and F. X. Roux, "Implicit Parallel Processing in Structural Mechanics," *Computational Mechanics Advances*, volume. 2, pp. 1 - 124 (1994)

[6.5]   D. T. Nguyen and P. Chen, "Automated Procedures for Obtaining Generalized Inverse for FETI Formulations," Structures Research Technical Note No. 03-22-2004, Civil & Env. Engr. Dept., Old Dominion University, Norfolk, VA 23529 (2004)

[6.6]   C. Farhat, M. Lesoinne, P. LeTallec, K. Pierson, and D. Rixen, "FETI-DP: A Dual-Primal Unified FETI Method- Part I: A Faster Alternative to the 2 Level FETI Method," *IJNME*, volume 50, pp. 1523 - 1544 (2001)

[6.7]   R. Kanapady and K. K. Tamma, "A Scalability and Space/Time Domain Decomposition for Structural Dynamics - Part I: Theoretical Developments and Parallel Formulations," Research Report UMSI 2002/ 188 (November 2002)

[6.8]   X. S. Li and J. W. Demmel, "SuperLU_DIST: A Scalable Distributed-Memory Sparse Direct Solver for Unsymmetric Linear Systems," *ACM Trans. Mathematical Software*, volume 29, no. 2, pp. 110 - 140 (June 2003)

[6.9]   A. D. Belegundu and T. R. Chandrupatla, *Optimization Concepts and Applications in Engineering*, Prentice-Hall (1999)

[6.10]   D. T. Nguyen and P. Chen, "Automated Procedures For Obtaining
          Generalized Inverse for FETI Formulation," Structures Technical
          Note [#] 03-22-2004, Civil & Env. Engr. Dept. ODU, Norfolk, VA 23529

[6.11]   M. Papadrakakis, S. Bitzarakis, and A. Kotsopulos, "Parallel
          Solution Techniques in Computational Structural Mechanics,"
          B. H. V. Topping (Editor), *Parallel and Distributed Processing
          for Computational Mechanics: Systems and Tools*, pp. 180 - 206,
          Saxe-Coburg Publication, Edinburgh, Scotland (1999)

[6.12]   L. Komzsik, P. Poschmann, and I. Sharapov, "A Preconditioning
          Technique for Indefinite Linear Systems," *Finite Element in
          Analysis and Design*, volume 26, pp. 253-258 (1997)

[6.13]   P. Chen, H. Runesha, D. T. Nguyen, P. Tong, and T. Y. P. Chang,
          "Sparse Algorithms for Indefinite System of Linear Equations,"
          pp. 712 - 717, *Advances in Computational Engineering Science*,
          edited (1997) by S. N. Atluri and G. Yagawa, Tech. Science Press,
          Forsyth, Georgia

[6.14]   D. T. Nguyen, G. Hou, H. Runesha, and B. Han, "Alternative Approach
          for Solving Sparse Indefinite Symmetrical System of Equations,"
          Advances in Engineering Software, volume 31 (8 - 9), pp. 581 - 584
          (2000)

[6.15]   J. Qin, D. T. Nguyen, T. Y. P. Chang, and P. Tong, "Efficient Sparse
          Equation Solver With Unrolling Strategies for Computational
          Mechanics", pp. 676 - 681, *Advances in Computational Engineering
          Science*, edited (1997) by S. N. Atluri and G. Yagawa, Tech. Science
          Press, Forsyth, Georgia

[6.16]   A. George and J. W. Liu, *Computer Solution of Large Sparse Positive
          Definite Systems*, Prentice-Hall (1981)

[6.17]   C. Farhat, M. Lesoinne, and K. Pierson, "A Scalable Dual-Primal
          Domain Decomposition Method," Numerical Linear Algebra with
          Applications, volume 7, pp. 687 - 714 (2000)

[6.18]   Nguyen, D. T., "Multilevel Structural Sensitivity Analysis,"
          Computers & Structures Journal, volume 25, no. 2, pp. 191 - 202,
          April 1987

[6.19]   S. J. Kim, C. S. Lee, J. H. Kim, M. Joh, and S. Lee, "ISAP: A High
         Performance Parallel Finite Element Code for Large-Scale Structural
         Analysis Based on Domain-wise Multifrontal Technique," proceedings
         of Super Computing, Phoenix, AZ (November 15 - 21, 2003)

[6.20]   J. H. Kim, and S. J. Kim, "Multifrontal Solver Combined with Graph
         Patitioners," *AIAA Journal*, volume 37, no. 8, pp. 964 - 970 (Aug. 1999)

[6.21]   I. Duff and J. Reid, "The Multifrontal Solution of Indefinite Sparse
         Symmetric Linear Systems," *Association for Computing Machinery
         Transactions Mathematical Software*, volume 9, pp. 302 - 325 (1983)

[6.22]   B. M. Iron, "A Frontal Solution Program for Finite Element Analysis,"
         IJNME, volume. 2, pp. 5 - 32 (1970)

[6.23]   F. J. Lingen, "A Generalized Conjugate Residual Method for the
         Solution of Non-Symmetric Systems of Equations with Multiple
         Right-Hand Sides," *IJNM in Engr.*, volume 44, pp. 641 - 656 (1999)

[6.24]   P. F. Fischer, "Projection Techniques for Iterative Solution
         of Ax = b with Successive Right-Hand Sides," ICASE Report # 93-90,
         NASA LaRC, Hampton, VA

[6.25]   S. Tungkahotara, D. T. Nguyen, W. R. Watson, and H. B. Runesha,
         "Simple and Efficient Parallel Dense Equation Solvers," 9th
         International Conference on Numerical Methods and Computational
         Mechanics, University of Miskolc, Miskolc, Hungary (July 15 - 19,
         2002)

Index

FINITE ELEMENT METHODS:
PARALLEL-SPARSE STATICS AND EIGEN-SOLUTIONS

Duc T. Nguyen

Dr. Duc T. Nguyen is the founding Director of the Institute for Multidisciplinary Parallel-Vector Computation and Professor of Civil and Environmental Engineering at Old Dominion University. His research work in parallel procedures for computational mechanics has been supported by NASA Centers, AFOSR, CIT, Virginia Power, NSF, Lawrence Livermore National Laboratory, Jonathan Corp., Northrop-Grumman Corp., and Hong Kong University of Science and Technology. He is the recipient of numerous awards, including the 1989 Gigaflop Award presented by Cray Research Incorporated, the 1993 Tech Brief Award presented by NASA Langley Research Center for his fast Parallel-Vector Equation Solvers, and Old Dominion University, 2001 A. Rufus Tonelson distinguished faculty award. Dr. Nguyen has been listed among the Most Highly Cited Researchers in Engineering in the world.